Human Anatomy
Laboratory Guide and Dissection Manual
Fourth Edition

Michael J. Timmons, M.S.

Moraine Valley Community College
Palos Hills, Illinois

Ralph T. Hutchings

Biomedical photographer, formerly Chief Medical Laboratory Scientific Officer,
Royal College of Surgeons of England

and

William C. Ober, M.D.

Medical illustrator

Claire W. Garrison, R.N.

Illustrator

D0024468

Prentice Hall
Upper Saddle River, New Jersey 07458

Editor in Chief: **Sheri Snavely**
Senior Acquisitions Editor: **Halee Dinsey**
Editorial Assistant: **Susan Zeigler**
Vice President of Production and Manufacturing: **David W. Riccardi**
Executive Managing Editor: **Kathleen Schiaparelli**
Production Editor: **James Buckley**
Assistant Managing Editor: **Dinah Thong**
Supplement Cover Management/Design: **Paul Gourhan**
Manufacturing Buyer: **Ilene Kahn**

Photographic facilities provided by WARD'S Natural Science Establishment, Inc.
5100 West Henrietta Road, Rochester, New York 14692-9012, 1-800-962-2660.

Use of 3B Models: Courtesy of Paul Binhold GmbH and American 3B Scientific

Technical assistance provided by WARD'S Natural Science Establishment, Inc.

Additional photography provided by Gervase Pevarnik,
WARD'S Natural Science Establishment, Inc.

Printed in the United States of America

13 12 11 10

ISBN 0-13-047547-5

Prentice-Hall International (UK) Limited, *London*
Prentice-Hall of Australia Pty. Limited, *Sydney*
Prentice-Hall Canada, Inc., *Toronto*
Prentice-Hall Hispanoamericana, S.A., *Mexico*
Prentice-Hall of India Private Limited, *New Delhi*
Prentice-Hall (*Singapore*) Pte. Ltd.
Prentice-Hall of Japan, Inc., *Tokyo*
Editora Prentice-Hall do Brasil, Ltda., *Rio de Janeiro*

Contents

15
The Nervous System: The Brain and Cranial Nerves 215

16
The Nervous System: Pathways and Higher-Order Functions 239

17
The Nervous System: Anatomic Division 245

18
The Nervous System: General and Special Senses 252

19
The Endocrine System 276

20
The Cardiovascular System: Blood 290

Dedication

To our students for their inspiration

To our children Molly, Kelly, Patrick, Katie, Sam and Isabel
for their constant support

To our dearest wives, Judy and Anne, for their devotion, encouragement,
and continued inspiration

Preface

Welcome to the new edition of this laboratory manual. Existing laboratory manuals and dissection guides typically contain extensive descriptions of textual narrative supplemented by poor quality line art and black-and-white photographs of specimens. This unrealistic, one-dimensional visual rendering of the human body is quite different from what the student will see in the anatomy laboratory and on the dissecting table. Instructors and students have been searching for a laboratory manual that offers an efficient and organized approach to the study of human anatomy. This manual, *Human Anatomy Laboratory Guide and Dissection Manual,* is designed to meet the needs of both the instructor and the student.

Two themes are conspicuous characteristics of this laboratory manual. The first is the portrayal of the human body as seen and studied in the laboratory. To accomplish this, the new design takes greater pedagogical advantage of the visual aspects of anatomy. This format offers distinct advantages. The illustration program emphasizes the three-dimensional relationships between anatomical structures and, wherever possible, anatomical structures are shown from multiple perspectives.

The second theme provides students with a framework to organize, interpret, and apply anatomy. Most undergraduate students (e.g., of nursing, occupational therapy, and allied health) and professional students (of physical therapy, pharmacy, chiropractic, and so on) learn introductory anatomy by viewing microscope slides to study organ histology, examining anatomical models, and dissecting preserved mammalian specimens and cadavers. The organized identification of anatomical structures portrayed by specimens and models is presented sequentially in this manual to facilitate linking of visual observations and identifications of anatomy in the laboratory with the interpretation and application of anatomical concepts. This pedagogical design facilitates student learning by creating a functional unit between lecture and laboratory settings.

Most students are good visual learners, but many have difficulty initially in identifying anatomical structures at both microscopic and gross levels. Additionally, some students have difficulty in organizing their laboratory study of anatomical structures. We understand the challenge students face in observing and conceptualizing the levels of anatomical architecture. *Human Anatomy Laboratory Guide and Dissection Manual* combines the traditional features of a laboratory manual with the features of a dissection guide and pictorial atlas.

Intended Use

This laboratory manual has been prepared to be used with any anatomy textbook and to accompany the fourth edition of **Human Anatomy** textbook by Martini, Timmons, and Tallitsch, published by Prentice Hall. The laboratory manual is designed specifically to meet the needs of undergraduate and selected professional students enrolled in either a one- or two-semester (quarter) human anatomy course that includes mammalian and cadaver dissection as part of the laboratory experience. *Human Anatomy Laboratory Guide and Dissection Manual* is designed to facilitate learning by creating a functional unit that connects lecture and laboratory. Identification of anatomical structures (by microscopic examination or study of skeletal elements, anatomical models, and cadaver and organ specimens) is presented in an organized and student-oriented manner. This laboratory manual is intended to act as an information bridge (both textual and visual), permitting the student to integrate the lecture descriptions with the laboratory observations and identifications.

FEATURES

1. The large format offers four distinct advantages: (1) more illustrations and photographs can be included, (2) the illustrations and photographs are enlarged, (3) a greater visual presentation of all anatomical structures, and (4) the enlarged images eliminate the need for students to purchase a separate pictorial, anatomical atlas.

2. The laboratory manual contains references to all appropriate **Human Anatomy** illustrations, specimen photographs, radiographs, and MRI images. The ordered and directed identification of anatomical structures is linked by the laboratory manual to the "visual anatomy" presented in the textbook, **Human Anatomy**. Two types of figure references are presented: (1) to the illustrations that appear in the laboratory manual and (2) to the links to the text and the illustrations (art and photographs) that appear in the **Human Anatomy** text.

3. All anatomical structures are listed by system for each area studied. The organization and sequence of topics presented in this laboratory manual matches the sequence presented in the **Human Anatomy** text.

4. Precise directions for the systematic observation of tissues, anatomical models, prosected cadaver specimens, and cat anatomy (as it relates to human

anatomy) are presented for all anatomical areas of study. Concise dissection directions have been tailored to current curricula for an efficient examination of both a prosected cadaver and cat anatomy.

5. Combines the traditional features of a laboratory manual with those of an atlas.

6. Color photographs of prosected cadaver specimens, anatomical models, and medical quality anatomical drawings appear in the laboratory manual. Color photographs of anatomical models aid in orientation of the specimen or body region and bridge the gap between the anatomical model studied in the laboratory and the human body. All color art and color photographs appear in a color folio insert, for easy access.

7. Icons identify microscopic and gross laboratory observations. Icons of a microscope and cadaver specimen appear as aids to organize the study sequence for anatomical structures.

8. Space is provided adjacent to each anatomical area of study for students to sketch or take notes.

9. Each chapter ends with two types of assessment questions for student review. Unlabeled drawings for students to label and color appear in the **Anatomical Identification Reviews** portion of each chapter. These drawings and micrographs are designed to be labeled and colored by the student during observation or review. Following the "Reviews" are **To Think About** questions. These short questions encourage critical thinking.

10. Twenty objective **Laboratory Review Questions** are arranged by chapter to assess student understanding of anatomical structure and function. The questions are designed to be turned into the instructor as part of the course assessment or for student self assessment, following the study of each chapter. These questions appear at the back of the book for convenient removal by students.

11. Cat anatomy art and concise dissection directions are presented for the systemic study of cat anatomy as it relates to human anatomy. The 27 drawings of cat anatomy serve as a visual primer to aid in the systematic observation of cat dissection as it relates to human anatomy.

A unique bar-code learning system by Timmons, the *LaserDisc with Bar Code Manual,* appears as a companion to this laboratory manual. Anatomical structures (e.g., specific text illustrations, histology, color specimen/cadaver photographs, cat dissection series) and video demonstrations of anatomical structures and dissection technique may be visualized immediately and directly by student or instructor accessing the Prentice-Hall Anatomy and Physiology LaserDisc either via the bar-code system or with Prentice-Hall's Multimedia Presenter software. Bar codes are found and accessed easily for all figures referenced in this laboratory manual. Laser-disc images

may be viewed from a TV monitor in the anatomy laboratory or learning resource center by either groups or individual students. This learning system will enhance student understanding of anatomical structures and architecture.

TO THE STUDENT

Icons appear to the left of two types of laboratory observations to aid you in the organized identification and study of anatomical structures.

- **Microscope Icon**: used to link specific microscope slides studied in the laboratory with illustrations in the laboratory manual or *Human Anatomy* textbook figures depicting this same specific tissue.
- **Cadaver Icon**: used to link the observation of cadaver specimens in the laboratory with illustrations in the laboratory manual or figures in the *Human Anatomy* textbook that depict specific anatomical structures seen in the cadaver.

The Illustrations and Photographs

- Illustrations depicting human anatomy: 125 labeled medical quality illustrations of human anatomy at different levels of dissection.
- Cadaver-specimen photographs: 175 labeled color and black-and-white photographs of cadaver and organ specimens.
- Photographs of anatomical models: 60 labeled color and black-and-white photographs of anatomical models used in teaching laboratories.
- Illustrations depicting cat anatomy: 27 illustrations show cat anatomy at different stages of dissection. This collection of labeled art depicting cat anatomy is arranged by system and will serve as a visual aid in your systematic observation of cat dissection as it relates to human anatomy. As a useful aid for reviewing dissected regions, color the appropriate drawings after your dissection and observations.

Illustrations to Be Colored and Labeled

Three-dimensional medical quality art and photographs of anatomical structures and tissues specimens appear at the end of each chapter as **Anatomical Identification Reviews.** These drawings or photographs are designed to be labeled and colored for review.

Figure References

Traditional laboratory manuals have figure references that connect the narrative with a specific illustration or photo. In addition to this traditional format, this dissection guide and manual goes a step further by providing you with a second set of figure references designed to link the "visual anatomy" presented by each figure in the **Human Anatomy** text to the specific anatomical area or structure under study. The images in the textbook provide you with an additional 1200 illustrations (art and photos) to help you "see" human anatomy from multiple perspectives.

Two types of figure references are presented in this lab manual:

- Figure references that are associated with the narrative of the laboratory manual and references to Color Plates appear in *italic*.
- Figures that reference the **Human Anatomy** text appear at the margins, preceded by a miniature icon of a muscular arm. A miniature chain link next to a figure or page number is used to reference the textbook. [∞ p277]

Use of Bar Codes

Printed bar codes appear in the companion to this laboratory manual, the *Laser Disc with Bar Code Manual* by Timmons. The Prentice-Hall anatomy and physiology LaserDisc contains animations depicting anatomical structure and function and brief video clips showing laboratory dissections. Still images of illustrations and photographs from your textbook, **Human Anatomy**, and from numerous other sources also appear on this laser disc. Images are arranged and grouped to permit easy access. Each animation sequence, video of laboratory dissection, and still image is titled and referenced to the **Human Anatomy** text. Images on the laser disc may be accessed in three ways: (1) manual entry, (2) "bar-code reader," and (3) software. Any still image, animation, or video laboratory dissection may be accessed and displayed on a TV monitor screen by using the appropriate function keys on the remote keypad unit of the video disc player or by entering a frame number manually. A quicker and easier way to view images is by "swiping" the bar-code reader across the printed bar codes that appear in the *Manual*. Ask your instructor to provide you with instructions for using the LaserDisc and bar-code system.

COMPANIONS AND SUPPLEMENTS TO *Human Anatomy Laboratory Guide and Dissection Manual*

Human Anatomy, 4th ed. (ISBN 0-13-061569-2) by Martini, Timmons, and Tallitsch

The fourth edition of **Human Anatomy** is the most comprehensive, visually oriented introductory anatomy text available today. The atlas-size format allows for larger, carefully crafted anatomical paintings, cadaver dissection photos, and photomicrographs to help students visualize the components of the human body. The lively, to-the-point writing style and student-friendly pedagogical aids transform this text into a trusted companion. The Text and Academic Authors Association presented **Human Anatomy** with their award for excellence. **Human Anatomy** has also received an Award of Excellence by The Association of Medical Illustrators for outstanding textbook medical illustration.

Instructor's Presentation CD-ROM for Human Anatomy (ISBN 0-13-066403-0)
This resource features images from the fourth edition of **Human Anatomy**, as well as an assortment of histology and cadaver dissection stills. The Power Point presentation software embedded on the CD allows instructors to customize lectures or create mini-tutorials for students by selecting and sorting images in the desired order of presentation. Users can also import files from word processing and presentational programs, making this an even more effective and versatile teaching tool. New features include electronic quizzes, enhanced Power Point presentations, fully interactive 3-D animations, and tutorials with audio for use in class and laboratory.

Primal Pictures 7 CD-ROM Set (ISBN 0-13-008904-4)
This award-winning 7 CD-ROM set from Primal Pictures™ is available to instructors free upon adoption of the Martini/Timmons **Human Anatomy** fourth edition textbook and the companion *Anatomy Laboratory Manual* by Timmons/Hutchings. This Anatomical and Clinical Pathology software features 3-D computer graphic models of human anatomy derived from scan data. The models can be rotated through 360 degrees, layers can be peeled away to reveal individual systems or regional structures. Movies illustrate muscle actions and slides highlight key pathology of anatomical structures. Anthroscopic materials provide fantastic views and stunning biomechanical animations demonstrate and describe the limits of motion and stresses. Radiology sections are also included. All materials can be imported into Power Point presentations.

A.D.A.M. Interactive Anatomy Dissection Manual (ISBN 0-13-082638-3) by Martha DePecol Sanner and Harry Greer
Offers step-by-step tutorials for use with the Digital Cadaver digital supplement. An online correlation guide is available for this supplement on the **Human Anatomy**, fourth edition website (www.prenhall.com/martini/ha4).

Correlation Guide to Prentice Hall's Digital Cadaver Anatomy Dissection Manual
This online supplement is the perfect complement to Prentice Hall's *Digital Cadaver* and Timmons/Hutchings *Human Anatomy Lab Manual*, fourth edition. It provides helpful cross-references of images between the Digital Cadaver and the Human Anatomy Lab Manual.

LaserDisc with Bar Code Manual (ISBN 0-13-341033-1 and 0-13-520032-6)
The LaserDisc and accompanying Bar Code Manual are companions to the Martini/Timmons **Human Anatomy** textbook (4th edition) and the Timmons/Hutchings *Human Anatomy Laboratory Guide and Dissection Manual.* The LaserDisc is organized by chapters that correspond to the chapters in the text and laboratory manual. The LaserDisc chapters contain text stills, histology, cadaver images, animations, and video of laboratory dissections. Custom lecture and laboratory multimedia presentations can be created easily and quickly using the *Bar Code Manual* as your directory and organizer of frame number addresses for all Laser-Disc images.

Instructor's Manual and Resource Guide For Human Anatomy Laboratory Guide and Dissection Manual The *Instructor's Guide* identifies all materials required for each laboratory observation, provides answer keys and catalog numbers for ordering anatomical models, microscope slides and supplies from Ward's Natural Science Establishment, Inc.

ACKNOWLEDGMENTS

We acknowledge our students, our colleagues, and the many reviewers who played a vital role in the development of this laboratory manual. Thank you Bill Ober, M.D., and Claire Garrison, R.N., for rendering the medical art that appears in this book. We appreciate the continued enthusiastic support of our entire book team at Prentice-Hall publishers. Many thanks to Mary Oreluk, R.N., for her assistance in organizing portions of the manuscript and diligence in typing and proofing the manuscript. James Buckley, our Production Editor, was responsible for the timely and accurate production of this book. A special note of thanks and deep appreciation to Sue Zeigler, our dynamic Editorial Assistant, for the key role she played in guiding the development of this book and overseeing its completion. It has been a pleasure to work with James and Sue, and we are grateful to them. Finally, we thank our Editor, Halee Dinsey, for her innovative, energetic, and visionary approach to improving the quality of our books. No author could ask for more.

In an effort to improve future editions, readers with pertinent information, suggestions, or comments concerning the organization or content of this manual are encouraged to send their remarks to the publisher: Halee Dinsey, Prentice Hall/Pearson Education, One Lake Street, Upper Saddle River, New Jersey 07458. The comments of colleagues and students will be deeply appreciated and carefully considered in the preparation of future editions of this laboratory manual and dissection guide.

Michael J. Timmons
Orland Park, IL

Ralph T. Hutchings
London, England

AN INTRODUCTION TO ANATOMY

Objectives

1. Apply anatomical terms to describe body areas, regions, and surfaces, body and organ sections, and relative positions of body structures.

2. Find and identify major body cavities and their locations relative to each other.

Superficial Anatomy

The language of anatomy is precise like the language of other sciences or mathematics. A strong foundation and continual, correct usage of the language of anatomy terms and phrases is crucial to building and maintaining an understanding of anatomical structures, references, and medical terminology. **Superficial** or **surface anatomy** describes general body form and structure: such studies are completed with the unaided eye. [∞ p13]

The human form is presented in the **anatomical position**, in which the subject stands with legs together and feet flat on the floor, hands at the sides, palms facing forward. Unless otherwise noted, all anatomical positions and descriptions given in this lab manual refer to the body in the anatomical position. [∞ p13]

Procedure

Identify the following **anatomical areas, regions,** and **structures** using a torso model and *Figures 1-1* and *1-2* for reference. Label each anatomical area, region, and structure in the illustration that is at the end of this chapter on *page 8*.

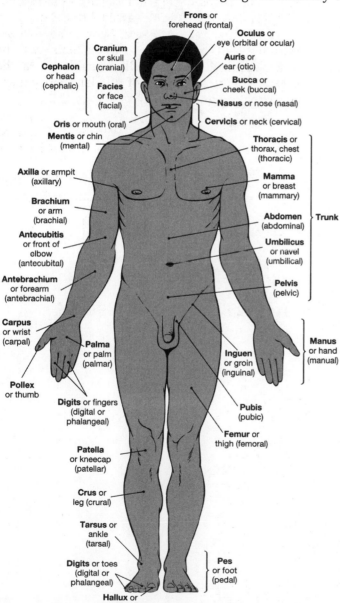

Frons or forehead (frontal)
Oculus or eye (orbital or ocular)
Cranium or skull (cranial)
Auris or ear (otic)
Cephalon or head (cephalic)
Facies or face (facial)
Bucca or cheek (buccal)
Nasus or nose (nasal)
Oris or mouth (oral)
Cervicis or neck (cervical)
Mentis or chin (mental)
Thoracis or thorax, chest (thoracic)
Axilla or armpit (axillary)
Mamma or breast (mammary)
Brachium or arm (brachial)
Abdomen (abdominal)
Antecubitis or front of elbow (antecubital)
Umbilicus or navel (umbilical)
Antebrachium or forearm (antebrachial)
Pelvis (pelvic)
Trunk
Carpus or wrist (carpal)
Manus or hand (manual)
Palma or palm (palmar)
Inguen or groin (inguinal)
Pollex or thumb
Digits or fingers (digital or phalangeal)
Pubis (pubic)
Femur or thigh (femoral)
Patella or kneecap (patellar)
Crus or leg (crural)
Tarsus or ankle (tarsal)
Digits or toes (digital or phalangeal)
Pes or foot (pedal)
Hallux or great toe

FIGURE 1-1 Anatomical landmarks.

LOCATE

See: Figs. 1-1, 1-2

Body Viewed from the Anterior

___ Cranium [*noun*] (cranial) [*adjective*]

___ Cephalon (cephalic)

___ Facies (face)

___ Frons or forehead (frontal)

___ Oculus or eye (orbital)

___ Auris or ear (otic)

___ Bucca or cheek (buccal)

___ Nasus (nasal)

___ Oris (oral)

___ Mentis (mental)

___ Cervicis (cervical) or neck

___ Axilla (axillary)

___ Brachium (brachial)

___ Antecubitis (antecubital)

___ Antebrachium (antebrachial)

___ Carpus (carpal)

___ Palma or palm (palmar)

___ Pollex

___ Fingers (digital or phalangeal)

___ Trunk

___ Thorax (thoracic)

___ Mamma (mammary)

___ Abdomen (abdominal)

___ Umbilicus (umbilical)

___ Pelvis (pelvic)

___ Manus (manual)

___ Pubis (pubic)

___ Inguen (groin)

___ Femur or thigh (femoral)

___ Anterior knee (patellar)

___ Crura or leg (crural)

___ Tarsus (tarsal)

___ Toes (digital or phalangeal)

___ Hallux

___ Pes (pedal)

Body Viewed from the Posterior

___ Head (cephalic)

___ Cervicis

___ Shoulder (acromial)

___ Dorsum (dorsal)

___ Lumbus or loin (lumbar)

___ Olecranon (olecranal)

___ Gluteus (gluteal)

___ Popliteus (popliteal)

___ Surus or calf (sural)

___ Calcaneus (calcaneal)

___ Plantus (plantar)

___ Upper limb (extremity)

___ Lower limb (extremity)

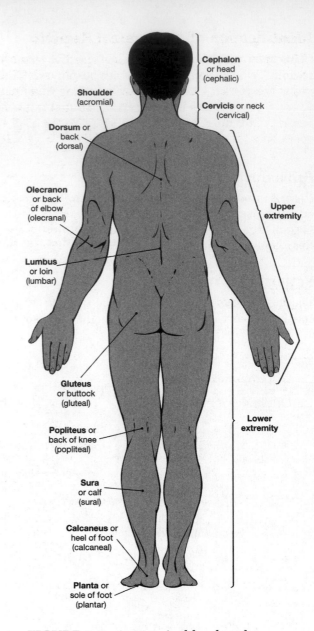

FIGURE 1-2 Anatomical landmarks.

Identification of Anatomical Regions

Clinicians and scientists often use specific terms to pinpoint an area or a location within the abdominal or pelvic region. Two different methods are employed. Clinicians favor the **abdominopelvic quadrant** method, which divides the region into four quarters using imaginary horizontal and vertical lines that intersect at the umbilicus (navel). Anatomists use more precise distinctions that describe the location and orientation of internal organs within the abdominopelvic region. This method subdivides the region into nine areas, which are formed by intersecting four parallel, imaginary lines (two horizontal and two vertical). [∞ p15]

Procedure

Identify **abdominal regions** by both the **quadrant** and the **anatomical method** using a *torso model, Color Plate 41,* and *Figures 1-3* and *1-4* for reference. Locate and identify the **abdominopelvic organs** that occupy these areas. Color each area and identify each internal organ shown in the accompanying illustrations. Use *Color Plate 102* for reference.

See: Figs. 1-3, 1-4

LOCATE

Quadrant Method

— Right upper quadrant (RUQ)

— Right lower quadrant (RLQ)

— Left upper quadrant (LUQ)

— Left lower quadrant (LLQ)

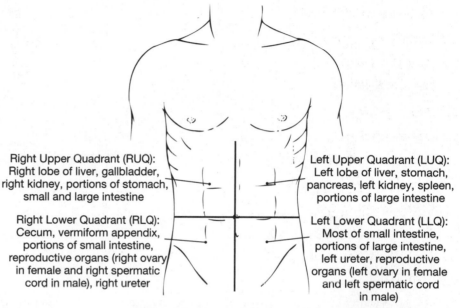

Right Upper Quadrant (RUQ):
Right lobe of liver, gallbladder, right kidney, portions of stomach, small and large intestine

Right Lower Quadrant (RLQ):
Cecum, vermiform appendix, portions of small intestine, reproductive organs (right ovary in female and right spermatic cord in male), right ureter

Left Upper Quadrant (LUQ):
Left lobe of liver, stomach, pancreas, left kidney, spleen, portions of large intestine

Left Lower Quadrant (LLQ):
Most of small intestine, portions of large intestine, left ureter, reproductive organs (left ovary in female and left spermatic cord in male)

FIGURE 1-3 Abdominopelvic quadrants and regions.

Anatomical Method

— Right hypochondriac region

— Epigastric region

— Left hypochondriac region

— Right lumbar region

— Umbilical region

— Left lumbar region

— Right inguinal region

— Hypogastric region

— Left inguinal region

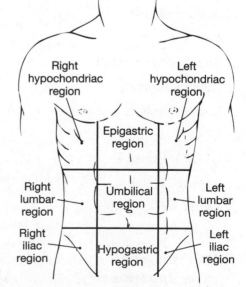

Description of Anatomical Directions

Terms describing anatomical regions and directions are the foundation for precise communication of anatomical structures, references, and medical procedures. It is important to remember that *right* and *left* refer to the subject's right and left. Principal regional and directional terms are described and examples of their use are presented in *Tables 1-1* and *1-2.* Review these terms before proceeding.

FIGURE 1-4 Abdominopelvic regions.

TABLE 1-1　Regions of the Human Body

Anatomical Name	Anatomical Region	Area Indicated
Cephalon	Cephalic	Area of head
Cervicis	Cervical	Area of neck
Thoracis	Thoracic	The chest
Abdomen	Abdominal	The abdomen
Pelvis	Pelvic	The pelvis (in general)
Pubis	Pubic	The anterior pelvis
Inguen	Inguinal	The groin (crease between thigh and trunk)
Lumbus	Lumbar	The lower back
Gluteus	Gluteal	The buttock
Brachium	Brachial	The segment of the upper limb closest to the trunk; the arm
Antebrachium	Antebrachial	The forearm
Carpus	Carpal	The wrist
Manus	Manual	The hand
Femur	Femoral	The thigh
Patella	Patellar	The kneecap
Crus	Crural	The leg, from knee to foot
Sura	Sural	The calf
Talus	Talan	The ankle
Pes	Pedal	The foot
Planta	Sole	Plantar region of foot

TABLE 1-2　Regional and Directional Terms

Term	Region or Reference	Example
Anterior	The front; before	The navel is on the *anterior* surface of the trunk.
Ventral	The bellyside (equivalent to anterior when referring to human body)	In humans, the navel is on the *ventral* surface.
Posterior	The back; behind	The shoulder blade is located *posterior* to the rib cage.
Dorsal	The back (equivalent to posterior when referring to human body)	The *dorsal* body cavity encloses the brain and spinal cord.
Cranial	The head	The *cranial*, or *cephalic*, border of the pelvis is toward the head rather than toward the thigh.
Cephalic	Same as cranial	
Superior	Above; at a higher level (in human body, toward the head)	The cranial border of the pelvis is *superior* to the thigh.
Caudal	The tail (coccyx in humans)	The hips are *caudal* to the waist.
Inferior	Below; at a lower level	The knees are *inferior* to the hips.
Medial	Toward the midline longitudinal axis of the body	The *medial* surfaces of the thighs may be in contact.
Lateral	Away from the midline longitudinal axis of the body	The thigh bone articulates with the *lateral* surface of the pelvis.
Proximal	Toward an attached base	The thigh is *proximal* to the foot.
Distal	Away from an attached base	The fingers are *distal* to the wrist.
Superficial	At, near, or relatively close to the body surface	The skin is *superficial* to underlying structures.
Deep	Toward the interior of the body; farther from the body surface	The bone of the thigh is *deep* to the surrounding skeletal muscles.

Procedure

Identify the **anatomical directions** using a *torso model* and *Figure 1-5* for reference.

LOCATE

___ Superior/Inferior

___ Cranial/Caudal

___ Lateral/Medial

___ Proximal/Distal

___ Anterior (ventral)/
    Posterior (dorsal)

See: Fig. 1-5

∞ *Fig. 1-10, p17*

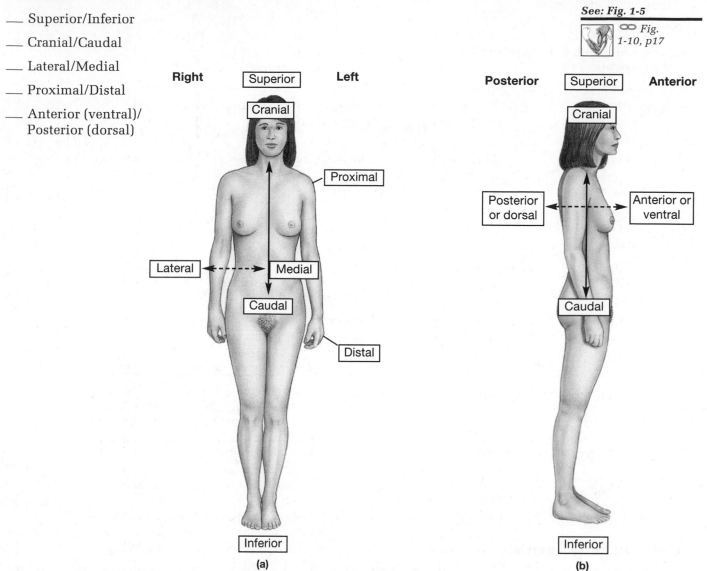

FIGURE 1-5   Directional references.

Description of Sectional Anatomy

A sectional view of an anatomical region or structure is sometimes the only way to view three-dimensional relationships between the parts of anatomical structures. Any slice through a body region or part describes one of three primary planes of section. Planes are flat surfaces formed by making a cut through the body or a part of it. The **transverse (horizontal) plane** lies at right angles to the long axis of the body, dividing it into superior and inferior sections. This plane can pass through the body at any level and is always parallel to the floor. It produces **transverse (cross) sections**. The **frontal (coronal) plane** extends from side to side, dividing the body into anterior and posterior sections. A **sagittal plane** that passes along the midline and divides the body into left and right halves is a **midsagittal section**. A sagittal plane that lies parallel to the midsagittal line but offset from it is a **parasagittal section**. [∞ p17]

Procedure

Identify the **three sectional planes** using a *torso model* and *Figure 1-6* for reference. Color each sectional plane in *Figure 1-6*.

LOCATE

___ Frontal (coronal plane)

___ Sagittal plane

___ Transverse plane or transverse (cross) section

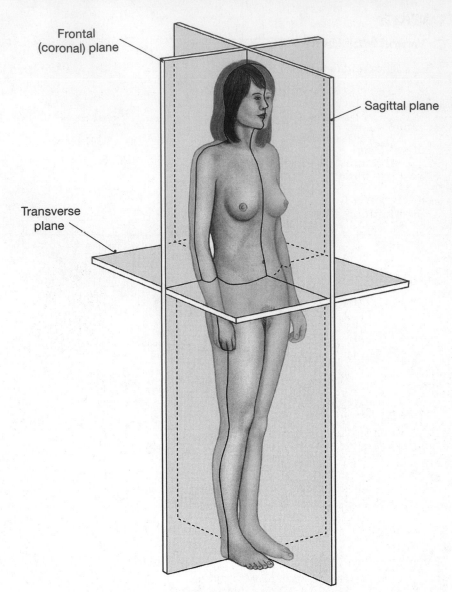

Frontal
(coronal) plane

Sagittal plane

Transverse
plane

FIGURE 1-6 Planes of section.

Location of Body Cavities

The human body is not solid. Many vital organs are suspended in internal chambers known as **body cavities**. These fluid-filled cavities serve both to protect delicate organs and to permit significant movement and growth of each organ. The **dorsal body cavity** contains the brain and spinal cord, while the much larger **ventral body cavity** contains the organs of the respiratory, cardiovascular, digestive, urinary, and reproductive systems. Each body cavity is subdivided. The dorsal body cavity is subdivided into a **cranial cavity** (enclosing the brain) and a **spinal (vertebral) cavity** (spinal cord). The ventral body cavity contains the organs of the respiratory, cardiovascular, digestive, urinary, and reproductive systems and is subdivided into a **thoracic cavity** and an **abdominopelvic cavity** by the muscular diaphragm. The thoracic cavity is further divided into **left** and **right pleural cavities** and a **pericardial cavity**. The pleural cavities enclose the lungs and the pericardial cavity encloses the heart. The abdominopelvic cavity is further subdivided into the **abdominal cavity** and the **pelvic cavity**. The abdomino-pelvic cavity is separated by an imaginary plane from the superior surface of the sacrum to the superior surface of the pubis into abdominal (superior) and pelvic (inferior) regions. Digestive organs and the kidneys occupy the abdominal region, while the urinary bladder, uterus, ovaries, and vagina occupy the pelvic region. [∞ p15]

Procedure

Identify the **ventral** and **dorsal body cavities** and their subdivisions using a *torso model* and *Figure 1-7* for reference. Label and color ventral and dorsal body cavities and their subdivisions in the illustration at the end of the chapter on *page 8*.

LOCATE

See: Fig. 1-7

__ **Ventral body cavity**

  __ Thoracic cavity

    __ Pleural cavities

    __ Pericardial cavity

    __ Mediastinum

  __ Abdominopelvic cavity

__ Peritoneal cavity

__ Abdominal cavity

__ Pelvic cavity

Dorsal body cavity

  __ Cranial cavity

  __ Spinal cavity

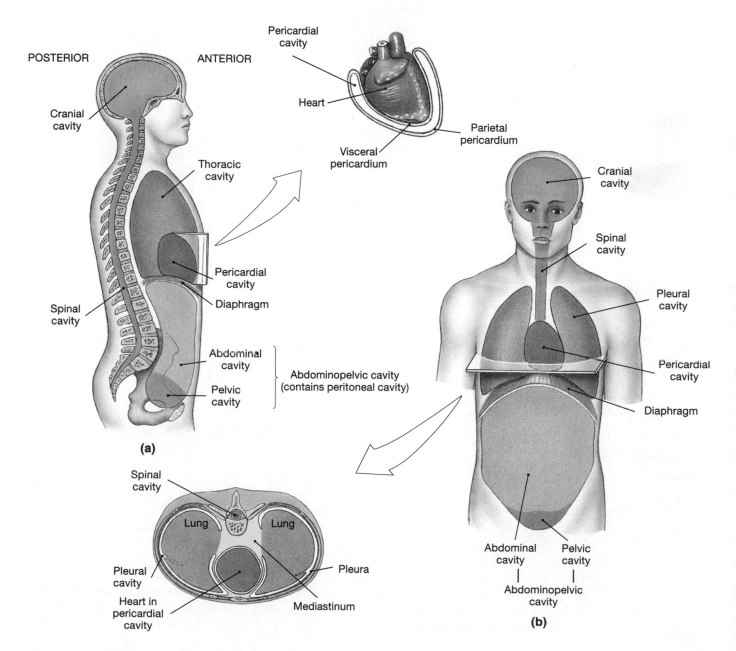

FIGURE 1-7 (a) Lateral view of dorsal and ventral body cavities (b) Anterior and sectional views of the ventral body cavity.

Anatomical Identification Review

On the accompanying drawing, label all body cavities and their divisions.

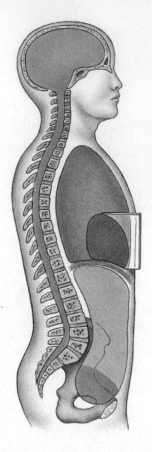

To Think About

1. Why is it important to be familiar with major anatomical landmarks?

2. How does the definition of anterior, as it is used in human terminology, differ from the usage with quadrupedal animals?

3. How does comparative anatomy contribute to the study of human anatomy?

4. How does the regional approach differ from the systemic approach in the study of human anatomy? What are the advantages of each approach?

2

THE CELL

Objectives

1. Identify the plasma membrane, cellular organelles, and cellular inclusions on a model of a cell.

2. Develop an understanding of cellular organelle structure and function.

Observation of Cellular Anatomy

The cell is the unit of living structure and the building block of the human body. Cells carry out all body activities and functions. **Cytology** is the study of the structure and function of individual cells. Most mammalian cells are microscopic. Using the light microscope, cellular structures can be observed when magnified up to 1,000 times. Different types of cells, such as epithelial cells, muscle cells, or neurons, exhibit specific shapes and characteristics, which can be distinguished both from each other and from large intracellular structures, such as collagen fibers. In the living state, individual cells are relatively transparent and difficult to distinguish from their neighbors. Staining tissue sections with dyes is the principal way of creating the contrast necessary for morphological study of membranes and/or organelles. A record of the study is made by taking a photograph, termed a **light micrograph** (LM), of the cell or structure. Another major approach to the structural and chemical characterization of cells and tissues is the isolation of cellular and tissue components by ultracentrifugation, followed by selected chemical analysis and electron microscopic examination. [∞ p27]

The plasma membrane surrounds the cytoplasm as the outer limiting boundary of the cell. The properties of the plasma membrane provide for control of substance passage across the membrane. The cytoplasm is composed of cytosol (fluid), organelles (functional units), and inclusions (storage units). Organelles are found in all body cells and each performs a specific and essential function. Organelles can be divided into two categories: *non-membranous* and *membranous.* The non-membranous organelles lack a membrane boundary and include the cytoskeleton, flagella, and ribosomes. As the name implies, membranous organelles are bounded by a limiting membrane. Membranous organelles include the nucleus: endoplasmic reticulum, mitochondria, Golgi apparatus, lysosomes, and peroxisomes.

Procedure

Identify the listed cellular structures on a *model* of a *typical cell.* Use *Color Plate 25* for reference. Label each cellular structure in the illustration that appears in the Anatomical Identification Review section on the next page. Briefly describe each structure and define its function as you label it.

LOCATE

See: Color Plate 25
∞ Fig. 2-3, p28

___ Centrosome

___ Centrioles

___ Cell membrane

___ Cilia

___ Microvilli

___ Cytosol

___ Cytoskeleton (microtubules, microfilaments)

___ Secretory vesicles

___ Lysosome

___ Golgi apparatus

___ Mitochondrion

___ Rough endoplasmic reticulum

___ Smooth endoplasmic reticulum

___ Ribosomes

___ Nucleus

___ Nuclear envelope

___ Nuclear pores

___ Nucleolus

Anatomical Identification Review

Organelle Identification   On the accompanying drawing, label the following organelles and describe their function in the table below.

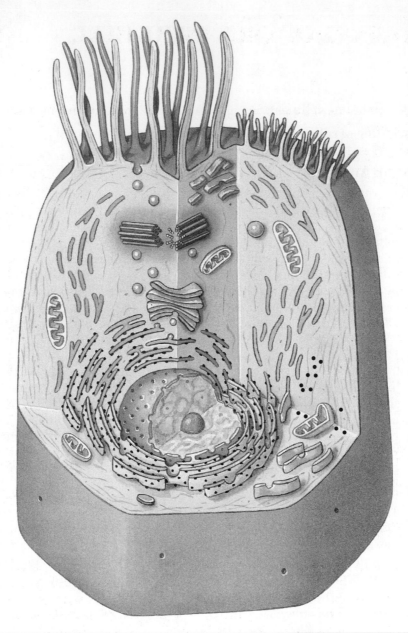

Organelle	Function
Cilia	
Nucleolus	
Ribosomes	
Golgi Apparatus	
Lysosome	
Mitochondrion	
Rough ER	
Centrioles	
Smooth ER	

To Think About

What is the advantage to having the nucleus of a cell enclosed within a membrane?

3

THE TISSUE LEVEL OF ORGANIZATION

Objectives

1. Understand the basis of the classification of body tissues into four major categories: epithelial, connective, muscle, and neural tissue.

2. Describe the function of each tissue type.

3. Identify types of epithelial tissue, connective tissues, muscle tissue, and the basic structure of neural tissue.

4. Recognize under the microscope the types of epithelial tissue, connective tissues, and muscular tissue and the basic structure of neural tissue.

Tissues

The human body contains trillions of cells, but there are only about 200 different types of cells in the body. These cell types combine to form **tissues**, which are collections of specialized cells and cell products that perform specific functions. [∞ p52] Tissues are classified into four **primary types**: epithelial, connective, muscle, and neural.

1. **Epithelial tissue** exists in several structural forms, but they all share a common role: to form continuous layers lining inner surfaces and covering outer surfaces throughout the body, or as glands to secrete various substances or both. [∞ p52] All epithelia have a free surface, termed *apical surface*, and a *basal surface* that is attached to an adhesive basement membrane. This membrane is composed of a basal lamina secreted by the epithelial cells and a reticular lamina secreted by connective tissue cells. When epithelial cells form a single layer, the structure is called **simple epithelium**. When multiple layers are formed, a **stratified epithelium** is present. Epithelial cells are held together by membrane specializations called **junctions** (i.e., tight and adhering junctions, and desmosomes). Epithelial tissues lack large amounts of intercellular matrix or blood vessels between adjacent cells. Because epithelial tissue is avascular, it must obtain nutrients from deeper tissues or from its exposed surfaces. Epithelial layers are always attached to underlying connective tissues and have a free surface exposed to either the external environment (i.e., skin) or to an internal chamber or passageway (i.e., lining of blood vessels).

2. **Connective tissues** are defined as the complex of cells and extracellular materials (matrix) that provides the internal supporting (i.e., bone and cartilage) and connecting (i.e., tendon) framework for all the other tissues of the body. [∞ p64] All connective tissues have three basic components: (a) *extracellular protein fibers*, (b) an *extracellular ground substance*, and (c) a population of *specialized cells* that synthesize these extracellular products. The extracellular fibers and ground substance constitute the **matrix** that surrounds the cells. Connective tissue can be classified into three categories based on the physical nature of the matrix: (a) the *connective tissue proper* (e.g., adipose [fat] tissue and tendons) has many cell types and extracellular fibers in a viscous ground substance, (b) the *fluid connective tissue* (e.g., blood and lymph) has a distinctive population of cells in an aqueous matrix, and (c) the *supporting connective tissue* (e.g., cartilage and bone) has a less diverse cell population and either a gel-like or hard matrix.

3. **Muscle tissue** is composed of individual, relatively long and slender cells (termed fibers), which are specialized for contraction. [∞ p77] Muscle fibers possess organelles and properties that distinguish them from other cells and participate in their excitation and contraction. Three types of muscle tissue are found in the body: (a) **skeletal**, (b) **cardiac**, and (c) **smooth**. Skeletal muscle (e.g., leg muscle) is chiefly responsible for voluntary movement or stabilization of the position of the skeleton. Cardiac muscle is found only in the heart and is responsible for the rhythmic, synchronized pumping action that drives the blood through vessels of the circulatory system. Smooth muscle occurs in the walls of all hollow organs

(e.g., small intestine) and some blood vessels (e.g., arteries and veins), and is responsible for involuntary movement to maintain the diameter of organ and vessel lumens.

4. **Neural tissue**, also called nervous or nerve tissue, is specialized for electrical impulse conduction from one region of the body to another. [∞ p79] Ninety-eight percent of neural tissue is located within the brain and spinal cord. This tissue contains two basic types of cells: **nerve cells** or **neurons** and various supporting cells called **neuroglia**. Neurons are the longest cells in the body, many reaching a meter in length. They communicate with other neurons, muscle, or gland cells, by transmitting impulses (electrical signals). Neuroglia provide structural and metabolic support for neurons and do not transmit electrical impulses.

Epithelial Tissue

Procedure

Identify the following **tissue types, cells,** and **associated structures** using the *microscope slides* provided. First view the sections under low power magnification and then under high power magnification. *As you recognize microscopic samples that compose the four primary tissues, sketch, label the black and white drawing, and highlight with colored pencils the appropriate cells and structures.* Use *Color Plates 1–22* for reference. These color light micrographs are only one example of the described tissue. Use the color photomicrographs and illustrations that appear in *Human Anatomy* as additional references for your microscopic study of tissues.

 Simple Squamous Epithelium   Cells are very thin, flat (often with the nuclear region creating a noticeable apical bulge), making them suitable for lining surfaces across which gases or metabolites must move rapidly. The lining of the terminal air spaces of the lungs, termed alveoli, is a simple squamous epithelium, as are the lining of both blood (endothelium) and lymphatic vessels and the lining (mesothelium) of the ventral body cavities.

   Slide # _____ Simple Squamous Epithelium

See: Color Plate 1

∞ *Fig. 3-4a, p56*

LOCATE

— Nucleus

— Cytoplasm

— Cell membrane of squamous epithelial cells

— Basement membrane

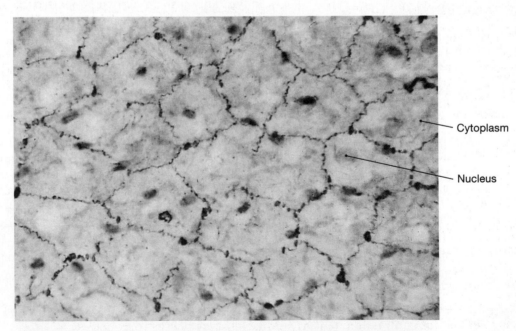

FIGURE 3-1   Simple squamous epithelium (mesothelium) × 263.

 Stratified Squamous Epithelium   This epithelium is composed of a basal layer of germinative cells, termed *stratum basale*, that are attached at the basement membrane to the underlying connective tissue. These basal cells give rise to several overlying layers of more differentiated cells, which are no longer in contact with the basement membrane. The cells lining the surface of this epithelium are very flat, the nuclear region creating a noticeable apical bulge. Stratified epithelia are named according to the shape of the cells at the apical surface. A stratified squamous epithelium forms the outer layer of skin (no nuclei in cells at apical surface) and the lining of the oral cavity, esophagus, rectum, anus, and vagina.

Slide # _____ Stratified Squamous Epithelium

LOCATE

__ Stratified squamous epithelial cells

__ Basement membrane

__ Loose connective tissue

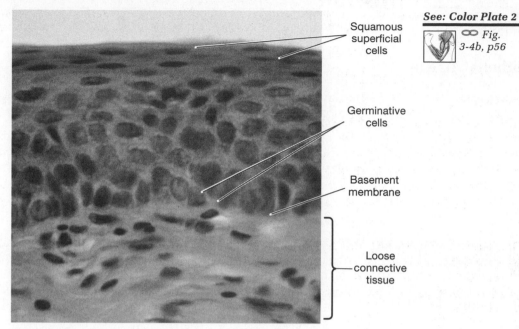

See: Color Plate 2

∞ *Fig. 3-4b, p56*

Squamous superficial cells

Germinative cells

Basement membrane

Loose connective tissue

FIGURE 3-2 Stratified squamous epithelium, skin × 310.

Simple Cuboidal Epithelium Cuboidal cells appear to be of equal height and width. They form some of the tubules and ducts of the kidney and also form ducts of glands; they cover the choroid plexus (capillary layers lining brain ventricles) and ciliary bodies (choroid layer of the eye to which the lens is attached) and are active in both absorption and secretion. Modified simple cuboidal epithelia are prevalent in glands. When glandular cells are grouped into "berry-like" clusters called acini, the simple cuboidal epithelial cells are pyramidal in appearance.

Slide # _____ Simple Cuboidal Epithelium

LOCATE

__ Cuboidal cells

__ Connective tissue

__ Lumen of duct

__ Basement membrane

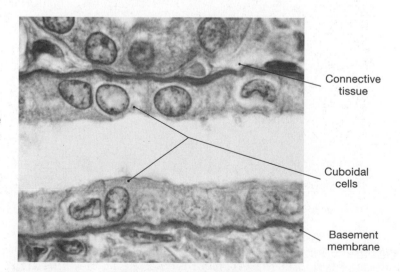

See: Color Plate 3

∞ *Fig. 3-5a, p58*

Connective tissue

Cuboidal cells

Basement membrane

FIGURE 3-3 Simple cuboidal epithelium, kidney tubule × 2000.

Stratified Cuboidal Epithelium This rare epithelium has two or more layers of cells with the apical surface having a cuboidal appearance. It is found along ducts of sweat glands and in the larger ducts of the mammary glands.

Slide # _____ Stratified Cuboidal Epithelium

See: Color Plate 4

 ∞ *Fig. 3-5b, p58*

LOCATE

___ Cuboidal cells

___ Connective tissue

___ Lumen of duct

___ Basement membrane

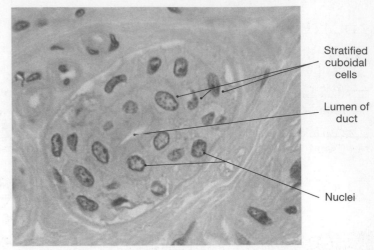

FIGURE 3-4 Stratified cuboidal epithelium, sweat gland duct × 729.

 **Transitional Epithelium** This epithelium is found in the urinary tract, where it lines the bladder, the ureters connecting the kidney to the bladder, and the renal pelvis of the kidney. The cells of this epithelium vary dramatically in shape, from a pyramidal or cuboidal-like to a squamous-like appearance, depending on how distended the structure is by the fluid it contains. In the empty bladder, a multilayered epithelium is found, but with maximal distension only two layers (a basal and a surface layer) are apparent.

Slide # _____ Transitional Epithelium

See: Color Plate 5

∞ *Fig. 3-5c, d, p58*

LOCATE

___ Transitional epithelial cells

___ Balloon-shaped squamous cells

___ Connective tissue and smooth muscle layers

___ Basement membrane

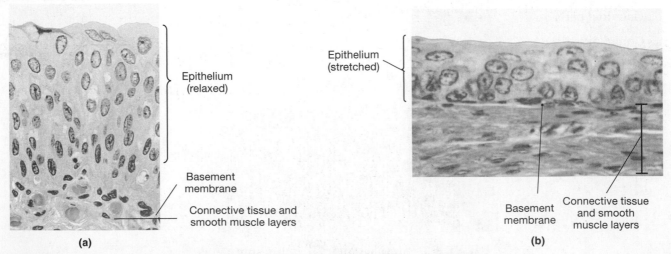

FIGURE 3-5 Transitional epithelium (a) empty bladder × 413 (b) full bladder × 486.

Simple Columnar Epithelium   Columnar cells are taller than they are wide and are the most strikingly polarized of the epithelia. Active in secretion, absorption or both, simple columnar epithelia form linings throughout the digestive tract, at various sites in the respiratory tract, in portions of the female reproductive system, in the urinary system, and in the major ducts of many glands. Goblet cells are single-celled exocrine glands that secrete mucin and are scattered among the columnar epithelial cells. Goblet cells are easily distinguished by their wine glass appearance.

> **Slide #** _____ Simple Columnar Epithelium

LOCATE

See: *Color Plate 6*

∞ Figs. 3-6a, p59

— Simple columnar epithelial cells

— Cytoplasm

— Nucleus

— Basement membrane

— Loose connective tissue

— Goblet cell

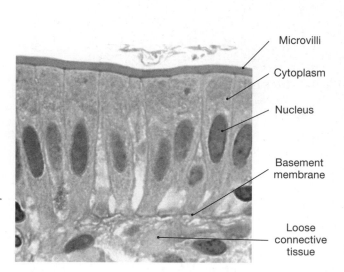

Microvilli

Cytoplasm

Nucleus

Basement membrane

Loose connective tissue

FIGURE 3-6   Simple columnar epithelium, small intestine × 550.

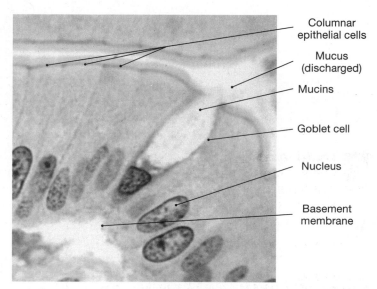

Columnar epithelial cells

Mucus (discharged)

Mucins

Goblet cell

Nucleus

Basement membrane

FIGURE 3-7   Goblet cells, small intestine × 1150.

 Stratified Columnar Epithelium   This epithelium has two or more layers of cells with the apical surface layer having a columnar appearance. Stratified columnar epithelia form parts of salivary gland ducts, small areas of the pharynx, urethra, anus, uterine tubes, and mammary glands.

> **Slide #** _____ Stratified Columnar Epithelium

LOCATE

See: *Color Plate 7*

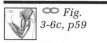

∞ Fig. 3-6c, p59

— Stratified columnar epithelium

— Nucleus of stratified columnar epithelial cell

— Basement membrane

— Cytoplasm

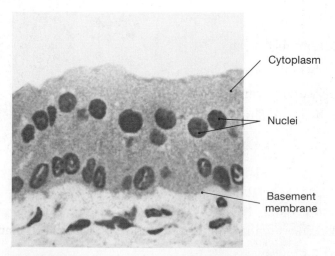

Cytoplasm

Nuclei

Basement membrane

FIGURE 3-8   Stratified columnar epithelium, salivary gland × 324.

Pseudostratified, Ciliated, Columnar Epithelium This epithelium is composed of a single layer of cells of differing heights. Nuclei are observed at different levels because the cells extend varied distances from the basement membrane. Pseudostratified epithelial cells have a multilayered appearance; however, it is not stratified because all of the cells maintain contact with the underlying basement membrane. The apical or free surface of the pseudostratified columnar epithelium characteristically possesses cilia and goblet cells. Pseudostratified columnar epithelia are found in the nasal cavity, trachea, bronchi, and large bronchioles.

Slide # _____ Pseudostratified Ciliated Columnar Epithelium With Goblet Cells

See: Color Plate 8

∞ *Fig.*
3-6b, p59

LOCATE

— Cilia

— Pseudostratified columnar epithelial cells

— Nuclei

— Loose connective tissue

— Basement membrane

— Goblet cells

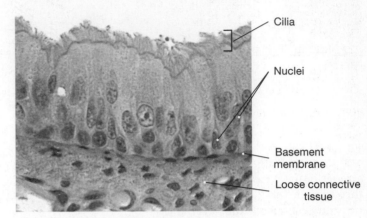

FIGURE 3-9 Pseudostratified epithelium, trachea × 324.

Connective Tissues

 Areolar (Loose) Connective Tissue Areolar connective tissue is characterized by a relative predominance of both connective tissue resident cells (e.g., fibroblasts, adipocytes, macrophages, and mast cells) and gel-like ground substance over fibrous elements. This soft, pliable tissue is widely and abundantly distributed in the body as a packing material between other tissues. Areolar connective tissue ensheathes blood vessels and nerves, surrounds glands, and cushions and attaches skin to underlying structures.

Slide # _____ Areolar Connective Tissue

See: Color Plate 9

∞ *Figs.*
3-11b, p65
3-13a, p67

LOCATE

— Fibers {Elastic / Collagen / Reticular

— Mast cell

— Macrophage

— Fibroblasts

— Lymphocyte

— Mesenchymal cell

— Fat cell (adipocyte)

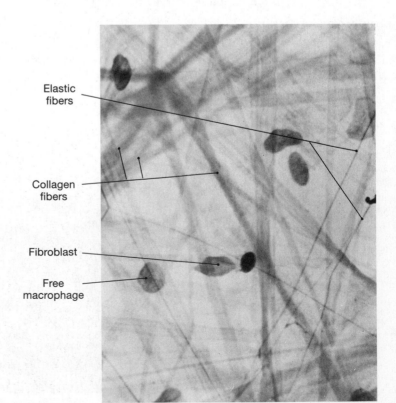

FIGURE 3-10 Loose (areolar) connective tissue, mesothelium × 520.

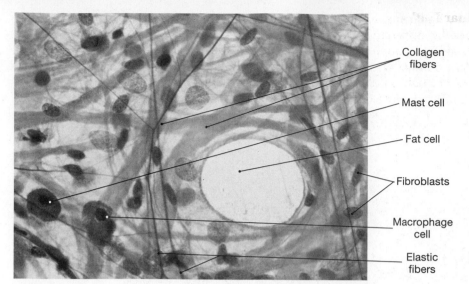

Collagen fibers

Mast cell

Fat cell

Fibroblasts

Macrophage cell

Elastic fibers

FIGURE 3-11 Areolar connective tissue, pleura × 380.

Adipose Tissue Adipose tissue, commonly called fatty tissue, is a type of connective tissue with a loose association of lipid-filled cells termed *adipocytes* in a sparse matrix of collagen fibers and viscous ground substance. The lipid is extracted during preparation, thus the adipocytes appear empty. Adipose tissue is located throughout the body and is especially prominent beneath the skin.

Slide # _____ Adipose Tissue

See: Color Plate 10

∞ *Fig. 3-13b, p67*

LOCATE

___ Adipocytes (fat cells)

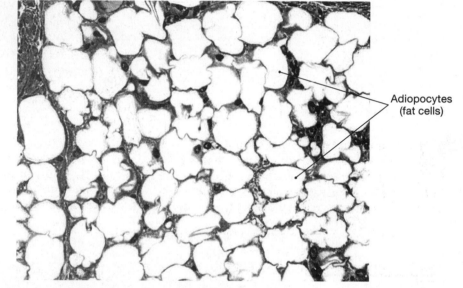

Adiopocytes (fat cells)

FIGURE 3-12 Adipose tissue, subcutaneous fat × 133.

Reticular Tissue Reticular tissue has a meshwork of delicate, branched reticular fibers. These prominently stained fibers support other cell types, such as fixed macrophages and fibroblasts. Reticular tissue forms a meshwork between the cells of the liver, spleen, lymph nodes, and bone marrow.

Slide # _____ Reticular Tissue

See: Color Plate 11

∞ *Fig. 3-13c, p67*

LOCATE

___ Reticular fibers

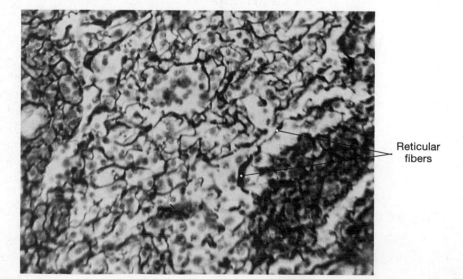

Reticular fibers

FIGURE 3-13 Reticular tissue, liver × 375.

Dense Regular Connective Tissue (Tendon)   In dense connective tissue, collagen fibers predominate over cells. The fibers are oriented preferentially in such dense regular tissue as tendons, ligaments, and aponeuroses, where the parallel arrangement of fibers helps to resist stress in one direction.

Slide # _____ White Fibrous (Dense Regular) Connective Tissue, Tendon

See: Color Plate 12

 ∞ *Fig.*
3-14a, p69

LOCATE

___ Skeletal muscle (may be present)

___ Tendon

___ Fibroblast cells

___ Collagen fibers

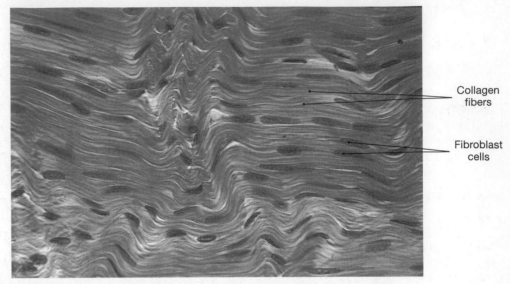

FIGURE 3-14   Dense regular connective tissue, tendon × 364.

 Elastic Ligaments   Elastic ligaments resemble tendons microscopically, but the ligaments contain large amounts of both elastic and collagen fibers that are packed less densely. Note the arrangement of collagen fibers and the significant number of elastic fibers that appear smaller and wavelike. Fibroblast cells are abundant between fibers. These ligaments are found along the vertebral column and are very important in stabilizing the positions of the vertebrae and yet permitting a limited amount of movement.

Slide # _____ Elastic Tissue

See: Color Plate 13

∞ *Fig.*
3-14b, p69

LOCATE

___ Bundles of elastic fibers

___ Fibroblast cells

___ Collagen fibers

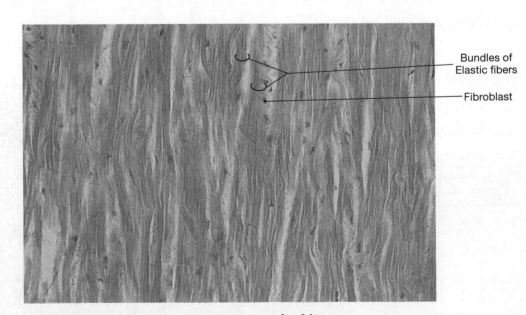

FIGURE 3-15   Elastic tissue in intervertebral ligament × 887.

Dense Irregular Connective Tissue   In dense connective tissue, collagen fibers predominate over cells. The fibers are not preferentially oriented in dense irregular connective tissue where the arrangement of fibers extends in all directions to help resist stress from any direction. Dense irregular connective tissue forms the connective tissue capsule around many internal organs, the dermis of the skin, and the wrapping sheath around muscle.

Slide # _____ Skin

See: Color Plate 14

∞ *Fig. 3-14c, p69*

LOCATE

— Fibroblast cells

— Collagen fibers

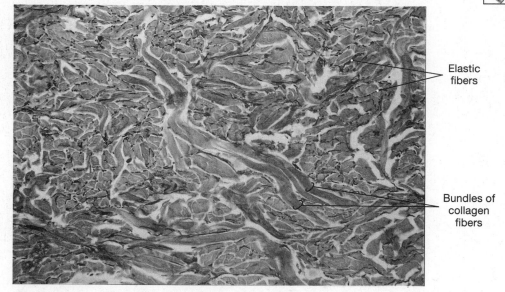

Elastic fibers

Bundles of collagen fibers

FIGURE 3-16   Dense irregular tissue, dermis × 111.

Hyaline Cartilage   Hyaline cartilage is a specialized form of supporting connective tissue composed of cells, fibrous macromolecules, and ground substance. The cartilage cells called **chondroblasts** synthesize **matrix** and, when mature, the adult cartilage cells or **chondrocytes** encase themselves within cavities called **lacunae**. Hyaline cartilage is present in the walls of the major respiratory passages (nose, larynx, trachea, and bronchi), the ventral ends of the ribs, and on the bone surfaces of joints. Fresh hyaline cartilage is translucent bluish-white. Unstained, hyaline cartilage matrix appears homogeneous, and fibers are not visible within the ground substance by conventional light microscopy.

Slide # _____ Hyaline Cartilage

See: Color Plate 15

∞ *Fig. 3-17a, p72*

LOCATE

— Chondroblasts

— Chondrocytes

— Lacuna

— Intercellular matrix

— Collagen fibers

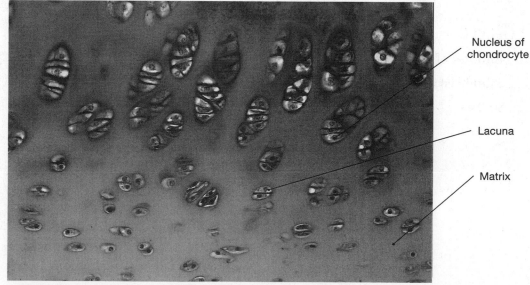

Nucleus of chondrocyte

Lacuna

Matrix

FIGURE 3-17   Hyaline cartilage, covering bone surfaces at shoulder joint × 455.

Elastic Cartilage Elastic cartilage is typically found in regions requiring a flexible form of support: the auricle of the ear (the fleshy moveable part of your ear), the walls of the external auditory canals and auditory tubes (tubes that link the middle ear cavity and the pharynx [throat]), the epiglottis, and some cartilages of the larynx. A preponderance of elastic fibers gives fresh elastic cartilage a yellow and more opaque look than hyaline cartilage. In section, the elastic fibers are highly branched and often obscure the ground substance. The **chondrocytes** occur within lacunae either singly or in cell nests.

Slide # _____ Elastic Cartilage

See: Color Plate 16

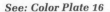 ∞ *Fig.
3-17b, p72*

LOCATE

— Chondrocytes

— Lacuna

— Matrix

— Elastic fibers
  in matrix

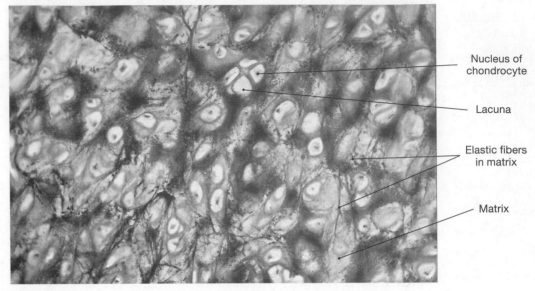

FIGURE 3-18 Elastic cartilage, external ear (pinna) × 320.

 Fibrocartilage Fibrocartilage is located in a few regions where firm support and tensile strength are necessary: the intervertebral discs (pads of fibrocartilage between the vertebrae), the pubic symphysis (pad of fibrocartilage that joins the left and right hipbones at the anterior), the attachments of tendons and ligaments, and the rims of certain articular cartilage (i.e. hip joint). It differs dramatically from hyaline cartilage in that the ground substance is sparse, and numerous collagenous fibers are visible as large irregular bundles between groups of **chondrocytes**. Occasionally, the chondrocytes are aligned in rows parallel to the collagen bundles.

Slide # _____ Fibrocartilage

See: Color Plate 17

∞ *Fig.
3-17c, p72*

LOCATE

— Chondrocytes

— Lacuna

— Fibrous matrix
  (collagen fibers
  in matrix)

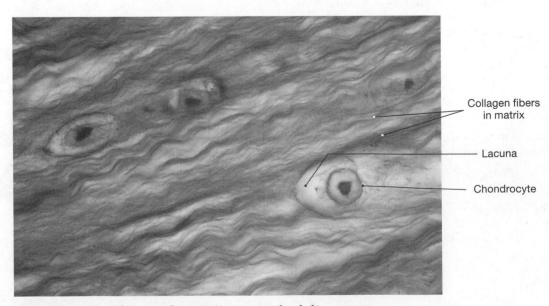

FIGURE 3-19 Fibrocartilage, in intervertebral disc × 750.

Bone, Compact   Bone differs from other connective tissues by its rigidity and hardness owing to the inorganic salts (calcium phosphate) impregnated into the collagen fibers of the **matrix**. On a gross level all bones are composed of two architectural structures: compact and cancellous. Adult bone is *lamellated* with a concentric ring pattern of organization. Each group of concentric rings is a structural and functional unit of mature compact bone termed **osteon (Haversian System)**. Within an osteon, bone cells called **osteocytes** are encased in cavities called **lacunae**. The osteocytes are arranged in lamellae around a **central canal** (Haversian canal) that contains one or more blood vessels (usually not visible). Compact bone forms the outer walls of all bones. Typically in observing this specimen, the intact matrix and central canals will appear off white, and the lacunae will appear black.

Slide # _____ Bone, Compact

LOCATE

__ Osteon (Haversian system)

__ Central canal (Haversian canal)

__ Osteocytes

__ Lacunae

__ Matrix

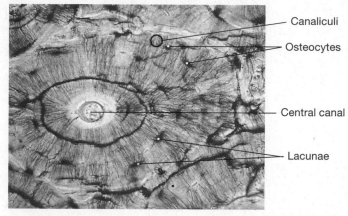

See: Color Plate 18

∞ Fig. 3-18, p73

FIGURE 3-20   Compact bone × 450.

Muscle Tissues

Skeletal Muscle Fibers   Skeletal muscle is associated, as the name implies, with the skeleton. The parallel muscle cells, usually called **muscle fibers**, are greatly elongated and each contains many peripheral nuclei. Because of conspicuous transverse striations or bands, owing to the precise arrangement of actin and myosin filaments, skeletal muscle is also referred to as *striated* muscle. It is controlled by the somatic nervous system and is often called voluntary muscle.

Slide # _____ Skeletal Muscle

LOCATE

__ Nucleus

__ Muscle fibers

__ Striations

__ Sarcolemma (cell membrane)

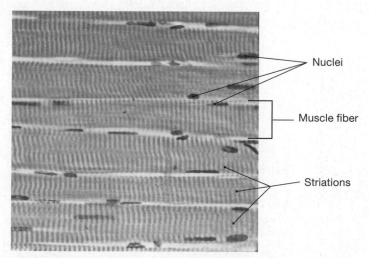

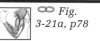

See: Color Plate 19

∞ Fig. 3-21a, p78

FIGURE 3-21   Skeletal muscle tissue × 181.

 **Smooth Muscle** Smooth muscle tissue is the simplest in appearance of the muscle types. It consists of narrow and relatively short tapering cells with no striations, each with a single centrally located nucleus. This type of muscle occurs in the walls of the blood vessels, and viscera (digestive, respiratory, urinary, and reproductive systems) and hence is often called *nonstriated involuntary muscle.*

Slide # _____ Smooth Muscle

See: Color Plate 20

∞ *Fig.
3-21c, p78*

LOCATE

___ Smooth muscle cells

___ Nucleus

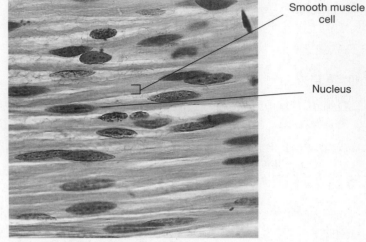

FIGURE 3-22 Smooth muscle tissue walls of digestive system × 235.

 **Cardiac Muscle** Cardiac muscle tissue is a highly specialized form of involuntary, striated muscle found only in the heart. The small cardiac muscle cell, or **cardiocyte**, is similar to skeletal muscle fibers in that the cells are transversely striated and multinucleated, but as in smooth muscle, the nuclei are centrally located.

Slide # _____ Cardiac Muscle

See: Color Plate 21

∞ *Fig.
3-21b, p78*

LOCATE

___ Cardiocytes

___ Intercalated discs

___ Nucleus

___ Striations

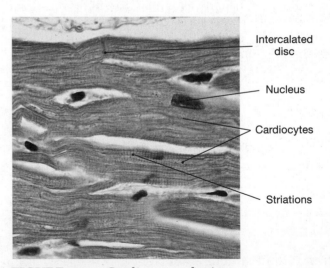

FIGURE 3-23 Cardiac muscle tissue, heart × 450.

Neural Tissue

Motor Neuron The most typical multipolar neurons are the motor cells, or motor neurons of the ventral horn of the spinal cord. The large **soma** (expanded part of the cell containing the prominent nucleus) usually gives rise to several large branching processes, termed **dendrites**, which radiate in all directions, and to a single **axon**. Dendrites receive incoming messages while axons conduct outgoing messages. Observe the small supporting cells, neuroglia, which are abundant in the background surrounding each neuron.

Slide # _____ Motor Neurons, Spinal Cord

LOCATE

___ Soma

___ Nucleus

___ Nucleolus

___ Axon

___ Dendrites

See: Color Plate 22

∞ *Fig. 3-22, p79*

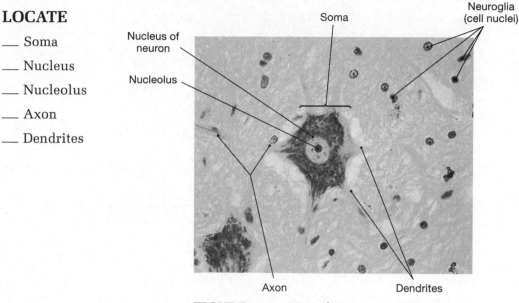

FIGURE 3-24 Neural tissue, spinal cord × 580.

To Think About

1. How does the structure of a ligament determine its function?

2. Why does pinching the skin of a body region not usually distort or damage the underlying muscles?

3. Contrast the structural differences between bone and cartilage in terms of strength.

4

THE INTEGUMENTARY SYSTEM

Objectives

1. Use a model of the skin to examine the organization and identify the relationship of epidermis, dermis, and hypodermis layers; the continuous cellular layers of the epidermis; and accessory structures and their location in the dermis.

2. Recognize under the microscope; the epidermis, dermis, and hypodermis layers; the five cell layers of the epidermis; and the accessory structures and their distribution in the dermis.

3. Use both the microscope and a skin model to describe the structure of hair, hair follicles, and their associated structures.

Description of Integumentary Structure

The skin, also called the **integumentary system** or *integument* is composed of two major components, the cutaneous membrane and accessory structures. The **cutaneous membrane** is an epithelial membrane organized into a stratified squamous epithelium (**epidermis**) and underlying connective tissues (**dermis**). The epidermis consists of four or five continuous strata or layers. [∞ p89] Underlying the epidermis is the dermis, which contains **accessory structures** derived from the epidermis (hair, nails, and multicellular exocrine glands), connective tissue cells and extracellular matrix, blood vessels, and receptors of sensory neurons. The dermis is well supplied with blood vessels that provide nutrients to the cutaneous membrane and sensory receptors that monitor the integument and its interface with the external world. [∞ p95] Under the dermis is the loose connective tissue of the **subcutaneous layer**, *superficial fascia*, or *hypodermis*. The hypodermis separates the integument from the underlying structures, such as muscles and bone, but is not part of the integument. The hypodermis is typically examined with integumentary structures. [∞ p95] This system functions partly to regulate temperature because of the extent of the blood supply to the dermis and the presence of numerous sweat glands, and as a sensory buffer for the immediate external environment of the body.

Microscopic Examination of the Skin

Procedure

Locate the components of the cutaneous membrane and accessory structures by viewing the microscope slides provided under either low power magnification or dissecting microscope and then under high power magnification. Identify microscopically epidermal and dermal layers and accessory structures. Sketch, label, and highlight with colored pencils the appropriate cell layers and structures in the black and white drawings. Use *Color Plates 14, 23* and *Figure 4-1* for reference. As additional reference, use the color photomicrographs and illustrations that appear in *Human Anatomy* for your study of the skin.

Skin Locate epidermis, dermis, and hypodermis regions on your specimen. The basement membrane is a good landmark because it identifies the interface between epidermal and dermal regions. Beginning at the basement membrane, identify each epidermal stratum, starting with the innermost, the stratum germinativum. This layer will be most evident in a specimen from the palm or sole (thick skin). Precise boundaries between strata are not always distinguished easily. The dermis contains both blood vessels and protein fibers of the extracellular matrix interwoven into the connective tissue.

Slide # _____ Skin

LOCATE

— Epidermis

— Epidermal ridge

— Stratum corneum

— Stratum lucidum

— Stratum granulosum

— Stratum spinosum

— Stratum germinativum

— Dermis

— Dermal papillae

See: Color Plates 14, 23, Fig. 4-1

 ∞ Figs.
4-1, p88
4-2, p89
4-3, p90
4-4, p92

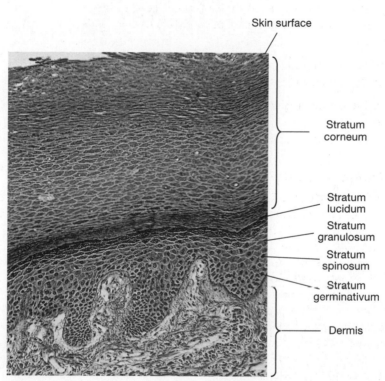

Skin surface

Stratum corneum

Stratum lucidum

Stratum granulosum

Stratum spinosum

Stratum germinativum

Dermis

FIGURE 4-1 The structure of the epidermis, skin × 200

Scalp Under low magnification, many hair and integumentary accessory structures, including hair follicles, hair shafts, arrector pili muscles, sweat and sebaceous glands, blood vessels, and sensory nerve endings (typically Tactile [Meissner's] corpuscles) will be easily observed. Notice the locations and relationships of the accessory structures. Compare the association and location of the arrector pili muscles, sebaceous glands, and hair shafts.

Slide # _____ Skin, Scalp

See: Fig. 4-2

∞ *Figs.*
4-9, p96
4-13, p100
4-14, p101

LOCATE

__ Epidermis

__ Dermis

__ Hypodermis

__ Hair shaft

__ Hair follicle

__ Arrector pili muscle

__ Sebaceous gland

__ Sweat gland

__ Sweat gland ducts

__ Tactile (Meissner's) corpuscles

__ Artery

__ Vein

__ Adipose cells

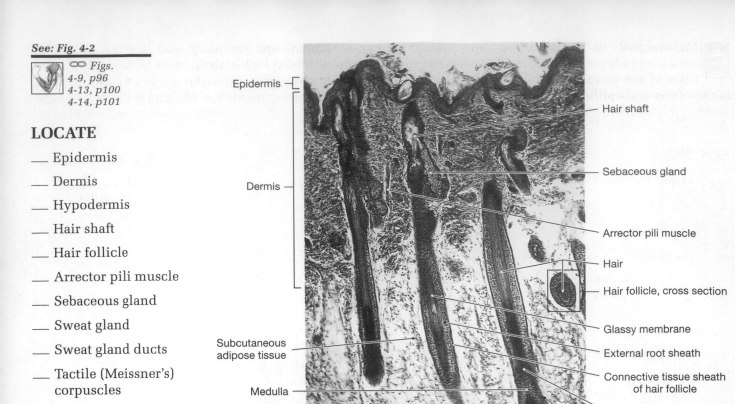

Epidermis

Dermis

Subcutaneous adipose tissue

Medulla

Papilla

Hair shaft

Sebaceous gland

Arrector pili muscle

Hair

Hair follicle, cross section

Glassy membrane

External root sheath

Connective tissue sheath of hair follicle

Cortex

Hair bulb

(a)

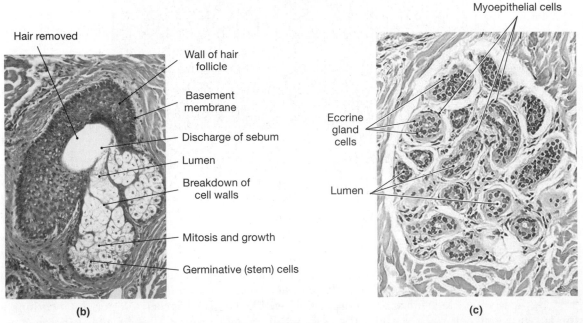

Hair removed

Wall of hair follicle

Basement membrane

Discharge of sebum

Lumen

Breakdown of cell walls

Mitosis and growth

Germinative (stem) cells

(b)

Myoepithelial cells

Eccrine gland cells

Lumen

(c)

FIGURE 4-2 Skin (a) Accessory structures × 66 (b) Sebaceous glands and follicles × 120 (c) Eccrine sweat gland × 243.

Melanocytes Skin color is determined by the amount of carotene and oxyhemoglobin, and the amount of melanin produced by melanocytes. Melanocytes are best observed at high magnification between epithelial cells of the stratum germinativum of thick skin specimens. Melanin pigments are found within both melanocytes and epithelial cells of the stratum germinativum (sometimes called stratum basale) layers.

Slide # _____ Pigmented Skin, Showing Melanocytes

LOCATE

See: Fig. 4-3

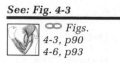 ∞ *Figs. 4-3, p90 4-6, p93*

— Stratum granulosum

— Stratum germinativum (stratum basale)

— Melanocytes

— Melanin pigment in epidermal cell and melanocytes

— Basement membrane

— Dermis

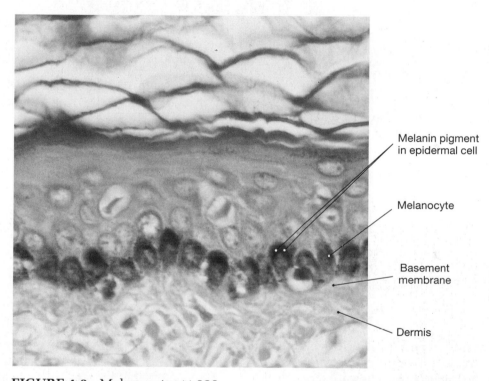

Melanin pigment in epidermal cell

Melanocyte

Basement membrane

Dermis

FIGURE 4-3 Melanocytes × 860.

Hair Structure   Compare and relate both frontal and cross sections through hair follicles and shafts. Begin with low magnification using longitudinal sections to observe the follicle and hair shaft structures derived from the overlying epidermis. Locate well-formed hair bulbs and hair shafts extending into the dermis. Begin your identification of hair structures at the hair bulb.

Slide # _____ Scalp

See: Figs. 4-4

∞ *Figs.*
4-9b, p96
4-10, p97
4-13, p100
4-14, p101

LOCATE

— Dermis

— Root hair plexus

— Hair follicle

— Hair bulb

— Arrector pili

— Cuticle

— Papilla

— Matrix

— Medulla

— Cortex

— Hair shaft

— Connective tissue sheath of follicle wall

— Glassy membrane

— External root sheath

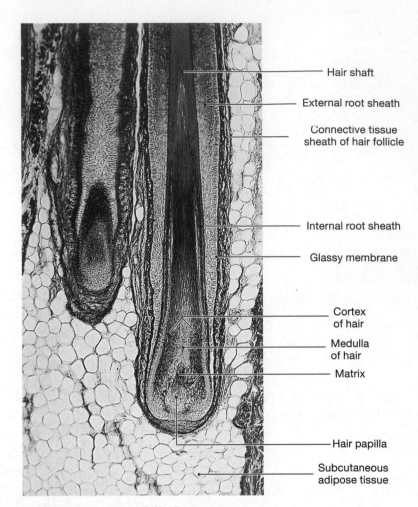

FIGURE 4-4   Hair follicles × 60.

Lamellated (Pacinian) Corpuscle   These deep touch receptors appear as lamellated corpuscles (concentric circles) that are sensitive to pulsating or high frequency vibrating stimuli. They are easily recognized by their concentric ring (lamellar) structures. They are large, oval to round receptors that have a bull's eye appearance.

Slide # _____ Pacinian Corpuscles, Skin

See: Fig. 4-5

∞ *Figs.*
4-2, p89
18-3f, p471

LOCATE

— Lamellated (Pacinian) corpuscles in dermis

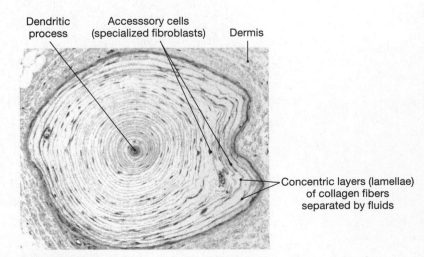

FIGURE 4-5   Pacinian corpuscle × 125.

Identification of Integumentary Structures

Identify the following integumentary structures using a *model of the skin* and *Color Plates 27* and *28* for reference.

LOCATE

See: *Color Plates 27, 28, Fig. 4-6*

∞ *Figs.*
4-2, p89
4-9b, p96

___ Epidermis

___ Dermis

___ Bundles of collagen fibers in dermis

___ Hypodermis (subcutaneous layer)

___ Hair shaft

___ Pore of sweat gland

___ Tactile (Meissner's) corpuscle

___ Sebaceous gland

___ Arrector pili

___ Sweat gland duct

___ Hair follicle

___ Sweat gland

___ Artery/vein (cutaneous plexus)

___ Adipose cells (adipocytes)

___ Lamellated (Pacinian) corpuscle

Anatomical Identification Review

1. Label the correct strata layers.

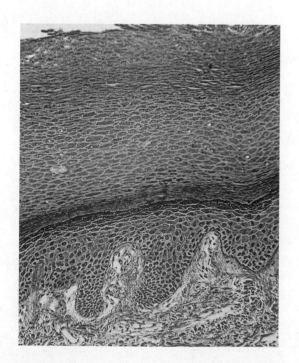

2. Write the correct labels next to the appropriate structures.

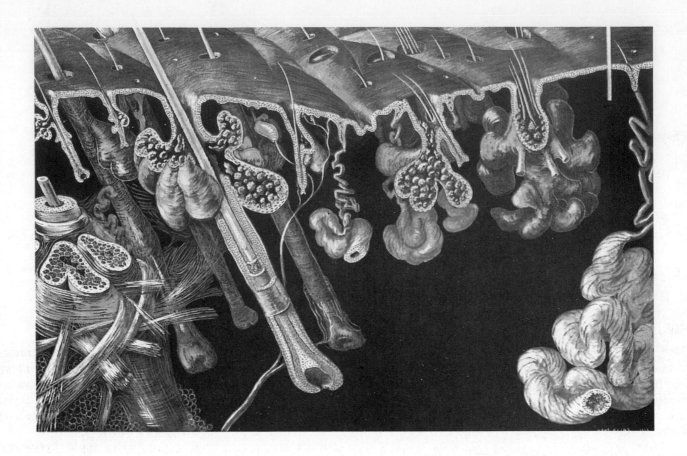

To Think About

1. What is the physiological function of hair on the scalp?

2. What is the function of melanocytes in skin? Why are fair-skinned people more likely to develop skin cancers in sun-belt regions (i.e., Florida)?

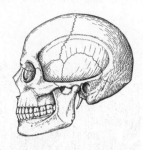

5

THE SKELETAL SYSTEM:
Osseous Tissue and Skeletal Structure

Objectives

1. Identify and describe similarities and differences between compact and spongy bone.

2. Observe and identify osteon component structures on a compact bone model.

3. Recognize under the microscope the organization of compact bone histology.

4. Recognize shapes and describe characteristics of the six anatomical classification categories of human bone specimens.

5. Identify bone markings and surface feature examples on the human skeleton.

Structure of Bone: Gross and Histology Organization

Bone is **osseous tissue**, a highly specialized type of supporting connective tissue that forms the skeletal framework. [∞ p113] Like other connective tissues, it contains specialized cells and an **extracellular matrix** consisting of both protein fibers and ground substance. Bone is distinguished from other types of connective tissue by the preponderance of solid extracellular material. Crystals of **hydroxyapatite**, $Ca_{10}(PO_4)_6(OH)_2$, form small plates that lie alongside collagen fibers [∞ p113]; the hydroxyapatite accounts for two-thirds of the matrix with the remaining one-third composed of collagen fibers and other calcium salts, such as calcium carbonate.

There are two types of osseous tissue: *compact* or *dense bone*, and *spongy or cancellous bone*. [∞ p113] Compact bone forms the walls and outer surfaces of bones. Compact bone is a dense, solid mass whose basic structural and functional unit is the **osteon**, or *Haversian system*. Osteons are cylindrical structures oriented parallel to the long axis of the bone. Compact bone thickness varies among individual bones. **Periosteum** is a connective tissue component of the deep fascia that covers compact bone surfaces everywhere except where *hyaline articular cartilage* covers articulating bone surfaces.

Spongy bone is formed by an open network of interconnecting and branching struts and plates (**trabeculae**). There are no osteons in spongy bone. The spicules and trabeculae create an open honeycomb network of bone with a sponge-like appearance. Both compact and spongy regions occur in bones of the skeleton. Spongy bone forms the internal structure where it surrounds loose connective tissue of the **marrow cavity**. [∞ p114]

Compact bone is thickest where stresses arrive from a limited range of directions. In long bones, stresses are normally applied along the tubular shaft, or **diaphysis**. Spongy bone abounds where bones are not heavily stressed, or where stress arrives from many directions. The expanded end regions of long bones are the **epiphyses**. [∞ p116]

Procedure

Identify the gross bone structures on either a *bone* or a *bone model*. Use the color photomicrographs and illustrations that appear in *Human Anatomy* as additional reference for your gross microscopic study of bone.

Bone Specimen

See: Fig. 5-1

∞ Fig. 5-3, p116

LOCATE

__ Compact bone

__ Spongy bone

__ Diaphysis

__ Epiphysis

__ Medullary (marrow) cavity

Spongy bone

Compact bone

Medullary (marrow) cavity

Epiphysis (head)

Diaphysis (shaft)

Epiphysis

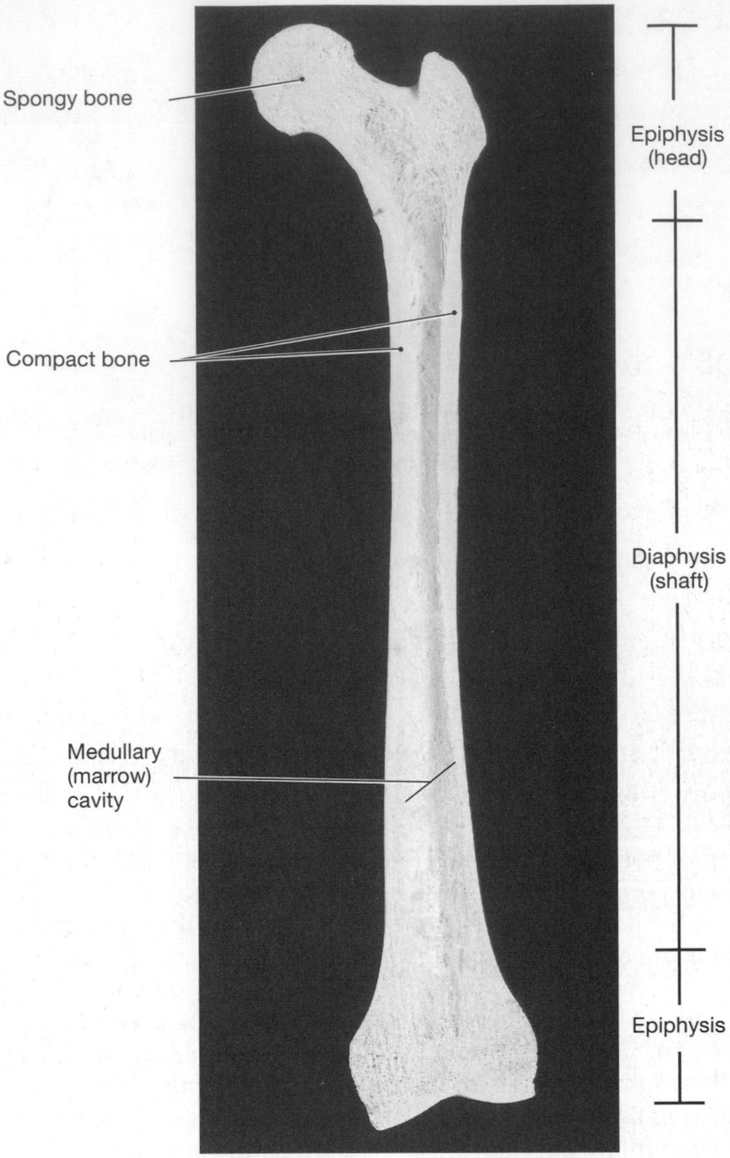

FIGURE 5-1 The femur as seen in a coronal section.

Observation of Osteon Model/Compact Bone, Ground Thin

Bone, Compact The structural and functional unit of compact bone is a cylindrical structure termed **osteon** (*Haversian System*), which is oriented parallel to the long axis of the bone. In cross section, the osteon displays a pattern of concentric rings (*lamellae*). This organization gives the osteon a bull's eye pattern in which the center of the pattern is occupied by the **central canal** (*Haversian canal*) containing one or more blood vessels. **Osteocytes** (bone cells) occupy spaces in the hard bone (**lacunae**) within the matrix between concentric lamellae. Lamellae appear as rings of the bull's eye. Miniature canals, termed **canaliculi**, enclose cytoplasmic projections of osteocytes and radiate between lacunae to provide a conduit for nutrient diffusion between neighboring osteocytes. The bone matrix is solid and primarily contains calcium phosphate crystals deposited upon collagen fibers and some ground substance. When viewed with the microscope, the hard bone matrix will appear off white, while the lacunae, canaliculi (microscopic channels between osteocytes), and sometimes the central canals appear black. Each osteon is connected to its neighbor through **perforating canals** (*Volkmanns canals*), which are oriented transverse to the long axis of the bone. The periosteum (connective tissue wrap around bones) is absent in specimens.

Procedure

Identify the components of **osteon structure** using a *model* depicting osteons (Haversian systems), *Color Plate 29* and *Figure 5-2*.

LOCATE

See: Color Plate 29,
Fig. 5-2

∞ Fig.
5-2b, p115

__ Periosteum

__ Osteon

__ Concentric lamellae

__ Central canal (Haversian)

__ Blood vessels

__ Lacunae

__ Osteocytes

__ Canaliculi

__ Matrix

__ Perforating canals (canals of Volkmann)

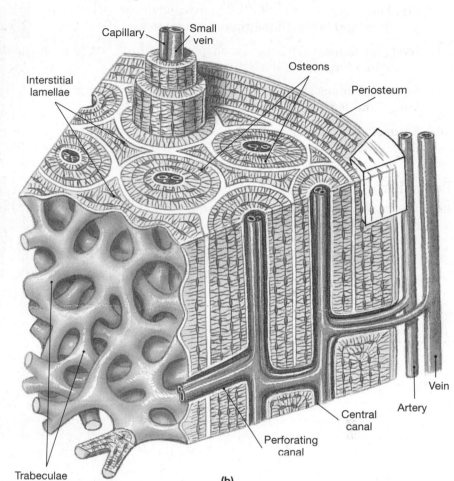

(a)

(b)

FIGURE 5-2 Structure of bone (a) thin section through compact bone × 272 (b) diagrammatic view.

Procedure

 Identify the following structures using the microscope slide provided by first viewing under low power magnification and then under high power magnification to see the details of the listed structures. As you identify the osteon structures, briefly write a description of their function next to their structure.

Slide # _____ Compact Bone, Ground Thin

See: Color Plate 18,
Figs. 5-2, 3-2

∞ Figs.
5-2, p115
3-18, p73

LOCATE

__ Osteon (Haversian system)

__ Canaliculi

__ Osteocytes

__ Central canal (Haversian canal)

__ Blood vessels within central canal

__ Lacunae

__ Matrix

__ Lamellae

Classification of Bones

The human skeleton contains approximately 206 bones. Human skeleton bones are classified into six general categories according to shape:

1. **Long bones** have distinct diaphysis (shaft) and epiphysis (end) regions and enclose a marrow cavity (examples: bones of the limbs).

2. **Short bones** are almost cube-shaped in appearance and lack both diaphysis and epiphysis portions (examples: bones of the wrists and ankles).

3. **Flat bones** are usually thin and composed of parallel plates of compact bone (examples: sternum, ribs, scapulae, and bones that form the roof of the skull).

4. **Irregular bones** have irregular shapes with short, flat-notched, or ridged surfaces (examples: the block-like bones that form the spinal column vertebrae).

5. **Sesamoid bones** are typically round, small, flat bones that develop inside tendons near joints in the hands, knees, and feet (example: kneecap).

6. **Sutural (Wormian) bones** are small, flat, oddly shaped bones located between the flat bones of the skull in the suture line. Sutural bones vary in number, shape, and size.

Procedure

Identify on a *human skeleton* examples of the following **anatomical classification of bone shapes**, using *Figure 5-3* and *Color Plates 30, 31* as reference.

See: Color Plates 30, 31, Fig. 5-3

LOCATE

___ Long bones

___ Short bones

___ Flat bones

___ Irregular bones

___ Sutural bones

___ Sesamoid bones

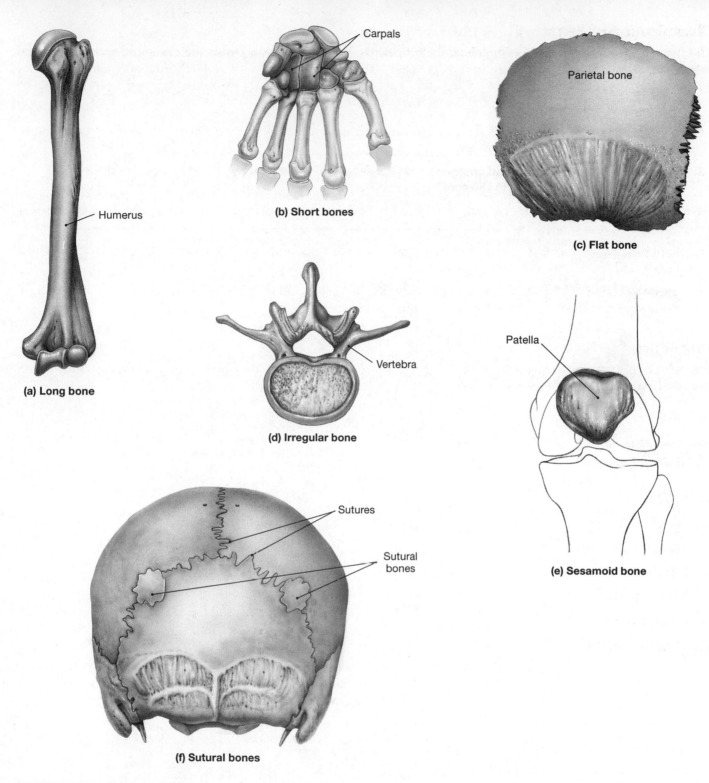

(a) Long bone

Humerus

(b) Short bones

Carpals

(c) Flat bone

Parietal bone

(d) Irregular bone

Vertebra

(e) Sesamoid bone

Patella

(f) Sutural bones

Sutures

Sutural bones

FIGURE 5-3   Shapes of bones.

Observation of Bone Markings (Surface Features)

Identify on a *human skeleton* examples of the following **common bone markings (surface features)**, using *Figure 5-4* and *Table 5-1* as reference. Bone markings occur either because of stress on a bone from attached muscles or because of articulations or passageways for other structures (such as blood vessels and nerves). Markings can be grouped into two types: (1) *elevations*, *projections*, and *processes*, or simply "outies," and (2) *depressions* and *openings*, or simply "innies."

TABLE 5-1   Common Bone Marking Terminology

General Description	Anatomical Term	Definition
Elevations and projections (general)	**Process**	Any projection or bump
	Ramus	An extension of bone making an angle to the rest of the structure
Processes formed where tendons or ligaments attach	**Trochanter**	A large, rough projection
	Tuberosity	A relatively smaller, rough projection
	Tubercle	A small, rounded projection
	Crest	A prominent ridge
	Line	A low ridge
	Spine	A pointed process
Processes formed for articulation with adjacent bones	**Head**	The expanded articular end of an epiphysis, separated from the shaft by a narrower neck
	Neck	A narrow connection between the epiphysis and diaphysis
	Condyle	A smooth, rounded or articular process
	Trochlea	A smooth, grooved articular process shaped like a pulley
	Facet	A small, flat articular surface
Depressions Openings	**Fossa**	A shallow depression
	Sulcus	A narrow groove
	Foramen	A rounded passageway for blood vessels and/or nerves
	Fissure	An elongate cleft
	Canal	A large-diameter passageway through the substance of a bone
	Sinus or antrum	A chamber within a bone, normally filled with air

See: Fig. 5-4

LOCATE

Elevations and Projections

___ Process

___ Ramus

Processes formed where tendons or ligaments attach

___ Trochanter

___ Tuberosity

___ Tubercle

___ Crest

___ Line

___ Spine

Processes formed for articulation with adjacent bones

___ Condyle

___ Trochlea

___ Facet

___ Head

Depressions

___ Fossa

Openings

___ Foramen

___ Fissure

___ Meatus (Canal)

___ Sinus (Antrum)

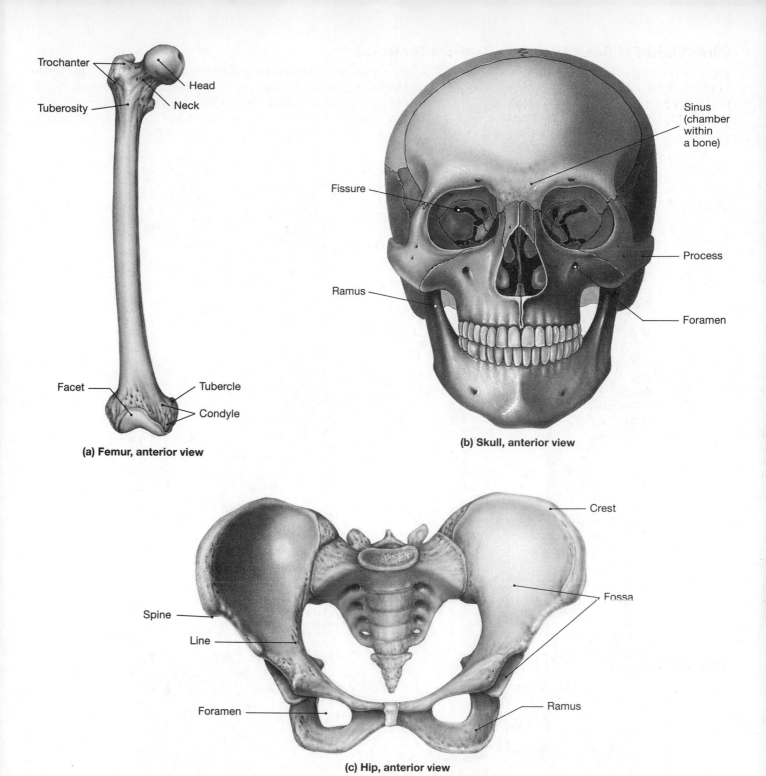

(a) Femur, anterior view

Trochanter
Head
Tuberosity
Neck
Facet
Tubercle
Condyle

(b) Skull, anterior view

Sinus (chamber within a bone)
Fissure
Process
Ramus
Foramen

(c) Hip, anterior view

Crest
Spine
Fossa
Line
Foramen
Ramus

FIGURE 5-4 Examples of bone markings (surface features).

Anatomical Identification Review

1. Identify in the photograph of the femur bone as many bone markings as you can.

2. Label on the photograph of the human skeleton two examples of each anatomical classification of bone.

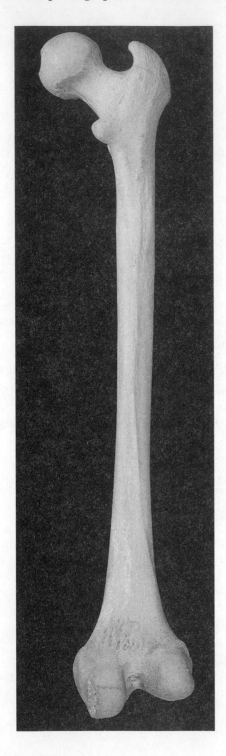

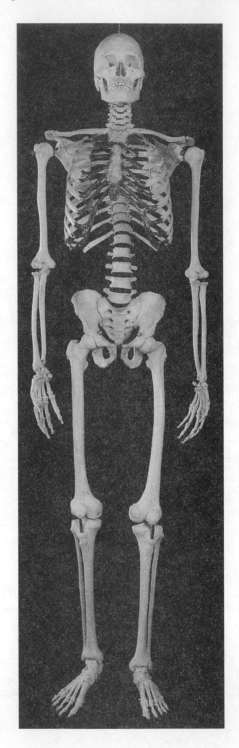

To Think About

1. How does an extensive exercise program affect the appearance of the skeleton? Why do these changes occur?

2. What is the significance of the orientation of the trabeculae that form the spongy bone of a long bone?

6

THE SKELETAL SYSTEM:
Axial Division

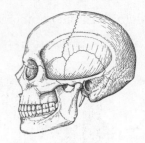

Objectives

1. Identify and name the bones of the axial skeleton and the skull (cranial and facial bones). For each bone, describe its function, explain the significance of markings, and identify prominent surface features.

2. Identify the structure of the nasal complex and the functions of the individual bone and cartilage elements.

3. Describe the structural differences between infant and adult skulls.

4. Identify and name the vertebral groups and the curvatures of the vertebral column. Identify the body (centrum), the vertebral arch and its components, and the articular processes on each vertebral group. Describe the structural and functional differences between the vertebral groups. Describe the functions of the curvatures.

5. Identify an intervertebral foramen and intervertebral disc.

6. Identify the surface features of the sternum and a typical rib: describe the significance of the articulations between the thoracic vertebrae, the ribs, and the sternum.

Description of the Skeletal System: The Axial Division

The approximately 206 bones that compose the skeletal system are conveniently divided into axial and appendicular divisions. The **axial skeleton** has 80 bones; 22 are skull bones and 7 are associated with the *skull*, 25 form the *thoracic cage*, and 26 form the *vertebral column*. The **appendicular skeleton** has 126 bones, including the bones of the limbs and the *pectoral* and *pelvic girdles* that attach the limbs to the trunk. [∞ p134]

Procedure

Identify the bones of the **axial** division, using a *human skeleton* and *Figures 6-1* and *6-2* for reference.

The Skull

Bones of the skull protect the brain and guard the entrances to the digestive and respiratory systems. The skull contains 22 bones: 8 form the **cranium** (braincase) and 14 form the **face**. Additionally, the hyoid bone and the auditory ossicles (3 within each ear) are associated with the skull. Bones of the skull are firmly attached together with ossified dense fibrous connective tissue to form immovable joints called **sutures**. The *coronal*, *sagittal*, *squamosal*, and *lambdoid* sutures are easily identified on the adult skull. [∞ p135-138]

Procedure

Examine the *skull* first by locating it and then comparing its position to those of other bones on an articulated skeleton. Hold your specimen in the anatomical position adjacent to the skull on the articulated skeleton. Identify **anterior/posterior, medial/lateral**, and **superior/inferior surfaces (borders)**. Now identify the **skull bones** and their **markings (surface features)** on your **specimen.** Use *Color Plates 30-33* and *Figures 6-2* to *6-12* for reference. Identify the skull bones first, using the *Color Plates*, then use the "Organizational Views" to locate and identify the bone markings.

LOCATE

___ Frontal bone

___ Parietal bones

___ Occipital bone

___ Sphenoid bone

___ Temporal bones

___ Maxillary bones (maxilla)

___ Zygomatic bones

___ Mandible

___ Nasal bones

___ Lacrimal bones

___ Ethmoid

See: Color Plates 30-33, Figs. 6-1 to 6-12

∞ *Figs.*
6-1, p134
6-2, p135
6-3, p136–139

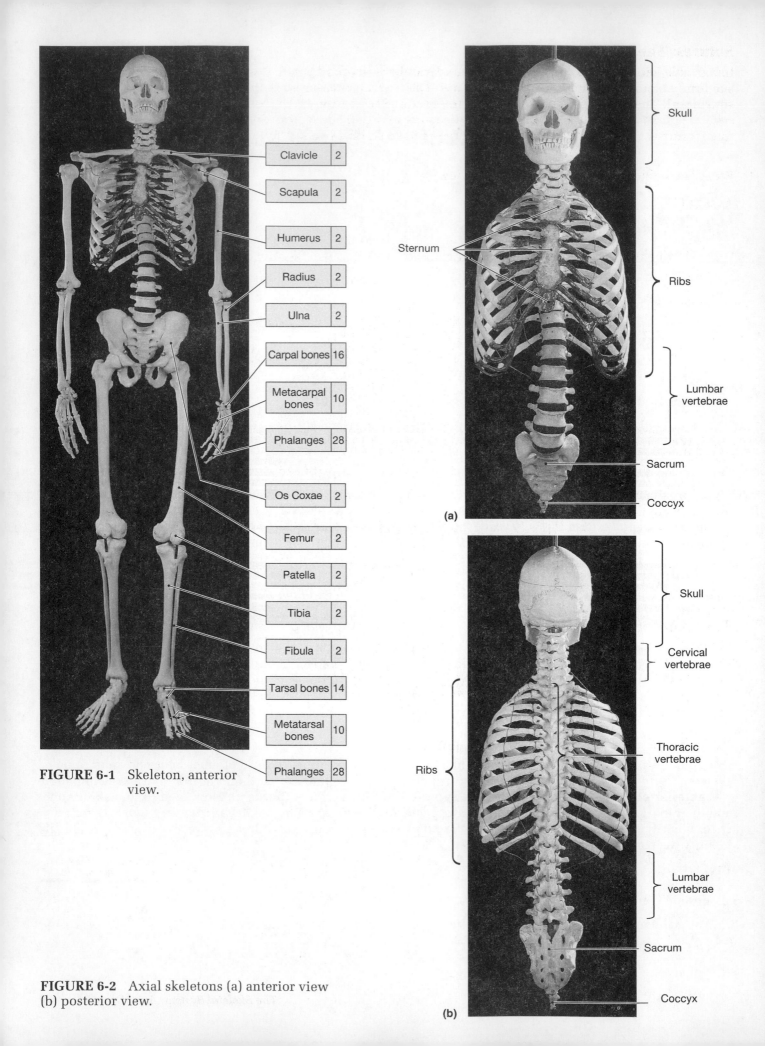

Clavicle	2
Scapula	2
Humerus	2
Radius	2
Ulna	2
Carpal bones	16
Metacarpal bones	10
Phalanges	28
Os Coxae	2
Femur	2
Patella	2
Tibia	2
Fibula	2
Tarsal bones	14
Metatarsal bones	10
Phalanges	28

FIGURE 6-1 Skeleton, anterior view.

FIGURE 6-2 Axial skeletons (a) anterior view (b) posterior view.

Sutures: The Adult Skull

Immovable joints called **sutures** form the boundaries between skull bones. Ossified dense fibrous connective tissue firmly bonds the bones together at a suture. The four major sutures are the **lambdoid, coronal, sagittal, and squamosal**. The lambdoid suture separates the occipital bone from the parietal bones. The sagittal suture separates the parietal bones. The coronal suture separates the frontal bone from the parietal bones. The squamosal suture identifies the boundary between the temporal bone and the parietal bone on each side of the skull.

From the Superior, Lateral, and Posterior View

LOCATE

___ Sagittal suture

___ Coronal suture

___ Squamosal suture

___ Lambdoid suture

See: Figs. 6-3 to 6-5

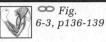 ∞ Fig. 6-3, p136-139

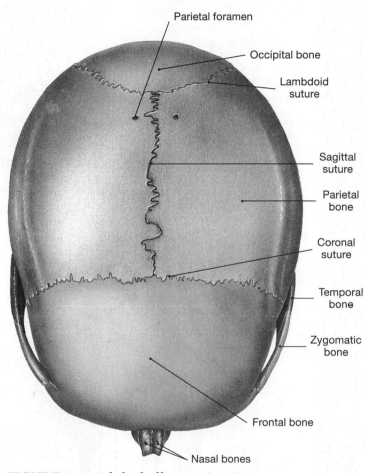

FIGURE 6-3 Adult skull, superior view.

See: Figs. 6-4 to 6-10

The Adult Skull

LOCATE

From the Posterior View

___ Sagittal suture

___ Parietal bone (paired bones)

___ Lambdoid suture

___ Occipital bone

___ External occipital protuberance

___ Occipital condyle

___ Squamous suture

___ Temporal bone (paired bone)

___ Mastoid process

___ Styloid process

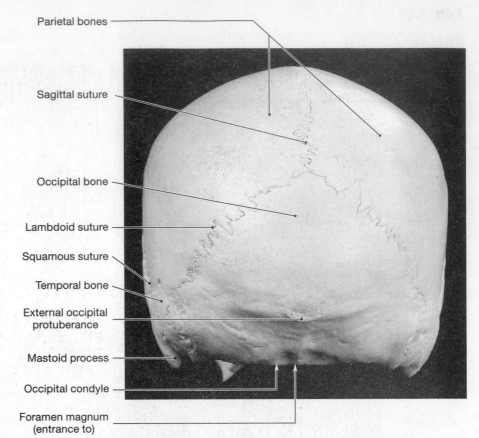

FIGURE 6-4 Adult skull, posterior view.

LOCATE

From the Superior View

___ Sagittal suture

___ Parietal bones

___ Lambdoid suture

___ Occipital bone

___ Coronal suture

___ Frontal bone

___ Nasal bones

___ Zygomatic bones (paired bones)

___ Temporal bones

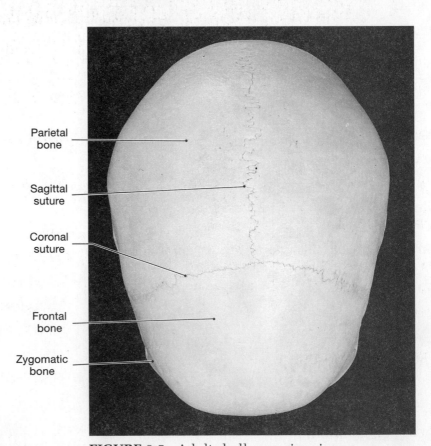

FIGURE 6-5 Adult skull, superior view.

LOCATE

From the Lateral View

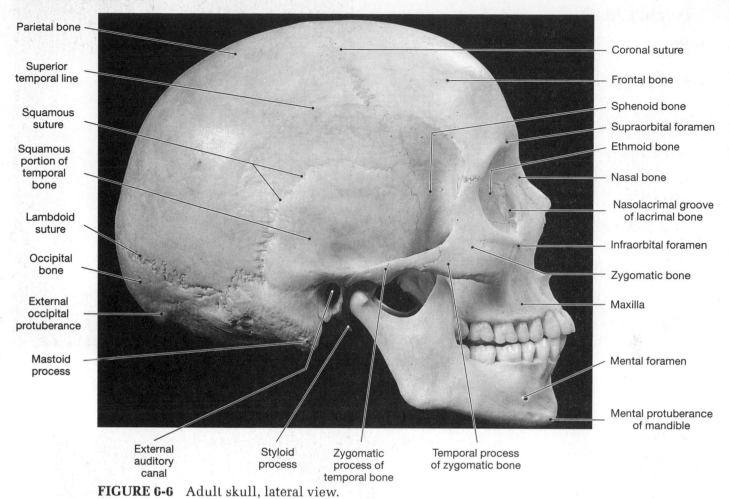

Parietal bone

Superior
temporal line

Squamous
suture

Squamous
portion of
temporal
bone

Lambdoid
suture

Occipital
bone

External
occipital
protuberance

Mastoid
process

Coronal suture

Frontal bone

Sphenoid bone

Supraorbital foramen

Ethmoid bone

Nasal bone

Nasolacrimal groove
of lacrimal bone

Infraorbital foramen

Zygomatic bone

Maxilla

Mental foramen

Mental protuberance
of mandible

External
auditory
canal

Styloid
process

Zygomatic
process of
temporal bone

Temporal process
of zygomatic bone

FIGURE 6-6 Adult skull, lateral view.

— Frontal bone

— Coronal suture

— Parietal bone

— Superior temporal line (see ∞ Fig. 6-6c, p143)
(not clearly defined on all specimens)

— Inferior temporal line (see ∞ Fig. 6-6c, p143)
(not clearly defined on all specimens)

— Squamous suture

— Lambdoid suture

— Occipital bone

— Supraorbital foramen

— Temporal bone

— External auditory meatus (canal)

— Mastoid process

— Styloid process

— Zygomatic bone

— Zygomatic arch { Zygomatic process
of temporal bone
Temporal process
of zygomatic bone

— Sphenoid bone

— Nasal bone

— Lacrimal bone

— Nasolacrimal groove of lacrimal bone

— Ethmoid bone

— Maxilla

— Infraorbital foramen

— Mandible

— Mental foramen

— Mental protuberance of mandible

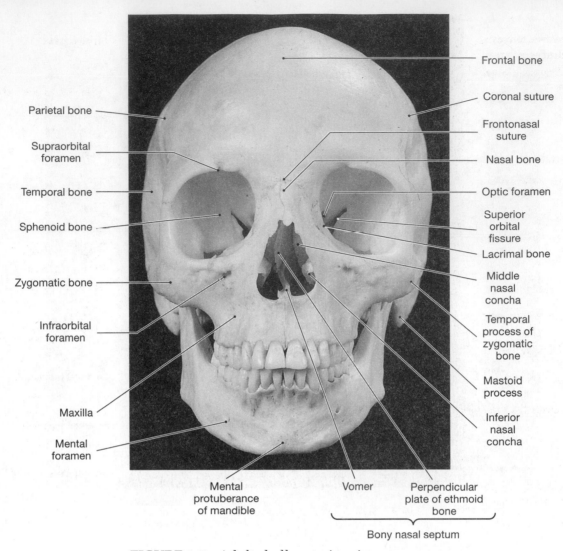

Frontal bone

Coronal suture

Frontonasal suture

Nasal bone

Optic foramen

Superior orbital fissure

Lacrimal bone

Middle nasal concha

Temporal process of zygomatic bone

Mastoid process

Inferior nasal concha

Parietal bone

Supraorbital foramen

Temporal bone

Sphenoid bone

Zygomatic bone

Infraorbital foramen

Maxilla

Mental foramen

Mental protuberance of mandible

Vomer

Perpendicular plate of ethmoid bone

Bony nasal septum

FIGURE 6-7 Adult skull, anterior view.

LOCATE

From the Anterior View

___ Coronal suture

___ Parietal bone

___ Frontal bone

___ Supraorbital foramen (or notch)

___ Temporal bone

___ Mastoid process

___ Zygomatic bone

___ Zygomaticofacial foramen

___ Temporal process of zygomatic bone

___ Ethmoid bone

___ Perpendicular plate of ethmoid bone

___ Vomer bone

___ Middle nasal concha

___ Inferior nasal concha

___ Lacrimal bone

___ Nasal bone

___ Frontonasal suture

___ Maxilla

___ Infraorbital foramen

___ Mandible

___ Mental protuberance of mandible

___ Mental foramen

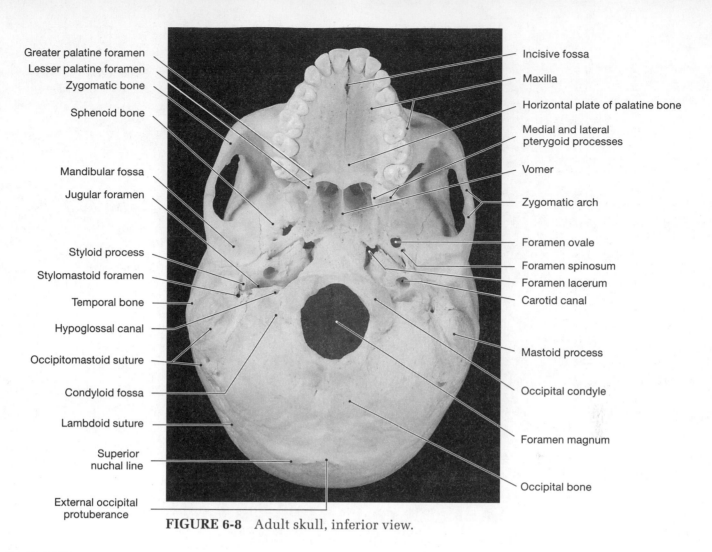

Greater palatine foramen
Lesser palatine foramen
Zygomatic bone

Sphenoid bone

Mandibular fossa
Jugular foramen

Styloid process
Stylomastoid foramen
Temporal bone
Hypoglossal canal
Occipitomastoid suture

Condyloid fossa

Lambdoid suture

Superior
nuchal line

External occipital
protuberance

Incisive fossa
Maxilla
Horizontal plate of palatine bone
Medial and lateral
pterygoid processes
Vomer
Zygomatic arch

Foramen ovale
Foramen spinosum
Foramen lacerum
Carotid canal

Mastoid process

Occipital condyle

Foramen magnum

Occipital bone

FIGURE 6-8 Adult skull, inferior view.

LOCATE

From the Inferior View

— Occipital bone

— External occipital protuberance

— Superior nuchal line
(see ∞ Fig. 6-6a, p143)

— Occipital condyle

— Fossa for cerebellum

— Occipitomastoid suture

— Lambdoid suture

— Stylomastoid foramen (mastoid foramen)

— Temporal bone

— Mastoid process

— Styloid process

— External auditory canal

— Jugular foramen

— Jugular fossa

— Carotid canal

— Foramen spinosum

— Foramen lacerum

— Foramen ovale

— Mandibular fossa

— Maxilla

— Incisive fossa

— Horizontal plate of palatine bone

— Palatine foramen (greater and lesser)

— Sphenoid bone

— Pterygoid processes (medial & lateral)

— Vomer

— Zygomatic arch

— Zygomatic bone

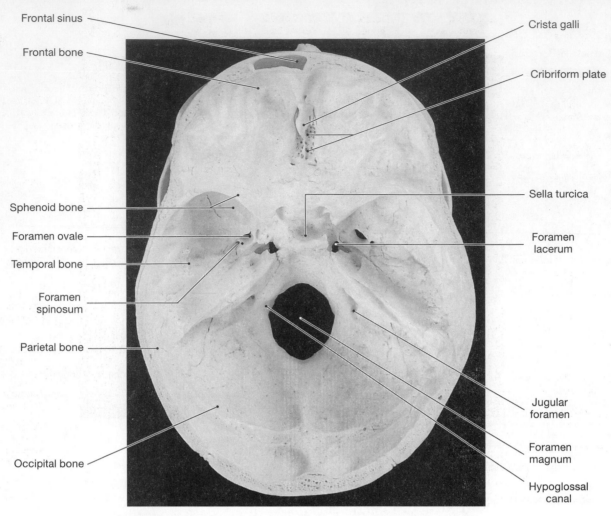

Frontal sinus

Frontal bone

Crista galli

Cribriform plate

Sphenoid bone

Foramen ovale

Temporal bone

Foramen spinosum

Parietal bone

Occipital bone

Sella turcica

Foramen lacerum

Jugular foramen

Foramen magnum

Hypoglossal canal

FIGURE 6-9   Adult skull, horizontal section.

Sectional Anatomy of the Skull

LOCATE

See: Color Plates 32, 33, Figs. 6-9 to 6-11

*Figs.
6-4, p140
6-5, p141
6-9, pp146-147
6-11, p149*

Horizontal Section

___ Frontal bone

___ Frontal sinus

___ Ethmoid bone

___ Crista galli

___ Cribriform plate

___ Sphenoid bone

___ Sella turcica

___ Optic canal (foramen)

___ Foramen rotundum

___ Foramen lacerum

___ Foramen ovale

___ Foramen spinosum

___ Carotid canal

___ Temporal bone

___ Petrous portion of temporal bone

___ Internal acoustic meatus

___ Foramen magnum

___ Jugular foramen

___ Hypoglossal canal

___ Parietal bone

___ Occipital bone

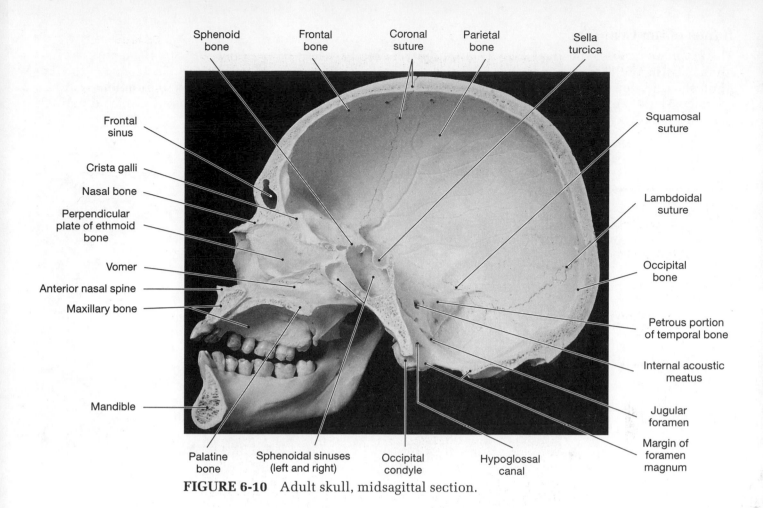

Sphenoid bone · Frontal bone · Coronal suture · Parietal bone · Sella turcica

Frontal sinus
Crista galli
Nasal bone
Perpendicular plate of ethmoid bone
Vomer
Anterior nasal spine
Maxillary bone
Mandible

Squamosal suture
Lambdoidal suture
Occipital bone
Petrous portion of temporal bone
Internal acoustic meatus
Jugular foramen
Margin of foramen magnum

Palatine bone · Sphenoidal sinuses (left and right) · Occipital condyle · Hypoglossal canal

FIGURE 6-10   Adult skull, midsagittal section.

Sagittal Section

___ Frontal bone

___ Frontal sinus

___ Coronal suture

___ Parietal bone

___ Lambdoid suture

___ Occipital bone

___ Margin of foramen magnum

___ Occipital condyle

___ Temporal bone

___ Styloid process

___ Petrous portion of temporal bone

___ Internal acoustic meatus

___ Jugular foramen

___ Hypoglossal canal

___ Sphenoid bone

___ Sphenoidal sinus

___ Hypophyseal fossa of sella turcica

___ Nasal bone

___ Ethmoid bone

___ Vomer bone

___ Maxillary bone

___ Palatine bone

___ Mandible

Bones of the Cranium

The *cranium* encloses the cranial cavity, a fluid-filled chamber that cushions and supports the brain. The cranium consists of the *frontal, parietal, occipital, temporal, sphenoid,* and *ethmoid* bones. Each bone of the cranium should be examined both individually to identify **additional ridges, depressions,** and **foramina,** and in a complete skull to identify relationships with adjoining bones. [∞ p142] The detail of these additional individual bone markings may be observed in the cited figures.

See: Fig. 6-11

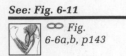

∞ *Fig. 6-6a,b, p143*

Occipital Bone

LOCATE

___ Markings shown in *Figure 6-11*

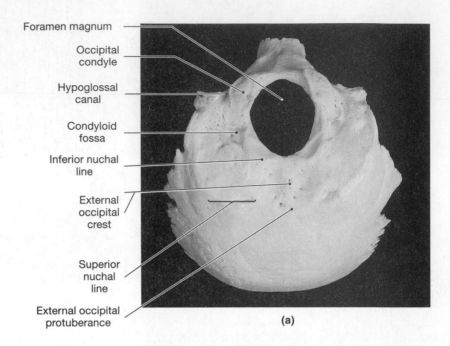

Foramen magnum
Occipital condyle
Hypoglossal canal
Condyloid fossa
Inferior nuchal line
External occipital crest
Superior nuchal line
External occipital protuberance

(a)

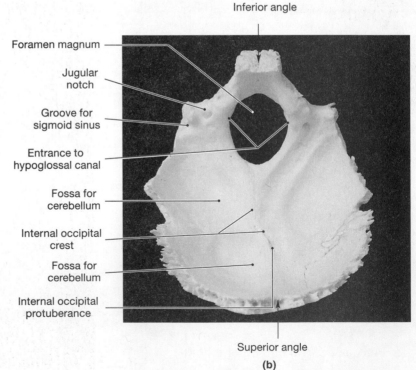

Inferior angle
Foramen magnum
Jugular notch
Groove for sigmoid sinus
Entrance to hypoglossal canal
Fossa for cerebellum
Internal occipital crest
Fossa for cerebellum
Internal occipital protuberance
Superior angle

(b)

FIGURE 6-11 Occipital bone (a) inferior (external) view (b) superior (internal) view.

See: Fig. 6-12

∞ *Fig. 6-6c, p143*

Parietal Bones

LOCATE

___ Markings shown in *Figure 6-12*

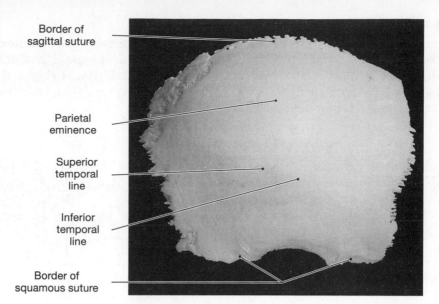

Border of sagittal suture

Parietal eminence

Superior temporal line

Inferior temporal line

Border of squamous suture

FIGURE 6-12   Parietal bone, lateral view.

See: Fig. 6-13

∞ *Fig. 6-7, p144*

Frontal Bone

LOCATE

___ Markings shown in *Figure 6-13*

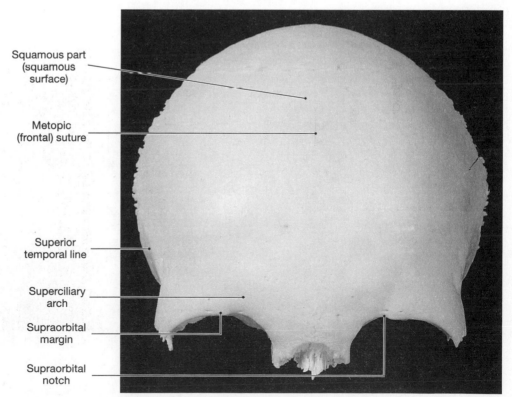

Squamous part (squamous surface)

Metopic (frontal) suture

Superior temporal line

Superciliary arch

Supraorbital margin

Supraorbital notch

FIGURE 6-13   Frontal bone, anterior view.

See: Fig. 6-14

 ∞ *Fig.*
6-8, p145

Temporal Bones

LOCATE

___ Markings shown
in *Figure 6-14*

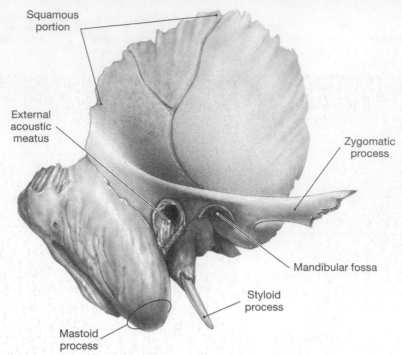

FIGURE 6-14 Temporal bone, lateral view.

Squamous portion

External acoustic meatus

Zygomatic process

Mandibular fossa

Styloid process

Mastoid process

See: Fig. 6-15, 6-16

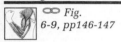 ∞ *Fig.*
6-9, pp146-147

Sphenoid Bone

LOCATE

___ Markings shown
in *Figures 6-15*
and *6-16*

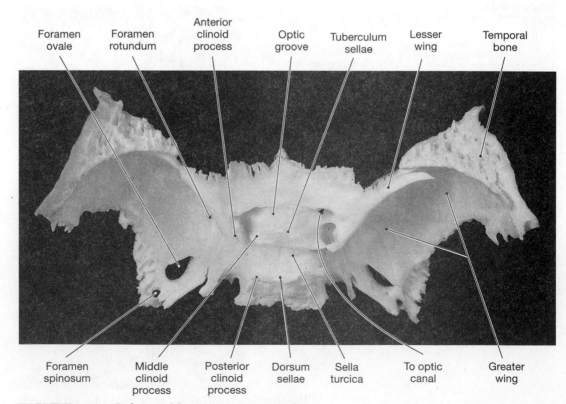

Foramen ovale | Foramen rotundum | Anterior clinoid process | Optic groove | Tuberculum sellae | Lesser wing | Temporal bone

Foramen spinosum | Middle clinoid process | Posterior clinoid process | Dorsum sellae | Sella turcica | To optic canal | Greater wing

FIGURE 6-15 Sphenoid bone, superior surface.

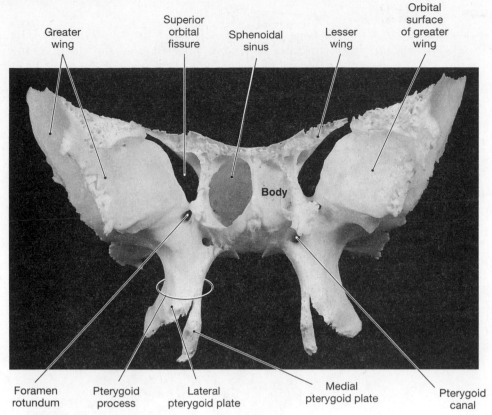

FIGURE 6-16   Sphenoid bone, anterior surface.

Greater wing

Superior orbital fissure

Sphenoidal sinus

Lesser wing

Orbital surface of greater wing

Body

Foramen rotundum

Pterygoid process

Lateral pterygoid plate

Medial pterygoid plate

Pterygoid canal

See: Fig. 6-17

∞ *Fig. 6-10. p148*

Ethmoid Bone

LOCATE

__ Markings shown in *Figure 6-17*

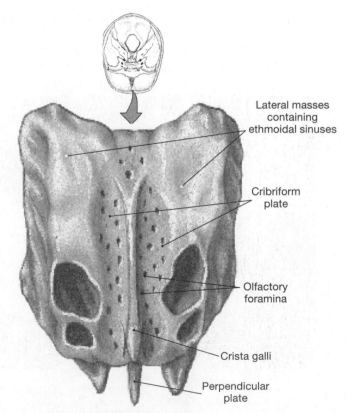

Lateral masses containing ethmoidal sinuses

Cribriform plate

Olfactory foramina

Crista galli

Perpendicular plate

FIGURE 6-17   Ethmoid bone, superior surface.

Bones of the Face

Facial bones protect and support the entrances to the digestive and respiratory tracts. The superficial facial bones, the *lacrimal, nasal, maxilla, zygomatic*, and *mandible*, provide areas for the attachment of muscles that control facial expressions and assist in the manipulation of food. Each *bone* of the *face* should be examined individually to identify **additional ridges, depressions**, and **foramina** and in a complete skull to identify relationships with adjoining bones. [∞ p150] The detail of these additional individual bone markings may be observed in the cited figures.

The Maxilla

See: Fig. 6-18

∞ *Fig. 6-12, p151*

LOCATE

___ Markings shown in *Figure 6-18*

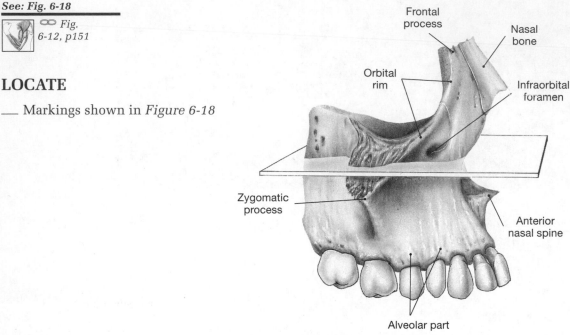

FIGURE 6-18   Maxilla, lateral surface.

See: Fig. 6-19

∞ *Fig. 6-13a, p151*

The Palatine Bone

LOCATE

___ Markings shown in *Figure 6-19*

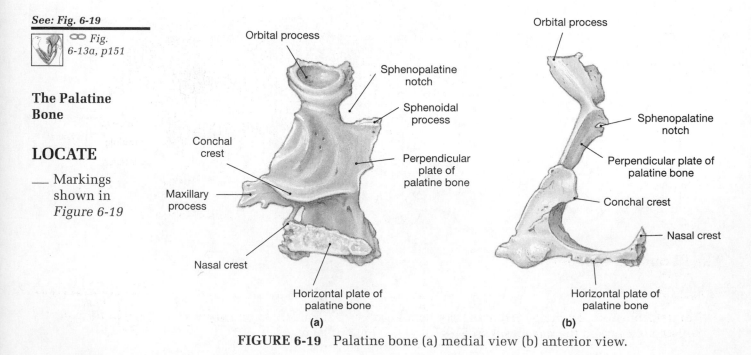

FIGURE 6-19   Palatine bone (a) medial view (b) anterior view.

The Nasal Bones

See: Fig. 6-10

 ∞ *Figs.*
6-2, p135
6-3, p137

LOCATE

___ Nasal bones shown in *Figures 6-6, 6-7,* and *6-18*

The Vomer Bone

See: Fig. 6-10

 ∞ *Figs.*
6-5, p141
6-16a, p154

LOCATE

___ Vomer bone

The Inferior Nasal Conchae

See: Color Plate 32,
Fig. 6-20

 ∞ *Fig.*
6-16b-d, p154

LOCATE

___ Nasal conchae
{
Superior
Middle
Inferior

See: Color Plate 33,
Fig. 6-20

 ∞ *Fig.*
6-16b-d, p154

The Nasal Complex

LOCATE

___ Markings shown in *Figure 6-20*

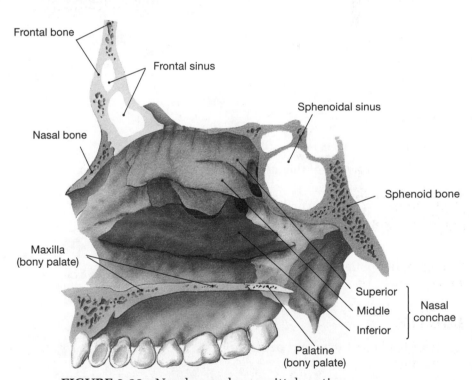

Frontal bone

Frontal sinus

Sphenoidal sinus

Nasal bone

Sphenoid bone

Maxilla
(bony palate)

Superior
Middle } Nasal
conchae
Inferior

Palatine
(bony palate)

FIGURE 6-20 Nasal complex, sagittal section.

The Paranasal Sinuses

LOCATE

___ Frontal, sphenoid, ethmoid, and maxillary bones each contain air-filled chambers collectively termed the **paranasal sinuses**

See: Fig. 6-10

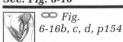 ∞ *Fig.*
6-16b, c, d, p154

The Zygomatic Bones

See: Fig. 6-21

⚭ *Fig.
6-15, p153*

LOCATE

___ Markings shown in
Figure 6-21

The Lacrimal Bones

See: Fig. 6-21

⚭ *Fig.
6-15, p153*

LOCATE

___ Markings shown in
Figure 6-21

The Orbital Complex

The *orbits* are the bony recesses that contain the eyes. Seven bones form the orbital complex for each orbit. The *frontal* bone forms the roof and the *maxilla* provides most of the floor. Other contributions to the orbit are made by the *lacrimal, ethmoid, sphenoid, zygomatic,* and *palatine* bones (see *Color Plates 30-31*). Examine the *orbit* first by locating its position on the skull and then by identifying the individual bones of the complex. [⚭ p152] Identify the following features of the **orbital complex**.

*See: Color Plates
30-31, Fig. 6-21*

⚭ *Fig.
6-15, p153*

LOCATE

___ Frontal bone forms roof of orbit

___ Maxilla form most of the orbital floor

___ Maxilla, lacrimal, and lateral mass of ethmoid bones form medial wall of orbit

___ Sphenoid bone forms posterior wall of orbit

___ Zygomatic bone forms lateral wall of orbit

___ Palatine bone forms part of inferior wall of orbit

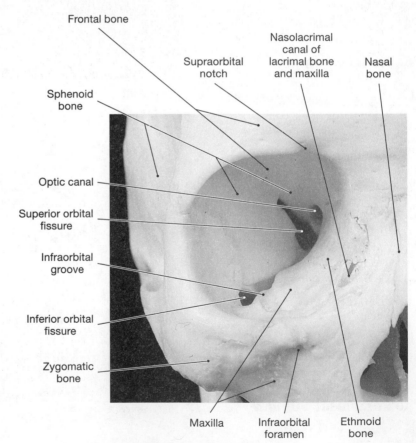

FIGURE 6-21   Orbital complex.

The Mandible

The *mandible* forms the entire lower jaw. It can be subdivided into the horizontal **body** and ascending **rami**. The **condylar processes** articulate with the **mandibular fossae** of the temporal bones at the temporomandibular joints (TMJ). Examine the *mandible* for the bone markings cited.

See: Fig. 6-22

∞ Fig. 6-14a, p152

From the Lateral View

LOCATE

— Body of mandible

— Mental foramen

— Mental protuberance

— Angle of mandible

— Ramus of mandible

— Mandibular notch

— Condylar process

— Coronoid process

— Alveolar part

— Mylohyoid line

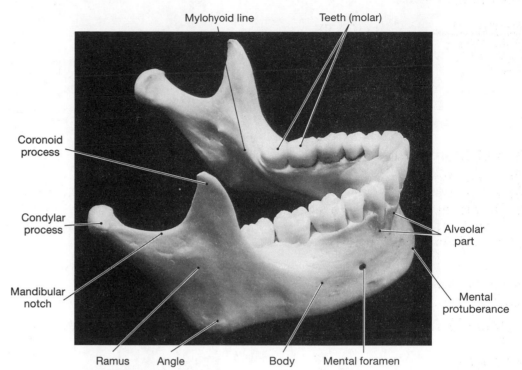

FIGURE 6-22 Mandible, lateral view.

From the Medial View

— Mandibular foramen

— Mylohyoid line

— Alveolar process

— Submandibular fossa

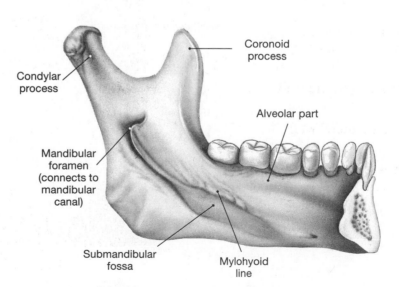

See: Fig. 6-23

∞ Fig. 6-14b, p152

FIGURE 6-23 Mandible, medial view.

The Hyoid Bone

See: Fig. 6-24

 Fig.
6-17, p155

LOCATE

___ Body of hyoid bone

___ Greater horn

___ Lesser horn

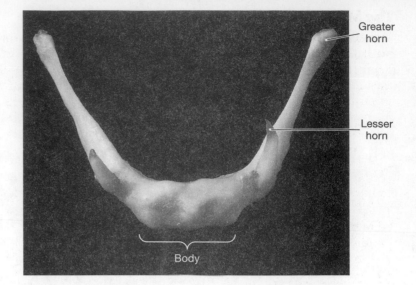

FIGURE 6-24 Hyoid bone, anterior view.

The Skull of an Infant

Procedure

The skulls of infants and adults differ in terms of the shape and structure of the bones. The differences account for variations in proportions as well as in size. The fibrous areas between the developing cranial bones are known as **fontanels**. [∞ p161] Using a *specimen* of the skull of an infant, identify **fontanels, sutures,** and **bone markings:**

See: Figs. 6-25, 6-26

∞ Fig.
6-18, p162

LOCATE

___ Frontal bone

___ Anterior fontanel

___ Coronal suture

___ Parietal bone

___ Sagittal suture

___ Posterior fontanel

___ Lambdoid suture

___ Occipital bone

___ Mastoid fontanel

___ Squamous suture

___ Temporal bone

___ Sphenoid fontanel

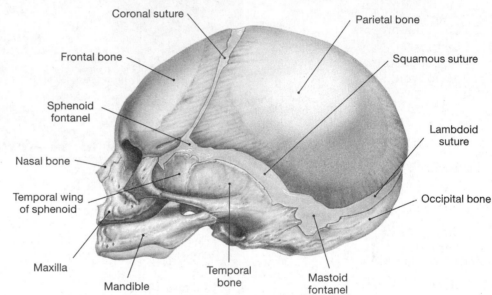

FIGURE 6-25
Skull of an infant, lateral view.

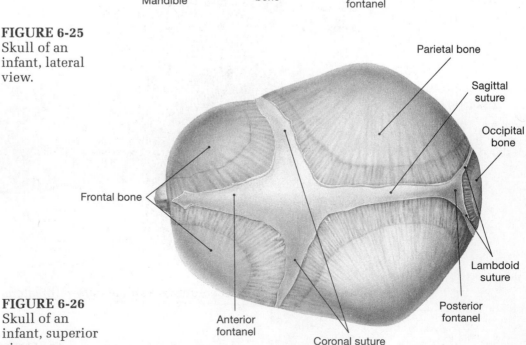

FIGURE 6-26
Skull of an infant, superior view.

The Vertebral Column

The adult **vertebral column** consists of 26 bones and is divided into five regions. Beginning at the base of the skull, the regions and their **vertebrae** are: *cervical* (7), *thoracic* (12), *lumbar* (5), *sacral* (1), and *coccygeal* (1). Typically, by age 25, an adult's sacral vertebrae and coccygeal vertebrae have fused into single bones, the **sacrum**, and the **coccyx**. [∞ p163] The cervical, thoracic, and lumbar vertebrae consist of 3 basic parts: (1) **body** or centrum, which transfers weight along the axis of the vertebral column, (2) a **vertebral arch**, which encloses the vertebral foramen that surrounds a portion of the spinal cord, and (3) **articular processes**, which have facets on their surfaces to permit articulation with adjacent vertebrae.

Procedure

Examine the *vertebral column* first by locating its position on an articulated skeleton. Then place your specimen in the anatomical position adjacent to the vertebral column of the articulated skeleton. Use *Figures 6-27* and *6-28* for reference. Identify **anterior/posterior, medial/lateral,** and **superior/inferior surfaces** or **borders**. Now identify the **vertebral regions, spinal curvatures,** and **associated structures**.

See: Figs. 6-27, 6-28

∞ Figs.
6-19, p163
6-21, p165

LOCATE

__ Cervical region/curvature

__ Thoracic region/curvature

__ Lumbar region/curvature

__ Sacral region/curvature

__ Intervertebral disc

__ Intervertebral foramen

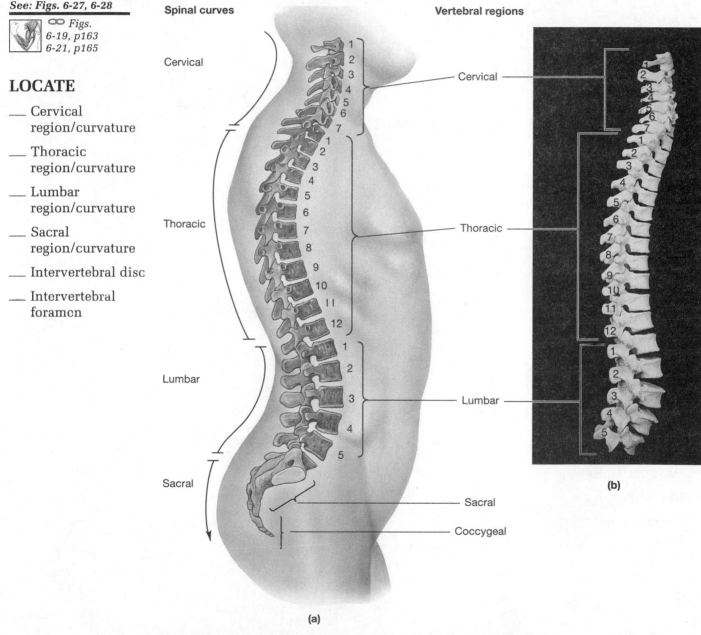

FIGURE 6-27 Vertebral column (a) spinal curves (b) lateral view.

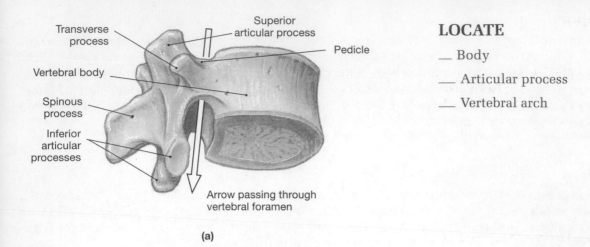

LOCATE
— Body
— Articular process
— Vertebral arch

(a)

Transverse process

Superior articular process

Pedicle

Vertebral body

Spinous process

Inferior articular processes

Arrow passing through vertebral foramen

Superior articular process

Lamina of vertebral arch

Intervertebral foramen

Intervertebral disc

Spinous process

Transverse process

Vertebral body

Inferior articular process

(b)

(c)

FIGURE 6-28 Vertebral column (a) lateral and inferior view (b) posterior view (c) lateral and sectional view.

Cervical Vertebrae

Procedure

Examine a typical *cervical vertebra* (C_3 to C_7), first by locating its position on an articulated skeleton. [∞ p166] Place your bone specimen in the anatomical position adjacent to that same vertebra on the articulated skeleton. Identify **anterior/posterior** and **superior/inferior surfaces** or **borders**. Now identify the three basic parts and the following **bone markings** on your **specimen**.

See: Fig. 6-29

 ∞ *Fig. 6-22, p167*

From Lateral and Superior Views

LOCATE

___ Vertebral body

___ Transverse process

___ Transverse foramen

___ Costal process

___ Vertebral foramen

___ Spinous process (often bifid)

___ Pedicle

___ Lamina

___ Vertebral arch

___ Superior articular process and facet

___ Inferior articular process and facet

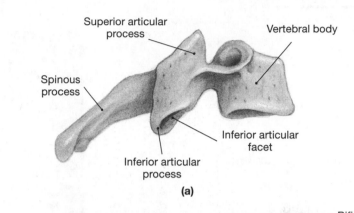

(a)

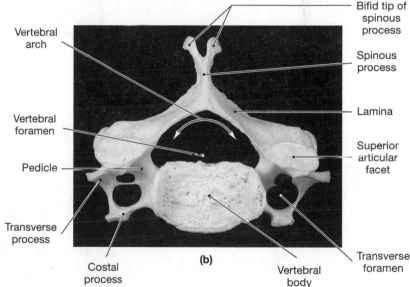

(b)

FIGURE 6-29 Cervical vertebra, (a) lateral view (b) superior view.

Cervical Vertebrae: Atlas and Axis

The **atlas** [C₁] holds up the head, articulating with the occipital condyles of the skull. This articulation permits nodding ["yes"] but prevents twisting. The **axis** [C₂] contains the prominent *dens* or *odontoid process*, which represents the fusion of the body of the atlas to the body of the axis during development. This "bowling pin" shaped process permits the formation of a pivot for rotation of the atlas and skull. Examine the *first two cervical vertebrae*, the *atlas* and the *axis* respectively, first by locating their positions on an articulated skeleton. Now identify the **common bone markings** of **cervical vertebrae** on the **atlas** and **axis**. Identify those bone markings that may be found only on these vertebrae.

See: Figs. 6-30 to 6-33

∞ *Fig. 6-23, p168*

LOCATE

Atlas

— Anterior tubercle

— Articular facet for dens of axis

— Posterior tubercle

— Posterior arch

— Vertebral foramen

— Transverse process

— Transverse foramen

— Costal process

— Superior/inferior articular processes

— Superior/inferior articular facets

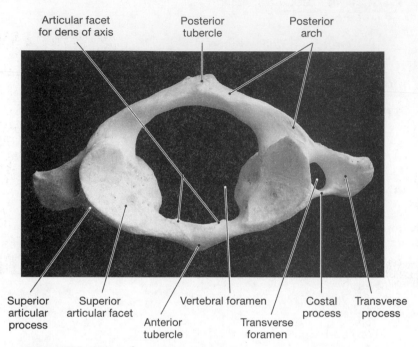

FIGURE 6-30 Atlas, superior view.

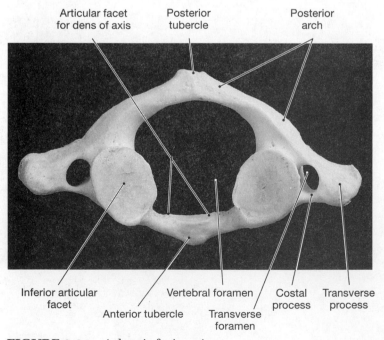

FIGURE 6-31 Atlas, inferior view.

Axis

— Articulation between atlas and axis

— Vertebral body

— Dens (odontoid process)

— Pedicle

— Superior articular process and facet

— Inferior articular process and facet

— Lamina

— Transverse process and transverse foramen

— Vertebral foramen

— Spinous process

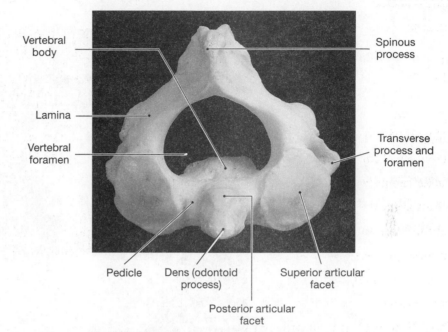

FIGURE 6-32 Axis, superior view.

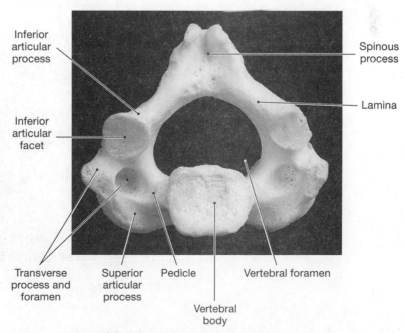

FIGURE 6-33 Axis, inferior view.

Thoracic Vertebrae

Procedure

Examine the *thoracic vertebrae* first by locating their positions on an articulated skeleton. Place your thoracic vertebra specimen in the anatomical position adjacent to that same vertebra on the articulated skeleton. Identify **anterior/posterior** and **superior/inferior views**. [∞ p167] Now identify the three basic parts and the following **bone markings** on your **thoracic specimen**.

See: Figs. 6-34, 6-35

 ∞ *Fig. 6-24, p169*

From Superior and Lateral Views

LOCATE

___ Vertebral body

___ Costal facets (superior and inferior)

___ Transverse process

___ Transverse costal facets

___ Vertebral foramen

___ Spinous process

___ Pedicle

___ Lamina

___ Vertebral arch

___ Superior articular process and facet

___ Inferior articular process and facet

___ Inferior vertebral notch

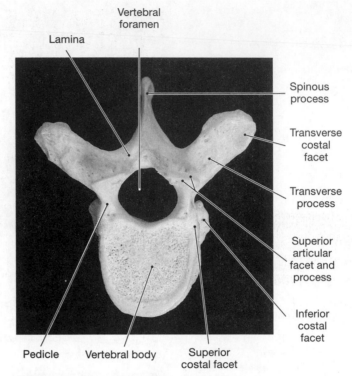

FIGURE 6-34 Thoracic vertebra, superior view.

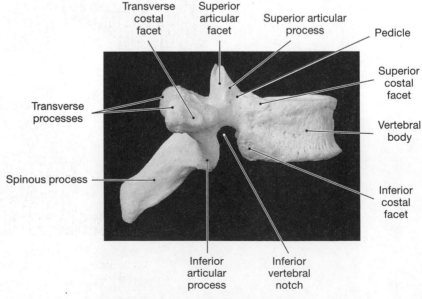

FIGURE 6-35 Thoracic vertebra, lateral view.

Lumbar Vertebrae

Procedure

Examine the *lumbar vertebrae* first by locating their positions on an articulated skeleton. Place your lumbar vertebra specimen in the anatomical position adjacent to that same lumbar vertebra on the articulated skeleton. Identify **anterior/posterior** and **superior/inferior surfaces** or **borders**. [∞ p170] Now identify the three basic parts and the following **bone markings** on your **lumbar specimen**.

See: Figs. 6-36, 6-37

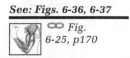 ∞ *Fig. 6-25, p170*

From Superior and Lateral Views

LOCATE

___ Vertebral body

___ Pedicle

___ Lamina

___ Spinous process

___ Transverse processes

___ Inferior articular process and facet

___ Superior articular process and facet

___ Vertebral foramen

___ Inferior vertebral notch

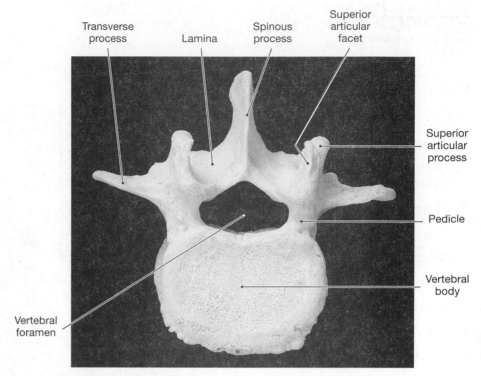

FIGURE 6-36 Lumbar vertebra, superior view.

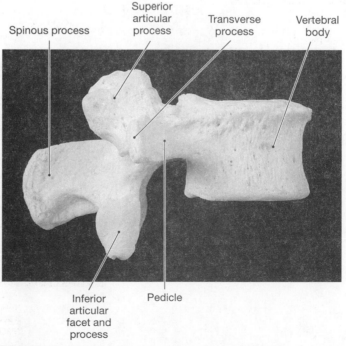

FIGURE 6-37 Lumbar vertebra, lateral view.

The Sacrum and Coccyx

Procedure

The *sacrum* consists of the fused components of five sacral vertebrae. The composite structure provides protection for reproductive, digestive, and excretory organs. Additionally, it provides for the attachment of the axial skeleton to the pelvic girdle of the appendicular skeleton. The *coccyx* consists of 3-5 fused coccygeal vertebrae that act as an attachment site for many ligaments and a muscle that constricts the anal opening. Examine the *sacrum* and *coccyx* first by locating their positions on an articulated skeleton. [∞ p134] Place your bone specimen(s) in the anatomical position adjacent to the sacrum/coccyx on the articulated skeleton. Identify **anterior/posterior, lateral**, and **superior/inferior surfaces** or **borders**. Now identify the following **bone markings** on your **specimens**.

See: Figs. 6-38 to 6-39

 ∞ Figs. 6-1, p134 6-26, p171

From Posterior, Lateral, and Anterior Views

LOCATE

__ Sacral canal

__ Superior articular process

__ Median sacral crest

__ Lateral sacral crest

__ Sacral foramina

__ Sacral cornua

__ Sacral hiatus

__ Coccygeal cornua

__ Sacral tuberosity

__ Auricular surface

__ Sacral curvature

__ Coccyx

__ Base

__ Sacral body

__ Sacral promontory

__ Ala

__ Apex

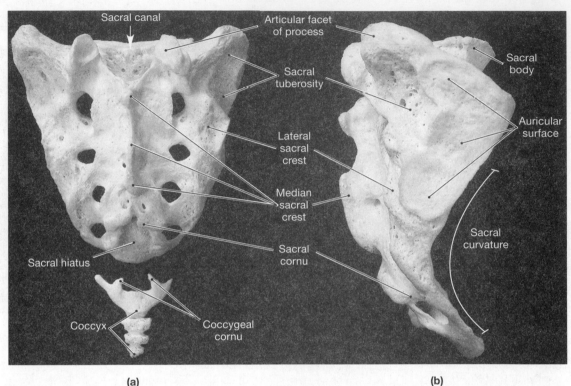

FIGURE 6-38 Sacrum and coccyx (a) posterior surface (b) lateral surface.

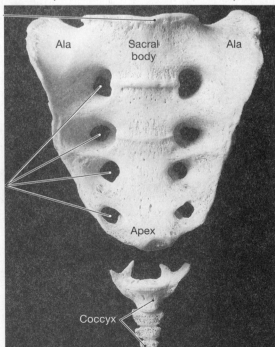

FIGURE 6-39 Sacrum and coccyx, anterior surface.

The Thoracic Cage

The skeleton of the **thoracic cage** consists of the *thoracic vertebrae*, the *ribs* (costae), and the *sternum*. Only the ribs and the sternum form the **rib cage**. There are 12 pairs of ribs. Ribs 1-7 are called **true** or **vertebrosternal ribs** because they reach the anterior body wall and are connected to the sternum by separate cartilaginous extensions, the **costal cartilages**. Ribs 8-12 are termed **false ribs**: ribs 8-10 are **vertebrochondral ribs** because their costal cartilages fuse together with the costal cartilage of rib 7, before reaching the sternum and therefore do not attach directly to the sternum, and ribs 11-12 are called **floating ribs** because they have no connection with the sternum. [∞ p174]

Procedure

Examine the **sternum** by first locating its position on an articulated skeleton. Place your specimen in the anatomical position adjacent to the sternum on the articulated skeleton. Use Figure 6-40 for reference. Identify **anterior/posterior, lateral**, and **superior/inferior surfaces** or **borders**. Now identify the following **bone markings** on your specimen.

See: Figs. 6-40, 6-41

∞ *Figs.*
6-1, p134
6-27a, c, p175

LOCATE

__ Sternum { Manubrium / Body / Xiphoid process

__ Jugular notch

__ Clavicular notch
(for articulation with first rib)

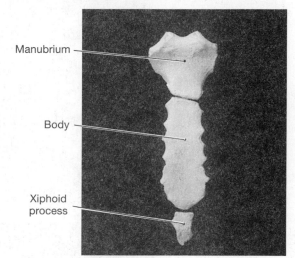

Manubrium

Body

Xiphoid process

FIGURE 6-40 Sternum, anterior view.

Ribs

Procedure

Ribs are elongated, curved, flattened bones that originate on or between the thoracic vertebrae and end in the wall of the thoracic cavity [see *Color Plates 68-72*]. Intercostal muscles attach to and move the ribs, affecting both the width and depth of the thoracic cage.

Examine the *ribs* first by locating their positions on an articulated skeleton. Place your bone specimen in the anatomical position adjacent to that same bone on the articulated skeleton. Identify **anterior/posterior, medial/lateral**, and **superior/inferior surfaces** or **borders**. Now identify the following **bone markings** on your specimens.

See: Fig. 6-41

∞ *Figs.*
6-1, p134
6-27, p175

LOCATE

__ True ribs

__ False ribs

___ Floating ribs

__ Costal cartilage

__ Head of rib

__ Neck of rib

__ Tubercle of rib

__ Articular facet of tubercle

__ Angle of rib

__ Transverse articular facet

__ Costal facets (superior and inferior)

__ Articular facets of head (superior and inferior)

__ Interarticular crest

__ Attachment to costal cartilage

__ Costal groove

__ Identify ribs 1-12 (from posterior and anterior views)

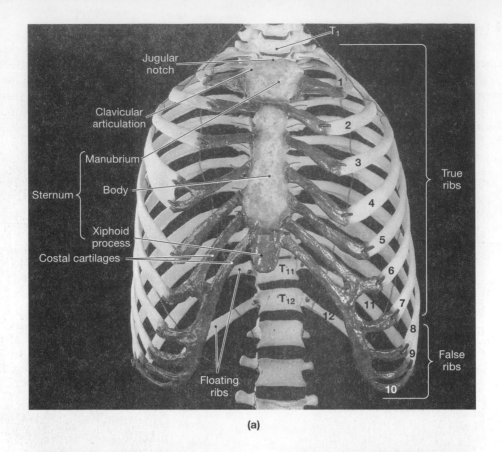

(a)

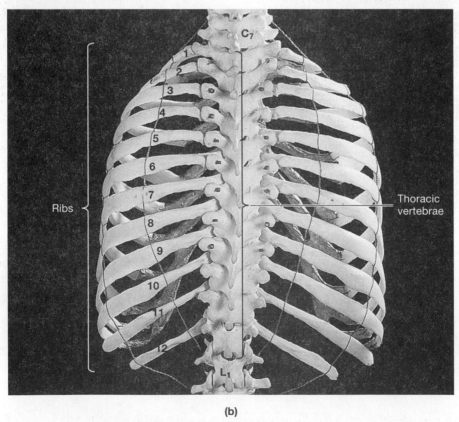

(b)

FIGURE 6-41 Thoracic cage (a) anterior view (b) posterior view.

Anatomical Identification Review

Label on the adjacent photo the individual bones that form the axial skeleton.

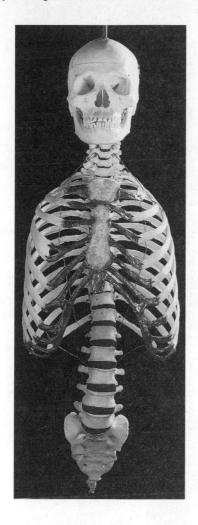
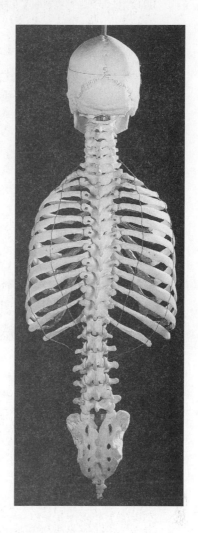

To Think About

1. Olivia has always hated her nose because she thinks it is too large and has a "bump" at the bridge. Therefore, for her sixteenth birthday, her parents give her a rhinoplasty as a present. What is she having done, and what anatomic structures are involved?

2. A series of X rays taken of a newborn infant whose head is increasing in size rapidly shows that some of the cranial bones are separating at the sutures. What is the likely cause and best treatment of this condition and what are the anatomic ramifications of not treating it?

3. A hero in an action adventure motion picture or television series is stabbed in the chest, inferior to the eighth rib. Although this injury slows him down slightly, he nevertheless manages to capture all of the bad guys. Had an individual in real life received this injury, would it have been serious, and what would have been the symptoms and the anatomical injuries?

4. A person with a serious case of osteoporosis has developed kyphosis. What anatomic structures are involved in this condition, and what are the functional implications?

5. A model is often complemented on her high cheekbones and large eyes. What is the anatomic basis for each of these features?

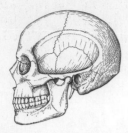

7

THE SKELETAL SYSTEM:
Appendicular Division

Objectives

1. Identify and name the bones of the appendicular skeleton, the pectoral girdle and upper extremity, and the pelvic girdle and lower extremity. For each bone, describe its function and identify prominent surface features.

2. Identify the anatomical differences between the female and the male pelvis.

3. Summarize the anatomical differences between the skeleton of a female and a male.

The **appendicular skeleton** includes the 126 bones of the upper and lower extremities and the supporting elements, called *girdles*, that connect the extremities to the trunk. The *pectoral girdles* and upper extremities are formed by 64 bones. The *pelvic girdles* and lower extremities are formed by 62 bones. [∞ p181] Use *Figure 6-1* for reference to aid you in the identification of the bones of the appendicular skeleton.

Description of the Pectoral Girdle and Upper Extremity

Each arm articulates with the trunk at the **pectoral girdle** or **shoulder**. It is formed by the **clavicle** (collar bone) and **scapula** (shoulder blade). The clavicle articulates with the manubrium of the sternum, but the scapula has no direct attachment to the thoracic cage and is held in position by skeletal muscles and by the clavicle. The clavicle and scapula position the shoulder joint and create the basis for arm movement. The bony ridges, processes, flanges, and flat surfaces of the scapula and clavicle are sites for muscle attachment. Muscles that originate both on the pectoral girdle and the axial skeleton help to move the upper extremity. [∞ p182]

The Clavicle

The smooth, S-shaped, compressed tubular **clavicle** originates at the craniolateral border of the manubrium of the sternum and extends until it articulates with the scapula. [∞ p182]

Procedure

Examine the *clavicle* first by locating its position on an articulated skeleton. Place your bone specimen in the anatomical position adjacent to the clavicle on the articulated skeleton. Identify **anterior/posterior, medial/lateral**, and **superior/inferior surfaces** or **borders**. Then identify the following **bone markings** on your specimen.

See: Fig. 7-1

∞ Figs.
7-1, p181
7-2, p182
7-3, p183
7-4, p183

LOCATE

— Acromial (lateral) end

— Sternal (medial) end

— Sternal facet for articulation with manubrium of sternum

— Costal tuberosity

— Facet for articulation with acromion of scapula

— Conoid tubercle

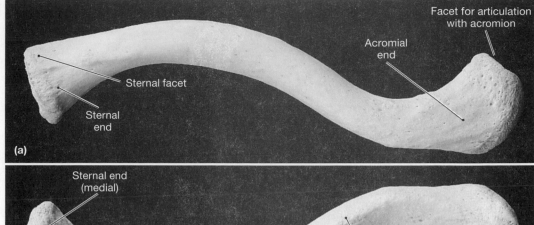

FIGURE 7-1 Right clavicle (a) superior view (b) inferior view.

The Scapula

The broad, flat, smooth, triangular **scapula** articulates with the proximal end of the arm bone, the humerus. The **acromion** of the scapula serves as the attachment point to the clavicle. [∞ p182] This articulation helps maintain the position of the scapula.

Procedure

Examine the *scapula* first by locating its position on an articulated skeleton. Place your bone specimen in the anatomical position adjacent to the scapula on the articulated skeleton. Identify **anterior/posterior, medial/lateral,** and **superior/inferior surfaces** or **borders**. Then identify the following **bone markings** on your specimen.

See: Figs. 7-2, 7-3

∞ Figs.
7-1, p181
7-5, p184

LOCATE

— Body of scapula

— Neck of scapula

— Superior border

— Superior angle

— Medial (vertebral) border

— Inferior angle

— Lateral (axillary) border

— Glenoid cavity

— Supraglenoid tubercle

— Infraglenoid tubercle

— Acromion

— Coracoid process

— Suprascapular notch

— Subscapular fossa

— Spine

— Supraspinous fossa

— Infraspinous fossa

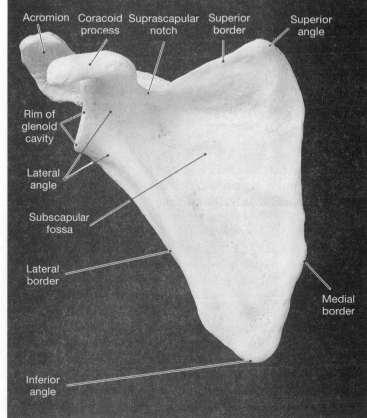

FIGURE 7-2 Scapula, anterior view.

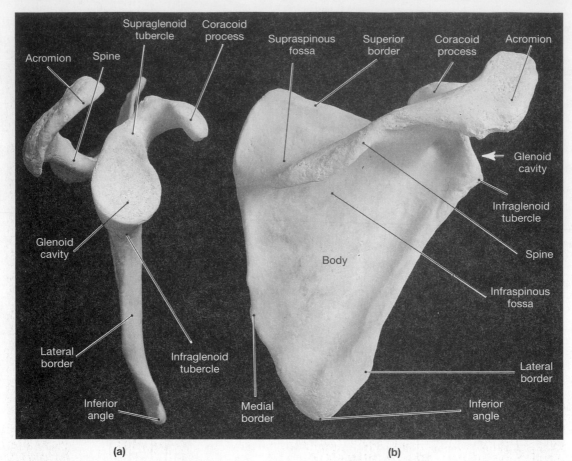

FIGURE 7-3 Scapula (a) lateral view (b) posterior view.

The Arm: Humerus

The **humerus** extends from the scapula to the elbow. The proximal end articulates with the scapula of the pectoral girdle. The distal end of the humerus articulates with the bones of the forearm. [∞ p185]

Procedure

Examine the *humerus* first by locating its position on an articulated skeleton. Place your bone specimen in the anatomical position adjacent to the humerus on the articulated skeleton. Identify **anterior/posterior, medial/lateral surfaces** or **borders**, and **proximal** and **distal ends**. Then identify the following **bone markings** on your specimen.

See: Figs. 7-4, 7-5

∞ *Figs.*
7-1, p181
7-6, pp186-187

LOCATE

From the Anterior View

___ Head

___ Necks { Anatomical / Surgical

___ Greater tubercle

___ Lesser tubercle

___ Shaft

___ Intertubercular groove

___ Deltoid tuberosity

___ Lateral epicondyle

___ Capitulum

___ Trochlea

___ Coronoid fossa

___ Medial epicondyle

___ Radial fossa

From the Posterior View

___ Head

___ Necks

___ Greater tubercle

___ Shaft

___ Groove for radial nerve

___ Olecranon fossa

___ Nutrient foramen

___ Trochlea

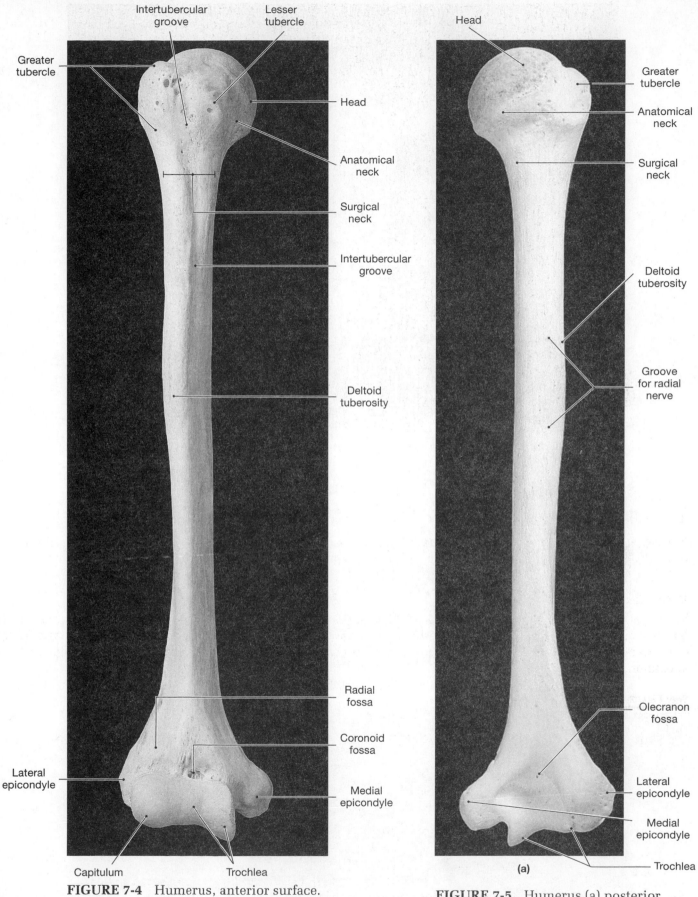

Intertubercular groove

Lesser tubercle

Greater tubercle

Head

Anatomical neck

Surgical neck

Intertubercular groove

Deltoid tuberosity

Radial fossa

Coronoid fossa

Lateral epicondyle

Medial epicondyle

Capitulum

Trochlea

FIGURE 7-4 Humerus, anterior surface.

Head

Greater tubercle

Anatomical neck

Surgical neck

Deltoid tuberosity

Groove for radial nerve

Olecranon fossa

Lateral epicondyle

Medial epicondyle

Trochlea

(a)

FIGURE 7-5 Humerus (a) posterior surface.

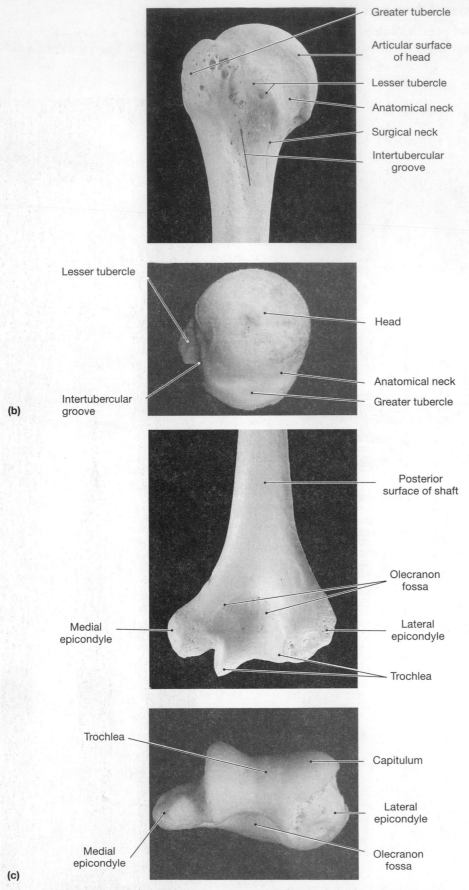

(b)

(c)

FIGURE 7-5 *(continued)*   Humerus (b) proximal end, anterior and superior surfaces (c) distal end, posterior and inferior surfaces.

Bones of the Forearm: Ulna and Radius

The **ulna** and **radius** are parallel bones of the forearm. The distal end of the humerus articulates with the proximal ends of these bones. Their distal ends articulate with the bones that form the wrist. [∞ p185]

Procedure

Examine the *ulna* and *radius* first by locating their positions on an articulated skeleton. Place each bone specimen in the anatomical position adjacent to that same bone on the articulated skeleton. In the anatomical position, the ulna lies medial to the radius. Identify **anterior/posterior, medial/lateral,** and **superior/inferior surfaces** or **borders,** and **proximal** and **distal ends.** Then identify the following **bone markings** on your specimen.

See: Figs. 7-6, 7-7

∞ *Figs.*
7-1, p181
7-7, p188-189

LOCATE

Ulna Markings

__ Olecranon process

__ Trochlear notch

__ Coronoid process

__ Ulnar tuberosity

__ Radial notch

__ Shaft

__ Head of ulna

__ Ulnar styloid process

__ Proximal and distal radioulnar joints

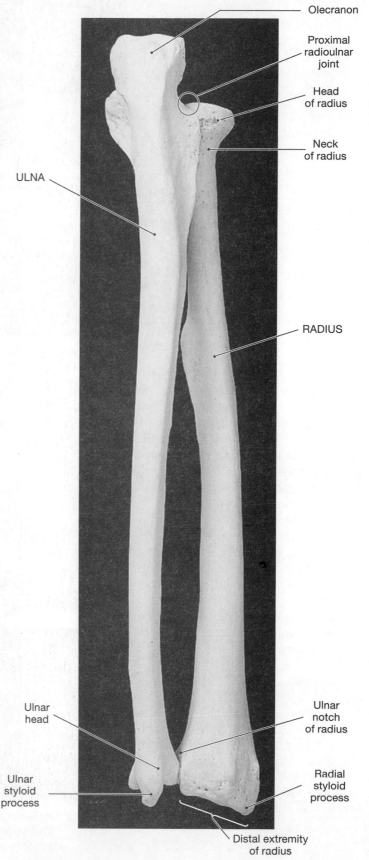

FIGURE 7-6 Radius and ulna, posterior view.

LOCATE

Radius Markings

— Head of radius

— Neck of radius

— Radial tuberosity

— Shaft of radius

— Attachment site for antebrachial interosseous membrane

— Ulnar notch

— Radial styloid process

Bones of the Wrist and Hand

The **wrist** or **carpus** is formed by eight **carpal bones** arranged in proximal and distal rows. The four *proximal carpal bones* are: **scaphoid, lunate, triangular (triquetrum)**, and **pisiform**. The four *distal carpal bones* are: **trapezium, trapezoid, capitate**, and **hamate**. [∞ p190]

Five **metacarpals** articulate with the distal carpal bones to form the palm of the hand. The metacarpals appear as miniature long bones. Roman numerals I-V identify each metacarpal. Metacarpal I is the lateral metacarpal that articulates with the trapezium and metacarpal V, the medial metacarpal, articulates with the hamate.

The proximal ends of the metacarpals articulate with the bones of the fingers, the **phalanges**. There are 14 phalangeal bones in each hand. Three bones form each finger, a *proximal*, a *middle*, and a *distal phalanx*. The thumb, termed **pollex**, contains only two bones---a proximal phalanx and a distal phalanx.

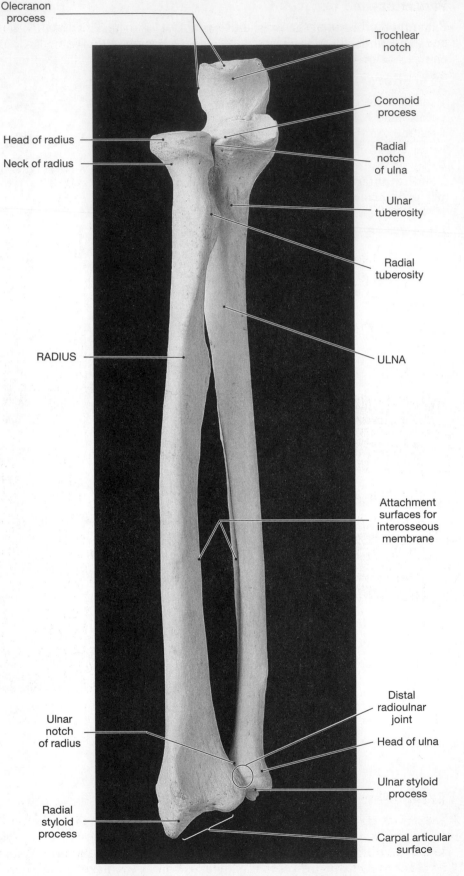

FIGURE 7-7 Radius and ulna, anterior view.

Olecranon process

Trochlear notch

Coronoid process

Head of radius

Neck of radius

Radial notch of ulna

Ulnar tuberosity

Radial tuberosity

RADIUS

ULNA

Attachment surfaces for interosseous membrane

Distal radioulnar joint

Head of ulna

Ulnar notch of radius

Ulnar styloid process

Radial styloid process

Carpal articular surface

Procedure

Examine the bones of the *wrist* and hand first by locating their positions on an articulated skeleton. Place your bone specimens in the anatomical position adjacent to that same hand on the articulated skeleton. Identify **anterior/posterior** and **medial/lateral surfaces**. When the radius and ulna are placed in the anatomical position, the hand is in the **supine position**. The reverse relationship of the radius and ulna places the hand in the **prone position**. Then identify the following **bone markings** on your specimen.

See: Fig. 7-8

∞ *Fig. 7-8, p190*

LOCATE

___ Phalanges

___ Distal phalanx

___ Middle phalanx

___ Proximal phalanx

___ Metacarpals (I to V)

___ Scaphoid bone

___ Lunate bone

___ Triquetrum

___ Pisiform bone

___ Trapezium bone

___ Trapezoid bone

___ Capitate bone

___ Hamate bone

___ Ulnar styloid process

___ Radial styloid process

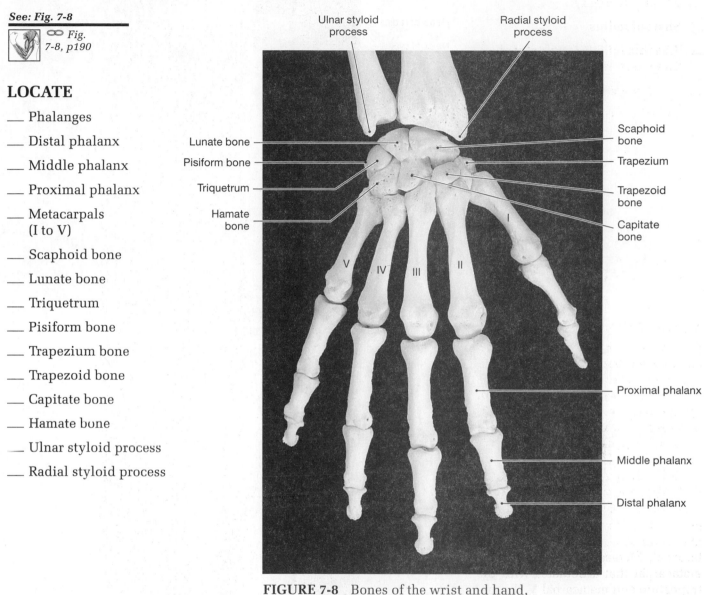

FIGURE 7-8 Bones of the wrist and hand, posterior view.

Description of the Pelvic Girdle and Lower Extremity

The **pelvic girdle** consists of two fused bones, the hip (**coxae**) or **innominate bones**. The hip bones, sacrum, and coccyx form a single bony structure, the **pelvis**. Each lower extremity, extending from the pelvis, consists of the **femur** (thigh bone), **tibia** and **fibula** (leg bones), **tarsals** (bones of the ankle), and bones of the foot. Notice the bones of the pelvic girdle and the lower extremity are heavier and more massive than those of the upper extremity. Bones of greater mass are required to support our body weight and permit movement. [∞ p194]

Hip Bone: Os Coxae and the Pelvis

Procedure

Examine the **os coxae**, or hip bone first by locating its position on an articulated skeleton. Determine if your hip bone specimen is a left or right by placing your bone specimen in the anatomical position adjacent to the hip bone on the articulated skeleton. Identify **anterior/posterior, medial/lateral**, and **superior/inferior surfaces** or **borders**. Then identify the following **bone markings** on your specimen.

See: Figs. 7-9 to 7-12

∞ *Figs.*
7-9, p194
7-10, p196-197
7-11, p198-199
7-12, p200

LOCATE

___ Ilium

___ Pubis

___ Ischium

___ Iliac crest

___ Anterior superior iliac spine

___ Anterior inferior iliac spine

___ Anterior gluteal line

___ Inferior gluteal line

___ Posterior gluteal line

___ Posterior superior iliac spine

___ Posterior inferior iliac spine

___ Greater sciatic notch

___ Ischial spine

___ Lesser sciatic notch

___ Ischial tuberosity

___ Ischial ramus

___ Obturator foramen

___ Inferior ramus of pubis

___ Superior ramus of pubis

___ Pubic tubercle

___ Pubic crest

___ Pubic symphysis

___ Acetabulum

___ Acetabular fossa

___ Lunate surface of acetabulum

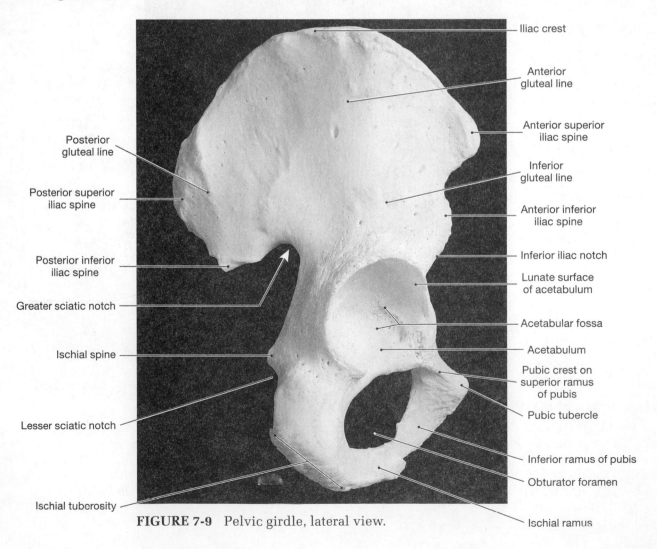

FIGURE 7-9 Pelvic girdle, lateral view.

From the Medial View

LOCATE

___ Same structures on lateral surface visible in medial view, except acetabular structures and gluteal lines

___ Auricular surface for articulation with sacrum

___ Iliac fossa

___ Iliac tuberosity

___ Arcuate line

___ Iliopectineal line

___ Obturator groove

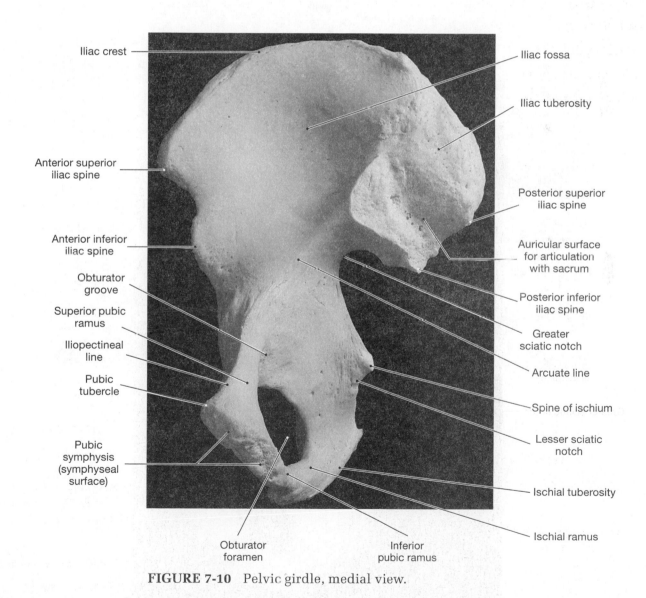

Iliac crest

Anterior superior iliac spine

Anterior inferior iliac spine

Obturator groove

Superior pubic ramus

Iliopectineal line

Pubic tubercle

Pubic symphysis (symphyseal surface)

Obturator foramen

Inferior pubic ramus

Iliac fossa

Iliac tuberosity

Posterior superior iliac spine

Auricular surface for articulation with sacrum

Posterior inferior iliac spine

Greater sciatic notch

Arcuate line

Spine of ischium

Lesser sciatic notch

Ischial tuberosity

Ischial ramus

FIGURE 7-10 Pelvic girdle, medial view.

LOCATE

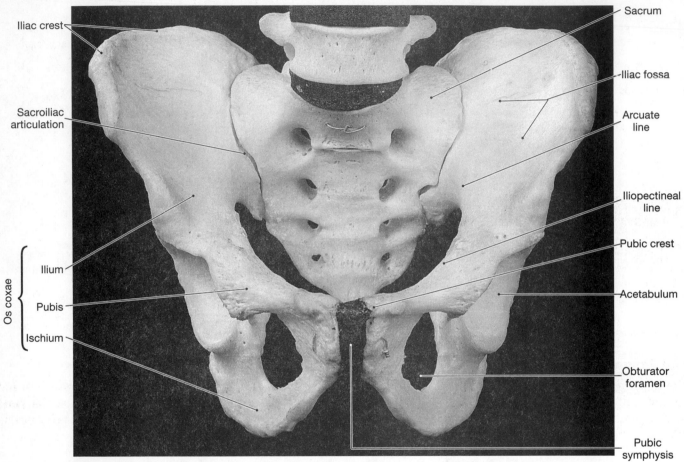

FIGURE 7-11 Male pelvis, anterior view. The pelvis consists of the two os coxae, the sacrum and the coccyx.

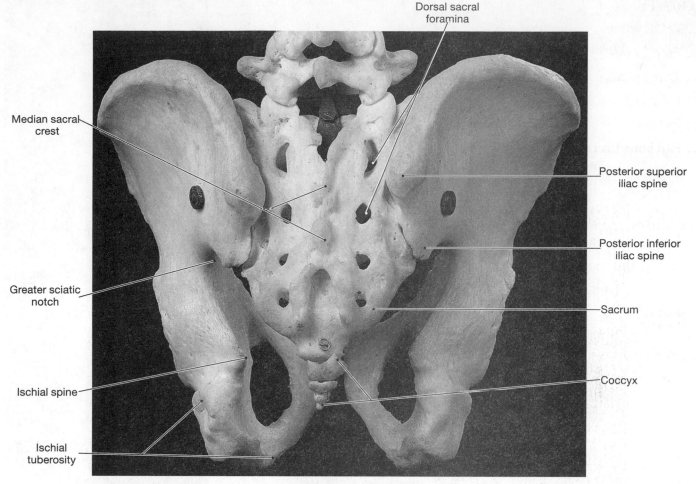

Dorsal sacral foramina

Median sacral crest

Posterior superior iliac spine

Posterior inferior iliac spine

Greater sciatic notch

Sacrum

Ischial spine

Coccyx

Ischial tuberosity

FIGURE 7-12   Male pelvis, posterior view.

The Thigh: Right Femur

The **femur** is the longest and heaviest bone in the body. The proximal portion of the femur articulates with the acetabulum of the hip bone, while the distal end articulates with the proximal portion of the tibia at the knee joint. Note the smooth anterior surface and the rough, raised posterior surface. The posterior surface serves as the attachment point for muscles that adduct and flex the leg. [∞ p201]

Procedure

Examine the *femur* first by locating its position on an articulated skeleton. Place your bone specimen in the anatomical position adjacent to the femur on the articulated skeleton. Identify **anterior/posterior, medial/lateral**, and **superior/inferior surfaces** and **proximal** and **distal ends**. Now identify the following **bone markings** on your specimen.

From the Anterior View

LOCATE

___ Head

___ Fovea for ligament of head

___ Neck

___ Greater trochanter

___ Lesser trochanter

___ Intertrochanteric line

___ Trochanteric fossa

___ Shaft (body)

___ Lateral condyle

___ Lateral epicondyle

___ Medial condyle

___ Medial epicondyle

___ Adductor tubercle

___ Patellar surface

From the Posterior View

LOCATE

___ Same structures on anterior surface visible except intertrochanteric line and patellar surface

___ Intertrochanteric crest

___ Gluteal tuberosity

___ Pectineal line

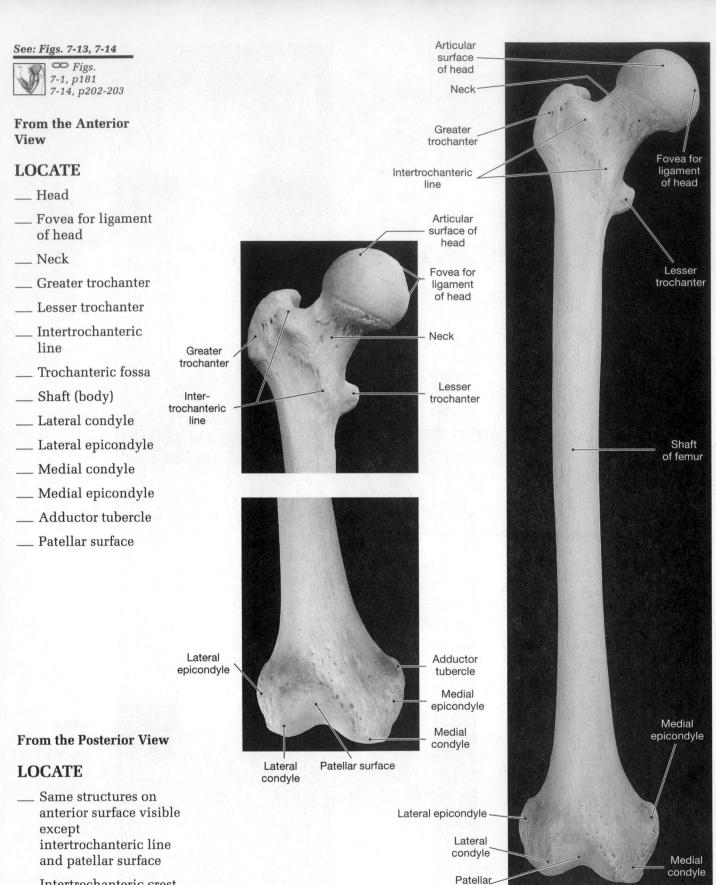

FIGURE 7-13 Right femur, anterior surface.

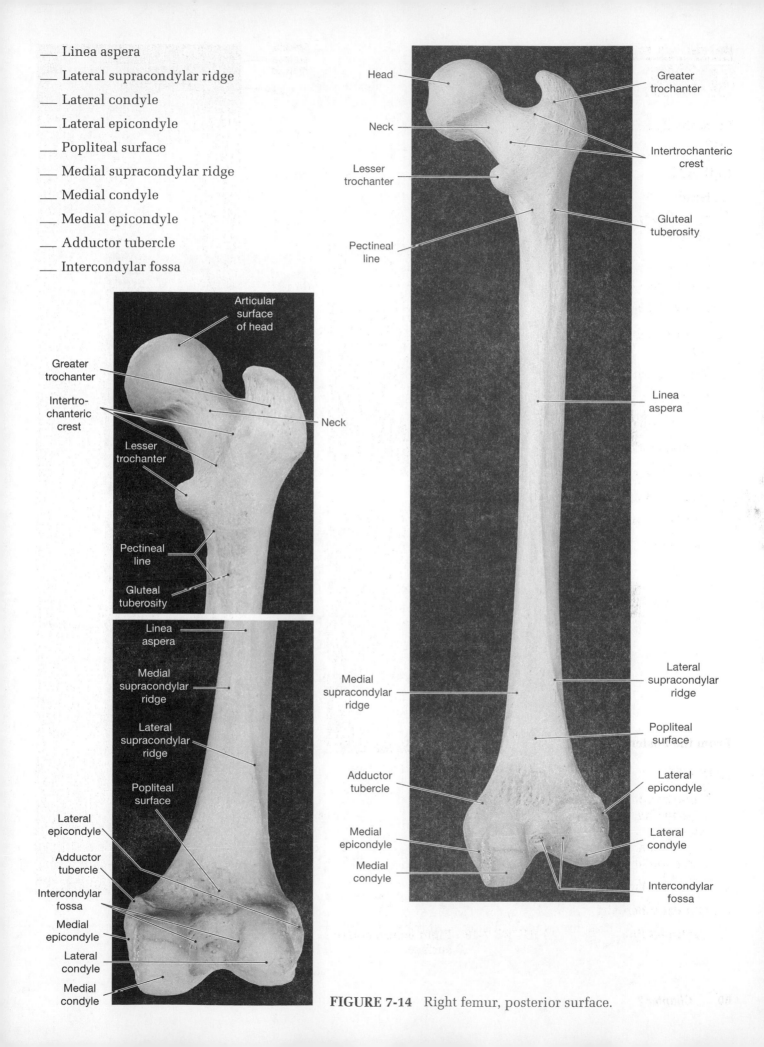

___ Linea aspera
___ Lateral supracondylar ridge
___ Lateral condyle
___ Lateral epicondyle
___ Popliteal surface
___ Medial supracondylar ridge
___ Medial condyle
___ Medial epicondyle
___ Adductor tubercle
___ Intercondylar fossa

Articular surface of head

Greater trochanter

Intertro-chanteric crest

Lesser trochanter

Neck

Pectineal line

Gluteal tuberosity

Linea aspera

Medial supracondylar ridge

Lateral supracondylar ridge

Popliteal surface

Lateral epicondyle

Adductor tubercle

Intercondylar fossa

Medial epicondyle

Lateral condyle

Medial condyle

Head

Neck

Lesser trochanter

Pectineal line

Greater trochanter

Intertrochanteric crest

Gluteal tuberosity

Linea aspera

Medial supracondylar ridge

Lateral supracondylar ridge

Popliteal surface

Adductor tubercle

Lateral epicondyle

Medial epicondyle

Lateral condyle

Medial condyle

Intercondylar fossa

FIGURE 7-14 Right femur, posterior surface.

Patella

The somewhat triangular-shaped **patella** (knee cap) is the largest sesamoid bone. The patella is located within the *tendon of the quadriceps femoris*. [∞ p204] The roughened anterior surface is convex, while the posterior surface is marked by a smooth concave *medial facet* and a *lateral facet*. These facets articulate respectively with the medial and lateral condyles of the femur. [∞ p204]

Procedure

Examine the *patella* by first locating its position on an articulated skeleton. Place your bone specimen in the anatomical position adjacent to that same patella on the articulated skeleton. Identify **anterior/posterior, medial/lateral**, and **superior/inferior surfaces** or **borders**. Now identify the following **bone markings** on your specimen.

See: Fig. 7-15

∞ *Figs.*
7-1, p181
7-15, p204

LOCATE

__ Base of patella

__ Apex of patella

__ Attachment site for quadriceps (tendon and patellar ligament-anterior surface)

__ Articular surface of patella { Medial facet / Lateral facet

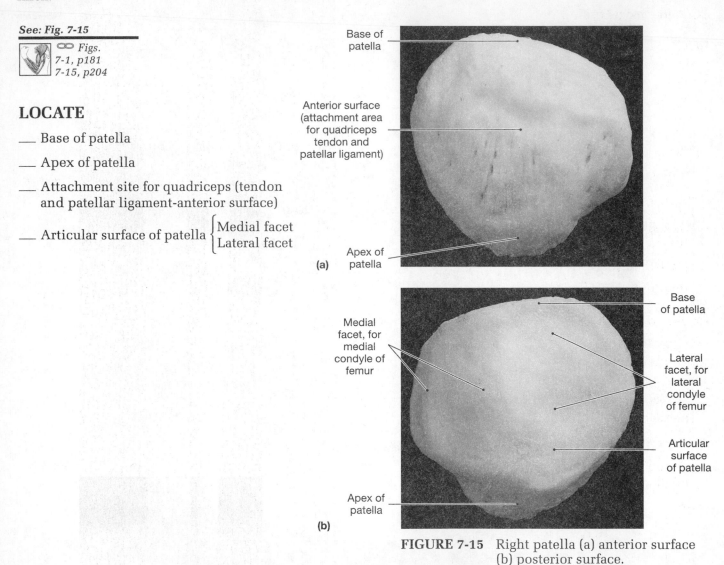

FIGURE 7-15 Right patella (a) anterior surface (b) posterior surface.

Bones of the Leg: Tibia and Fibula

The large medial bone of the leg is the **tibia**. The medial and lateral condyles of the femur articulate with the *medial* and *lateral tibial condyles* of the proximal end of the tibia. The tibia transfers body weight to the ankle and foot and serves as attachment for muscles that move the foot and toes. Paralleling the tibia is the long thin sharp-shafted **fibula** bone. The fibula is not part of the knee joint nor does it transfer weight to the ankle and foot. It is an important stabilizer of the leg and site for attachment of those muscles that move the ankle, foot, and toes. [∞ p204]

Procedure

Examine the *tibia* and *fibula* by first locating their positions on an articulated skeleton. Place your bone specimens in the anatomical position adjacent to the tibia and fibula on the articulated skeleton. Identify **anterior/posterior, medial/lateral borders**, and **proximal** and **distal ends**. Now identify the following **bone markings** on your specimen.

See: Figs. 7-16, 7-17

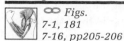

∞ *Figs.
7-1, 181
7-16, pp205-206*

From the Anterior View

LOCATE

Tibia

___ Shaft

___ Medial condyle

___ Lateral condyle

___ Tibial tuberosity

___ Anterior margin (border)

___ Interosseous border of tibia

___ Medial malleolus

___ Inferior articular surface

Fibula

___ Fibula { Head
Neck
Shaft

___ Interosseous crest (border) of fibula

___ Lateral malleolus

___ Tibiofibular { Superior
joints { Inferior

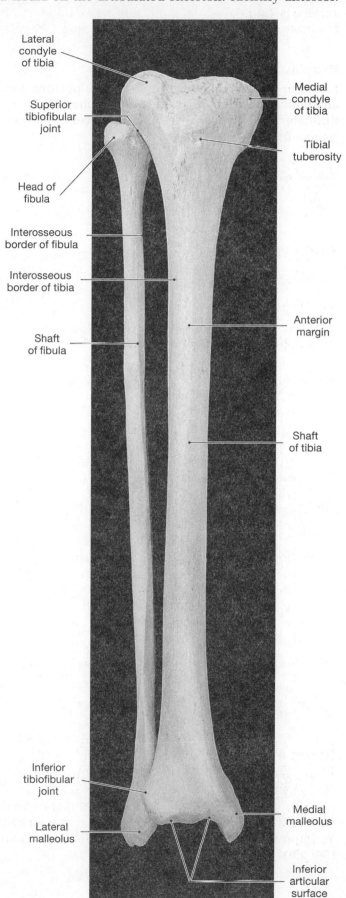

FIGURE 7-16   Tibia and fibula, anterior view.

From the Posterior View

LOCATE

___ Same structures on anterior surface visible except tibial tuberosity, anterior crest (border)

___ Intercondylar eminence

___ Medial and lateral tubercles of intercondylar eminence

___ Popliteal line

___ Articular surfaces of tibia and fibula

Bones of the Ankle and Foot

The ankle, termed **tarsus**, consists of seven bones called **tarsals**. The tarsal bones are the **calcaneus**, **talus**, **cuboid**, **navicular**, and three **cuneiforms**. The calcaneus, or heel bone, is the largest tarsal bone. The articular surface of the tibia transmits the weight of the body to the talus, then to the calcaneus, which in turn, transmits the weight anteriorly to the toes and then to the ground. The three cuneiforms are named according to their location: *medial cuneiform, intermediate cuneiform,* and *lateral cuneiform.* The distal surfaces of the cuneiforms and the cuboid articulate with the bones that form the sole of the foot. [∞ p207]

Five long **metatarsal bones** form the sole of the foot. Like the metacarpals, the metatarsals are also identified by Roman numerals I–V. The medial metatarsal is identified by Roman numeral I and the lateral by numeral V. The distal portion of each metatarsal articulates with the toes.

The skeletal organization of the toes, termed **phalanges**, is the same as that of the fingers. Each toe consists of a **proximal, medial,** and **distal phalanx**. The great toe, termed **hallux**, has only a proximal phalanx and a distal phalanx.

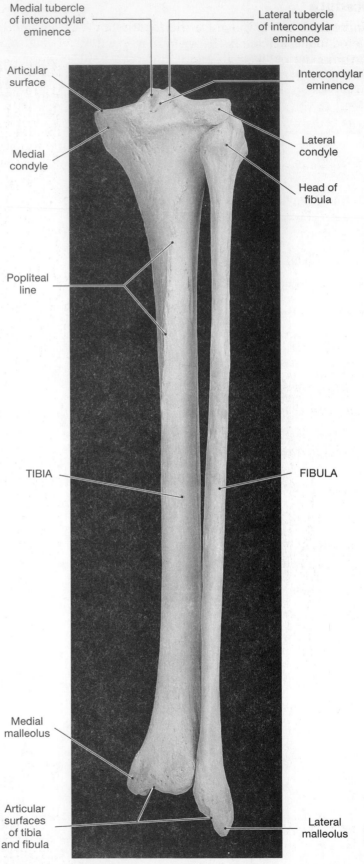

FIGURE 7-17 Tibia and fibula, posterior view.

Procedure

Examine the bones of the *ankle* and *foot* first by locating their positions on an articulated skeleton. Place your articulated foot specimen in the anatomical position adjacent to that same foot on the articulated skeleton. Identify **anterior/posterior, medial/lateral,** and **superior/inferior surfaces** or **borders**. Now identify the following **bone markings** on your specimen.

See: Fig. 7-18

∞ *Figs.*
7-1, p181
7-17, p207
7-18, p208

LOCATE

___ Distal phalanx

___ Middle phalanx

___ Proximal phalanx

___ Hallux (great toe)

___ Metatarsal bones (I to V) { Base Shaft Head

___ Cuboid bone

___ Cuneiform bones { 1st (medial) 2nd (intermediate) 3rd (lateral)

___ Navicular bone

___ Talus bone

___ Trochlea of talus

___ Calcaneus

___ Transverse arch

___ Longitudinal arch

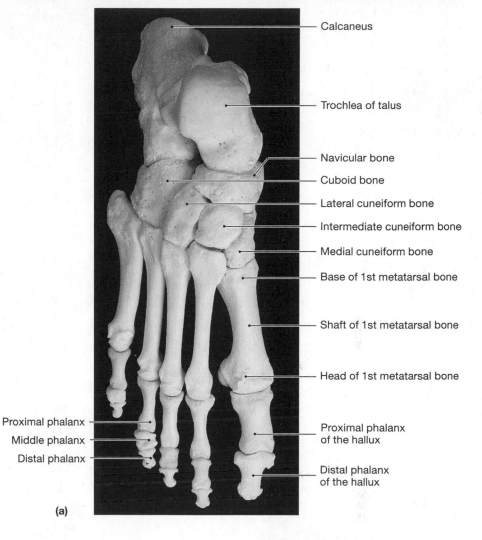

Calcaneus

Trochlea of talus

Navicular bone

Cuboid bone

Lateral cuneiform bone

Intermediate cuneiform bone

Medial cuneiform bone

Base of 1st metatarsal bone

Shaft of 1st metatarsal bone

Head of 1st metatarsal bone

Proximal phalanx

Middle phalanx

Distal phalanx

Proximal phalanx of the hallux

Distal phalanx of the hallux

(a)

FIGURE 7-18 Bones of the ankle and foot (a) dorsal view (b) sectional view.

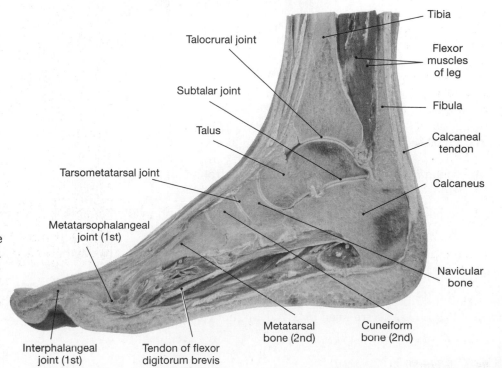

Talocrural joint

Subtalar joint

Talus

Tarsometatarsal joint

Metatarsophalangeal joint (1st)

Interphalangeal joint (1st)

Tendon of flexor digitorum brevis

Metatarsal bone (2nd)

Cuneiform bone (2nd)

Tibia

Flexor muscles of leg

Fibula

Calcaneal tendon

Calcaneus

Navicular bone

(b)

Anatomical Identification Review

Label the individual bones that form the appendicular skeleton.

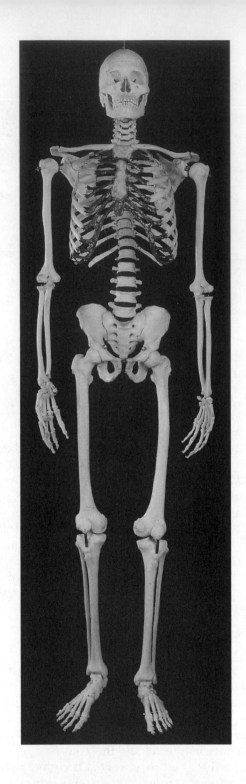

To Think About

1. What is the advantage of the lack of strong bone to bone articulations between the girdle of the upper extremity and the axial skeleton?

2. What is the significance of the strong bone to bone attachments of the girdle of the lower extremity to the axial skeleton?

3. How would a forensic scientist decide whether a partial skeleton found in the forest after a plane crash is that of a male or female?

4. Which skeletal features can be used by a forensic scientist to help determine the identity of an individual?

5. Which joints are used by a star pitcher for a major league baseball team when throwing a baseball?

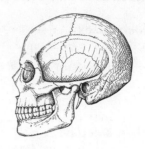

8

THE SKELETAL SYSTEM:
Articulations

Objectives

1. Classify joints (articulations) into categories based on the range of motion permitted and provide examples for each category.

2. Identify, name, and describe the range of motion permitted for the anatomical components of articulations between vertebrae in the vertebral column.

3. Identify, name, and describe the range of motion permitted by each joint for anatomical structures in articulations of both the upper extremity (shoulder, elbow, wrist, and hand) and the lower extremity (hip, knee, ankle, and foot).

Classification of Joints

Joints, or **articulations**, exist wherever two bones meet. The function of each joint depends on its anatomical design. Some joints are interlocking and completely prohibit movement, others permit slight movement, while others permit extensive movement. Joints are classified either on the basis of their structure (bony fusion, fibrous, cartilaginous, or synovial) or on the range of motion they permit. [∞ p214] An immovable or rigid joint, termed **synarthrosis**, permits no movement because the bony surfaces are held together firmly or even interlocked (bones of the skull). A slightly movable joint, termed **amphiarthrosis**, permits a very limited range of motion (adjacent vertebrae of vertebral column). This type of joint is characterized by bands of ligaments binding bones together. An exception to this restricted movement is observed in the moveable joint between the radius and ulna. A freely movable joint, called **diarthrosis**, permits a wide range of free movements (shoulder). This type of joint is characterized by: (1) a **fibrous joint capsule**, (2) a **synovial membrane** that produces **synovial fluid**, (3) **articular cartilages** that cap the articular surfaces of the bones forming the joint, and (4) **accessory structures**—pads of fat or cartilage (menisci or articular discs), intracapsular and extracapsular ligaments, tendons, and bursae.

Intervertebral Articulations

The articulations between the superior and inferior articular processes of adjacent vertebrae are gliding joints that permit small movements associated with flexion, extension, and rotation of the vertebral column. [∞ p223] Pads of fibrocartilage, called **intervertebral discs**, separate and cushion adjacent vertebrae and form amphiarthrotic joints. The size of the disc is variable, but conforms to the size of the vertebral body. One-fourth of the length of the adult vertebral column is due to the intervertebral discs. Intervertebral discs are not found between the first two cervical vertebrae, the atlas and axis, nor within the sacrum or coccyx. Numerous ligaments are attached to the bodies and processes of all vertebrae to bind them together and stabilize the vertebral column. Neighboring vertebrae are connected and stabilized by the following six ligaments:

1. **anterior longitudinal ligament** connects the anterior surfaces of each vertebral body and disc from occipital bone to sacrum.

2. **posterior longitudinal ligament** connects the posterior surfaces of each vertebral body and disc from C_2 to sacrum.

3. **ligamentum flavum** interconnects the laminae of adjacent vertebrae.

4. **interspinous ligament** connects the spinous processes of adjacent vertebrae.

5. **supraspinous ligament** connects the ends of the spinous processes from C_7 to the sacrum.

6. **ligamentum nuchae** extends longitudinally from the external occipital protuberance of the skull to C_7.

Procedure

✓ Quick Check

Before you begin to identify the articulations between adjacent vertebrae, review the articulation between lumbar vertebrae on the articulated skeleton (*Figure 6-2, p40*).

Identify the articulations between adjacent vertebrae first by observing a *model* or *skeleton* and then a *cadaver specimen.* Start your observation first by reviewing the **structures** of a **typical cervical, thoracic,** or **lumbar vertebra**. Begin your examination of the cadaver back in the lower thoracic and lumbar regions of the vertebral column, then move cranially to the cervical region. Reflect the *erector spinae muscles* [*Figure 10-8,* ∞ Figure 10-12] to see all of the above ligaments, except the anterior longitudinal ligament. This ligament can be observed in the lumbar region by first reflecting the abdominal viscera that covers the posterior abdominal wall. The vertebral column is still hidden by the inferior vena cava and descending abdominal aorta (*Color Plate 87* and *Figure 22-6, p323*). These vessels must be separated and the underlying fascia may need to be cleaned.

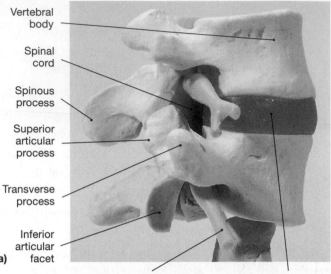

(a)

Vertebral body
Spinal cord
Spinous process
Superior articular process
Transverse process
Inferior articular facet
Spinal nerve emerging from intervertebral foramen
Intervertebral disc

See: Fig. 8-1

∞ Fig. 8-8, p223

LOCATE

— Vertebral body

— Intervertebral disc

— Anterior longitudinal ligament

— Spinous process

— Supraspinous ligament

— Interspinous ligament

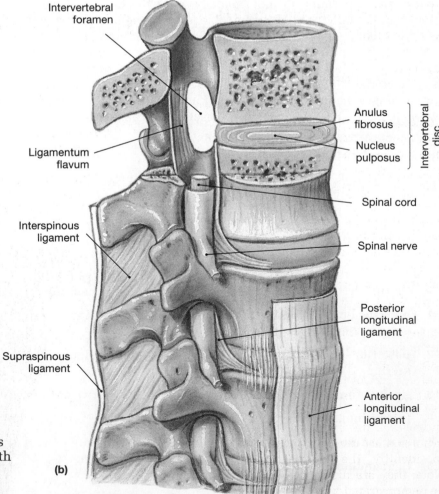

Intervertebral foramen
Ligamentum flavum
Interspinous ligament
Supraspinous ligament
Anulus fibrosus
Nucleus pulposus
Intervertebral disc
Spinal cord
Spinal nerve
Posterior longitudinal ligament
Anterior longitudinal ligament

(b)

FIGURE 8-1 Intervertebral articulations (a) articulation model of 4th and 5th lumbar vertebrae (b) vertebral ligaments, lateral views.

Description of the Shoulder Joint (Glenohumeral)

The **shoulder joint** (glenohumeral) permits the greatest range of motion of all the joints of the body. [∞ p223] The head of the humerus articulates with the glenoid fossa of the scapula to form this ball-and-socket type joint. The fibrocartilaginous **glenoid labrum** is attached to the margin of the glenoid fossa (cavity), completely encircling and deepening it. The stability of the shoulder joint is provided both by ligaments between the scapula and either the humerus or the clavicle and by the surrounding muscles and associated tendons. The major ligaments involved in stabilizing the joint are:

1. **glenohumeral ligaments** lie in the anterior region of the articular capsule, and are difficult to see well.

2. **coracohumeral ligament** originates at the coracoid process of the scapula and inserts on the greater tubercle of the humerus, and is difficult to see.

3. **coracoacromial ligament** lies superior to the capsule and bridges the gap between coracoid process and acromion.

4. **acromioclavicular ligament** binds the acromion to the clavicle to restrict clavicular movement at the acromion.

5. **coracoclavicular ligament** binds the clavicle to the coracoid process to prevent the clavicle from being pulled away from the scapula and to provide major support for the acromioclavicular joint.

6. **transverse humeral ligament** straps down the tend from the long head of the biceps brachii muscle into the intertubercular groove of the humerus.

Muscles and the underlying shoulder articular capsule are protected by a number of **bursae**. These are small, membranous sacs, which function to reduce friction and act as shock absorbers. They are filled with synovial fluid produced by the synovial membrane. The prominent **subacromial bursa** and the **subcoracoid bursa** prevent the acromion and the coracoid process of the scapula from coming in contact with the shoulder articular capsule.

✓ Quick Check

Before you begin to identify the structures of the shoulder joint, review the relative positions and articulations of the clavicle, scapula, and humerus on the articulated skeleton (*Figures 6-1, p40* and *7-1 to 7-5, pp69-72*).

Identify the following **joint structures** using first a *model* of the *shoulder joint* and then a *cadaver specimen*. To observe the shoulder joint in the cadaver, the deltoid and trapezius muscles must first be removed from the acromion and the lateral end of the clavicle. The tendon of long head of biceps brachii muscle and the head of the humerus are good landmarks to begin your examination. View the joint first from the anterior. The large subacromial bursa lie superior to the tendon and humeral head; the subcoracoid and subscapular bursae lie inferior to the coracoid process. Now locate the shoulder articular capsule as it attaches to the glenoid fossa and glenoid labrum proximally and to the anatomical neck of the humerus distally.

The interior of the shoulder articular capsule and joint cavity is best viewed from the posterior. Clean from the posterior of the capsule any remaining muscle, ligaments, and fascia. Cut through the posterior part of the capsule to expose the interior of the joint capsule. Locate first the articular surface of the head of the humerus and the glenoid labrum and use these as landmarks from which to identify the glenohumeral ligaments. Note: they are difficult to identify and may be demonstrated by the instructor.

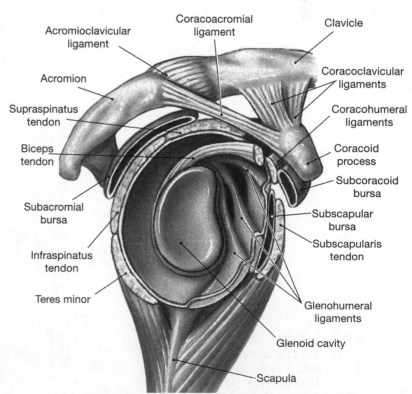

FIGURE 8-2 Shoulder joint, lateral view of pectoral girdle.

See: Figs. 8-2 to 8-4

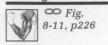 ∞ *Fig. 8-11, p226*

LOCATE

__ Coracoid process

__ Acromion

__ Head of humerus

__ Glenoid fossa

__ Clavicle

__ Acromioclavicular ligament

__ Subacromial bursa

__ Articular capsule

__ Glenohumeral ligaments

__ Coracoacromial ligament

__ Coracoclavicular ligaments

__ Tendon of biceps brachii muscle

__ Subcoracoid bursa

__ Subscapular bursa

__ Subscapularis muscle

__ Deltoid muscle

__ Pectoralis major muscle

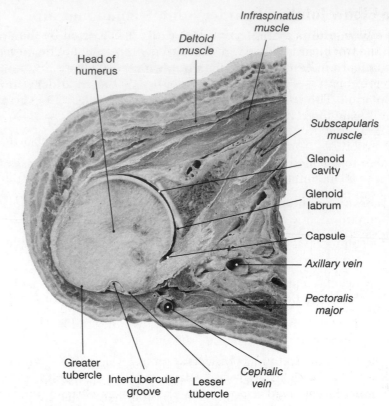

FIGURE 8-3 Shoulder joint, superior view, horizontal section.

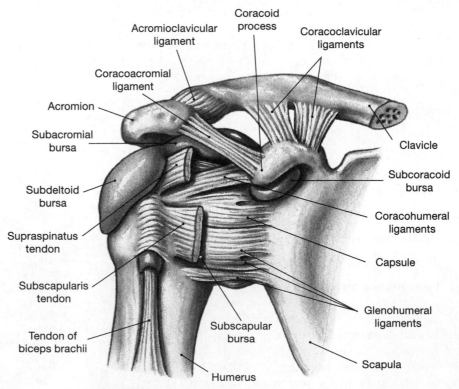

FIGURE 8-4 Right shoulder joint, anterior view.

The Elbow Joint

The **elbow joint** is a hinge joint. [∞ p227] It is formed by the articulation of both the trochlea of the humerus with the trochlear notch of the ulna and the capitulum of the humerus with the head of the radius. Viewed in the anatomical position, the articular capsule extends from the medial epicondyle of the humerus to the anterior coronoid process. From the posterior, the articular capsule extends from the capitulum to the olecranon process of the ulna. The anatomical structure of the elbow joint is very stable. The stability of the elbow joint results from (1) the interlocking of the trochlea with the trochlear notch, (2) a thick articular capsule, and (3) capsule strengthening provided by the *ulnar collateral ligament, annular ligament*, and *radial collateral ligament* accessory structures. The **ulnar collateral ligament** extends from the anterior and posterior surfaces of the medial epicondyle to the medial surface of the coronoid process and olecranon. The **annular ligament** wraps around the radial head and attaches to the radial notch of the ulna. The **radial collateral ligament** extends from the inferior border of the lateral epicondyle to the lateral surface of the ulna and the annular ligament.

Procedure

✓ **Quick Check**

Before you begin to identify the structure of the elbow joint, review the articulation of the humerus, radius, and ulna on the articulated skeleton (*Figures 6-1, p40* and *7-5 to 7-7, p71-74*).

Identify the following **joint structures**, using first a *model* and then a *cadaver specimen* of the humeroulnar (elbow) *joint*. To observe the humeroulnar joint capsule in the cadaver, the muscles that surround it first must be removed and the underlying fascia cleaned and removed. By cutting through the anterior part of the capsule, the interior of the joint and articular surfaces may be viewed.

See: Figs. 8-5, 8-6

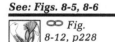
∞ *Fig. 8-12, p228*

LOCATE

___ Humerus

___ Ulna/radius

___ Articular capsule

___ Articular cartilage of olecranon

___ Articular cartilage of capitulum

___ Annular ligament

___ Biceps brachii tendon

___ Triceps muscle and tendon

___ Interosseous membrane

___ Ulnar collateral ligament

___ Radial collateral ligament

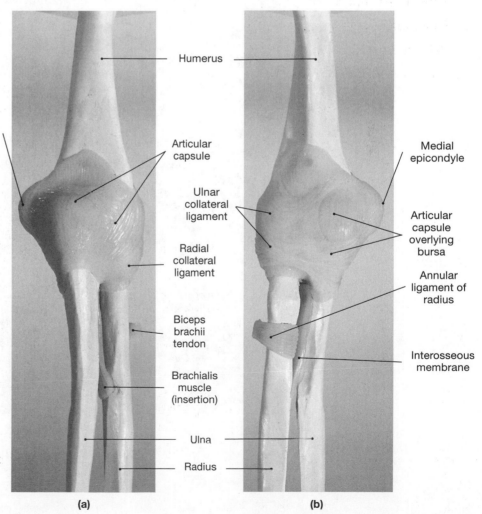

FIGURE 8-5 Model of elbow joint
(a) anterior view
(b) posterior view.

(a)   (b)

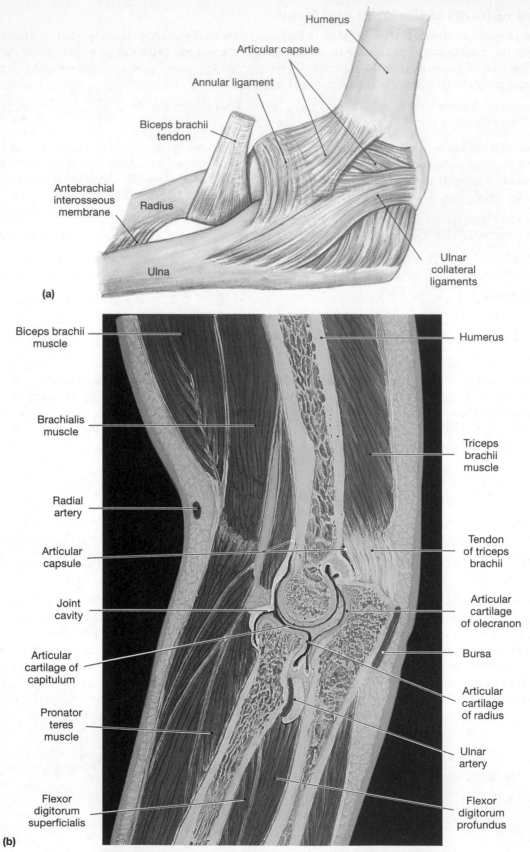

(a)

Humerus

Articular capsule

Annular ligament

Biceps brachii
tendon

Antebrachial
interosseous
membrane

Radius

Ulna

Ulnar
collateral
ligaments

(b)

Biceps brachii
muscle

Brachialis
muscle

Radial
artery

Articular
capsule

Joint
cavity

Articular
cartilage of
capitulum

Pronator
teres
muscle

Flexor
digitorum
superficialis

Humerus

Triceps
brachii
muscle

Tendon
of triceps
brachii

Articular
cartilage
of olecranon

Bursa

Articular
cartilage
of radius

Ulnar
artery

Flexor
digitorum
profundus

FIGURE 8-6 Elbow joint (a) superficial view, medial
aspect (b) coronal section.

Description of Joints of the Wrist and Hand

The wrist, or carpus, contains the **wrist joint**, which consists of three articulations: (1) the **distal radioulnar articulation**, (2) the **radiocarpal articulation**, and (3) the **intercarpal articulations**. [∞ p229] Wrist stability is maintained by a strong connective tissue capsule, numerous tendons of flexor and extensor muscles, and ligaments. The major ligaments that stabilize the wrist joint are:

1. **flexor retinaculum** is a thickening of the deep fascia attached to the superficial aponeurosis that arches across the anterior (palmar) surface of the carpals and forms the carpal tunnel.

2. **extensor retinaculum** is a broad ligament band that arches across the posterior (dorsum) of the wrist and is attached to the styloid process of the ulna, the triangular and pisiform bones, and the radius; immediately under the retinaculum, six canals are present to permit the passage of tendons for the dorsal extensor muscles of the hand.

3. **palmar radiocarpal ligament** connects the distal portion of the radius to the anterior surfaces of the scaphoid, lunate, and triangular carpals (it is not easily separated from other wrist ligaments).

4. **dorsal radiocarpal ligament** connects the distal portion of the radius to the posterior surface of the scaphoid, lunate, and triangular carpals (it is not easily separated from other wrist ligaments).

5. **ulnar collateral ligament** connects the styloid process of the ulna to the medial surface of the triangular carpal.

6. **radial collateral ligament** connects the styloid process of the radius to the lateral surface of the scaphoid bone.

The carpal bones articulate with the metacarpals of the palm to form **carpometacarpal joints**. Metacarpal I articulates with the trapezium bone to form the carpometacarpal joint of the thumb. Metacarpal II articulates with both trapezium and trapezoid bones. Metacarpal III articulates with capitate bone, and metacarpals IV and V articulate with the hamate bone. **Interosseous metacarpal ligaments** bind and stabilize the proximal ends of all the metacarpals by attaching to the medial and lateral surfaces of the bases of each metacarpal.

Procedure

✓ **Quick Check**

Before identifying the articulations of the wrist and hand, review bones of the wrist and hand on the articulated skeleton (*Figure 7-8, p75*).

Identify **structures** of the **wrist** and **hand**, using first a *model* and then a *cadaver specimen* of the *wrist* and *hand*. To view the carpus and wrist joint in the cadaver, the extensor and flexor tendons of the forearm must be removed distal to the styloid processes of the ulna and the radius. To observe the joint surfaces of the bones from the palmar surface, the wrist joint must be forced open in extension. To observe the joint surface of the bones from the dorsal surface, the wrist joint must be forced open in flexion. Identify first the articular surfaces of the radius and ulna before locating the scaphoid, lunate, and triangular bones. Should you wish to observe the joints of the hand, your instructor will provide you with directions to display and examine these joints.

See: Figs. 8-7, 8-8

∞ Fig.
8-13, p229

LOCATE

__ Tendons from muscles of forearm

__ Tendon sheath of flexor pollicis longus (radial bursa)

__ Tendon sheath of flexor carpi radialis

__ Common sheath of digital flexor tendons (ulna, bursa)

__ Radial collateral ligament

__ Palmar radiocarpal ligament

__ Ulnar collateral ligament

__ Carpometacarpal joint of little finger

__ Intercarpal ligaments

__ Carpometacarpal ligaments

__ Interosseous metacarpal

__ Distal radioulnar joint

__ Digitocarpal ligaments

__ Interosseous metacarpal ligaments

__ Midcarpal joint

__ Carpometacarpal joint of thumb

__ Carpal tunnel

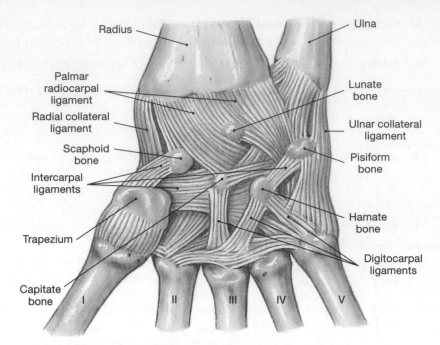

FIGURE 8-7 Ligaments of the wrist.

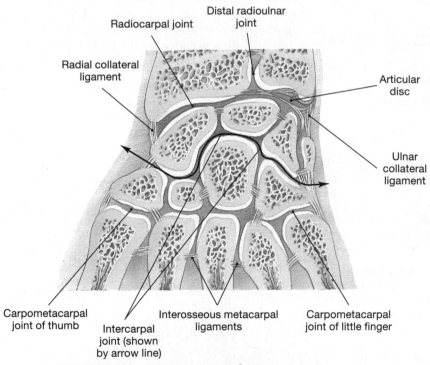

FIGURE 8-8 Wrist, sectional view.

Observation of the Hip Joint

The **hip joint** is categorized as a ball-and-socket type of joint, formed by the head of the femur articulating with the acetabulum of the hip bone. [∞ p230] A dense, fibrous articular capsule extends from the rim of the acetabulum to the intertrochanteric line in front to neck of the femur in back. (Note, the capsule is not attached to the intertrochanteric crest.) The articular capsule is strengthened and stabilized by the following ligaments:

1. **iliofemoral ligament** extends from the anterior inferior iliac spine to the intertrochanteric line as a Y-shaped ligament.

2. **pubofemoral ligament** extends from the superior pubic ramus and merges with the fibers of the iliofemoral ligament.

3. **ischiofemoral ligament** extends from the posterior ischium inferior and posterior to the acetabulum to merge with the fibers of the articular capsule at the junction of the neck and greater trochanter of the femur.

4. **transverse acetabular ligament** bridges the acetabular notch to complete the formation of the acetabulum.

5. **ligament of** (the) **head of** (the) **femur** *ligamentum capitis femoris* originates from acetabular notch and transverse acetabular ligament to attach to the fovea of the femur.

Procedure

✓ **Quick Check**

Before you begin to identify the structure of the hip joint, review the articulation of the head of the femur with the acetabulum of the hip bone (*Figures 7-9, p76* and *7-13, 7-14, p80-81*).

Identify the following **structures** of the **hip joint**, using first a *model* and then a *cadaver specimen* of the *hip joint*. To observe the hip joint capsule in the cadaver, the muscles that surround it must first be removed and the underlying fascia cleaned and removed. Observe from the anterior and locate first the prominent iliofemoral and pubofemoral ligaments. From the posterior observe the ischiofemoral ligament.

To view both the interior of the joint and its articular surfaces, cut through the iliofemoral and pubofemoral ligaments from the anterior. Now dislocate the joint by pulling the femur from the acetabular fossa. In order, locate first the lunate surface, acetabular labrum, and then the transverse ligament. The *ligament of the femoral head* has been severed during the dislocation process, but a portion should still be prominent within the acetabular fossa.

See: Figs. 8-9 to 8-11

∞ *Figs.*
8-14, p231
8-15, p232

LOCATE

___ Acetabulum   ___ Acetabular labrum

___ Lunate surface   ___ Acetabular fossa

___ Articular capsule

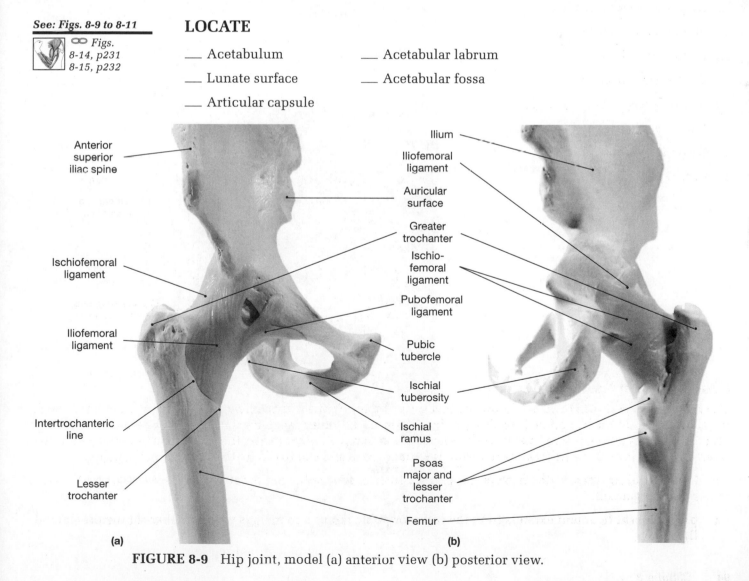

FIGURE 8-9 Hip joint, model (a) anterior view (b) posterior view.

___ Fovea for ligament of the femoral head

___ Acetabular notch

___ Transverse acetabular ligament

___ Greater trochanter

___ Iliofemoral ligament

___ Ischiofemoral ligament

___ Pubofemoral ligament

___ Ischial tuberosity

___ Femur

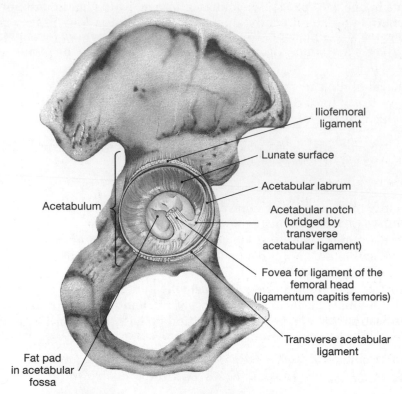

Iliofemoral ligament

Lunate surface

Acetabular labrum

Acetabular notch (bridged by transverse acetabular ligament)

Fovea for ligament of the femoral head (ligamentum capitis femoris)

Transverse acetabular ligament

Acetabulum

Fat pad in acetabular fossa

FIGURE 8-10   Right hip joint, lateral view.

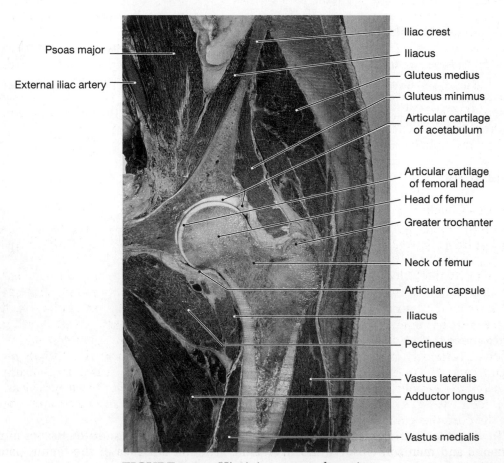

Psoas major

External iliac artery

Iliac crest

Iliacus

Gluteus medius

Gluteus minimus

Articular cartilage of acetabulum

Articular cartilage of femoral head

Head of femur

Greater trochanter

Neck of femur

Articular capsule

Iliacus

Pectineus

Vastus lateralis

Adductor longus

Vastus medialis

FIGURE 8-11   Hip joint, coronal section.

Description of the Knee Joint

The **knee joint** is the most complex joint in the body. It is categorized as a hinge joint, but is actually three separate joints in one. [∞ p231] Two joints are formed by the femur and tibia (medial condyle to medial condyle and lateral condyle to lateral condyle) and the third joint is formed between the patellar surface of the femur and patella. The femur-tibia relationship permits flexion, extension, and slight rotation when the leg is flexed. The femur-patella relationship creates a sliding up and a sliding down of the patella when locking and unlocking of the joint occurs.

A dense, fibrous articular capsule extends from the femur to the tibia and is strengthened and stabilized by connective tissue (of the iliotibial tract), muscles, tendons, and ligaments. The articular capsule is strengthened and stabilized by the following ligaments:

1. **patellar ligament** contains the patella and holds it in place; provides support as it passes over the anterior surface of the joint, and attaches to the anterior tibia.

2. **popliteal ligaments** extend between the femur and the heads of the tibia and fibula to reinforce the posterior knee joint; helps to prevent extension and medial rotation of the leg.

3. **anterior cruciate ligament** extends from lateral femoral condyle to anterior intercondylar eminence of tibia; serves to prevent the slipping forward of the tibia on the femur because the ligament is tense in all positions of the joint; most important of all the ligaments because it prevents lateral rotation of the leg and extension.

4. **posterior cruciate ligament** extends from medial femoral condyle to posterior intercondylar eminence of tibia and the posterior portion of the lateral meniscus; prevents posterior displacement of tibia on femur.

5. **tibial collateral ligament** extends from the medial epicondyle of the femur to the medial condyle and shaft of the tibia; serves to reinforce the medial surface of the knee joint and prevents abduction of the leg.

6. **fibular collateral ligament** extends from the lateral condyle of the femur to the head of fibula; serves to reinforce the lateral side of the knee joint and prevents adduction of the leg.

Between the femoral and tibial articular surfaces are a pair of fibrocartilaginous pads, the *medial* and *lateral menisci*. The crescent-shaped **medial meniscus** attaches at its anterior to the area anterior to the intercondylar eminence of the tibia and at its posterior to the area posterior to the intercondylar eminence of the tibia between the attachments of the posterior cruciate ligament and lateral meniscus. It is located anterior to the anterior cruciate ligament. The almost circular **lateral meniscus** attaches at its anterior end anterior to the intercondylar eminence of the tibia and lateral and posterior to the anterior cruciate ligament. At its posterior end, it is attached posterior to the intercondylar eminence of the tibia and anterior to the posterior end of the medial meniscus.

The articular capsule and synovial membrane of the knee are the largest and most complex in the body. The synovial membrane extends both superior to and inferior to the patella and lines the non-articular surfaces of the interior of the joint. The knee contains 12 bursae, positioned around the joint in groups (anterior, posterior, medial, and lateral). They are named according to the structures with which they are associated, such as *suprapatellar* and *infrapatellar bursae*.

Procedure

✓ **Quick Check** ───

Before you begin to identify the structure of the knee joint, review the articulation of the femur, patella, and tibia on the articulated skeleton (*Figures 7-13, p80* and *7-15, p82*).

Identify the following **structures of the knee**, using first a *model* and then a *cadaver specimen* of the *knee*. Observe the knee in the extended position from anterior, lateral, medial and posterior views. To observe the knee joint capsule in the cadaver, the thigh muscles superior to the patella and the leg muscles inferior to the patella must be separated at their tendons, and the underlying fascia cleaned and removed. The thigh muscles [∞ Figures 11-13 and 11-14b] superior to the patella should be cut, separated, and reflected just superior to their tendons. Posteriorly, in the popliteal region, note the origin on the femur of the *gastrocnemius muscle* (*Figures 11-18, p170* and *11-20, p172*). The gastrocnemius muscle must be cut distal to its origin and reflected. Observe the blood vessels and nerves in the popliteal region. Divide and separate these vessels and nerves, but do not remove them. Identify the *popliteus muscle* (*Figure 11-18, p170*). Remove and clean fascia from the popliteal fossa to expose the posterior surface of the articular capsule of the knee joint.

To view the interior, the capsule of the knee joint and all surrounding tissues must be removed to reveal the ligaments and menisci. Observe the ligaments, articular surfaces of the femur, patella, and tibia, the menisci from all sides, and, if possible, the joint in both flexed and extended positions.

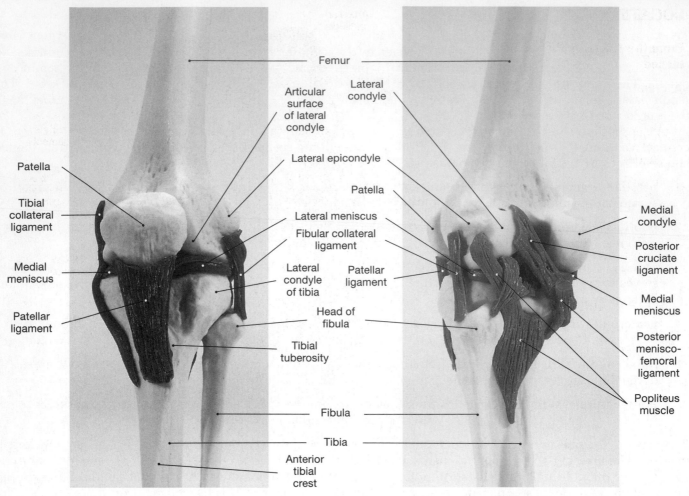

FIGURE 8-12 Knee joint, model.

See: Figs. 8-12 to 8-15

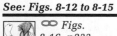 ∞ *Figs.*
8-16, p233
8-17, p234

LOCATE

From the Posterior View, Knee Extended

___ Femur

___ Condyles { Medial
Lateral

___ Posterior cruciate ligament

___ Tibia

___ Fibula

___ Lateral meniscus

___ Fibular collateral ligament

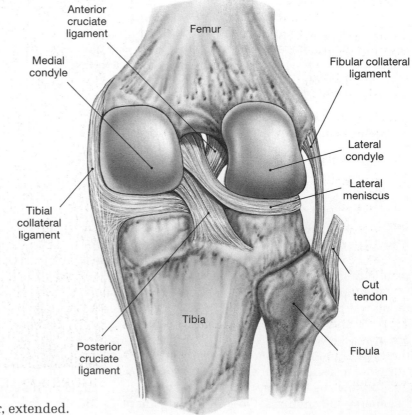

FIGURE 8-13 Knee joint, posterior, extended.

From the Anterior View, Knee Flexed

___ Femur

___ Fibula

___ Tibia

___ Articular cartilage

___ Patellar surface

___ Medial condyles

___ Fibular collateral ligament

___ Medial meniscus

___ Anterior cruciate ligament

___ Tibial collateral ligament

___ Lateral condyles

___ Lateral meniscus

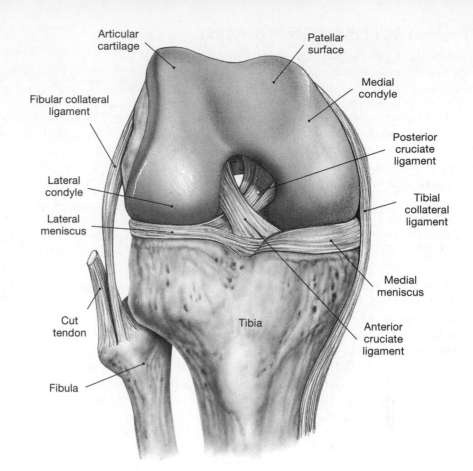

FIGURE 8-14 Knee joint, anterior, flexed.

LOCATE

From the Lateral View, Sagittal Section

___ Femur (medial and lateral condyles)

___ Tendon of quadriceps femoris muscle

___ Suprapatellar bursa

___ Patella

___ Fat pad

___ Articular cartilage of femur/tibia/patella

___ Patellar ligament

___ Popliteal ligament

___ Lateral meniscus

___ Anterior/posterior cruciate ligaments

___ Tibia

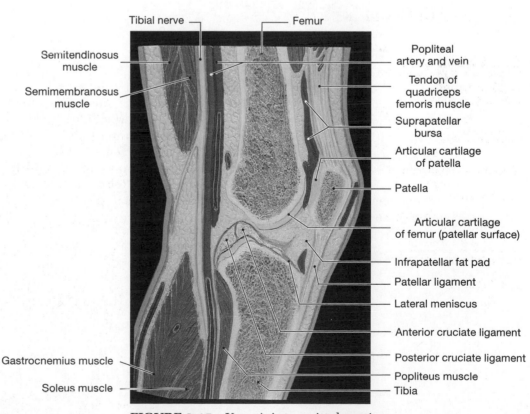

FIGURE 8-15 Knee joint, sagittal section.

Observation of the Joints of the Ankle and Foot

The ankle, or **tibiotalar joint**, is formed by the medial malleolus of the tibia, the lateral malleolus of the fibula, and the distal articular surface of the tibia resting upon the talus bone. [∞ p235] The talus sits on the anterior portion of the calcaneus, but the calcaneus is not one of the bones forming the tibiotalar (talocrural) joint. The ankle is a hinge type of joint, which permits limited dorsiflexion (extension) and plantar flexion (flexion) movements. The articular capsule of the ankle joint extends from the distal surfaces of the tibia, medial malleolus, and the lateral malleolus, to the talus. The stability of the ankle is maintained by a strong connective tissue capsule, numerous tendons of muscles acting on the ankle joint and joints in the foot, and ligaments. The ligaments that cover the medial and lateral surfaces of the capsule are exceptionally strong. The major ligaments involved in stabilizing the ankle joint are:

1. **deltoid ligament** consists of four medial ligaments that collectively bind the medial malleolus of the tibia to the navicular, talus, and calcaneal bones, thereby limiting the amount of dorsiflexion.

2. **lateral ligaments** consist of three ligaments: *anterior talofibular*, *posterior talofibular*, and *calcaneofibular*, which help to prevent side-to-side sliding.

The foot contains four types of synovial joints. **Intertarsal joints** are formed by tarsal to tarsal articulation and are gliding joints that permit limited sliding and twisting movements. **Tarsometatarsal joints** are formed by tarsal to metatarsal articulation and are also gliding joints that permit limited sliding and twisting movements. Cuneiforms I, II, and III articulate respectively with metatarsals I, II, and III. The cuboid articulates with metatarsals IV and V. **Metatarsophalangeal joints** are formed between the heads of the metatarsals and the bases of the proximal phalanges. These are ellipsoidal joints and permit flexion/extension and adduction/abduction movements. **Interphalangeal joints** are hinge joints formed between the head and base portions of adjacent phalanges, permitting flexion and extension movements. Ligaments bind and stabilize the joints of the foot by attaching to plantar, dorsal, medial, and lateral joint surfaces.

Procedure

✓ **Quick Check**

Before you begin to identify the structures of the ankle and foot joints, review the bones of the ankle and foot on the articulated skeleton (*Figures 7-16* to *7-18, pp83-85*)

Identify the **structures** of the **ankle** and **foot**, using first a *model* and then *cadaver specimen* of the *ankle* and *foot*. To view the ankle joint in the cadaver, the tendons and muscles must be removed superior to the medial malleolus of the tibia and the lateral malleolus of the fibula. Note the thickness of the articular capsule on medial and lateral sides compared to anterior and posterior sides. To view the interior of the capsule of the ankle, all surrounding tissues must be removed. Enter the joint cavity by cutting with scissors through anterior, lateral, and posterior surfaces of the capsule. The deltoid (medial) ligament is kept intact by this approach.

Observe the lateral ligaments and articular surfaces of the tibia and talus. If the talus is removed, the articular surfaces of calcaneus and navicular may be observed along with their relationships in forming the *subtalar joint*. Should you wish to observe the joints of the foot, your instructor will provide you with directions to display and examine these joints.

See: Figs. 8-16 to 8-18

∞ *Figs.*
8-18, p236
8-19, p237

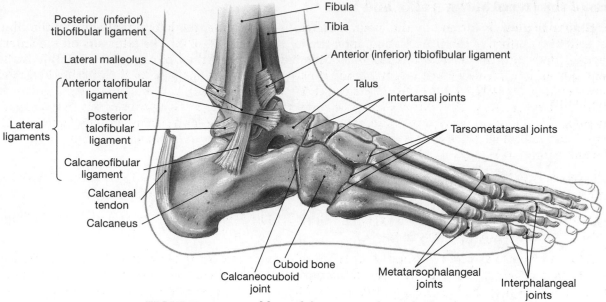

FIGURE 8-16 Ankle and foot joints, lateral view.

LOCATE

From the Lateral and Sectional Views

___ Fibula

___ Tibia

___ Posterior tibiofibular ligament

___ Anterior talofibular ligament

___ Anterior tibiofibular ligament

___ Lateral malleolus

___ Posterior talofibular ligament

___ Calcaneal tendon

___ Calcaneofibular ligament

___ Calcaneus

___ Cuboid bone

___ Calcaneocuboid joint

___ Intertarsal joints

___ Tarsometatarsal joints

___ Metatarsophalangeal joints

___ Interphalangeal joints

___ Talus bone

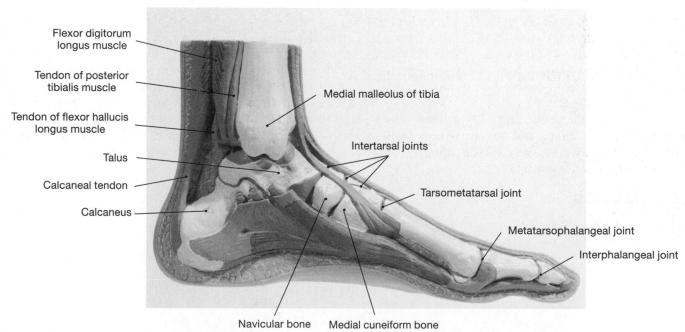

FIGURE 8-17 Normal foot, medial view.

From Posterior and Sectional Views

___ Tibia

___ Fibula

___ Medial malleolus

___ Lateral malleolus

___ Tibiotalar (talocrural) joint

___ Subtalar joint

___ Deltoid ligament

___ Talocalcaneal ligament

___ Talus

___ Calcaneus

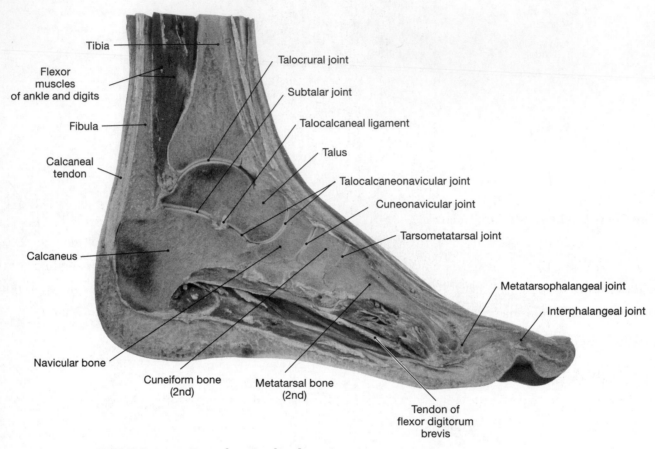

FIGURE 8-18 Foot, longitudinal section.

Anatomical Identification Review

Label on the accompanying knee joint photograph the following structures which characterize a diarthrosis joint: fibrous joint capsule, synovial membrane, articular cartilages and accessory structures (e.g. bursa and tendons).

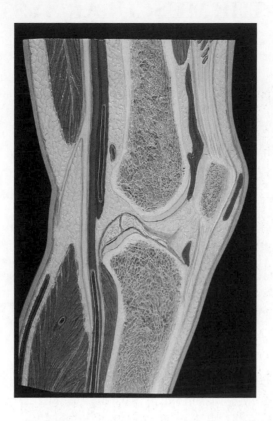

To Think About

1. How does the condition of rheumatoid arthritis destroy the small joints of the body?

2. Which joint in the body is most often injured? Which joint(s) in the body are rarely injured? Explain your answers.

3. Why is "clipping" worth a 15-yard penalty?

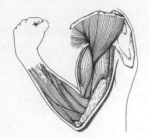

9

THE MUSCULAR SYSTEM:
Skeletal Muscle Tissue

Objectives

1. Identify the three types of muscle tissue and provide an example of where each may be found in the body.

2. Describe the organization of skeletal muscle from both the macroscopic and the microscopic level.

3. Describe the components of a motor unit.

4. Identify under the microscope the components of the neuromuscular junction.

Muscle tissue consists chiefly of elongated cells, termed **muscle fibers**, which are capable of contracting along their longitudinal axis when stimulated. The body contains three types of muscle tissue: *skeletal muscle, cardiac muscle*, and *smooth muscle*. The contraction of **skeletal muscle tissue**, acting directly or indirectly upon the skeleton, leads to body and/or limb movements, and maintenance of posture and balance. The contraction of **cardiac muscle tissue**, found exclusively in the heart, provides a means to propel blood through the body's blood vessels. The contraction of **smooth muscle tissue** propels foods and liquids through the digestive tract, changes the diameters of blood vessels and bronchial passageways, and performs other functions as well. [∞ p245]

Organization of Skeletal Muscles

The structural organization of muscle is based on its association with connective tissue. The **epimysium** is a dense, irregular connective tissue layer, composed chiefly of collagen fibers that surround the entire skeletal muscle. The epimysium is part of the deep fascia. Within each muscle, a connective tissue layer, termed **perimysium**, subdivides that muscle into numerous distinct compartments. Surrounded by the perimysium, each compartment contains a group or bundle of muscle fibers called a **fascicle**. Within every fascicle, each skeletal muscle fiber is surrounded by a loose connective tissue network of reticular fibers, termed the **endomysium**. It serves to bind each muscle fiber to neighboring muscle fibers.

　　Typically, at the ends of each muscle, the fibers of the epimysium form a tubular band called a **tendon**. Tendons attach skeletal muscle to bone. Sometimes tendons are formed into broad flat sheets, known as **aponeuroses**, which serve to attach superficial muscles to other muscles or bone.

Procedure

Identify the following structures of a **muscle fiber**, using a *model* of a *skeletal muscle fiber*. Use *Figure 9-1* for reference.

See: Figs. 9-1

∞ Figs.
9-1, p246
9-3, p248

LOCATE

__ Endomysium	__ Sarcolemma	__ Perimysium
__ Myofibril	__ Mitochondria	__ Blood vessels
__ Nucleus	__ Actin (thin filament)	__ Muscle fiber

Microscopic Identification of the Neuromuscular Junction

 Skeletal muscle contracts upon stimulation by motor neurons of the central nervous system. The axon (nerve fiber) endings of a motor neuron penetrate the endomysium of the skeletal muscle to stimulate the cell membrane of individual muscle fibers. Stimulation occurs through chemical communication from the

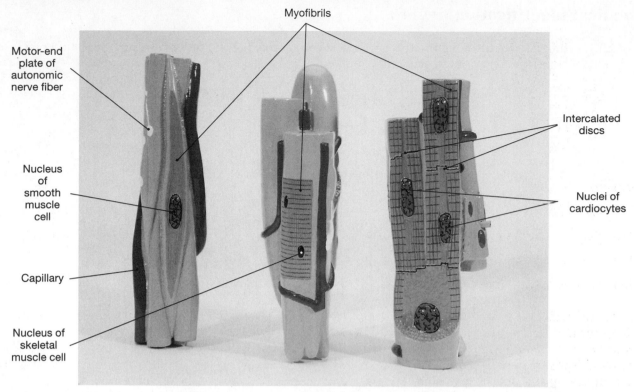

FIGURE 9-1   Model of muscle fiber types: smooth, cardiac, and skeletal.

synaptic knob at the axon terminal of the motor neuron to the *motor end plate* surface of the skeletal muscle cell. This specialized intercellular connection is termed the **neuromuscular (myoneural) junction**.

Procedure

Identify the following structures using the *microscope slide* provided, first by viewing the slide under low power magnification and then under high power magnification to see the details of both **muscle fibers** and the **neuromuscular junction**.

　　　Slide # _____ Motor End Plate (whole mount)

See: Fig. 9-2

 ∞ *Fig. 9-2a, p247*

LOCATE

__ Striated skeletal muscle fibers

__ Nuclei of skeletal muscle fibers

__ Nerve

__ Axons of motor neurons

__ Neuromuscular (myoneural) junction

__ Motor unit

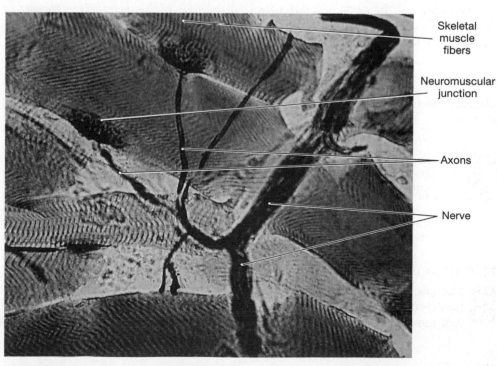

FIGURE 9-2   Neuromuscular junction × 230.

Anatomical Identification Review

Label on the accompanying drawing the three wraps of connective tissue that encase each skeletal muscle, separates muscle fascicles, and surrounds each muscle fiber.

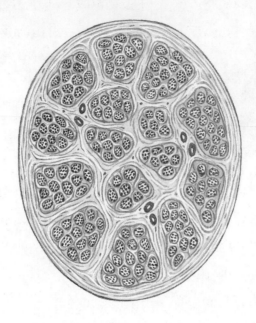

To Think About

1. How do the different connective tissue layers of skeletal muscle relate to one another?

2. Describe the arrangement of filaments within a myofibril.

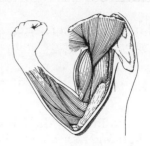

10

THE MUSCULAR SYSTEM:
Muscle Organization and the Axial Musculature

Objectives

1. Locate, identify, name, and describe the function of head and neck muscles (facial expression muscles, extrinsic eye muscles, mastication muscles, tongue muscles, and pharyngeal muscles).

2. Locate, identify, name, and describe the function of anterior neck muscles.

3. Locate, identify, name, and describe the function of muscles of the spine.

4. Locate, identify, name, and describe the function of abdominal wall muscles (including diaphragm).

5. Locate, identify, name, and describe the function of muscles of the pelvic floor.

Skeletal muscles compose the **muscular system** of the human body. The human muscular system includes more than 700 skeletal muscles. Based on muscle location(s) on the axial and appendicular skeleton and the movements or stabilization these muscles provide, the muscular system is separated into axial and appendicular musculatures. [∞ p267] The **axial musculature** originates and inserts on the axial skeleton, and the **appendicular musculature** moves or stabilizes limbs or parts of limbs of the appendicular skeleton. The axial musculature is the focus of our study in this chapter and the appendicular muscles are examined in Chapter 11.

The appearance of each skeletal muscle provides clues to its function, location, and even its name. The organization of fascicles gives each skeletal muscle its characteristic appearance. Four different organizational patterns are formed: *parallel muscles, convergent muscles, pennate muscles*, and *circular muscles*. [∞ p257] Each muscle begins at an **origin**, the stationary end of the muscle, ends at an **insertion**, the movable portion of the muscle, and contracts to produce a specific **action**. [∞ p260] Muscles are grouped into one of three categories based on the primary action of their contraction: (1) *agonists*, or *prime movers*, are responsible for the motion observed, (2) *synergists* assist the prime movers or help stabilize the origin of the prime mover, and (3) *antagonists* oppose the motion of the prime mover, thereby helping to control the speed of movement and to ensure its smoothness. [∞ p261] Skeletal muscles are named on the basis of various features. Names may be based on fascicle organization, structure, relative position, size, shape, origin, insertion, or actions. Typically, one or more of these features are combined to create the muscle name. The terminology used to name most muscles is presented in *Table 9-2* and should be reviewed at this time. [∞ p260]

The axial musculature can be divided into five groups:

1. muscles of the head and neck.

2. muscles of the anterior neck.

3. muscles of the spine.

4. abdominal wall muscles.

5. muscles of the pelvic floor.

The muscles of the head and neck can be further divided as presented. [∞ p269]

Observation of the Muscles of the Head and Neck

The muscles of the head and neck can be organized into the following groups: *muscles of facial expression, muscles of mastication, muscles of the tongue*, and *muscles of the pharynx*. These muscles are responsible for facial expression, chewing, movements of the tongue, and movements of the larynx for swallowing. [∞ p269]

Procedure

✓ **Quick Check**

Before you begin to examine the muscles of the head and neck, review the external bone markings of the adult skull and features of the axial skeleton. [*Figures 6-4 to 6-8, pp42-45*; ∞ Figure 6-3, p136-139] Keep a skull or articulated skeleton nearby to aid you in locating points of muscle attachment as you identify muscles of the head and neck.

Locate and identify the following **head** and **neck muscles** and **associated structures**, using a *torso model* or *prosected cadaver specimen.* The superficial **platysma** muscle covers most of the anterior and some of the lateral surface of the neck and is typically removed in the prosected cadaver specimen in order to see the underlying muscles.

See: Color Plates 34 to 36, Fig. 10-1

 ∞ *Fig. 10-3, p269*

LOCATE

___ Epicranial aponeurosis

___ Frontal belly of occipitofrontalis

___ Temporoparietalis

___ Orbicularis oculi

___ Corrugator supercilii

___ Procerus

___ Zygomaticus major

___ Levator labii superioris

___ Orbicularis oris

___ Platysma

___ Masseter

___ Buccinator

___ Depressor anguli oris

___ Depressor labii inferioris

___ Mentalis

___ Sternocleidomastoid

___ Trapezius

Location of the Muscles of Facial Expression

Muscles of facial expression do just what the term implies. This includes wrinkling or elevating the brow, and moving the mouth and cheeks. The muscles of facial expression are unusual because they arise from bony surfaces of the skull and insert into the skin. These muscles are grouped according to their location (mouth, eye, nose, ear, scalp and neck) and their associated action(s). [∞ p269] See *Table 10-1* for a detailed summary of the muscles of facial expression. Use *Color Plates 34, 35*, and *Figure 10-1* for reference.

Procedure

Locate and identify the following **muscles** of **facial expression** and **associated structures**, using a *torso model* or *prosected cadaver specimen.* Observe the muscles of facial expression on a model first before proceeding to the cadaver. Begin in the frontal area of the head and identify the **frontal belly of occipitofrontalis** muscle, the connective tissue sheet termed **epicranial aponeurosis**, the **occipital belly of occipitofrontalis,** the **orbicularis oculi**, and the **orbicularis oris** muscles. These muscles serve as landmarks for observing other facial muscles. Palpate for the zygomatic arch and mandible to locate the **buccinator** and **masseter** muscles. In the cadaver, the buccinator muscle lies under fat and under part of the masseter muscle.

Since a number of blood vessels and nerves are located in this area, be careful during your examination not to damage these structures. All of the muscles of facial expression are innervated by the seventh cranial nerve, the facial nerve (N VII). These vessels and nerves need to remain intact for future investigations. Check with your instructor before proceeding.

TABLE 10-1 Muscles of Facial Expression

Region/Muscle	Origin	Insertion	Action	Innervation
Mouth				
Buccinator	Alveolar processes of maxilla and mandible	Blends into fibers of orbicularis oris	Compresses cheeks	Facial nerve (N VII)
Depressor labii inferioris	Mandible between the anterior midline and the mental foramen	Skin of lower lip	Depresses lip	As above
Levator labii superioris	Inferior margin of orbit, superior to the infraorbital foramen	Orbicularis oris	Raises upper lip	As above
Mentalis	Incisive fossa of mandible	Skin of chin	Elevates and protrudes lower lip	As above
Orbicularis oris	Maxilla and mandible	Lips	Compresses, purses lips	As above
Risorius	Fascia surrounding parotid salivary gland	Angle of mouth	Draws corner of mouth to the side	As above
Depressor anguli oris	Anterolateral surface of mandibular body	Skin at angle of mouth	Depresses corner of mouth	As above
Zygomaticus major	Zygomatic bone near the zygomatic maxillary suture	Angle of mouth	Retracts and elevates corner of mouth	As above
Zygomaticus minor	Zygomatic bone posterior to zygomaticotemporal suture	Upper lip	Retracts and elevates corner of mouth	As above
Eye				
Corrugator supercilii	Medial margin of orbital rim of frontal bone near frontonasal suture	Eyebrow	Pulls skin inferiorly and anteriorly; wrinkles brow	As above
Levator palpebrae superioris	Tendinous band around optic foramen	Upper eyelid	Raises upper lid	Oculomotor nerve (N III)[a]
Orbicularis oculi	Medial margin of orbit	Skin around eyelids	Closes eye	Facial nerve (N VII)
Nose				
Procerus	Nasal bones and lateral nasal cartilages	Aponeurosis at bridge of nose and skin of forehead	Moves nose, changes position, shape of nostrils	As above
Nasalis	Maxilla and alar cartilage of nose	Bridge of nose	Compresses bridge, depresses tip of nose; elevates corners of nostrils	As above
Ear (extrinsic)				
Temporoparietalis	Fascia around external ear	Epicranial aponeurosis	Tenses scalp, moves pinna of ear	As above
Scalp (Epicranius)				
Occipitofrontalis				
Frontal belly	Epicranial aponeurosis	Skin of eyebrow and bridge of nose	Raises eyebrows, wrinkles forehead	As above
Occipital belly	Superior nuchal line	Epicranial aponeurosis	Tenses and retracts scalp	As above
Neck				
Platysma	Superior thorax between cartilage of second rib and acromion of scapula	Mandible and skin of cheek	Tenses skin of neck, depresses mandible	As above

[a]This muscle originates in association with the extrinsic oculomotor muscles, so its innervation is unusual.

LOCATE

___ Epicranial aponeurosis

___ Frontal belly of occipitofrontalis

___ Occipital belly of occipitofrontalis

___ Temporoparietalis

___ Temporalis

___ Orbicularis oculi

___ Corrugator supercilii

___ Procerus

___ Nasalis

___ Buccinator

___ Depressor labii inferioris

___ Levator labii superioris

___ Mentalis

___ Orbicularis oris

___ Risorius

___ Depressor anguli oris

___ Zygomaticus major

___ Platysma

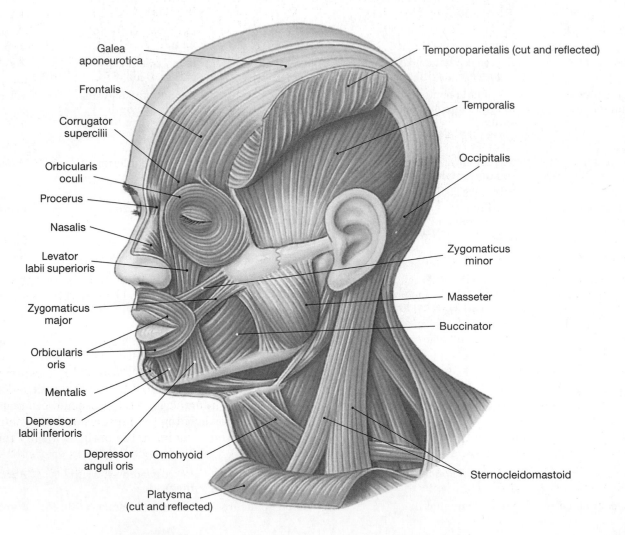

FIGURE 10-1 Muscles of facial expression, lateral view.

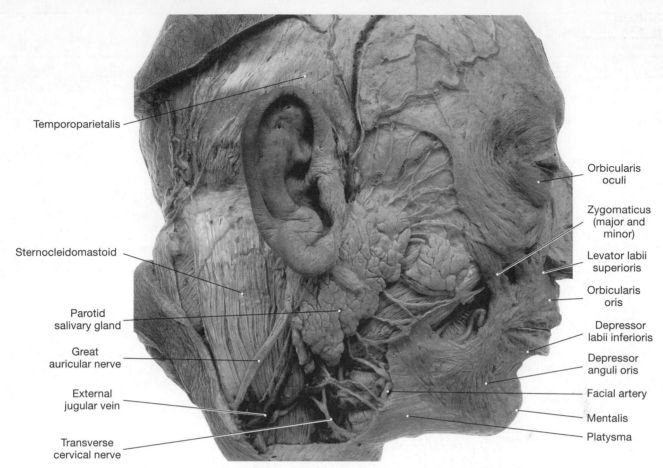

Temporoparietalis

Sternocleidomastoid

Parotid
salivary gland

Great
auricular nerve

External
jugular vein

Transverse
cervical nerve

Orbicularis
oculi

Zygomaticus
(major and
minor)

Levator labii
superioris

Orbicularis
oris

Depressor
labii inferioris

Depressor
anguli oris

Facial artery

Mentalis

Platysma

FIGURE 10-2 Muscles of the face, lateral view of the right side.

Observation of the Extraocular (Extrinsic) Eye Muscles

The **extraocular**, or **extrinsic eye muscles** are outside the eyeball and rotate it within the bony orbit. The contractions of six extrinsic eye muscles produce all movements associated with the eyeball. [∞ p272] See *Table 10-2* for a detailed summary of the extraocular eye muscles.

Procedure

Locate and identify the following **extraocular eye muscles** and **associated structures**, using a *model* or *specimen* of the *eye within the orbit.* Observe the extrinsic eye muscles on a model first before proceeding to observations on the cadaver. Use *Figures 10-3* and *18-13 (p268)* for references. In the prosected cadaver the eyeball is often partially collapsed, but this will not prevent you from examining the muscles and associated structures. The muscles and nerves associated with the eye are best viewed superior to the orbit. Typically the roof of the orbit (orbital portion of frontal bone and lesser wing of the sphenoid) is dissected open to create a "window" through which these structures may be viewed undisturbed. Using *Figure 18-14 (p269),* which is a view of the eye from superior to the orbit, try to identify as many structures as possible.

TABLE 10-2 Extraocular Eye Muscles

Muscle	Origin	Insertion	Action	Innervation
Inferior rectus	Sphenoid bone around optic foramen	Inferior, medial surface of eyeball	Eye looks down	Oculomotor nerve (N III)
Lateral rectus	As above	Lateral surface of eyeball	Eye rotates laterally	Abducens nerve (N VI)
Medial rectus	As above	Medial surface of eyeball	Eye rotates medially	Oculomotor nerve (N III)
Superior rectus	As above	Superior medial surface of eyeball	Eye looks up	As above
Inferior oblique	Maxilla at anterior portion of orbit	Inferior, lateral surface of eyeball	Eye rolls, looks up and to the medial	As above
Superior oblique	Sphenoid bone around optic foramen	Superior, medial surface of eyeball	Eye rolls, looks down and to the medial	Trochlear nerve (N IV)

See: Figs. 10-3, 18-13

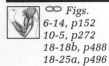

∞ *Figs.*
6-14, p152
10-5, p272
18-18b, p488
18-25a, p496

LOCATE

— Orbit (identify those bones that compose its roof, floor & walls, see *Figure 6-21, p54*)

— Superior oblique muscle

— Superior rectus muscle

— Medial rectus muscle

— Lateral rectus muscle

— Inferior rectus muscle

— Inferior oblique muscle

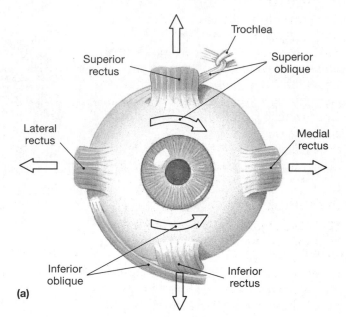

(a)

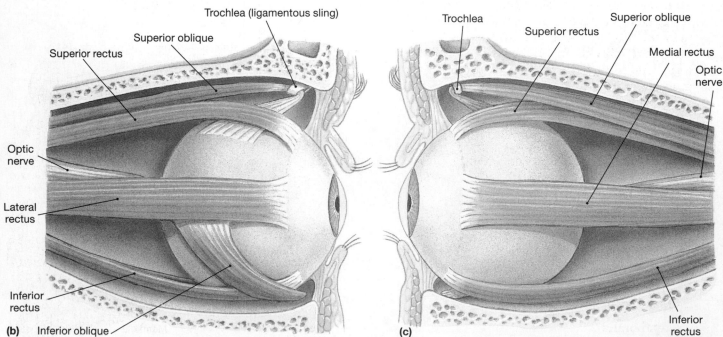

FIGURE 10-3 Extraocular muscles of the right eye (a) anterior view (b) lateral surface (c) medial surface.

TABLE 10-3 Muscles of Mastication

Muscle	Origin	Insertion	Action	Innervation
Masseter	Zygomatic arch	Lateral surface of mandibular ramus	Elevates mandible and closes the jaw	Trigeminal nerve (N V), mandibular branch
Temporalis	Along temporal lines of skull	Coronoid process of mandible	Elevates mandible and closes the jaw	As above
Pterygoids (medial and lateral)	Lateral pterygoid plate and processes	Medial surface of mandibular ramus	*Medial*: Elevates the mandible and closes the jaws, or moves mandible from side to side	As above
			Lateral: Opens jaws, protrudes mandible, or moves mandible from side to side	As above

Identification of the Muscles of Mastication

The power to grind and chew our food is provided by the muscles of mastication. The strongest of these muscles are the *masseter*, which forms part of each cheek, and the *temporalis*, which forms part of the temple. The *pterygoid muscles* produce side-to-side motion and are strong closers of the mouth. [∞ p273] A detailed summary of the muscles of mastication is presented in *Table 10-3*.

Procedure

 Locate and identify the **muscles** of **mastication** and **associated structures**, using a *model* or *prosected cadaver specimen*. Observe the muscles of mastication on a model first before proceeding to the cadaver. Use *Color Plate 34-37* for reference. Locate first the **temporalis muscle**, then the **masseter muscle**. To view the **lateral** and **medial pterygoid muscles** in the cadaver, perform the following: (1) the overlying muscles, chiefly the masseter, must be removed or severed and then reflected, (2) remove the zygomatic process of the temporal bone, and (3) remove most of the anterior border of the mandibular ramus.

See: Color Plates 34-37, Fig. 10-4

∞ *Figs.*
10-3, p269
10-6, p273

LOCATE

___ Masseter

___ Temporalis

___ Lateral and medial pterygoid muscles

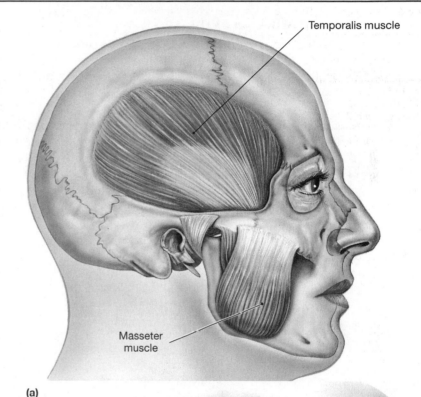

(a)

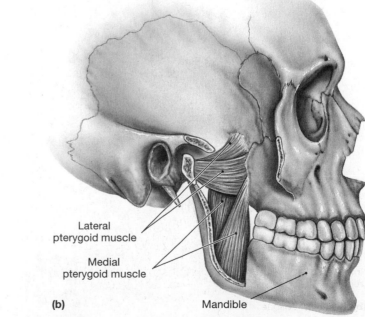

(b)

FIGURE 10-4 Muscles of mastication (a) lateral view (b) lateral view pterygoid muscle exposed.

TABLE 10-4 Muscles of the Tongue

Muscle	Origin	Insertion	Action	Innervation
Genioglossus	Medial surface of mandible around chin	Body of tongue, hyoid bone	Depresses and protracts tongue	Hypoglossal nerve (N XII)
Hyoglossus	Body and greater cornu of hyoid bone	Side of tongue	Depresses and retracts tongue	As above
Palatoglossus	Anterior surface of soft palate	As above	Elevates tongue, depresses soft palate	Cranial branch of accessory nerve (N XI) via vagus nerve (N X)
Styloglossus	Styloid process of temporal bone	Via side to the tip and base of tongue	Retracts tongue, elevates sides	Hypoglossal nerve (N XII)

Observation of the Muscles of the Tongue

The muscles of the tongue perform the movements necessary for speech and for working with food in the oral cavity prior to swallowing. Any muscle name ending with "-glossus" inserts in the tongue. The first part of the name gives the origin of the muscle. [∞ p274] A detailed summary of the muscles of the tongue is presented in *Table 10-4*.

 Locate and identify the **muscles** of the **tongue** and their **bony attachments**, using a *model* or *prosected cadaver specimen*. Use *Color Plate 35* and *91* for reference. Observe the muscles of the tongue on a model first before proceeding to the cadaver. In the cadaver, identify the *digastric* and *mylohyoid muscles* (see Anterior Muscles of the Neck). If not previously cut, then cut the mylohyoid in the midline from anterior to posterior (to hyoid bone) and reflect laterally the cut ends. The **genioglossus** and **hyoglossus** muscles may now be observed. The hyoglossus is a key landmark because blood vessels and nerves lie deep and superficial to this muscle.

See: Color Plates 35, 91, Fig. 10-5

∞ Fig. 10-8, p274

LOCATE

___ Mandible

___ Hyoid bone

___ Styloid process

___ Genioglossus

___ Hyoglossus

___ Styloglossus

___ Palatoglossus

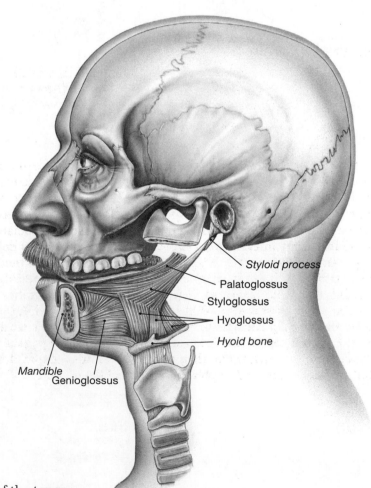

FIGURE 10-5 Muscles of the tongue.

TABLE 10-5   Muscles of the Pharynx

Muscle	Origin	Insertion	Action	Innervation
PHARYNGEAL CONSTRICTORS			Constrict pharynx to propel bolus into esophagus	Branches of pharyngeal plexus (N IX & X)
Superior constrictor	Pterygoid processes of sphenoid, medial surfaces of mandible	Median raphe attached to occipital bone		N X
Middle constrictor	Cornua of hyoid	Median raphe		N X
Inferior constrictor	Cricoid and thyroid cartilages of larynx	Median raphe		N X
LARYNGEAL ELEVATORS[a]			Elevate larynx	Branches of pharyngeal plexus (N X)
Palatopharyngeus	Soft palate	Thyroid cartilage		N X
Salpingopharyngeus	Cartilage around the inferior portion of the auditory tube	Thyroid cartilage		N X
Stylopharyngeus	Styloid process of temporal bone	Thyroid cartilage		N IX
PALATAL MUSCLES			Elevate soft palate	
Levator veli palatini	Petrous portion of temporal bone and tissues around the auditory tube	Soft palate		Cranial branch of accessory nerve (N XI) via vagus (N X)
Tensor veli palatini	Spine of sphenoid, tissues around the auditory tube	Soft palate		Mandibular branch of trigeminal nerve (N V)

[a]Assisted by the thyrohyoid, geniohyoid, stylohyoid, and hyoglossal muscles.

Observation of the Muscles of the Pharynx

The muscles of the pharynx participate in the process of moving foods and liquids into the esophagus. Any muscle name that includes "pharyngo-" or "palat-" in any form is located in the throat and associated with swallowing. [∞ p275] *Table 10-5* presents a detailed summary of the muscles of the pharynx.

Procedure

Locate and identify the following **muscles** of the **pharynx** and **associated structures**, using a *model* or *prosected cadaver specimen*. Use *Figures 10-6* and *10-7* for reference. Observe the muscles of the pharynx on a model first before proceeding to the cadaver. In the cadaver, the muscles of the pharynx are best observed by partially removing the pharynx, larynx, and trachea. Especially strong fascia connects the pharynx to the base of the skull. With the posterior wall of the pharynx removed, view the muscles associated with the pharynx from the posterior. The **superior, middle**, and **inferior pharyngeal constrictor muscles** collectively form a circular outer layer of muscle. Each is a fan-shaped muscle, with the large part of the fan extending posteriorly. The constrictor muscles lie under and overlap each other, with inferior overlapping the middle and middle overlapping the superior. Each constrictor has the same insertion (median raphe) but a different origin: superior pterygoid processes of the sphenoid, medial surfaces of the mandible; middle cornua of the hyoid bone; and inferior cricoid cartilage of the larynx. Identify the middle constrictor first. Use this muscle as a landmark to then observe the fibers of the superior and inferior constrictors. The fibers of the small **stylopharyngeus muscle** pass from the styloid process over the superior constrictor muscle and then under the middle constrictor. The **tensor veli palatini** can be observed on the medial surface of the medial pterygoid plate of the sphenoid. Try to observe as many pharyngeal muscles as possible.

See: Fig. 10-6

∞ *Fig.
10-9, p275*

LOCATE

___ Tensor veli palatini

___ Levator veli palatini

___ Stylopharyngeus

___ Pharyngeal constrictors { Superior / Middle / Inferior

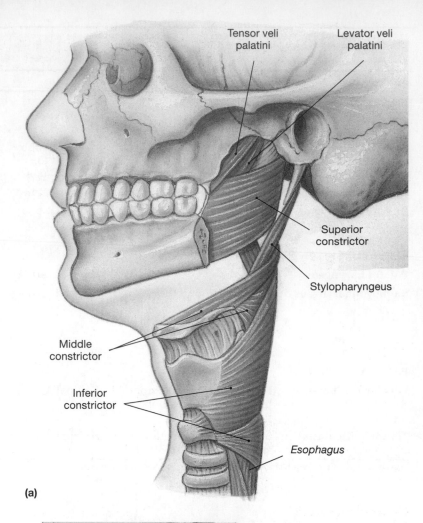

(a)

Observation of the Anterior Muscles of the Neck

The anterior muscles of the neck are responsible for depressing the mandible, positioning the larynx during breathing and swallowing, and positioning and turning the head towards the shoulder. [∞ p276] *Table 10-6* presents a detailed summary of the anterior muscles of the neck.

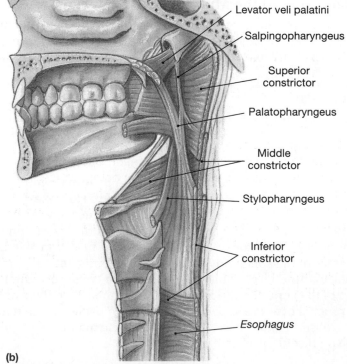

(b)

FIGURE 10-6 Muscles of the pharynx (a) lateral view (b) midsagittal view.

TABLE 10-6 Anterior Muscles of the Neck

Muscle	Origin	Insertion	Action	Innervation
Digastric	Two bellies: *posterior* to mastoid region of temporal; *anterior* to inferior surface of mandible at chin	Hyoid bone	Depresses mandible and/or elevates larynx	Facial nerve (N VII) to posterior belly; trigeminal nerve (N V), mandibular branch, to anterior belly
Geniohyoid	Medial surface of mandible at chin	Body of hyoid bone	As above and pulls hyoid anteriorly	Cervical spinal nerves, via hypoglossal nerve (N XII)
Mylohyoid	Mylohyoid line of mandible	Median connective tissue band (raphe) that runs to hyoid bone	Elevates floor of mouth, elevates hyoid, and/or depresses mandible	Trigeminal nerve (N V), mandibular branch
Omohyoid	Central tendon attaches to clavicle and 1st rib	Two bellies: *superior* attaches to body of hyoid bone; *inferior* to superior margin of scapula	Depresses hyoid and larynx	Cervical spinal nerves (ansa cervicalis, C_2–C_3)
Sternohyoid	Clavicle and manubrium	Inferior surface of the body of the hyoid bone	As above	As above
Sternothyroid	Dorsal surface of manubrium and 1st costal cartilage	Thyroid cartilage of larynx	As above	As above
Stylohyoid	Styloid process of temporal bone	Lesser cornu of hyoid bone	Elevates larynx	Facial nerve (N VII)
Thyrohyoid	Thyroid cartilage of larynx	Inferior border of the greater cornu of the hyoid bone	Elevates thyroid, depresses hyoid	Cervical spinal nerves via hypoglossal nerve (N XII)
Sternocleidomastoid	Two bellies: *clavicular head* attaches to sternal end of clavicle; *sternal head* attaches to manubrium of sternum	Mastoid process of skull	Together they flex the neck; alone, one side bends head toward shoulder and turns face to opposite side	Accessory nerve (N XI) and cervical spinal nerves

Procedure

Locate and identify the following **anterior muscles** of the **neck** and **associated structures**, using a *torso model* or *prosected cadaver specimen*. Use *Color Plates 34-36* and *Figure 10-7* for reference. Observe the anterior neck muscles on a model first before proceeding to the cadaver. In the cadaver, locate the body of the mandible, hyoid bone, and thyroid cartilage before proceeding to identify the anterior muscles of the neck. From the chin and on either side of the midline, identify the **digastric** muscles. These muscles overlie the broad, flat **mylohyoid** muscle. Reflect the mylohyoid muscle at the midline. Inferior to the mylohyoid on either side of the midline, observe four belt-like muscles, the **omohyoid, sternohyoid, sternothyroid**, and **thyrohyoid**, which insert on the hyoid bone (except sternothyroid, which inserts on the thyroid cartilage of the larynx) and descend inferiorly towards the manubrium of the sternum. The omohyoid and sternohyoid muscles should be located first, as the sternothyroid and thyrohyoid muscles lie deep to these muscles. Try to observe as many anterior muscles of the neck as possible.

∞ *Figs.*
10-3, p269
10-11, p277

LOCATE

___ Digastric
(anterior &
posterior bellies)

___ Cartilages of
the larynx

___ Mylohyoid

___ Hyoid bone

___ Geniohyoid

___ Stylohyoid

___ Thyrohyoid

___ Sternohyoid

___ Omohyoid

___ Sternum

___ Clavicle

Sternocleidomastoid
{ Clavicular head
Sternal head

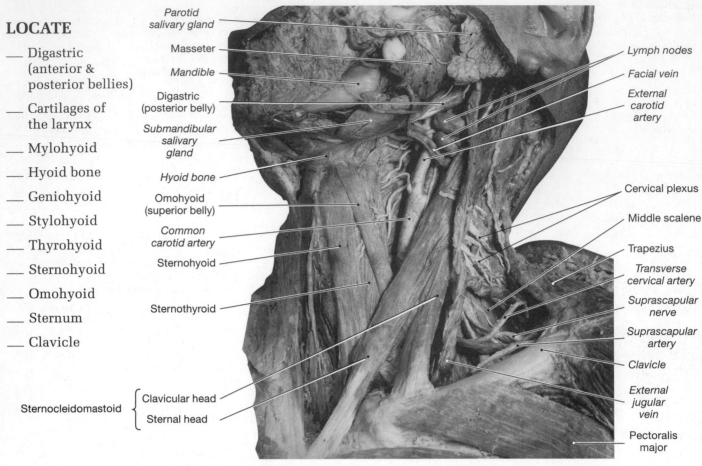

Parotid
salivary gland
Masseter
Mandible
Digastric
(posterior belly)
Submandibular
salivary
gland
Hyoid bone
Omohyoid
(superior belly)
Common
carotid artery
Sternohyoid
Sternothyroid

Lymph nodes
Facial vein
External
carotid
artery

Cervical plexus
Middle scalene
Trapezius
Transverse
cervical artery
Suprascapular
nerve
Suprascapular
artery
Clavicle
External
jugular
vein
Pectoralis
major

FIGURE 10-7   Anterior muscles of the neck, anterior view.

Observation of the Muscles of the Vertebral Column

The muscles of the vertebral column primarily function in postural activities. [∞ p278] They are covered by the more superficial back muscles and work together to extend the vertebral column. When only the left or right side contracts, the vertebral column is flexed laterally. Their names tell much about the action or position of these muscles. *Table 10-7* presents a detailed summary of the muscles of the vertebral column.

Procedure

✓ **Quick Check**

Before you begin to examine the muscles of the vertebral column, review both the external bone markings of the posterior region of the adult skull (occipital and temporal bones), [*Color Plate 31, Figures 6-4, 6-6,* and *6-8, pp42-45*] and posterior features of the vertebral column and pelvis. [∞ p163 and ∞ p195] Keep an articulated skeleton nearby to aid you in locating points of muscle attachment as you identify muscles of the column.

TABLE 10-7   Muscles of the Vertebral Column

Group/Muscle	Origin	Insertion	Action	Innervation
SUPERFICIAL LAYER				
Splenius (Splenius capitis, splenius cervicis)	Spinous processes and ligaments connecting lower cervical and upper thoracic vertebrae	Mastoid process, occipital bone of skull, and upper cervical vertebrae	The two sides act together to extend head; either alone rotates and tilts head to that side	Cervical spinal nerves
Erector Spinae				
Spinalis group				
Spinalis cervicis	Inferior portion of ligamentum nuchae and spinous process of C_7	Spinous process of axis	Extends neck	Cervical spinal nerves
Spinalis thoracis	Spinous processes of lower thoracic and upper lumbar vertebrae	Spinous processes of upper thoracic vertebrae	Extends vertebral column	Thoracic and lumbar spinal nerves
Longissimus group				
Longissimus capitis	Processes of lower cervical and upper thoracic vertebrae	Mastoid process of temporal bone	The two sides act together to extend head; either alone rotates and tilts head to that side	Cervical and thoracic spinal nerves
Longissimus cervicis	Transverse processes of upper thoracic vertebrae	Transverse processes of middle and upper cervical vertebrae	As above	As above
Longissimus thoracis	Broad aponeurosis and at transverse processes of lower thoracic and upper lumbar vertebrae; joins iliocostalis to form "sacrospinalis"	Transverse processes of higher vertebrae and inferior surfaces of lower 10 ribs	Extends and/or bends spinal column to one side	As above
Iliocostalis group				
Iliocostalis cervicis	Superior borders of vertebrosternal ribs near the angles	Transverse processes of middle and lower cervical vertebrae	Extends or bends neck, elevates ribs	As above
Iliocostalis thoracis	Superior borders of lower 7 ribs medial to the angles	Upper ribs and transverse process of last cervical vertebra	Stabilizes thoracic vertebrae in extension	Thoracic spinal nerves
Iliocostalis lumborum	Sacrospinal aponeurosis and iliac crest	Inferior surfaces of lower 7 ribs near their angles	Extends spine, depresses ribs	Lumbar spinal nerves

Locate and identify the following **muscles** of the **vertebral column** and **associated structures**, using a *torso model* or *prosected cadaver specimen*. Use *Color Plates 37, 42* and *Figures 10-8, 10-9* for reference. Observe the muscles of the vertebral column on a torso model first before proceeding to the cadaver. To observe the muscles of the spine, the prosected cadaver must be face down, in prone position, and if possible, with the arms abducted.

The **splenius muscles** are the most superficial muscles of the spine found in the neck region. The **splenius capitis** is a flat, wide muscle that arises in the midline from spinous processes of C_7 and T_{1-4} and runs superiorly and laterally to insert at the base of the skull. The **splenius cervicis** is a narrow muscle that arises at about the T_3–T_6 spinous processes and runs along the lateral border of the splenius capitis to about C_2–C_3.

TABLE 10-7   Muscles of the Vertebral Column (Cont.)

Group/Muscle	Origin	Insertion	Action	Innervation
DEEP LAYER (TRANSVERSOSPINALIS)				
Semispinalis group				
Semispinalis capitis	Processes of lower cervical and upper thoracic vertebrae	Occipital bone, between nuchal lines	Together, the two sides act to extend head; alone, each extends and tilts head to that side	Cervical spinal nerves
Semispinalis cervicis	Transverse processes of T_1–T_5/T_6	Spinous processes of C_2–C_5	Extends vertebral column and rotates toward opposite side	As above
Semispinalis thoracis	Transverse processes of T_6–T_{10}	Spinous processes of C_5–T_4	As above	Thoracic spinal nerves
Multifidus	Sacrum and transverse processes of each vertebra	Spinous processes of the third or fourth more superior vertebrae	Extends vertebral column and rotate toward opposite side	Cervical, thoracic, and lumbar spinal nerves
Rotatores (cervicis, thoracis, and lumborum)	From the articular processes of cervical, from transverse processes of thoracic vertebrae, from mamillary processes of lumbar vertebrae	Spinous process of adjacent, more superior vertebra	Extends vertebral column and rotate toward opposite side	As above
Interspinales	Spinous processes of each vertebra	Spinous processes of preceding vertebra	Extends vertebral column	As above
Intertransversarii	Transverse processes of each vertebra	Transverse process of preceding vertebra	Lateral flexion of the vertebral column	As above
SPINAL FLEXORS				
Longus capitis	Transverse processes of cervical vertebrae	Base of the occipital bone	Together, the two sides act to bend head forward; alone, each rotates head to that side	Cervical spinal nerves
Longus colli	Anterior surfaces of cervical and upper thoracic vertebrae	Transverse processes of upper cervical vertebrae	Flexes and/or rotates neck; limits hyperextension	As above
Quadratus lumborum	Iliac crest and iliolumbar ligament	Last rib and transverse processes of lumbar vertebrae	Together they depress ribs; each side alone flexes spine laterally	Thoracic and lumbar spinal nerves

In the cervical, thoracic, and upper lumbar regions, the splenius and erector spinae muscle groups are covered by other muscles. In order to view the splenius and erector spinae groups, the overlying muscles must be reflected. Working from C_7 to L_5 and from superficial to deep, reflect the following upper limb muscles: *trapezius, latissimus dorsi, rhomboideus minor, rhomboideus major*, and *serratus posterior (inferior)* muscles (see *Chapter 11, p145*). The tough connective sheet, *thoracolumbar fascia*, covers the erector spinae group of muscles. In the thoracic region, the fascia is thin, but in the lumbar region, it is thick and tough.

From the base of the skull to the pelvis, three long columns of muscle should be visible on either side of the vertebral column. These are the superficial spinal extensors of the erector spinae group. From the spinous processes, proceed laterally and identify the most medial column as the **spinalis group**, the middle column as the **longissimus group**, and the most lateral column as the **iliocostalis group**. Reflection of the spinalis group reveals the *multifidus* group of muscles.

See: Fig. 10-8

∞ Figs.
10-2, p268
10-12, p279

LOCATE

**Superficial and Deep Muscles
of the Erector Spinae Group**

___ Trapezius, cut and
reflected

___ Splenius { Capitis
Cervicis

___ Anterior scalene (slips)

___ Semispinalis capitis

___ Spinalis thoracis

___ Longissimus { Capitis
Cervicis
Thoracis

___ Iliocostalis { Cervicis
Thoracis
Lumborum

___ Quadratus lumborum

**Muscles Arising from
the Anterior Surfaces
of the Vertebrae**

___ Longus capitis

___ Longus colli

___ Anterior scalene (slips)

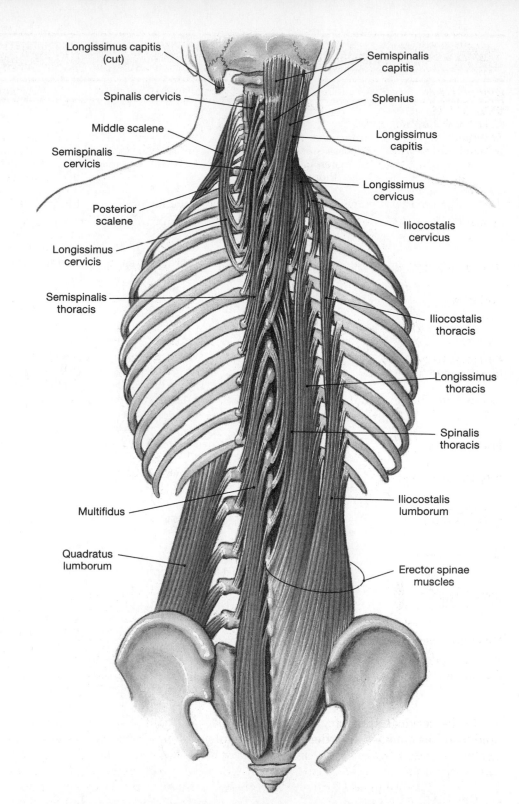

Longissimus capitis (cut)
Spinalis cervicis
Middle scalene
Semispinalis cervicis
Posterior scalene
Longissimus cervicis
Semispinalis thoracis
Multifidus
Quadratus lumborum

Semispinalis capitis
Splenius
Longissimus capitis
Longissimus cervicus
Iliocostalis cervicus
Iliocostalis thoracis
Longissimus thoracis
Spinalis thoracis
Iliocostalis lumborum
Erector spinae muscles

FIGURE 10-8 Muscles of the vertebral column. Superficial (right) and
deep (left) muscles of the erector spinae group.

Observation of the Oblique and Rectus Muscles

The oblique and rectus muscles form the abdominal wall and compress the underlying structures of the abdominal cavity or rotate the vertebral column [∞ p278]. The abdominal wall muscles are thin, but layered in such a way to provide strength. The muscles of the oblique and rectus groups lie between the vertebral spines and the ventral midline. These muscles can be separated into cervical, thoracic, and abdominal groups. See *Table 10-8* for a detailed summary of the oblique and rectus muscles.

TABLE 10-8 Oblique and Rectus Muscles

Group/Muscle	Origin	Insertion	Action	Innervation
OBLIQUE GROUP **Cervical region** Scalenes (anterior, middle, and posterior)	Transverse and costal processes of cervical vertebrae C_2–C_7	Superior surfaces of first two ribs	Elevate ribs, and/or flex neck; one side bends neck and rotates head and neck toward opposite side	Cervical spinal nerves
Thoracic region External intercostals	Inferior border of each rib	Superior border of the next rib, inferiorly	Elevate ribs	Intercostal nerves (branches of thoracic spinal nerves)
Internal intercostals	Superior border of each rib	Inferior border of the previous rib, superiorly	Depress ribs	As above
Transversus thoracis	Posterior surface of sternum	Cartilages of ribs	As above	As above
Abdominal region External oblique	External and inferior borders of ribs 5–12	External oblique aponeurosis extending to linea alba and iliac crest	Compresses abdomen; depresses ribs; flexes, bends to side, or rotates spine	Intercostal, iliohypogastric, and ilioinguinal nerves
Internal oblique	Lumbodorsal fascia and iliac crest	Inferior surfaces of ribs 9–12, costal cartilages 8–10, linea alba, and pubis	As above	As above
Transversus abdominis	Cartilages of inferior ribs (6–12), iliac crest, and lumbodorsal fascia	Linea alba and pubis	Compresses abdomen	As above
Serratus posterior (inferior)	Aponeurosis from spinous processes T_{10}–L_3	Inferior borders of ribs 8–12	Pulls ribs inferiorly; also pulls outward, opposing diaphragm	Thoracic nerves T_9–T_{12}
RECTUS GROUP **Cervical region**	*See muscles in Table 10-6 (except sternocleidomastoid)*			
Thoracic region Diaphragm	Xiphoid process, ribs 7–12 and associated costal cartilages, and anterior surfaces of lumbar vertebrae	Central tendinous sheet	Contraction expands thoracic cavity, compresses abdominopelvic cavity	Phrenic nerves
Abdominal region Rectus abdominis	Superior surface of pubis around symphysis	Inferior surfaces of costal cartilages (ribs 5–7) and xiphoid process of sternum	Depresses ribs, flexes vertebral column, and compresses abdomen	Intercostal nerves (T_7–T_{12})

Procedure

✓ Quick Check

Before you begin to examine the oblique and rectus muscles, identify rib borders, costal margin, xiphoid process [*Figure 6-40, p66*], and bone markings of the superior surface of the pelvis. [*Figures 7-9 to 7-12, pp76-79*] Keep an articulated skeleton nearby to aid you in locating points of muscle attachment as you identify abdominal wall muscles.

Locate and identify the following **oblique** and **rectus muscles** and **associated structures**, using a *torso model* or *prosected cadaver specimen*. Use *Color Plates 41, 42* and *Figures 10-9, 10-10* for reference. Observe these muscles on a torso model first before proceeding to the cadaver. The oblique group of muscles include the *scalenes*, in the cervical region, and the *intercostals* and *transversus*, in the thoracic region. Bend slightly the head of the cadaver towards the shoulder and observe anteriorly and laterally the **scalenes** (anterior, middle, and posterior muscle portions), which appear deep to the *jugular vein* and *sternocleidomastoid muscle*. In the thoracic region, the **external intercostal** muscles run obliquely (inferiorly and anteriorly) from the inferior border of the superior rib to the superior border of the inferior rib. The **internal intercostal** muscles run deep to and at right angles to those of the external intercostals (inferiorly and posteriorly) from the inferior border of the superior rib to the superior border of the inferior rib. The intercostals are important muscles in moving the ribs for respiration. The **transversus thoracis** are small muscles found on the inner chest wall arising at oblique angles from the sternum and costal cartilages to insert on the inferior borders of ribs 4–6. If the chest plate can be removed intact, the intercostals and thoracis muscles are easily observed.

Three large sheets of muscle cover the anterior and lateral walls of the abdominopelvic surface: an outer **external oblique**, a middle **internal oblique**, and an inner **transversus abdominis**. Identify on either side of the **linea alba** (midline) the **rectus abdominis** and note the horizontal connective tissue **tendinous inscriptions** of the rectus abdominis muscles.

See: Color Plates 41, 42, Figs. 10-9, 10-10

∞ Figs.
10-1, p267
10-12, p279
10-13, p281

LOCATE

Anterior View of the Trunk

___ External intercostals

___ Internal intercostals

___ External oblique

___ Xiphoid process

___ Rectus abdominis

___ Tendinous inscriptions of rectus abdominis

___ Iliac crest

___ Pubic tubercle

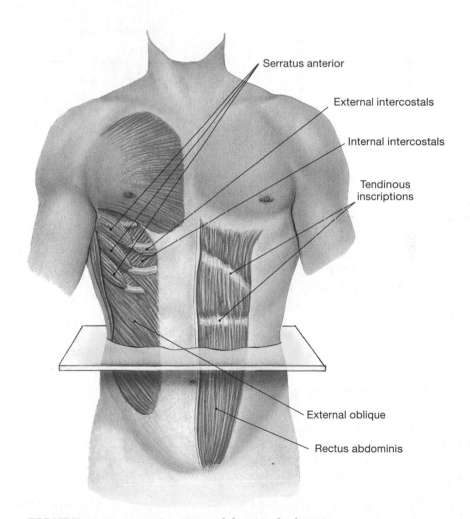

FIGURE 10-9 Anterior view of the trunk showing oblique and rectus muscles.

Abdominal Region

___ Linea alba

___ Rectus abdominis

___ Tendinous inscriptions

___ Transversus abdominis

___ Internal oblique

___ External oblique

___ Lumbar vertebra

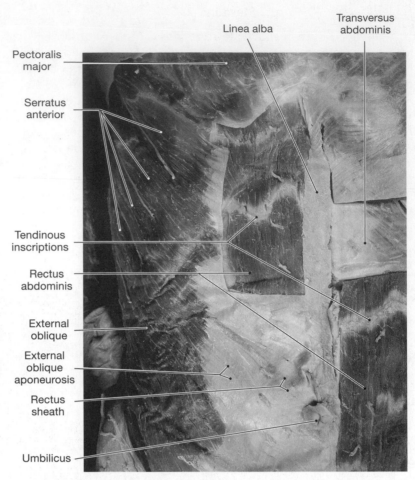

Pectoralis major

Serratus anterior

Tendinous inscriptions

Rectus abdominis

External oblique

External oblique aponeurosis

Rectus sheath

Umbilicus

Linea alba

Transversus abdominis

FIGURE 10-10 Oblique and rectus muscles, anterior view.

Examination of the Diaphragm

The dome-shaped **diaphragmatic muscle**, or **diaphragm**, serves not only as a partition between thoracic and abdominopelvic cavities, but is a major muscle of respiration. When contracted, the diaphragm flattens toward the abdominal cavity and expands the volume of the thoracic cavity. [∞ p278]

Procedure

Locate and identify the **position** and **parts** of **attachment** of the **diaphragmatic muscle** and **associated structures**, using a *torso model* or *prosected cadaver specimen*. Use *Color Plates 68, 72, 109, 120*, and *Figure 10-11* for reference. Observe the diaphragm on a torso model first before proceeding to the cadaver. On the torso model, the diaphragm is easily observed by removing the thoracic and abdominal viscera. To observe in the prosected cadaver, view first the superior surface of the diaphragm, then the inferior surface of the diaphragm. Both thoracic and abdominal viscera must be manipulated and reflected in order to view superior and inferior surfaces of the diaphragmatic muscle. The dome-shape and skeletal muscle fibers of the diaphragm are most obvious on the superior surface. To observe the inferior surface, reflect the liver and note the **impression for liver**, reflect the stomach and note the **impression for stomach**. The diaphragmatic muscle contains openings for the *inferior vena cava, esophagus,* and *aorta* to pass from thorax to abdomen. These structures are easily observed. The **right** and **left crura** (pl. form of crus) are muscular ligaments that arise from L_3 and L_4 vertebral bodies to attach to the **central tendon of diaphragm**, aiding in the support of the diaphragm from inferior. The right crus is larger than the left.

See: Color Plates 68, 72, 109, 120, Fig. 10-11

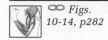

Figs. 10-14, p282

LOCATE

___ Sternum

___ Xiphoid process

___ Costal cartilages

___ Diaphragmatic muscle

___ Central tendon of diaphragm

___ Inferior vena cava

___ Impression for stomach

___ Impression for liver

___ Esophagus in esophageal hiatus

___ Aorta

___ 12th rib

___ Lumbar vertebra (1st and 2nd)

___ Crus (left/right)

___ Arcuate ligaments (medial and lateral)

___ Psoas muscle

___ Quadratus lumborum muscle

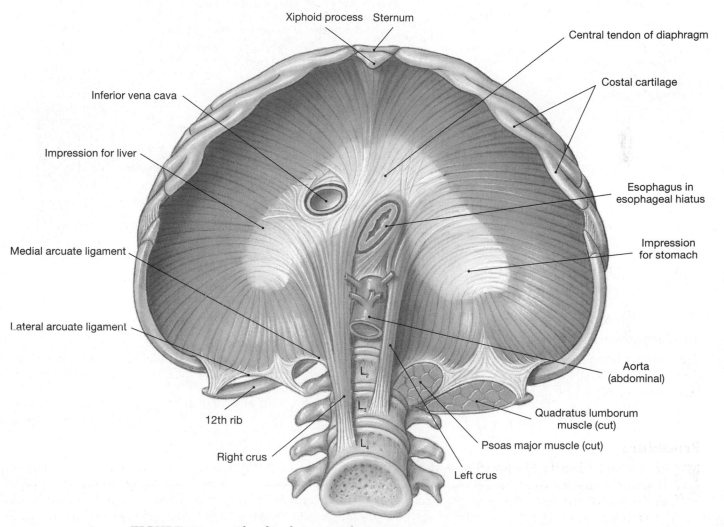

FIGURE 10-11   The diaphragm, inferior view.

Observation of the Muscles of the Pelvic Floor

The muscles of the pelvic floor extend from the sacrum and coccyx to the ischium and pubis. [∞ p283] The pelvic floor forms a muscular sling to support and prevent the abdominopelvic viscera from falling out of the body. These muscles also control the movement of materials through the urethra and anus. See *Table 10-9* for a detailed summary of the muscles of the pelvic floor.

TABLE 10-9 Muscles of the Pelvic Floor

Group/Muscle	Origin	Insertion	Action	Innervation
UROGENITAL TRIANGLE **Superficial muscles** Bulbospongiosus (or *bulbocavernosus*):				
Male	Collagen sheath at base of penis; fibers cross over urethra	Median raphe and central tendon of perineum	Compresses base, stiffens penis, ejects urine or semen	Pudendal nerve, perineal branch
Female	Collagen sheath at base of clitoris; fibers run on either side of urethral and vaginal openings	Central tendon of perineum	Compresses and stiffens clitoris, narrows vaginal opening	As above
Ischiocavernosus	Inferior ramus and tuberosity of ischium	Symphysis pubis anterior to base of penis or clitoris	Compresses and stiffens penis or clitoris, helping to maintain erection	As above
Superficial transverse perineal	Ischial ramus	Central tendon of perineum	Stabilizes central tendon of perineum	As above
Deep muscles: urogenital diaphragm Deep transverse perineal	Ischial ramus	Median raphe of urogenital diaphragm	Stabilizes central tendon of perineum	Pudendal nerve, perineal branch
Urethral sphincter:				
Male	Ischial and pubic rami	To median raphe at base of penis; inner fibers encircle urethra	Closes urethra, compresses prostate and bulbourethral glands	As above
Female	Ischial and pubic rami	To median raphe; inner fibers encircle urethra	Closes urethra, compresses vagina and greater vestibular glands	As above
ANAL TRIANGLE **Pelvic diaphragm** Coccygeus	Ischial spine	Lateral, inferior borders of the sacrum and coccyx	Flexes coccyx and coccygeal vertebrae	Lower sacral nerves (S_4–S_5)
External anal sphincter	Via tendon from coccyx	Encircles anal opening	Closes anal opening	Pudendal nerve, inferior rectal branch
Levator ani: Iliococcygeus	Ischial spine, pubis	Coccyx and median raphe	Tenses floor of pelvis, supports pelvic organs, flexes coccyx, elevates and retracts anus	Pudendal nerve and lower sacral nerves
Pubococcygeus	Inner margins of pubis	Coccyx and median raphe	As above	As above

Procedure

✓ **Quick Check**

Before you begin to examine the muscles of the pelvic floor, identify the bone markings on the articulated pelvis. (*Figures 7-11* and *7-12, pp78-79*) Orient an articulated pelvis (female pelvis if studying a female cadaver, male pelvis if studying a male cadaver) so that its position matches the cadaver to aid you in identifying points of muscle attachment.

Locate and identify the **muscles** of the **pelvic floor** and **associated structures**, using both male and female *torso models* or *prosected cadaver specimens*. The boundaries of the **perineum** are established by the inferior margins of the pelvis. On the articulated pelvis, draw an imaginary line between the ischial tuberosities, thereby dividing the diamond-shaped perineum into two triangular portions. [∞ p283] Anterior to this line, the small triangular area formed is termed the anterior or **urogenital triangle**, and contains the urethra. In the female, the opening to the vagina is also contained in the anterior triangle. [∞ p749] Posterior to the line, a larger triangular area is formed, the posterior or **anal triangle**, which contains the *rectum* and *anal sphincter*. [∞ p687]

Observe the muscles of the pelvic floor on a torso model (male and female) before proceeding to the cadaver. Use *Color Plates 120, 127, 130,* and *Figures 10-12, 10-13* for reference. To observe some of these muscles in the torso model, remove the abdominopelvic viscera and observe the pelvic floor from the abdominopelvic cavity. To observe in the prosected cadaver, one of two positions may be used. One method is to place the cadaver in the supine position with the thighs widely spread. Keeping the thighs widely spread will require other students to aid in the abducting of the lower limbs. Props may be needed to maintain the limbs in this position. A second positioning method is to examine the pelvic floor by placing the body in the prone position and abducting the lower limbs. The posterior triangle is more easily observed from this position. An alternative to either of these positioning methods is to disarticulate one lower limb at the hip joint to allow for better visibility of the perineum. This might be accomplished after the lower limb has been studied. Check with your instructor for the cadaver position you are to use.

The **pelvic diaphragm** is a funnel-shaped, muscular sheet that is attached to the pelvic walls and occupies the space within the anterior and posterior triangles. The pelvic diaphragm separates the pelvic cavity from the perineum. The rectum is located in the conical portion of the muscular pelvic diaphragm and should be used as a landmark for locating structures. Most of the pelvic diaphragm is formed by the **levator ani** muscle, which is located in the anterior triangle and part of the posterior triangle. The levator ani muscle consists of three portions, a pair of large thick anterior **pubococcygeus** muscles, a middle pair of smaller **puborectalis** muscles, and a pair of smaller **iliococcygeus** muscles located posteriorly. These muscles are found on the left and right sides of the midline. Immediately external and posterior to the iliococcygeus is the small, thin, triangular **coccygeus** muscle, which forms the remaining posterior portion of the pelvic diaphragm. The **sacrotuberous ligament** lies external to the coccygeus muscle. Lateral to the midline lies the feather-shaped **bulbocavernosus** muscle. In the female, this muscle surrounds the vagina. In the male, this muscle wraps around the base of the penis and a portion of the penile shaft. The **ischiocavernosus** muscle is easily observed in males. Additionally, you may observe a large, superficial muscle that forms the bulk of the buttock mass, the *gluteus maximus*. It is a major extensor of the thigh and does not participate in forming the pelvic floor.

As you examine the pelvic diaphragm, note the passage of blood vessels, nerves, and other structures. Be careful not to damage these vessels and structures, as they will be identified in future chapters. However, your instructor may wish to do so at this point. Check with your instructor before proceeding.

See: Color Plates 120, 127, 130, Figs. 10-12, 10-13

 ∞ Figs.
10-15, p284
27-9d, p736
27-20a, p749

LOCATE

Superficial Muscles

___ Ischiocavernosus

___ Bulbocavernosus

___ Superficial transverse perineal

___ Anus, vagina, urethra, urethral sphincter (male and female)

___ External anal sphincter

___ Pubococcygeus

___ Gluteus maximus

Deep Muscles

___ Deep transverse perineal

___ Levator ani (pubococcygeus, puborectalis, and iliococcygeus)

___ Sacrotuberous ligament

___ Coccygeus

___ Urogenital triangle

___ Anal triangle

___ No difference between deep musculature in male and female

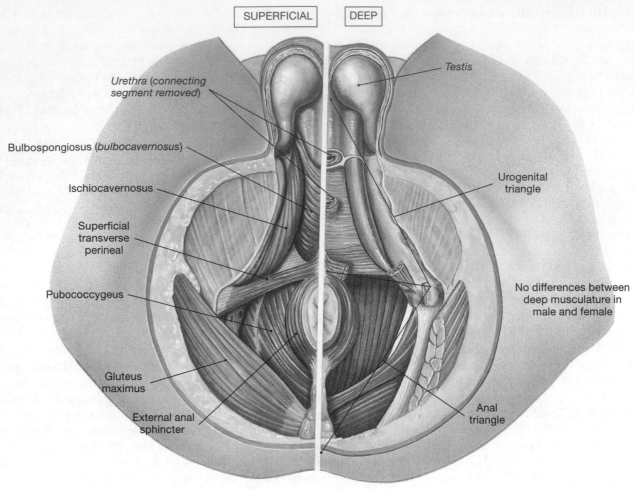

Urethra (connecting segment removed)

Bulbospongiosus (bulbocavernosus)

Ischiocavernosus

Superficial transverse perineal

Pubococcygeus

Gluteus maximus

External anal sphincter

Testis

Urogenital triangle

No differences between deep musculature in male and female

Anal triangle

FIGURE 10-12 Muscles of the male pelvic floor.

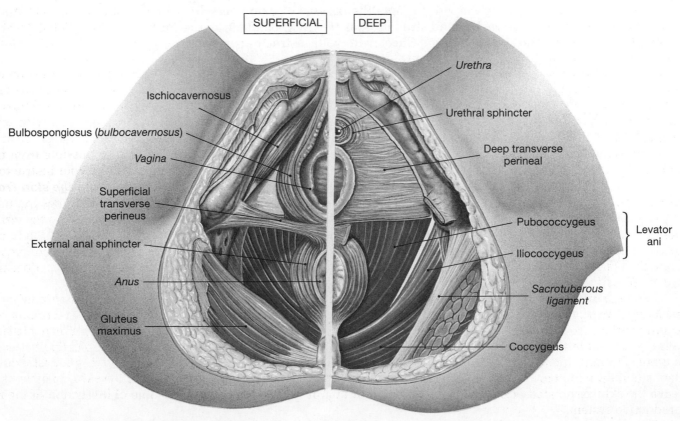

Ischiocavernosus

Bulbospongiosus (bulbocavernosus)

Vagina

Superficial transverse perineus

External anal sphincter

Anus

Gluteus maximus

Urethra

Urethral sphincter

Deep transverse perineal

Pubococcygeus

Iliococcygeus

} Levator ani

Sacrotuberous ligament

Coccygeus

FIGURE 10-13 Muscles of the female pelvic floor.

Dissection of Cat Musculature

Introduction

This exercise is intended to complement the study of human muscles. Some muscles found in the cat are lacking in humans. Other muscles that are fused in humans occur as separate muscles in animals. Muscle fibers within a skeletal muscle form are present in bundles termed *fascicles*. The fibers within a single fascicle are arranged in parallel lines, but the organization of the fascicles can vary, as can the relationship between the fascicles and the associated tendon. Four different patterns of fascicle organization produce *parallel, convergent, pennate*, and *circular muscles*. Each muscle is usually attached at its ends by connective tissue, which forms either a band-like *tendon* or a flat sheet-like *aponeurosis*. Skeletal muscles are named in a systematic manner. Skeletal muscles have an origin, the stationary end of the muscle, usually more proximal, and an insertion, the movable part of the muscle, usually more distal. Muscles are grouped into categories based on the actions produced by their contraction: *agonists* are prime movers; *synergists* assist the prime movers; and *antagonists* oppose the action of the prime movers.

Preparing the Cat for Dissection and Removal of the Skin

All students must practice the highest levels of laboratory safety. Be extremely careful when using a scalpel or other sharp instrument. Direct cutting motions away from you. Always point scissors away from you when cutting tissue and insert with the blunt end into the tissue. Before cutting any tissue, make sure that the tissue to be cut is freed from underlying or adjacent tissue, so that it will not be severed accidentally.

Read this entire section and familiarize yourself completely with the exercise before proceeding. You must exercise care to prevent muscle damage during skin removal. The attachment of the skin to the underlying hypodermis will be variable: in some places, skin is loosely attached to the body with subcutaneous fat between the skin and the underlying muscle (e.g., abdominopelvic region), whereas in other places the skin is held tightly to underlying muscle by tough fascia (e.g., thigh). Additionally, the cat has some large, thin masses of muscle in the area of the neck and face, which attach directly to and move the skin. The muscle is termed the *platysma*, but in humans it is thinner.

Procedure

Remove the intact skin piece. It should be saved to wrap the body for storage. The skin wrapping keeps the body moist and prevents excessive drying of the muscles. Carefully wrap the body with the skin prior to storage. Never rinse your cat with water, as this will remove the preservative and promote the growth of mold. You may need to supplement the skin wrap with some paper towels. Store the cat in the plastic bag provided, seal the bag, attach a name tag with your name and your lab partner's name to the cat, and place in the assigned storage area.

When you are working with the specimen, place the cat ventral side down on a dissecting tray before you begin removing the skin. With your scalpel, make a short, shallow incision in the midline just anterior to the tail. Insert a blunt probe or your finger to separate the skin from underlying muscle and fascia. Extend your incision anteriorly by cutting the separated skin with the scissors. Continue to separate with the probe and cut with scissors until you have an incision that extends up to the neck. Remember to place the blunt-ended blade of the scissors under the skin. On either side of this mid-dorsal incision line, separate as much skin as possible from the underlying muscle and fascia. Using bone forceps, cut off the tail and discard it, if requested by your instructor.

From the dorsal surface, make several incision lines. ***NOTE, always make sure to separate the skin from the underlying tissues using a blunt probe or your finger prior to cutting the skin.*** The incision lines include: (1) a complete encirclement of the neck, (2) the lateral side of each forelimb from the dorsal incision to the wrist; completely cutting the skin around the wrists, (3) the lateral side of each hind leg from the dorsal incision to the ankle; completely cutting the skin around the ankle, and (4) from the dorsal surface of the tail an incision encircling the anus and the genital organs. Loosen the skin as much as possible using your fingers. If the skin does not come off easily, it must be freed by cutting with the scalpel.

Use your fingers to free the skin from the loose connective tissue that attaches the skin to the underlying structures. Work from the dorsal surface toward the ventral surface at the posterior of the cat, then work on the ventral surface from the posterior end of the cat toward its neck. Depending on your specimen, you may observe some or all of the following: thin, rubber-band-like red or blue latex-injected blood vessels projecting between the muscles and the skin; in female cats, mammary glands between the skin and the underlying muscle; cutaneous nerves, which are small, white, cord-like structures extending from the muscles to the skin. In male cats, leave the skin associated with the male external genitalia. It will be removed at the time of dissection of the reproductive system.

If at any time you are confused about the nature of any of the tissue, always check with your instructor before removing anything.

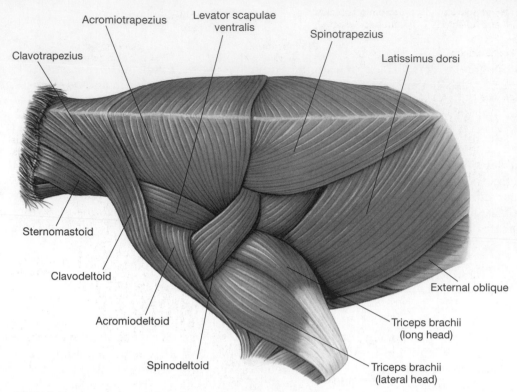

FIGURE 10-14 Superficial muscles of back and upper forelimb (lateral view).

Superficial Muscles of Back and Upper Forelimb (Lateral View)

Begin your dissection of the upper back region between the two forelimbs. Be sure that the skin has been removed from the neck up to the base of the ears.

1. Locate the **trapezius group** of muscles, which covers the dorsal surface of the neck and all of the scapula. The three cat muscles are homologous with the single trapezius muscle of the human. Their prefix denotes site of insertion.

   a. The **spinotrapezius** is the most posterior of the trapezius muscles. This triangular-shaped muscle should be separated from the acromiotrapezius (anteriorly) and the latissimus dorsi (posteromedially). It originates from spinous processes of the posterior thoracic vertebrae and inserts on the scapular spine. It pulls the scapula dorsocaudad.

   b. Anterior to the spinotrapezius is the **acromiotrapezius**, a large, almost square muscle that originates from spinous processes of cervical and anterior thoracic vertebrae and inserts into the scapular spine. It holds the scapula in place.

   c. The **clavotrapezius** is a broad muscle anterior to the acromiotrapezius. It originates from the lambdoidal crest and axis and inserts on the clavicle. It draws the scapula craniad and dorsad.

2. The large, flat muscle posterior to the trapezius group is the **latissimus dorsi**. Its origin is on the spines of thoracic and lumbar vertebrae, and it inserts on the medial aspect of the humerus. The latissimus dorsi acts to pull the forelimb posteriorly and dorsally.

3. A flat, strap-like muscle, the **levator scapulae ventralis**, lies on the scapular spine between the clavotrapezius and the acromiotrapezius. The occipital bone and atlas is the origin, and the vertebral border of scapula is the insertion. This muscle pulls the scapula forward.

4. Locate the **deltoid group** of muscles, which are shoulder muscles lateral to the trapezius group. The following three cat muscles are homologous with the single deltoid muscle in the human.

   a. The **spinodeltoid** is the most posterior of the deltoid group. It originates on the scapula ventral to insertion of the acromiotrapezius, and inserts on the proximal humerus. The action of this muscle is to flex the humerus and rotate it laterally.

   b. The middle muscle of the deltoid group is the **acromiodeltoid**. It originates from the acromion process of the scapula deep to the levator scapulae ventralis, and inserts on the proximal end of the humerus. The acromiodeltoid flexes the humerus and rotates it laterally.

   c. The **clavodeltoid** (clavobrachialis) originates on the clavicle and inserts on the ulna. It is a continuation of the clavotrapezius beyond the clavicle and extends down the arm from the clavicle. It functions to flex the forelimb.

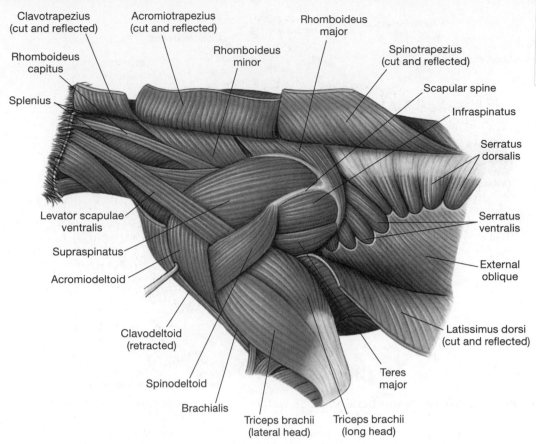

FIGURE 10-15 Deep muscles of back and upper forelimb (lateral view).

Deep Muscles of Back and Upper Forelimb (Lateral View)

To expose the deep muscles of the shoulder and back, carefully bisect (divide into two equal parts) both the acromiotrapezius and spinotrapezius; reflect (fold back) these muscles. Repeat the procedure with the latissimus dorsi.

5. Deep to the acromiotrapezius, the **supraspinatus** occupies the lateral surface of the scapula in the supraspinous fossa. It originates on the scapula and inserts on the humerus. It functions to extend the humerus.

6. On the lateral surface of the scapula, the **infraspinatus** occupies the infraspinous fossa. It originates on the scapula and inserts on the humerus, and causes the humerus to rotate laterally.

7. The **teres major** occupies the axillary border of the scapula, where it has its origin. It inserts on the proximal end of the humerus, and acts to rotate and draw the humerus posteriorly.

8. The **rhomboideus group** connects the spinous processes of cervical and thoracic vertebrae with the vertebral border of the scapula. They hold the dorsal part of the scapula to the body.

   a. The posterior muscle of this group is the **rhomboideus major**. This fan-shaped muscle originates from the spinous processes and ligaments of the posterior cervical and anterior thoracic vertebrae, and inserts on dorsal posterior angle of the scapula. It draws the scapula dorsally and anteriorly.

   b. The larger muscle anterior to the rhomboideus major is the **rhomboideus minor**. Its origin is on the spines of posterior cervical and anterior thoracic vertebrae. This muscle inserts along the vertebral border of the scapula. It functions to draw the scapula anteriorly and dorsally.

   c. The most anterolateral muscle of the rhomboideus group, is a narrow, thin, ribbon-like muscle, termed the **rhomboideus capitis**. It originates from the spinous nuchal line, and inserts on the vertebral border of the scapula. It elevates and rotates the scapula. It is *not found in humans*.

9. Deep to the rhomboideus capitis, a broad, flat, and thin **splenius** muscle covers most of the lateral surface of the cervical and thoracic vertebrae. This muscle has its origin on spines of the thoracic vertebrae, and its insertion on the superior nuchal line of the occipital bone. It acts to turn or raise the head.

10. To observe the muscle that occupies the subscapular fossa, the **subscapularis**, position the elbows to touch in the midline. It originates from the subscapular fossa and inserts on the humerus. It is an adductor of the humerus.

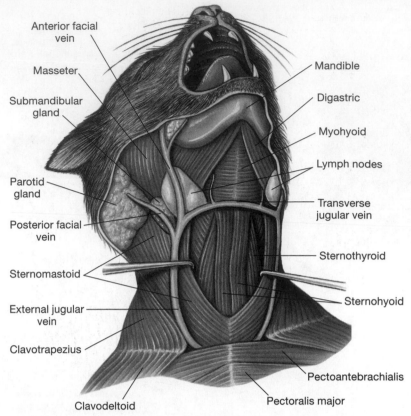

FIGURE 10-16   Superficial muscles of neck (ventral view).

Superficial Muscles of Neck (Ventral View)

Remove any remaining skin up to the ear from the left side of the neck. Be careful not to sever the *external jugular vein*. This is the large vein (blue-colored) that lies on the ventral surface of the neck. Free this vein from the underlying muscles, then clean off the connective tissue from both the back of the left shoulder and from the ventral and lateral surfaces of the left neck. Do not remove the fascia in the midline of the back.

11. A large V-shaped muscle between the sternum and head, the **sternomastoid**, originates from the manubrium of the sternum and passes obliquely around the neck to insert on the superior nuchal line and on the mastoid process. It turns and depresses the head.

12. The **sternohyoids** are a narrow pair of muscles (may be fused in the midline), that lie over the larynx, along the mid-ventral line of the neck. Their origin is the costal cartilage of the first rib, while their insertion is the hyoid bone. They act to depress the hyoid bone.

13. Locate the pair of superficial muscles extending along the inner surface of the mandible, the **digastrics**. Each originates on the occipital bone and mastoid process, and functions as a depressor of the mandible.

14. The **mylohyoid** is a superficial muscle running transversely in the midline and passing deep to the digastrics. It originates on the mandible and inserts on the hyoid bone. The mylohyoid raises the floor of the mouth.

15. The large muscle mass anteroventral to the parotid gland at the angle of each jaw is the **masseter**. This cheek muscle originates from the zygomatic bone. It inserts on the posterolateral surface of the dentary bone, and elevates the mandible.

Deep Muscles of Neck (Ventral View)

16. The **sternothyroids** lie deep and lateral under the cut ends of sternohyoid. They originate from the first costal cartilage and insert on the thyroid cartilage. These muscles act to pull the larynx posteriorly.

17. Anterior to the sternothyroid, deep and lateral under the cut ends of sternohyoid, lie the **thyrohyoids**. These small muscles have their origin at the thyroid cartilage and their insertion on the hyoid bone. They help to elevate the larynx.

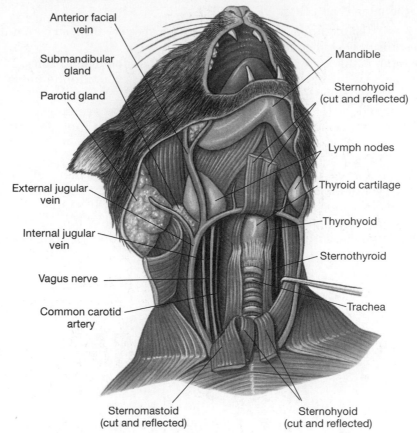

FIGURE 10-17 Deep muscles and vessels of neck (ventral view).

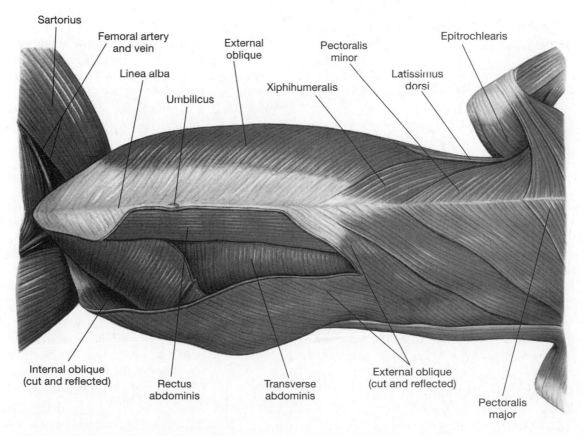

FIGURE 10-18 Superficial muscles of abdomen (ventral view).

The Muscular System: Muscle Organization and the Axial Musculature **133**

Superficial Muscles of Abdomen (Ventral View)

Three layers of muscle form the abdominal wall. These muscles are very thin, so care must be taken to separate them. Collectively, these muscles act to compress the abdomen.

18. The **external oblique** is the most superficial of the lateral abdominal muscles. It originates on posterior ribs and lumbodorsal fascia, and inserts on the linea alba from the sternum to the pubis. Its fibers run from anterodorsal to posteroventral, and it acts to compress the abdomen.

19. Deep to the external oblique lies the **internal oblique**, with its fibers running perpendicular to those of the external oblique in a posterodorsal to anteroventral orientation. It originates from the pelvis and lumbodorsal fascia, and inserts on the linea alba, where it functions to compress the abdomen. Reflect the external oblique to view the internal oblique.

20. Deep to the internal oblique, the **transverse abdominis**, with fibers that run transversely across the abdomen, forms the deepest layer of the abdominal wall. It originates from the posterior ribs, lumbar vertebrae and the ilium, and inserts on the linea alba. It acts to compress the abdomen.

21. The **rectus abdominis** is a long, ribbon-like muscle in the midline of the ventral of the abdomen. It originates from the pubic symphysis and inserts on the costal cartilage. It compresses internal organs of the abdomen.

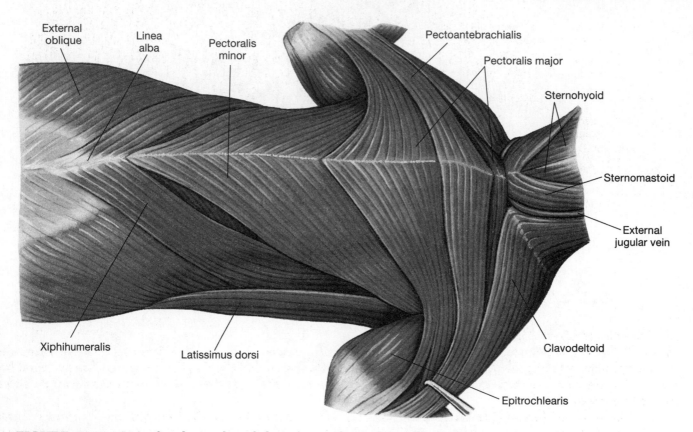

FIGURE 10-19 Superficial muscles of chest (ventral view).

Superficial Muscles of Chest (Ventral View)

22. The **pectoralis group** consists of the large muscles covering the ventral surface of the chest. They arise from the sternum and attach, for the most part, to the humerus. There are four subdivisions in the cat, but only two in man. In the cat, the relatively great degree of fusion gives the cat's pectoral muscles the appearance of a single muscle. This fusion makes it rather difficult to dissect as they do not separate from each other easily.

   a. The most superficial of the chest muscles, the **pectoantebrachialis**, has its origin on the manubrium. It inserts on the fascia of the forearm. This muscle adducts the forelimb. It is *not found in humans.*

   b. Posterior to the pectoantebrachialis is the much broader and triangular-shaped **pectoralis major**. Its origin is on the sternum and insertion on the posterior humerus. It functions to adduct the forelimb.

c. Posterior to the pectoralis major lies the larger **pectoralis minor**. This is the broadest and thickest muscle of the group. It extends posterior to the pectoralis major. It originates on the sternum and inserts near the proximal end of the humerus. It adducts the forelimb.

23. The thin **xiphihumeralis** is posterior to the posterior edge of the pectoralis minor. It originates on the xiphoid process of the sternum and inserts by narrow tendon on the humerus. It adducts and helps rotate the forelimb. It is *not found in humans*.

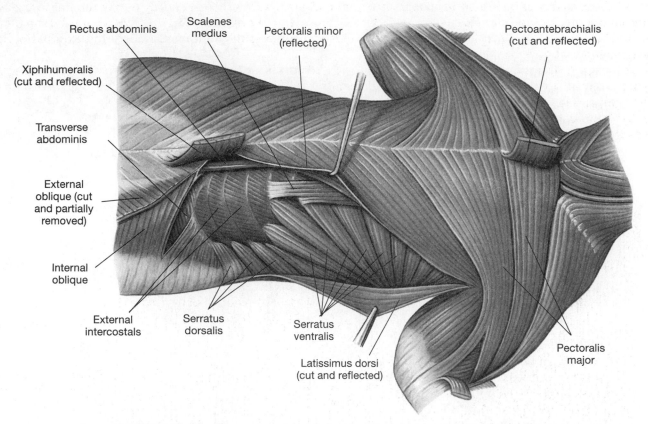

FIGURE 10-20 Deep muscles of chest and abdomen (ventral view).

Deep Muscles of Chest and Abdomen (Ventral View)

24. On the lateral thoracic wall, just ventral to the scapula, a large, fan-shaped muscle, the **serratus ventralis**, originates by separate slips from the ribs. It passes ventral to the scapula, and inserts on the vertebral border of the scapula. It is homologous to the serratus anterior of humans, and functions to pull the scapula anteriorly and ventrally.

25. The **serratus dorsalis** is a serrated muscle that lies medial to the serratus ventralis and overlies the anterior divisions of the sacrospinalis. Its origin is mid-dorsal cervical, thoracic, and lumbar raphe, and it inserts on the ribs. It acts to pull the ribs craniad and laterally.

26. On the lateral surface of the trunk is a three-part muscle, the **scalenus**, which unites anteriorly. It originates on the ribs and inserts on the transverse processes of the cervical vertebrae. It acts to flex the neck and draw the ribs anteriorly.

27. Deep to the external oblique, muscle fibers of the **external intercostals** run obliquely from the posterior border of one rib to the anterior border of the next rib. Their origin is on the caudal border of one rib, and their insertion is on the cranial border of the next rib. They lift the ribs during inspiration.

28. Toward the midline and deep to the external intercostals, observe the **internal intercostals**. Bisect one external intercostal muscle to expose an internal intercostal. The fibers of the intercostals run at oblique angles. The internal from median to lateral, the external from lateral to medial. Their origin is on the superior border of the rib below, and insert on the inferior border of the rib above. They draw adjacent ribs together and depress ribs during active expiration.

Muscles of Upper Forelimb (Medial View)

29. Observe the **epitrochlearis**, which is a broad, flat muscle covering the medial surface of the upper forelimb. It appears to be an extension of the latissimus dorsi, and originates from the fascia of the latissimus dorsi. It inserts on the olecranon process where it acts to extend the forelimb. It is *not found in humans*.

30. The **biceps brachii** is a convex muscle interior to the pectoralis major and minor on the ventromedial surface of the humerus. It originates on the scapula and inserts on the radial tuberosity near the proximal end of the radius. It functions as a flexor of the forelimb.

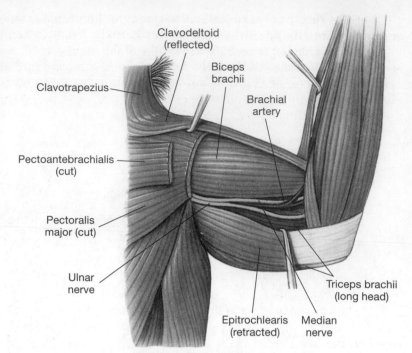

FIGURE 10-21 Muscles of upper forelimb (medial view).

Muscles of Forelimb (Medial View)

31. Observe a ribbon-like muscle on the lateral surface of the humerus. This is the **brachioradialis**. Its origin is on the mid-dorsal border of the humerus and its insertion on the distal end of the radius. It supinates the forelimb paw.

32. Deep to the brachioradialis, locate the **extensor carpi radialis**. There are two parts to this muscle: a shorter, triangular extensor carpi radialis brevis, and deep to it the extensor carpi radialis longus. It originates from the lateral surface of the humerus dorsal to the lateral epicondyle and inserts on the bases of the second and third metacarpals. It causes extension at the carpal joints.

33. The narrow muscle next to the extensor carpi radialis is the **pronator teres**. It runs from its point of origin on the medial epicondyle of the humerus and declines in size as it approaches to insertion on the radius. It rotates the radius for pronation.

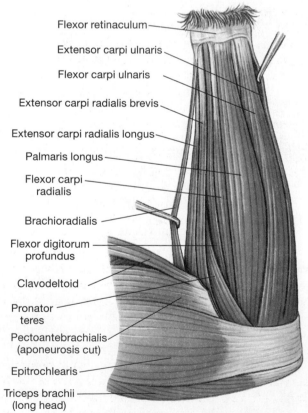

FIGURE 10-22 Muscles of forelimb (medial view).

34. The **flexor carpi radialis** is adjacent to the pronator teres. It originates on the distal end of the humerus and inserts into the second and third metacarpals. It acts to flex the wrist.

35. The large flat muscle in the center of the medial surface of the forelimb is the **palmaris longus**. Its origin is on the medial epicondyle of the humerus, and it inserts into all digits. It flexes the digits.

36. The flat muscle on the posterior edge of the forelimb is the **flexor carpi ulnaris**. It arises from a two-headed origin (medial epicondyle of the humerus, and the olecranon process) and inserts by a single tendon on the ulnar side of the carpals. It is a flexor of the wrist.

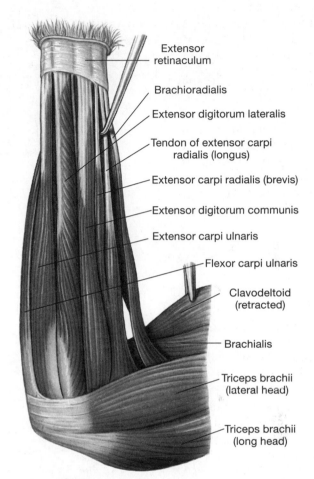

Extensor retinaculum

Brachioradialis

Extensor digitorum lateralis

Tendon of extensor carpi radialis (longus)

Extensor carpi radialis (brevis)

Extensor digitorum communis

Extensor carpi ulnaris

Flexor carpi ulnaris

Clavodeltoid (retracted)

Brachialis

Triceps brachii (lateral head)

Triceps brachii (long head)

Muscles of Left Forelimb (Lateral View)

37. Locate the **brachialis** on the ventrolateral surface of the humerus. It originates on the lateral side of the humerus and inserts on the proximal end of the ulna. It functions to flex the forelimb.

38. On the lateral and posterior surfaces of the forelimb, observe the largest superficial muscle in the upper forelimb, the **triceps brachii**. It arises from three heads:

  a. The **long head** is the large muscle mass on the posterior surface, originating on the lateral border of the scapula.

  b. The **lateral head** lies next to the long head on the lateral surface, originating on the deltoid ridge of the humerus.

  c. The small **medial head** lies deep to the lateral head, originating on the shaft of the humerus. All heads have a single insertion into the olecranon process of the ulna. Its function is to extend the forelimb.

FIGURE 10-23 Muscles of left forelimb (lateral view).

39. Distal and deep to the triceps medial and long heads is the L-shaped muscle surrounding the elbow, the **anconeus**. This superficial muscle has fibers that originate near the lateral epicondyle of the humerus and inserts near the olecranon. It is sometimes a darker color than its neighbors. It keeps the articular capsule of the elbow joint taut and assists in the extension of the forelimb.

40. Adjacent to the extensor carpi radialis on the forelimb is the **extensor digitorum communis**, a block-like muscle that originates from the lateral surface of the humerus dorsal to the epicondyle. It inserts on the second to fifth digits and acts to extend the digits.

41. Lateral to the extensor digitorum communis is the **extensor digitorum lateralis**, another block-like muscle that originates from the lateral surface of the humerus dorsal to the lateral epicondyle. It inserts with the tendons of the extensor digitorum communis on the third to fifth digits and functions to extend the digits.

42. The block-like muscle lateral to the extensor digitorum lateralis is the **extensor carpi ulnaris**. It originates from the lateral epicondyle of the humerus and inserts on the proximal end of the fifth metacarpal. It extends the ulnar side of the wrist and the fifth digit.

Superficial Muscles of Thigh (Ventromedial View)

43. Identify a wide, superficial muscle that covers the anterior half of the medial aspect of the thigh, the **sartorius**. It originates on the ilium and inserts on the tibia. The sartorius adducts and rotates the femur, and it extends the tibia.

44. The broad muscle that covers the posterior portion of the medial aspect of the thigh is the **gracilis**. It originates from the ischium and pubic symphysis and inserts on the medial surface of the tibia. It adducts the thigh and draws it posteriorly.

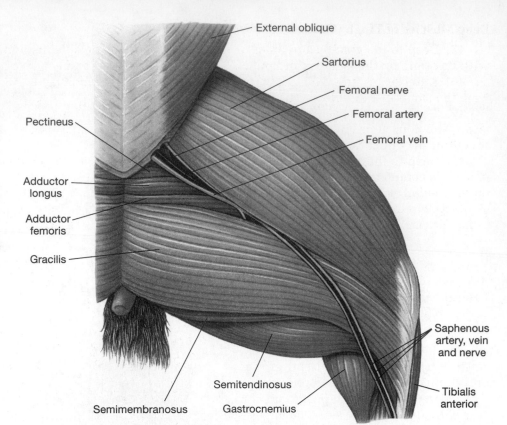

FIGURE 10-24 Superficial muscles of thigh (ventromedial view).

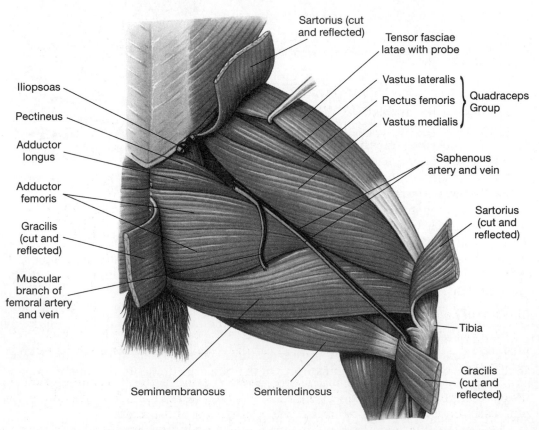

FIGURE 10-25 Deep muscles of thigh (ventromedial view).

Deep Muscles of Thigh (Ventromedial View)

45. Deep to the gracilis and anterior to the semimembranosus, is the **adductor femoris**. It is a large muscle with its origin on the ischium and pubis. It inserts on the femur and acts to adduct the thigh.

46. Anterior to the adductor femoris is the **adductor longus** muscle. It is a narrow muscle that originates from the ischium and pubis, and inserts on the proximal surface of the femur. It adducts the thigh.

47. The **pectineus** is anterior to the adductor longus. It is a deep, small muscle posterior to both the femoral artery and vein. It originates on the anterior border of the pubis and inserts on the proximal end of the femur. It functions to adduct the thigh.

48. Deep to the *femoral artery, vein* and *nerve*, and lateral to the pectineus muscle, is the **iliopsoas**. In humans it is composed of two muscles - iliacus and psoas major. It originates from the lumbar vertebrae and inserts on the medial aspect of the proximal femur. This muscle flexes and laterally rotates the thigh.

49. On the posterior aspect of the thigh, a strap-like muscle, the **semitendinosus**, originates from the ischial tuberosity and inserts on the medial side of the tibia. It flexes the knee.

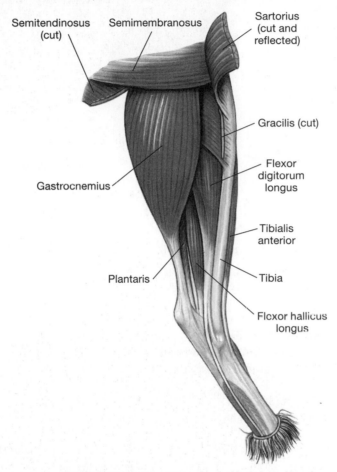

FIGURE 10-26   Muscles of left hindlimb

Muscles of Left Hindlimb (Medial View)

50. The **gastrocnemius** or calf muscle is on the posterior of the lower hindlimb. It has two heads of origin, a medial and a lateral (the knee fascia and the distal end of the femur), which unite in the calcaneal tendon to insert on the calcaneus. It extends the foot.

51. Between the gastrocnemius and the tibial bone is the **flexor digitorum longus**. This muscle has two heads of origin (the distal end of the tibia and the head and shaft of the fibula), and it inserts by four tendons onto bases of the terminal phalanges. It acts as a flexor of the digits.

52. On the anterior surface of the tibial bone, the **tibialis anterior** originates on the proximal ends of the tibia and fibula. It inserts on the first metatarsal and functions to flex the foot.

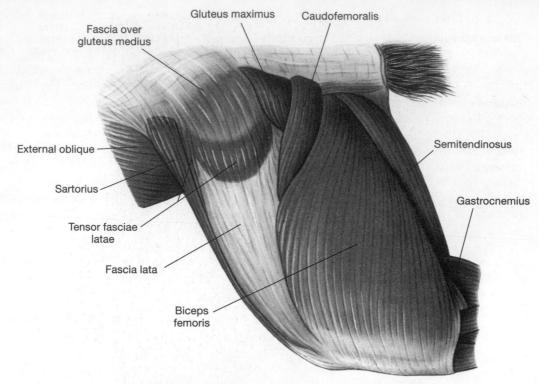

Fascia over
gluteus medius

Gluteus maximus

Caudofemoralis

External oblique

Sartorius

Tensor fasciae
latae

Fascia lata

Biceps
femoris

Semitendinosus

Gastrocnemius

FIGURE 10-27 Superficial muscles of left hindlimb
(posterolateral view).

Superficial Muscles of Left Hindlimb (Posterolateral View)

53. Observe the large, broad muscle covering the lateral region of the thigh. This is the **biceps femoris**. It originates on the ischial tuberosity and inserts on the tibia. The biceps femoris abducts the thigh and flexes the knee.

54. Posterior to the sartorius, locate a triangular-shaped muscle mass, the **tensor fasciae latae**. This muscle originates from the crest of the ilium and inserts into the fascia lata. The tensor fasciae latae extends the thigh.

55. Dorsal to the tensor fasciae latae lies the **gluteus medius**. This is the largest of the gluteus muscles, and it has its origin on both the ilium and the transverse processes of the last sacral and first caudal vertebrae. It inserts on the femur and acts to abduct the thigh.

56. Just posterior to the gluteus medius, locate the **gluteus maximus**, a small triangular hip muscle. It originates from the transverse processes of the last sacral and first caudal vertebrae, and inserts on the proximal femur. It abducts the thigh.

57. Anterior to the biceps femoris and posterior to the gluteus maximus, locate the **caudofemoralis**. It originates from the transverse processes of second and third caudal vertebrae, and inserts on the patella. The caudofemoralis abducts the thigh and extends the knee. It is *not found in humans*.

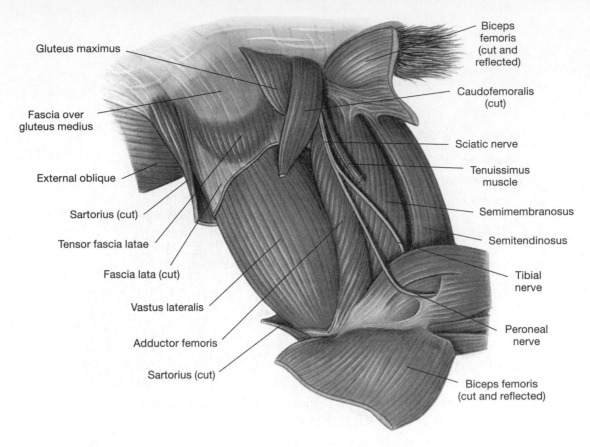

FIGURE 10-28 Deep muscles of left hindlimb (posterolateral view).

Deep Muscles of Left Hindlimb (Posterolateral View)

58. Observe a large, posterior muscle that lies deep to the gracilis and medial to the semitendinosus. This is the **semimembranosus**. It originates from the ischium and inserts on the medial epicondyle of the femur and medial surface of the tibia. It extends the thigh.

59. Locate the group of four large muscles that compose the **quadriceps femoris**. The quadriceps femoris covers about one-half the area of the thigh. Collectively, these muscles insert into the patellar ligament and act as a powerful extender of the knee. Bisect the sartorius, and free both borders of the tensor fasciae latae. Reflect these and observe that the muscles of the quadriceps femoris converge and insert on the patella.

   a. The large muscle on the anterolateral surface of the thigh is the **vastus lateralis**. It originates along the entire length of the lateral surface of the femur.

   b. The large muscle on the medial surface of the femur is the **vastus medialis**. It originates on the shaft of the femur.

   c. The large, cylindrical **rectus femoris** lies between the medialis and lateralis muscles. It originates on the acetabulum.

   d. The **vastus intermedius** lies deep to the rectus femoris. Bisect the rectus femoris and reflect it to expose the intermedius. It originates on the shaft of the femur.

60. Deep to the biceps femoris locate the **sciatic nerve** and observe that it divides into the **common peroneal** and **tibial nerves.**

61. Adhering to the superior surface of the sciatic nerve is a long thin ribbon of muscle, the **tenuissimus**. It originates on the transverse process of the second caudal vertebrae and inserts into the fascia of the biceps femoris. The tenuissimus aids in abducting the thigh and it is *not found in humans.*

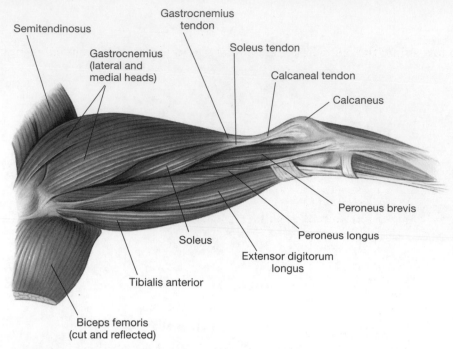

Semitendinosus

Gastrocnemius
tendon

Gastrocnemius
(lateral and
medial heads)

Soleus tendon

Calcaneal tendon

Calcaneus

Peroneus brevis

Peroneus longus

Extensor digitorum
longus

Soleus

Tibialis anterior

Biceps femoris
(cut and reflected)

FIGURE 10-29 Muscles of left hindlimb (posterolateral view).

Muscles of Left Hindlimb (Posterolateral View)

62. Deep to the gastrocnemius, but visible on the lateral surface of the calf is the **soleus**. The soleus originates on the fibula and inserts on the calcaneus. It extends the foot.

63. Deep to the soleus on the posterior and lateral surfaces, locate the three **peroneus** muscles.

a. The **peroneus brevis** lies deep to the tendon of the soleus, originating from the distal portion of the fibula and inserting on the base of fifth metatarsal.

b. The **peroneus longus** is a long, thin muscle that lies on the lateral surface, originating from the proximal portion of the fibula and inserting by a tendon that passes through a groove on the lateral malleolus and turns medially to attach to the bases of metatarsals.

c. The **peroneus tertius** lies along the tendon of the peroneus longus, originating on the fibula and inserting on the base of the fifth metatarsal. The peroneus longus and tertius flex the foot, while the peroneus brevis extends it.

64. On the anterolateral border of the tibia, the **extensor digitorum longus** originates from the lateral epicondyle of the femur. It inserts by long tendons on each of five digits, and functions to extend the digits.

To Think About

1. Why are there so few spinal flexors, when there are a large number of different extensor groups?

2. Identify the muscles involved in normal quiet breathing.

3. Which muscles make up the anal triangle and what are their function?

4. How do the muscles of the tongue function in mastication?

5. What muscles are involved in controlling the position of the head on the vertebral column?

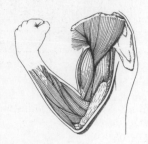

11

THE MUSCULAR SYSTEM:
The Appendicular Musculature

Objectives

1. Locate, identify, name, and describe the action of the muscles that position the pectoral girdle.

2. Locate, identify, name, and describe the action of the muscles of the upper limb that move: (a) the arm, (b) the forearm and hand, (c) and fingers.

3. Locate, identify, name, and describe the action of the muscles of the lower limb that move: (a) the thigh, (b) the leg, (c) the foot and toes.

The appendicular musculature is responsible both for stabilizing the pectoral and pelvic girdles and for moving the upper and lower limbs. [∞ p291] This musculature is separated into two distinct groups of muscles that stabilize and move the extremities: (1) the muscles of the upper limb (shoulder and upper limb), and (2) the muscles of the lower limb (pelvis and lower limb).

Muscles of the Upper Limb

There are four discrete groups of muscles of the upper limb: (1) muscles that position the pectoral girdle, (2) muscles that move the arm, (3) muscles that move the forearm and hand, and (4) muscles that move the fingers. [∞ p291] These groupings are merely generalizations and not universally accepted, because not all muscles fit conveniently into these selected categories. For example, many muscles that move the fingers also move the hand.

Muscles that Position and Move the Pectoral (Shoulder) Girdle

The *trapezius muscle* is the largest muscle involved in shoulder stabilization. The *rhomboideus muscles* and the *levator scapulae* lie deep to the trapezius and help stabilize the shoulder joint. Anteriorly, the shoulder is stabilized by both the *serratus anterior* and *pectoralis minor*. The trapezius and the serratus anterior also have an important function in rotating the scapula during some arm movements. [∞ p292]

Procedure

✓ **Quick Check** ──

Before you begin to examine the muscles of the upper limb, review the bone markings of the posterior of the skull (*Figure 6-4 p42*), cervical and thoracic vertebrae (*Figures 6-34* and *35 p62*), and clavicle (*Figure 7-1 p69*). Keep an articulated skeleton nearby to aid you in identifying sites of muscle attachment as you locate and identify muscles of the upper limb.

Locate and identify the **muscles** that position the shoulder girdle, using a *torso model* or *prosected cadaver specimen*. Use *Color Plates 42 to 44* and *Figure 11-1* for reference. The origin, insertion, and action of these muscles is presented in *Table 11-1* and should be reviewed and referred to as you proceed in your observations. Observe these muscles first on the torso and then on the cadaver. The cadaver must be in the prone position for observation of these muscles.

TABLE 11-1 Muscles That Move the Pectoral Girdle

Muscle	Origin	Insertion	Action	Innervation
Levator scapulae	Transverse processes of first 4 cervical vertebrae	Vertebral border of scapula near superior angle	Elevates scapula	Dorsal scapular nerve
Pectoralis minor	Ventral surfaces of ribs 3–5	Coracoid process of scapula	Depresses and protracts shoulder; rotates scapula so glenoid cavity moves inferiorly (downward rotation); elevates ribs if scapula is stationary	Medial pectoral nerve
Rhomboideus major	Spinous processes of upper thoracic vertebrae	Vertebral border of scapula from spine to inferior angle of scupula	Adducts and performs downward rotation	Dorsal scapular nerve
Rhomboideus minor	Spinous processes of vertebrae C_7–T_1	Vertebral border of scapula near spine	As above	As above
Serratus anterior	Anterior and superior margins of ribs 1–9	Anterior surface of vertebral border of scapula	Protracts shoulder, rotates scapula so glenoid cavity moves superiorly (upward rotation)	Long thoracic nerve
Subclavius	First rib	Clavicle	Depresses and protracts clavicle and shoulder	Subclavian nerve
Trapezius	Occipital bone, ligamentum nuchae, and spinous processes of thoracic vertebrae	Clavicle and scapula (acromion and scapular spine)	Depends on active region and state of other muscles; may elevate, retract, depress, or rotate scapula upward and/or elevate clavicle; can also	Accessory nerve (N XI) and cervical spinal nerves

The back is covered by two superficial muscles, *trapezius* and *latissimus dorsi.* The neck and most of the upper back is covered by the superficial, diamond-shaped **trapezius muscle**. It originates both at the nuchal lines and in the midline from C_1 through T_{12} spinous processes and inserts on the clavicle, acromion, and scapular spine. The inferior and medial edges of the trapezius muscle overlie the superior and medial edge of the latissimus dorsi muscle. Expose the borders of the trapezius and latissimus dorsi muscles. Note the location and position of these muscles and the fiber direction.

Reflect the trapezius muscle and observe the **rhomboideus major** and **rhomboideus minor muscles**, each resembling the shape of a collapsed square box. They originate on the spinous processes of the cervical and thoracic vertebrae and insert laterally and inferiorly on the vertebral border of the scapula. Deep to the reflected trapezius and superior to the rhomboideus minor lies the **levator scapulae muscle**. Also deep to the completely reflected trapezius, the splenius muscle overlies the superior border of the levator scapulae.

Return the cadaver to the supine position and observe the **pectoralis minor muscle**, which lies immediately deep to the *pectoralis major muscle*. The pectoralis minor originates on the ventral surfaces of ribs 3–5 and inserts on the coracoid process of the scapula. Superior to the pectoralis minor observe the **subclavius muscle**. It originates from the first rib and inserts on the inferior surface of the clavicle.

Position the arm at a right angle to the body. Reflect the pectoralis major superiorly towards the arm and view the pectoralis minor, serratus anterior, and subclavius muscles.

See: Color Plates 42 to 44, Fig. 11-01

∞ *Figs.*
11-1, p291
11-2, p292
11-4, p294

LOCATE

Viewed from the Posterior—Superficial and Deep Muscles

— Trapezius

— Levator scapulae

— Rhomboideus minor

— Rhomboideus major

Viewed from the Anterior—Deep Muscles

— Sternocleidomastoid

— Subclavius

— Pectoralis minor

— Serratus anterior

SUPERFICIAL   **DEEP**

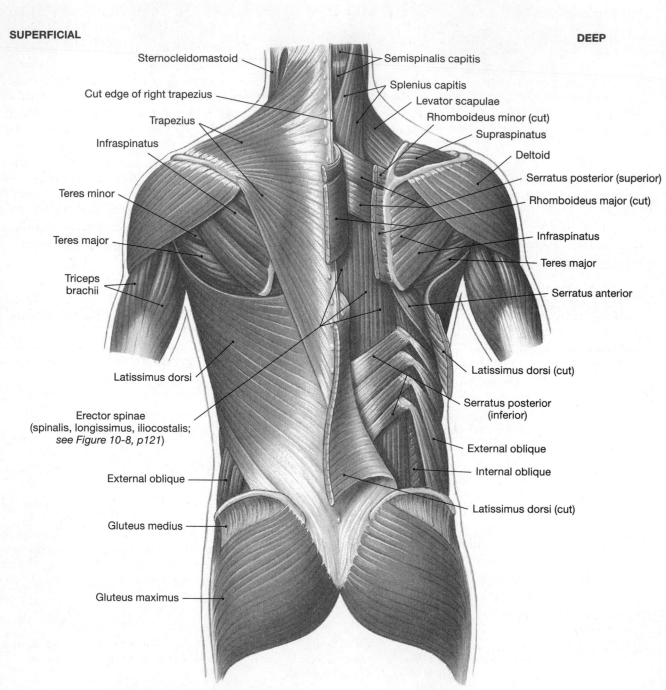

Sternocleidomastoid

Cut edge of right trapezius

Trapezius

Infraspinatus

Teres minor

Teres major

Triceps brachii

Latissimus dorsi

Erector spinae
(spinalis, longissimus, iliocostalis;
see Figure 10-8, p121)

External oblique

Gluteus medius

Gluteus maximus

Semispinalis capitis

Splenius capitis

Levator scapulae

Rhomboideus minor (cut)

Supraspinatus

Deltoid

Serratus posterior (superior)

Rhomboideus major (cut)

Infraspinatus

Teres major

Serratus anterior

Latissimus dorsi (cut)

Serratus posterior (inferior)

External oblique

Internal oblique

Latissimus dorsi (cut)

FIGURE 11-1 Superficial and deep muscles of the neck, shoulder, back, and gluteal regions, posterior view.

Muscles that Move the Arm

The muscles that move the arm lie on the chest and upper back. These can be grouped into muscles visible from the posterior and muscles visible from the anterior. Collectively, they provide the support for the loose glenohumeral joint. [∞ p295] The tendons of the *teres minor, subscapularis, infraspinatus*, and *supraspinatus* combine to form the *rotator cuff*, which adds immense strength and support to the shoulder.

Procedure

✓ Quick Check

Before you begin to examine the muscles that move the arm, review the bone markings of the ribs (*Figure 6-41 p66*), clavicle (*Figure 7-1 p69*), scapula (*Figure 7-2 p69*), and humerus. (*Figures 7-4* and *7-5 pp71-72*) Keep an articulated skeleton nearby to aid you in identifying sites of muscle attachment as you locate and identify muscles of the upper limb.

Locate and identify the following **arm muscles**, using an *arm torso model* or *prosected cadaver specimen*. Use *Color Plates 42* to *44*, and *Figures 11-2* and *11-3* for reference. The origin, insertion, and action of these muscles are presented in *Table 11-2* and should be reviewed and referred to as you proceed. Observe these muscles first on the torso before proceeding to the cadaver. To examine these muscles, the cadaver must be in prone position.

The **deltoid muscle** caps the shoulder and is the major abductor of the arm. Identify its triangular shape. Note its borders, as the muscle originates on the clavicle and scapula and inserts on the deltoid tuberosity of the humerus. The **latissimus dorsi muscle** is the wide superficial muscle of the lower chest and back. It originates on the *lumbodorsal fascia* of the spinous processes of the thoracic and lumbar vertebrae and inserts on the humerus. Note the location of the latissimus dorsi and its relationship to the trapezius muscle. Along with the trapezius, these two muscles occupy most of the back; however, remember the trapezius is a neck muscle and the latissimus dorsi is a powerful arm muscle, not a back muscle.

TABLE 11-2 Muscles that Move the Arm

Muscle	Origin	Insertion	Action	Innervation
Coracobrachialis	Coracoid process	Medial margin of shaft of humerus	Adducts and flexes arm	Musculocutaneous nerve
Deltoid	Clavicle and scapula (acromion and adjacent scapular spine)	Deltoid tuberosity of humerus	Abducts arm	Axillary nerve
Supraspinatus	Supraspinous fossa of scapula	Greater tubercle of humerus	Abducts arm	Suprascapular nerve
Infraspinatus	Infraspinous fossa of scapula	Greater tubercle of humerus	Rotates arm laterally	Suprascapular nerve
Subscapularis	Subscapular fossa of scapula	Lesser tubercle of humerus	Rotates arm medially	Subscapular nerve
Teres major	Inferior angle of scapula	Medial lip of the intertubercular groove of the humerus	Extends, adducts, and medially rotates arm	Lower subscapular nerve
Teres minor	Lateral (axillary) border of scapula	Greater tubercle of humerus	Lateral rotation of humerus	Axillary nerve
Triceps brachii (long head)	*See Table 11-3*			
Latissimus dorsi	Spinous processes of lower thoracic vertebrae, ribs 8–12, the spines of lumbar vertebrae, and the lumbodorsal fascia	Floor of the intertubercular groove of the humerus	Extends, adducts, and medially rotates arm	Thoracodorsal nerve
Pectoralis major	Cartilages of ribs 2–6, body of sternum, and inferior, medial portion of clavicle	Crest of greater tubercle of humerus (lateral lip of intertubercular groove)	Flexes, adducts, and medially rotates arm	Pectoral nerves

Reflect the trapezius muscle superiorly to permit easy observation of a group of muscles: the **supraspinatus, infraspinatus, subscapularis**, and **teres minor muscles**, which collectively compose the **rotator cuff**. The first three muscles are named according to their origin on the scapula. The small, round teres minor muscle lies inferior and lateral to the infraspinatus. The larger more triangular-shaped **teres major muscle** occupies most of the lateral area inferior to the teres minor. The tendons of these muscles are not easily separated.

Return the cadaver to the supine position to view the following muscles. The **pectoralis major muscle** is the large, triangular superficial chest muscle that covers most of the anterior chest wall. This muscle is a strong adductor and flexor of the arm. The pectoralis major overlies the pectoralis minor, serratus anterior, and subclavius muscles. These four muscles compose the muscles of the *pectoral region.* Review the underlying muscles on the supine cadaver. Position the arm at a right angle to the body, and reflect the pectoralis major superiorly towards the arm to permit observation of the underlying muscles.

The **coracobrachialis muscle** is very important in stabilizing the arm close to the body when lifting heavy objects. It originates from the coracoid process of the scapula and inserts on the medial surface of the humerus. This muscle lies medial to the *short head of biceps brachii* and can be easily traced from its point of insertion by maintaining the arm at a right angle to the body.

See: Color Plates 41 to 44, Figs. 11-2, 11-3

∞ *Figs. 11-1, p291 11-4, p294 11-5, p296*

LOCATE

Viewed from the Anterior

___ Deltoid

___ Pectoralis major

___ Coracobrachialis

___ Biceps brachii

Viewed from the Posterior

___ Deltoid

___ Latissimus dorsi

___ Rotator cuff muscles ⎰ Supraspinatus / Infraspinatus / Subscapularis / Teres minor

___ Teres major

___ Triceps brachii (long head)

___ Tendons of rotator cuff muscles

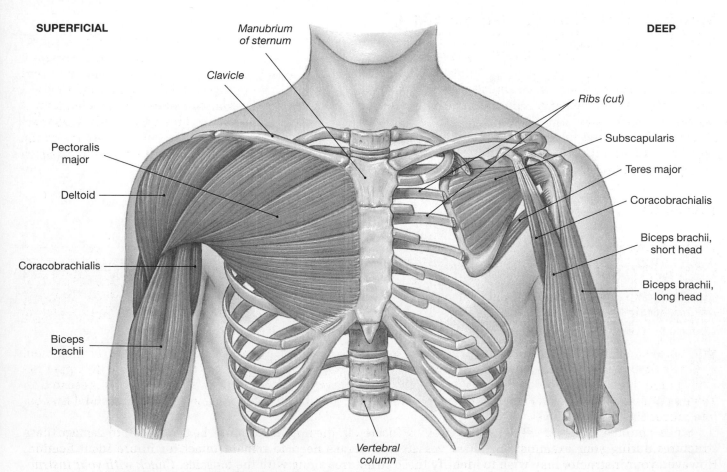

FIGURE 11-2 Muscles that move the arm, anterior view.

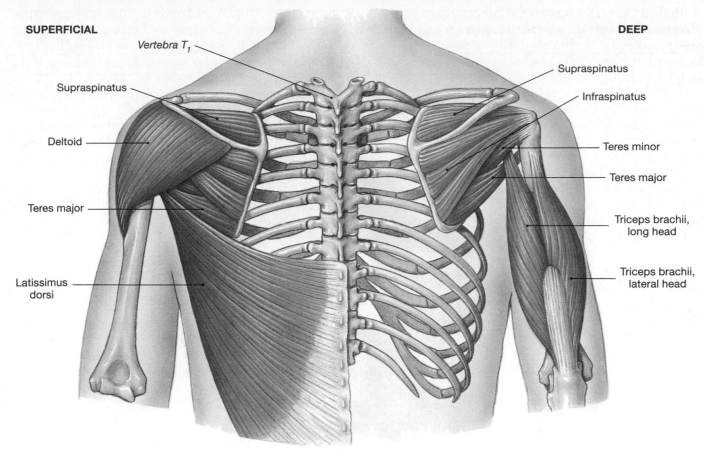

SUPERFICIAL

DEEP

Vertebra T₁

Supraspinatus

Supraspinatus

Infraspinatus

Deltoid

Teres minor

Teres major

Teres major

Triceps brachii, long head

Latissimus dorsi

Triceps brachii, lateral head

FIGURE 11-3 Muscles that move the arm, posterior view.

Muscles that Move the Forearm and Hand

Most of the muscles that move the forearm and hand originate on the humerus. [∞ p297] The exceptions to this are the *biceps brachii* and the *long head of the triceps brachii*. Both superficial and deep muscles of the forearm perform flexion and extension of the fingers. In the anatomical position, the flexors and the pronators insert and lie on the anterior surface of the forearm, while the extensors and supinators insert and lie on the posterior surface of the forearm. Flexor muscles will be much larger because they are stronger than extensors. The tendons of the flexor and extensor muscles must pass across the wrist to reach the fingers. Two broad flat bands of connective tissue, the *extensor retinaculum* and the *flexor retinaculum*, maintain the position of the tendons to the wrist. Just like a bracelet or wrist band, the retinacula aid in keeping the tendons in position as they perform flexion and extension movements of the hand and fingers.

Procedure

✓ **Quick Check**

Before you begin to examine the muscles that move the forearm and hand, you may wish to review the bone markings of the scapula (*Figure 7-2 p69*), humerus (*Figures 7-4* and *7-5 pp71-72*), ulna and radius. (*Figures 7-6* and *7-7 pp73-74*) Keep an articulated skeleton nearby to aid you in identifying sites of muscle attachment as you locate and identify muscles of the forearm and hand.

Locate and identify the following **arm muscles**, using an *arm torso model* or *prosected cadaver specimen*. Use *Color Plates 40, 42* to *44* and *Figures 11-4* to *11-7* for reference. The *arm* is that portion of the upper extremity between the shoulder and elbow. The origin, insertion, and action of arm muscles is presented in *Table 11-3* and should be reviewed and referred to as you proceed. Observe these muscles on the model first before proceeding to the cadaver.

Since a number of blood vessels and nerves are located in the upper extremity, be careful not to damage these structures during your examination. These vessels and nerves need to remain intact for future identification. However, your instructor may wish to identify these structures along with the muscles. *Check with your instructor before proceeding.*

TABLE 11-3   Muscles That Move the Forearm and Hand

Muscle	Origin	Insertion	Action	Innervation
PRIMARY ACTION AT THE ELBOW **Flexors**				
Biceps brachii	*Short head* from the coracoid process; *long head* from the supraglenoid tubercle (both on the scapula)	Tuberosity of radius	Flexes and supinates forearm; flexes arm	Musculocutaneous nerve
Brachialis	Anterior, distal surface of humerus	Tuberosity of ulna	Flexes forearm	As above
Brachioradialis	Lateral epicondyle of humerus	Lateral aspect of styloid process of radius	As above	Radial nerve
Extensors				
Anconeus	Posterior surface of lateral epicondyle of humerus	Lateral margin of olecranon on ulna	Extends forearm, moves ulna laterally during pronation	As above
Triceps brachii				
lateral head	Superior, lateral margin of humerus	Olecranon process of ulna	Extends forearm	As above
long head	Infraglenoid tubercle of scapula	As above	Extends and adducts arm	As above
medial head	Posterior surface of humerus inferior to radial groove	As above	Extends forearm	As above
PRONATORS/ SUPINATORS				
Pronator quadratus	Medial surface of distal portion of ulna	Anterolateral surface of distal portion of radius	Pronates forearm	Median nerve
teres	Medial epicondyle of humerus and coronoid process of ulna	Distal lateral surface of radius	As above	As above
Supinator	Lateral epicondyle of humerus and ulna	Anterolateral surface of radius distal to the radial tuberosity	Supinates forearm	Radial nerve
PRIMARY ACTION AT THE HAND **Flexors**				
Flexor carpi radialis	Medial epicondyle of humerus	Bases of 2nd and 3rd metacarpal bones	Flexes and abducts hand	Median nerve
Flexor carpi ulnaris	Medial epicondyle of humerus; adjacent medial surface of olecranon and anteromedial portion of ulna	Pisiform, hamate, and base of 5th metacarpal bone	Flexes and adducts hand	Ulnar nerve
Palmaris longus	Medial epicondyle of humerus	Palmar aponeurosis and flexor retinaculum	Flexes hand	Median nerve
Extensors				
Extensor carpi radialis, longus	Lateral supracondylar ridge of humerus	Base of 2nd metacarpal	Extends and abducts hand	Radial nerve
brevis	Lateral epicondyle of humerus	Base of 3rd metacarpal	As above	As above
Extensor carpi ulnaris	Lateral epicondyle of humerus; adjacent dorsal surface of ulna	Base of 5th metacarpal	Extends and adducts hand	Deep radial nerve

*See: Color Plates 40, 42
to 44, Fig. 11-4*

 ∞ *Fig.
11-6, p298*

Arm

With the cadaver in the supine position, reflect both the pectoralis major and deltoid muscles. The bulk of the anterior surface of the arm is formed by the **biceps brachii muscle**. Explore medially and observe the **short head** of the biceps, arising from the coracoid process of the scapula. On the lateral side, identify the **long head** of the biceps, which originates on the supraglenoid tubercle. Note the relationship of the *brachial artery* [∞ pp367, 584-585] and *median nerve* [∞ pp367-369] as they pass along the medial surface of the biceps. The **brachialis muscle** lies immediately deep to the biceps brachii and can be identified by separating it from the distal end of the biceps brachii, at the biceps tendon. Your fingers work well to perform this task. Work superiorly and note how the superficial fibers of the brachialis connect with the inferior fibers of the deltoid muscle. With the arm in the anatomical position, return to the biceps insertion tendon and locate on the lateral (thumb) side the **brachioradialis muscle**. The *radial nerve* [∞ p367-369] may be observed at this point lying deep between the brachialis and brachioradialis muscles.

On the posterior surface of the arm, identify the large **triceps brachii muscle**, which is named for its number of origins. The **long head** originates from the infraglenoid tubercle of the scapula. Observe how the long head extends between the teres minor and teres major muscles. The **lateral head** originates from the lateral surface of the proximal portion of the humerus and the **medial head** originates from the posterior margin of the humerus. With your fingers, separate the long and lateral heads. Be careful not to damage the *radial nerve* and *brachial artery*, which pass along the lateral head of the triceps muscle. The medial head is best observed by reflecting the cut edges of the lateral head (cut the belly at an oblique angle if necessary) of triceps brachii.

To understand the relation among muscles, blood vessels, and nerves, it is helpful to view these structures in transverse section. If a transverse section of the arm is available for viewing, these relations can easily be observed.

Forearm

The *forearm* is that portion of the upper extremity between the elbow and wrist. The forearm contains flexor and pronator muscles on the anterior side and extensor and supinator muscles on the posterior side. The tendons of most of these muscles (except for the brachioradialis, pronators, and supinators) cross the wrist (carpal bones) and enter the hand.

To Be Observed On Posterior Surface Of Forearm

The superficial extensor muscles of the posterior forearm are best observed and identified by working from the lateral to the medial side. Use your fingers as a probe for separating and examining the muscles. With the cadaver in the supine position, rotate the upper extremity to observe the posterior surface of the arm and forearm with the hand pronated. Begin your identification with the **brachioradialis muscle**. Use this muscle as a landmark from which to identify the superficial

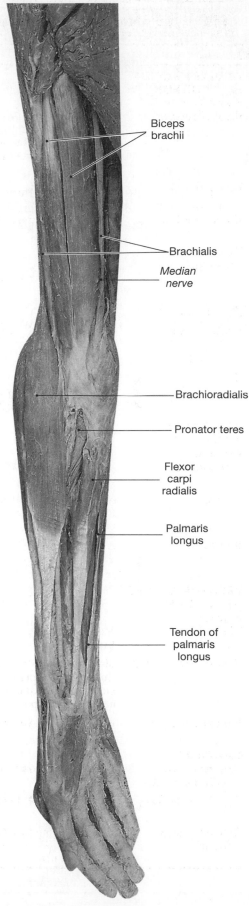

FIGURE 11-4 Forearm muscles, anterior view.

posterior muscles of the forearm. From the brachioradialis, work in a lateral direction to identify, in order, the **extensor carpi radialis longus, extensor carpi radialis brevis, extensor digitorum, extensor digiti minimi, extensor carpi ulnaris**, and the **anconeus**. Verify the extensor carpi ulnaris by tracing its tendon to the little finger side. The radialis pair is on the thumb side. Trace the course of each muscle and its tendon to the wrist and note the **extensor retinaculum**.

The deep muscle group of the posterior of the forearm may now be examined. Identify the **supinator muscle** in the area of the *cubital fossa* (*Figure 12-9 p189*) by separating the extensor carpi radialis brevis and extensor digitorum muscles. Now identify in order the **abductor pollicis longus, extensor pollicis longus, extensor pollicis brevis muscles**, and **extensor indicis**.

To Be Observed On The Anterior Surface Of The Forearm

Now rotate the upper limb to the anatomical position, viewing the anterior surface of the arm and forearm with the hand supinated. The superficial flexor muscles of the anterior forearm are best observed and identified by working from lateral to medial. Begin your identification at your landmark muscle, the brachioradialis, and work medially.

Identify in order the **pronator teres, flexor carpi radialis, palmaris longus** (not present in about 15% of the population), **flexor digitorum superficialis**, and **flexor carpi ulnaris muscles**. Trace the course of each muscle and tendon to the wrist and note the **flexor retinaculum**. Beginning in the area of the retinaculum, each tendon passes through a hollow tubular bursa, as it passes over the wrist and into the hand. The tendon and its bursa sleeve is termed a **tendon sheath** (or synovial sheath). Observe the four tendons of the flexor digitorum superficialis as they pass deep to the flexor retinaculum into the hand.

TABLE 11-4	Muscles That Move the Hand and Fingers			
Muscle	*Origin*	*Insertion*	*Action*	*Innervation*
Abductor pollicis longus	Proximal dorsal surfaces of ulna and radius	Lateral margin of 1st metacarpal bone	Abducts thumb	Deep radial nerve
Extensor digitorum	Lateral epicondyle of humerus	Posterior surfaces of the phalanges, fingers 2–5	Extends fingers and hand	As above
Extensor pollicis brevis	Shaft of radius distal to origin of adductor pollicis longus	Base of proximal phalanx of thumb	Extends thumb, abducts hand	As above
Extensor pollicis longus	Posterior and lateral surfaces of ulna and interosseous membrane	Base of distal phalanx of thumb	Extends thumb, abducts hand	As above
Extensor indicis	Posterior surface of ulna and interosseous membrane	Posterior surface of phalanges of little (5th) finger, with tendon of extensor digitorum	Extends and adducts little finger	As above
Extensor digiti minimi	Via extensor tendon to lateral epicondyle of humerus, and from intermuscular septa	Posterior surface of proximal phalanx of little finger	Extends little finger	As above
Flexor digitorum superficialis	Medial epicondyle of humerus; adjacent anterior surfaces of ulna and radius	Midlateral surfaces of middle phalanges of fingers 2–5	Flexes fingers, specifically middle phalanx on proximal; flexes hand	Median nerve
Flexor digitorum profundus	Medial and posterior surfaces of ulna, medial surface of coronoid process, and interosseous membrane	Bases of distal phalanges of fingers 2–5	Flexes distal phalanges and to a lesser degree the other phalanges and hand	Palmar interosseous nerve, from median nerve and ulnar nerve
Flexor pollicis longus	Anterior shaft of radius and interosseous membrane	Base of distal phalanx of thumb	Flexes thumb	Median nerve

The deep muscle group of the anterior forearm may now be examined. Identify in order, **flexor digitorum profundus, flexor pollicis longus,** and **pronator quadratus.** Trace the course of each muscle and tendon to the flexor retinaculum. (You will not be able to trace the pronator quadratus to the retinaculum.) Observe how the flexor digitorum profundus forms four tendons, just like the flexor digitorum superficialis. The flexor pollicis longus lies immediately lateral to the profundus. Return to the wrist and reflect the tendons of the flexor digitorum profundus in order to identify the pronator quadratus. At the cubital fossa, the *brachial artery* branches into the *radial artery* and *ulnar artery* [∞ pp583-585], as shown in *Color Plates 44* and *82* to *84.*

*See: Color Plates 42
to 44, 82 to 84, Figs.
11-5 to 11-7*

  ∞ *Figs.
11-1, p291
11-4, p294
11-6, p298-299
11-7, p300-301
11-8, p302*

LOCATE

Muscles (Superficial) Viewed on the Anterior Surface

___ Biceps brachii (long and short heads)

___ Brachialis

___ Brachioradialis

___ Pronator teres

___ Flexor carpi radialis

___ Palmaris longus

___ Flexor digitorum superficialis

___ Flexor carpi ulnaris

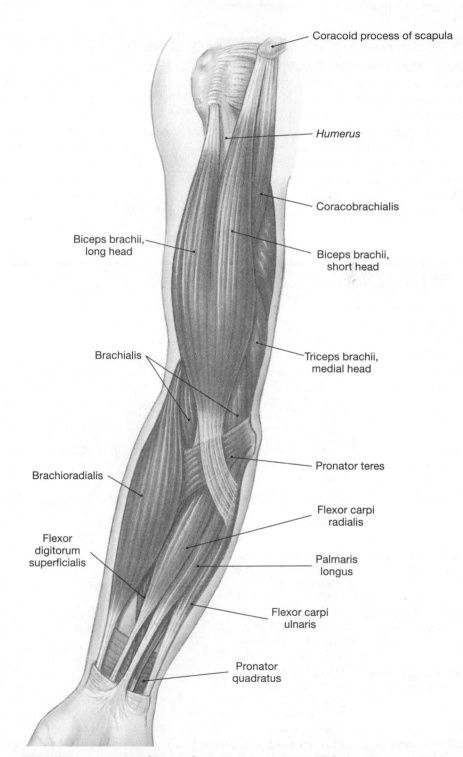

FIGURE 11-5   Muscles on the anterior aspects of the arm.

Muscles (Deep) Viewed on the Anterior Surface

___ Flexor digitorum profundus

___ Flexor pollicis longus

___ Pronator quadratus

Muscles (Deep) Viewed on the Posterior Surface

___ Anconeus

___ Supinator

___ Abductor pollicis longus

___ Extensor pollicis longus

___ Extensor pollicis brevis

___ Extensor indicis

___ Tendons of extensor digitorum

___ Tendons of extensor digiti minimi

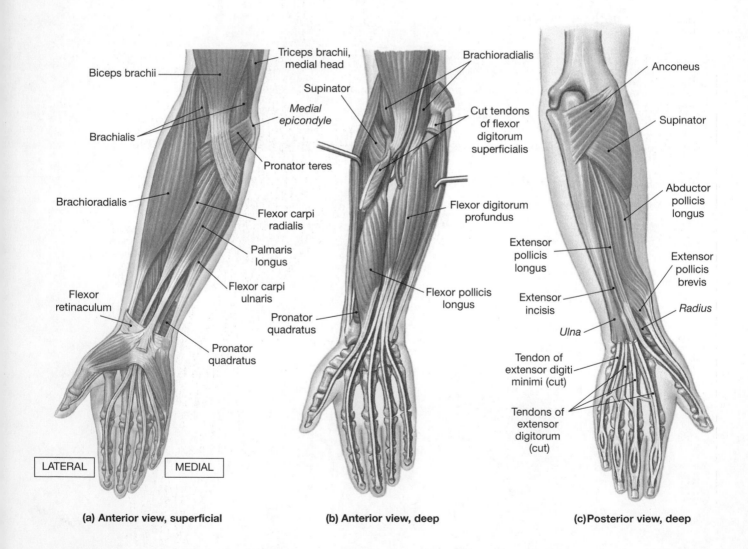

(a) Anterior view, superficial

(b) Anterior view, deep

(c) Posterior view, deep

FIGURE 11-6 Muscles that move the hand and fingers.

Muscles (Superficial) Viewed on the Posterior Surface

___ Triceps brachii, lateral head

___ Triceps brachii, long head

___ Brachioradialis

___ Extensor carpi radialis longus

___ Extensor carpi radialis brevis

___ Anconeus

___ Extensor digitorum

___ Extensor carpi ulnaris

___ Extensor digiti minimi

___ Abductor pollicis longus

___ Flexor carpi ulnaris

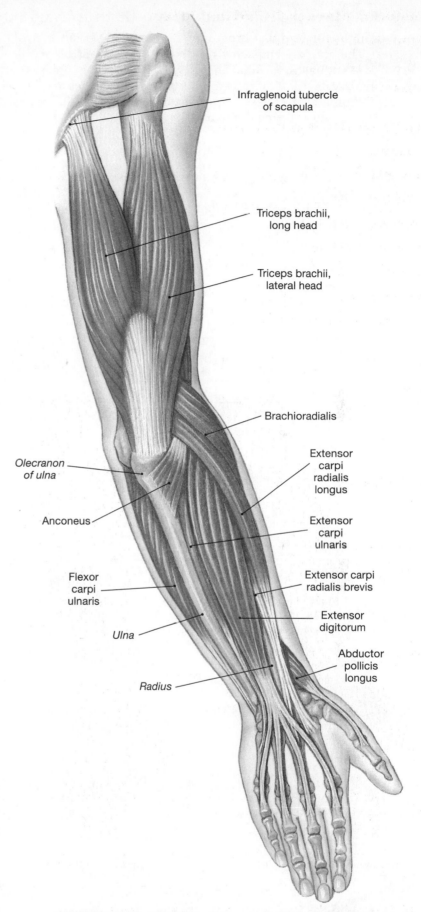

Infraglenoid tubercle of scapula

Triceps brachii, long head

Triceps brachii, lateral head

Brachioradialis

Olecranon of ulna

Anconeus

Extensor carpi radialis longus

Extensor carpi ulnaris

Extensor carpi radialis brevis

Flexor carpi ulnaris

Extensor digitorum

Abductor pollicis longus

Ulna

Radius

FIGURE 11-7 Muscles of the posterior aspect of the arm.

Muscles that Move the Hand and Fingers (Extrinsic and Intrinsic Groups)

The muscles of the forearm that provide strength and the basic control of the hand and fingers are termed *extrinsic muscles*. The *intrinsic muscles* provide the fine control of the hand and originate on the carpals and metacarpals. [⚭ p299] Superficial and deep muscles of the forearm perform flexion and extension of the fingers (as previously described). Again, flexor and pronator muscles are located on the anterior side of the forearm, and extensor and supinator muscles on the posterior side of the forearm with the tendons of most of these muscles crossing the wrist and entering the hand.

Procedure

Identify the following **intrinsic hand muscles**, using an *arm/hand torso model* or *prosected cadaver specimen*. Use *Color Plates 45* to *48* and *Figures 11-10, 11-11* for reference. The origin, insertion, and action of these muscles is presented in *Table 11-5* and should be reviewed and referred to as you proceed in your observation. View these muscles on the model first before proceeding to the cadaver.

To Be Observed On Anterior (Palmar Surface) Of Hand

Observe the muscles and tendons of the hand on the same side as you observed the forearm muscles. With the hand supinated (anterior view), return to the flexor retinaculum and note its medial and lateral attachments. At

TABLE 11-5	Intrinsic Muscles of the Hand			
Muscle	Origin	Insertion	Action	Innervation
Adductor pollicis	Metacarpal and carpal bones	Proximal phalanx of thumb	Adducts thumb	Ulnar nerve, deep branch
Opponens pollicis	Trapezium	First metacarpal bone	Opposition of thumb	Median nerve
Palmaris brevis	Palmar aponeurosis	Skin of medial border of hand	Moves skin on medial border toward midline of palm	Ulnar nerve, superficial branch
Abductor digiti minimi	Pisiform bone	Proximal phalanx of little finger	Abducts little finger and flexes its proximal phalanx	Ulnar nerve, deep branch
Abductor pollicis brevis	Transverse carpal ligament, scaphoid bone, and trapezium	Radial side of the base of the proximal phalanx of the thumb	Abducts the thumb	Median nerve
Flexor pollicis brevis*	Flexor retinaculum, trapezium, capitate bone, and ulnar side of first metacarpal bone	Ulnar side of the proximal phalanx of the thumb	Flexes and adducts the thumb	Branches of median and ulnar nerves
Flexor digiti minimi brevis	Hamate bone	Proximal phalanx of little finger	Flexes little finger	Ulnar nerve, deep branch
Opponens digiti minimi	Hamate bone	Fifth metacarpal	Opposition of fifth metacarpal bone	Ulnar nerve, deep branch
Lumbricals (4)	Tendons of flexor digitorum profundus	Tendons of extensor digitorum	Flexes metacarpophalangeal joints, extends middle and distal phalanges	#1 and #2 by median nerve, #3 and #4 by ulnar nerve, deep branch
Dorsal interossei (4)	Each originates from opposing faces of two metacarpal bones (I and II, II and III, III and IV, IV and V)	Bases of proximal phalanges of fingers 2–4	Abduct fingers 2–4 away from the midline axis of the middle finger (3), flex metacarpophalangeal joints, extend fingertips	Ulnar nerve, deep branch
Palmar interossei (3–4)*	Sides of metacarpal bones II, IV, and V	Bases of proximal phalanges of fingers 2, 4, and 5	Adduct fingers 2, 4, and 5 toward the midline axis of the middle finger (3), flex metacarpophalangeal joints, extend fingertips	Ulnar nerve, deep branch

*The portion of the flexor pollicis brevis originating on the first metacarpal is sometimes called the *first palmar interosseus* muscle.

the level of the proximal row of carpal bones and deep to the retinaculum, again identify the tendons of the flexor digitorum profundus, flexor digitorum superficialis, and flexor pollicis longus muscles, and the median nerve. Collectively, they comprise the contents of the **carpal tunnel**. The **median nerve** at this point lies superficial to the scaphoid bone and sandwiched between the flexor carpi radialis tendon and flexor digitorum superficialis tendons. Notice how the flexor retinaculum forms the "roof of the tunnel" through which these structures pass. (see *Color Plates 45, 46,* and *Figure 11-10 p159*)

To observe the muscles and tendons of the palm of the hand, the fascia and tough aponeurosis covering these structures must be removed. *If this has not been removed, check with your instructor before proceeding.* At the base of the thumb, reflect medially the abductor pollicis brevis to identify the underlying **opponens pollicis muscle**. Between thumb and index finger, identify the **adductor pollicis muscle**.

Starting at the most medial edge of the palm, identify the **abductor digiti minimi** muscle. Lateral to it, identify the **flexor digiti minimi brevis** muscle. Deep to the flexor is the **opponens digiti minimi** muscle. The thin **palmaris brevis** may need to be reflected, as it lies over the proximal borders of the abductor digiti minimi and flexor digiti muscles.

Using the tendons of the flexor digitorum profundus as landmarks, identify the four **lumbrical muscles** arising from these tendons. Deep to the lumbricals lie the three **palmar interossei muscles**: 1st medial to index finger, 2nd and 3rd lateral to ring and little fingers. Also deep to the lumbricals lie the four **dorsal interossei muscles**: 1st and 2nd lateral to index and middle fingers, 3rd and 4th medial to middle fingers.

To Be Observed On Posterior (Dorsum) Of Hand

Use the same hand to examine the muscles and tendons of the dorsum of the hand as was used to observe the palm of the hand structures. Use *Color Plate 47* and *Figure 11-11* for reference. With the hand supinated (posterior view), return to the extensor retinaculum and note its medial and lateral attachment points. Now identify, on the dorsum of the hand, the tendons of the extensor muscles previously observed and identified on the forearm. Identify the tendons of the extensor digitorum muscle as they insert on the dorsal surface of the phalanges. At the wrist in the area of the base of the thumb, identify the tendon of the extensor pollicis longus muscle. Lying under the pollicis tendon, at the wrist, are the tendons of the extensor carpi radialis longus and brevis muscles.

See: Color Plates 6, 45–47,
Figs. 11-7 to 11-11

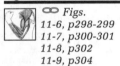 ∞ *Figs.*
11-6, p298-299
11-7, p300-301
11-8, p302
11-9, p304

LOCATE

Extrinsic Muscles of the Hand (Anterior and Posterior Views)

___ Flexor digitorum superficialis

___ Flexor digitorum profundus

___ Flexor pollicis longus

___ Abductor pollicis longus

___ Extensor digitorum

___ Extensor retinaculum

___ Flexor retinaculum

___ Extensor pollicis brevis

___ Extensor pollicis longus

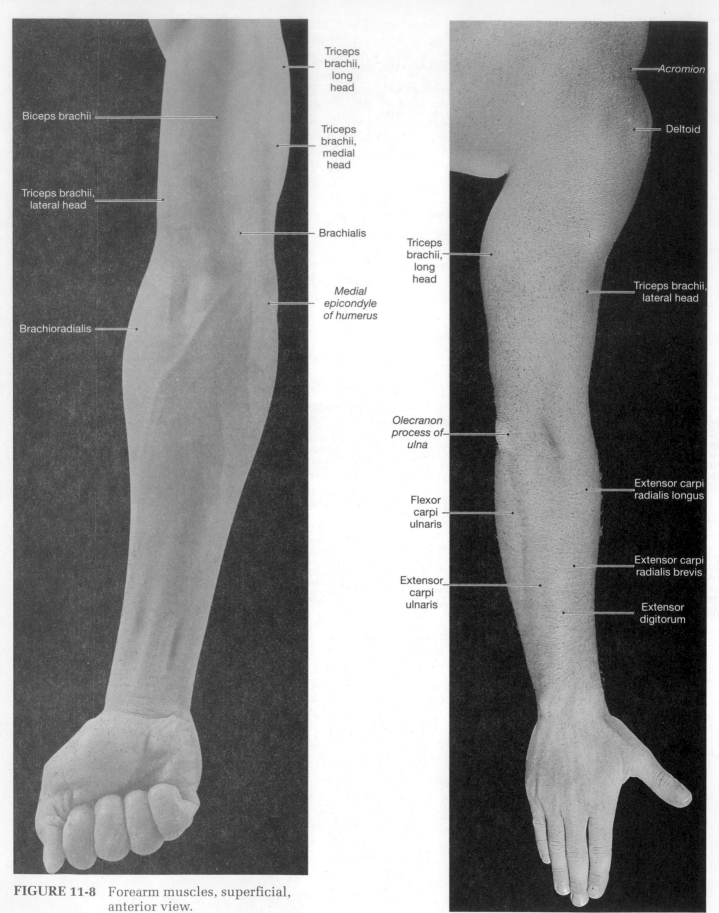

Biceps brachii

Triceps brachii, lateral head

Brachioradialis

Triceps brachii, long head

Triceps brachii, medial head

Brachialis

Medial epicondyle of humerus

FIGURE 11-8 Forearm muscles, superficial, anterior view.

Acromion

Deltoid

Triceps brachii, long head

Triceps brachii, lateral head

Olecranon process of ulna

Flexor carpi ulnaris

Extensor carpi ulnaris

Extensor carpi radialis longus

Extensor carpi radialis brevis

Extensor digitorum

FIGURE 11-9 Forearm muscles, superficial, posterior view.

Intrinsic Muscles of the Hand (Anterior and Posterior Views)

___ Abductor pollicis brevis

___ Adductor pollicis

___ Opponens pollicis

___ Palmaris brevis

___ Abductor digiti minimi

___ Flexor digiti minimi brevis

___ Opponens digiti minimi

___ Lumbricals (4)

___ Dorsal interossei (4)

___ Palmar interossei (3)

___ Flexor pollicis brevis

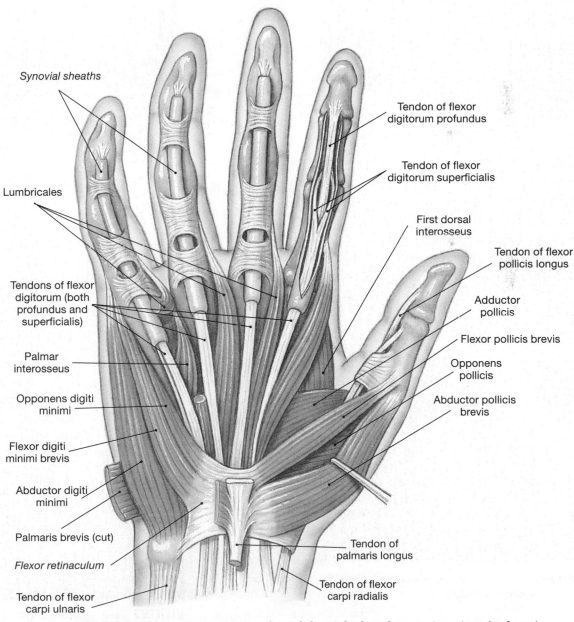

FIGURE 11-10 Intrinsic muscles of the right hand, anterior view (palmer).

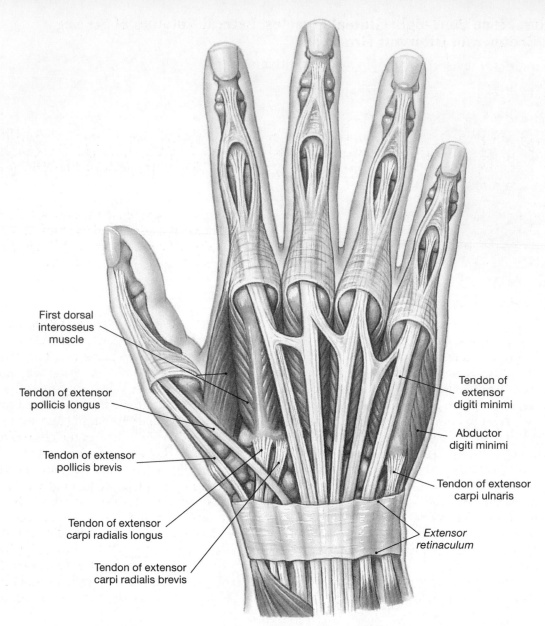

First dorsal interosseus muscle

Tendon of extensor pollicis longus

Tendon of extensor pollicis brevis

Tendon of extensor carpi radialis longus

Tendon of extensor carpi radialis brevis

Tendon of extensor digiti minimi

Abductor digiti minimi

Tendon of extensor carpi ulnaris

Extensor retinaculum

FIGURE 11-11   Intrinsic muscles of the right hand, posterior view.

Muscles of the Lower Limb

The lower limb is divided into regions and parts; namely, *gluteal region* or buttock, *thigh, knee, leg,* and *foot.* (*Figures 1-1* and *1-2 pp1-2*) The muscles of the lower limb [∞ p305] can be divided into four groups: (1) muscles that move the thigh, (2) muscles that move the leg, (3) muscles that move the foot and toes, and (4) muscles of the foot. Muscles originating on the pelvis move the thigh. These muscles can be conveniently grouped according to their orientation around the hip joint. The point of origin on the pelvis, as well as the insertion on the femur, will determine the movement permitted by each of these muscles. The muscles that move the leg are arranged similarly to those of the upper limb: extensors on the anterior surface (but on upper limb on the posterior surface) and flexors on the posterior surface (except for sartorius muscle).

Three muscular compartments surround the femur. The *medial compartment* adducts the thigh at the hip. The *anterior compartment* extends the knee, collectively called the *quadriceps femoris.* The posterior compartment contains the muscles to flex the knee, called the *hamstring muscles.*

Since a number of blood vessels and nerves are located in the lower limb, be careful not to damage these structures during your examination. These vessels and nerves need to remain intact for future identification. However, your instructor may wish to identify these structures along with the muscles. *Check with your instructor before proceeding.*

Muscles that Move the Thigh: Gluteal Muscles, Lateral Rotators of the Leg, Adductor Group, and Iliopsoas Group

Procedure

✓ Quick Check

Before you begin to examine the muscles of the thigh, review the bone markings of the coxae (*Figures 7-9 and 7-10 on pp76-77*), pelvis (*Figures 7-11* and *7-12 pp78-79*), and femur. (*Figures 7-13* and *7-14 pp80-81*) Keep an articulated skeleton nearby to aid you in identifying points of muscle attachment as you proceed in identifying muscles of the thigh.

Identify the following **thigh muscles**, using a *leg torso model* or *prosected cadaver specimen*. Use *Color Plates 39* to *41, 49* to *54,* and *Figures 11-12* to *11-17.* The origin, insertion, and action of the muscles that move the thigh are presented in *Tables 11-6* and *11-7* and should be reviewed and referred to as you proceed in your observation. Observe the muscles on the torso first before proceeding to the cadaver.

To Be Observed On The Posterior Surface Of The Thigh

With the cadaver in the prone position, the muscles that rotate the thigh can be examined. (see *Table 11-6*) Begin your observation at the gluteal region or buttock. The *gluteal muscles* can be observed to fill most of the gluteal surface and insert in various directions onto the femur. Use *Color Plate 42* and *Figure 11-17* for reference.

Identify the **gluteus maximus muscle** and trace its fibers to the borders as it arises from the lumbodorsal fascia, sacrum, and iliac crest. The iliac crest represents its most superior border. The maximus is the largest extender and lateral rotator of the thigh. Reflect the maximus (cut at the belly into medial and lateral portions) to expose underlying muscles and structures. Observe the **gluteus medius muscle** deep and superior to the maximus and note the oblique angle at which the fibers run. Deep to the belly of the medius lies the **gluteus minimus muscle**. The minimus muscle abducts and medially rotates the thigh, but the medius only abducts. The **piriformis muscle** is easily observed as it overlies the minimus just inferior to the medius, running horizontally from sacrum to greater trochanter of the femur. Inferior to the piriformis is the small somewhat triangular tendon of the **obturator internus muscle**, which arises on the internal surface of the obturator foramen to insert on the medial surface of the greater trochanter of the femur. The piriformis and obturator are lateral rotators of the thigh.

Identify on the lateral surface of the thigh the **tensor fasciae latae muscle**. The tensor lies anterior and lateral to the gluteus medius and arises from the anterior superior iliac spine and crest. Observe how the fibers of the tensor merge with those of the gluteus maximus, with both muscles inserting into the **iliotibial tract**. Several layers of tough dense fascia form the iliotibial tract, which runs vertically down the lateral thigh terminating just inferior to the lateral epicondyle of the tibia. The tract appears as a wide side stripe, like the stripe on tuxedo trousers.

To Be Observed On The Medial And Anterior Surfaces Of The Thigh

With the cadaver in the supine position, the adductor and flexor muscles of the thigh can be examined. (See *Table 11-6*) Use *Color Plates 39, 40, Figures 11-16* and *22-7* for reference. Identify the **gracilis muscle**, a long belt-like muscle that runs vertically along the medial surface of the thigh. It arises from the inferior rami of pubis and ischium to insert on the medial surface of the tibia. At the medial border of the femoral triangle, identify the wedge-shaped **pectineus muscle**, which arises from the superior border of the pubis. [*Fig 22-7 p324*; ∞ p592] Between the pectineus and gracilis muscles lies the **adductor longus muscle**. Reflect the adductor longus to identify the adductor brevis muscle. Reflect the adductor longus and the adductor brevis in order to identify the **adductor magnus muscle**. This muscle lies deep to the longus and lateral to the gracilis and is the largest adductor muscle.

The **iliopsoas** is a muscle of the groin and is formed at its origin inside the pelvis by two separate muscles, the **iliacus muscle** and **psoas muscle**. These muscles are clearly visible within the abdomen. The iliacus muscle can be observed easily in the thigh just medial to the *sartorius* and deep to the *femoral nerve.*

The **femoral triangle** is an area (*Figure 22-7 p324*) formed by three sides, a floor, and a roof: (1) at the superior side by the boundary of the inguinal ligament; (2) laterally, by the medial border of the sartorius muscle; (3) medially, by the lateral border of the adductor longus muscle; (4) the roof or anterior wall is formed by the fascia lata; and (5) the floor or posterior wall is formed by the adductor longus, pectineus and iliopsoas muscles. [∞ p592] Observe within the femoral triangle the **femoral nerve, femoral artery**, and **femoral vein**. Use *Color Plates 41* and *51* for reference. Begin your identification at the femoral nerve and proceed medially to identify the femoral artery and vein.

TABLE 11-6 Muscles That Move the Thigh

Group/Muscle	Origin	Insertion	Action	Innervation
Gluteal Group				
Gluteus maximus	Iliac crest of ilium, sacrum, coccyx, and lumbodorsal fascia	Iliotibial tract and gluteal tuberosity of femur	Extends and laterally rotates thigh	Inferior gluteal nerve
Gluteus medius	Anterior iliac crest of ilium, lateral surface between posterior and anterior gluteal lines	Greater trochanter of femur	Abducts and medially rotates thigh	Superior gluteal nerve
Gluteus minimus	Lateral surface of ilium between inferior and anterior gluteal lines	Greater trochanter of femur	Abducts and medially rotates thigh	As above
Tensor fasciae latae	Iliac crest and lateral surface of anterior superior iliac spine	Iliotibial tract	Flexes, abducts, and medially rotates thigh; tenses fascia lata, which laterally supports the knee	As above
Lateral Rotator Group				
Obturators (externus and internus)	Lateral and medial margins of obturator foramen	Trochanteric fossa of femur (externus); medial surface of greater trochanter (internus)	Laterally rotates thigh	Obturator nerve (externus) and special nerve from sacral plexus (internus)
Piriformis	Anterolateral surface of sacrum	Greater trochanter of femur	Laterally rotates and abducts thigh	Branches of sacral nerves
Gemelli (superior and inferior)	Ischial spine and tuberosity	Medial surface of greater trochanter	Laterally rotates thigh	Nerves to obturator internus and quadratus femoris
Quadratus femoris	Lateral border of ischial tuberosity	Intertrochanteric crest of femur	Laterally rotates thigh	Special nerve from sacral plexus
Adductor Group				
Adductor brevis	Inferior ramus of pubis	Linea aspera of femur	Adducts, medially rotates, and flexes thigh	Obturator nerve
Adductor longus	Inferior ramus of pubis anterior to brevis	As above	Adducts, flexes, and medially rotates thigh	As above
Adductor magnus	Inferior ramus of pubis posterior to adductor brevis and ischial tuberosity	Linear aspera and adductor tubercle of femur	Adducts thigh; superior portion flexes and medially rotates thigh, inferior portion extends and laterally rotates thigh	Obturator and sciatic nerves
Pectineus	Superior ramus of pubis	Pectineal line inferior to lesser trochanter of femur	Flexes, medially rotates, and adducts thigh	Femoral nerve
Gracilis	Inferior ramus of pubis	Medial surface of tibia inferior to medial condyle	Flexes leg, adducts, and medially rotates thigh	Obturator nerve
Iliopsoas Group				
Iliacus	Iliac fossa of ilium	Femur distal to lesser trochanter; tendon fused with that of psoas	Flexes hip and/or lumbar spine	Femoral nerve
Psoas major	Anterior surfaces and transverse processes of vertebrae T_{12}–L_5	Lesser trochanter in company with iliacus	As above	Branches of the lumbar plexus

See: Color Plates 39 to 41, 49 to 54, Figs. 11-12 to 11-15

 Figs.
11-1, p291
11-4, p294
11-10, p307
11-11, p308
11-12, p309
11-15, p314

LOCATE

Gluteal Muscle Group (Posterior and Lateral Views)

___ Gluteal group { Gluteus maximus   ___ Tensor fasciae latae
Gluteus medius   ___ Iliotibial tract (band)
Gluteus minimus

Lateral Rotator Group (Anterior and Posterior Views)

___ Obturators (internal and external)   ___ Piriformis

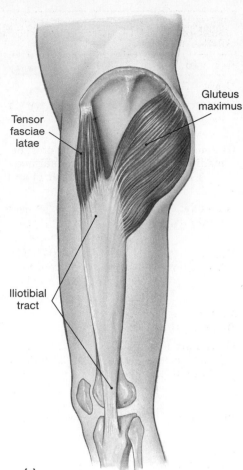

(a)

FIGURE 11-12 The gluteal muscles and lateral rotators of the thigh (a) lateral view, gluteal muscles (b) posterior view, deep muscles.

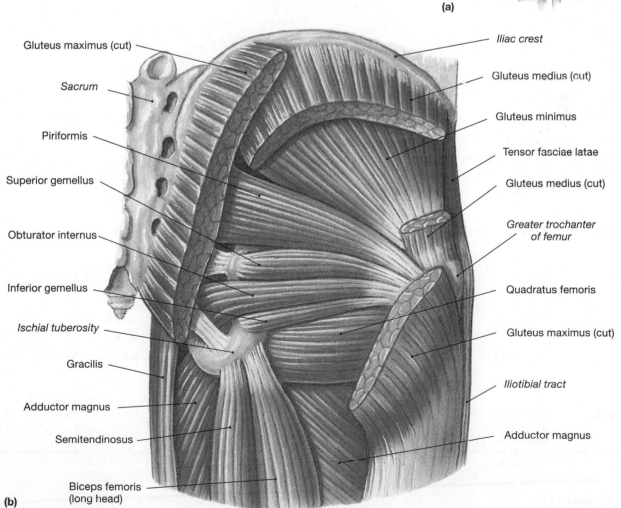

(b)

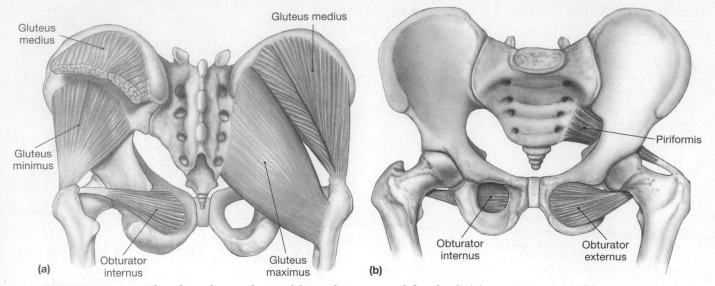

FIGURE 11-13 The gluteal muscles and lateral rotators of the thigh (a) posterior view (b) anterior view.

Adductor Group (Anterior and Posterior Views)

___ Adductor brevis

___ Adductor longus

___ Adductor magnus

___ Pectineus

___ Gracilis

___ Iliopsoas muscle ⎰ Iliacus
    group ⎱ Psoas major

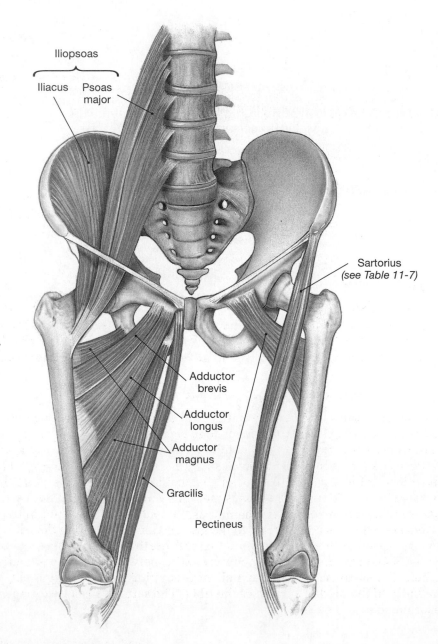

FIGURE 11-14 Adductor group, anterior view.

Muscles that Move the Leg: Flexors and Extensors

The muscles that move the leg originate on the thigh. [∞ p309] The flexor muscles of the leg are located on the posterior surface of the femur, with the exception of the *sartorius* muscle. The sartorius muscle originates on the anterior surface of the pelvis. The knee extensor muscles are collectively referred to as the *quadriceps femoris*. [∞ p309] The extensor group is composed of the *vastus lateralis, vastus medialis, rectus femoris*, and the *vastus intermedius*. All four of these muscles insert on the tibia through the **patellar ligament**.

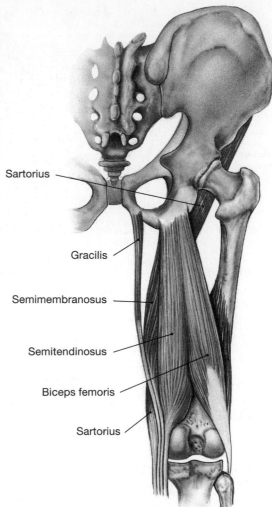

Sartorius

Gracilis

Semimembranosus

Semitendinosus

Biceps femoris

Sartorius

FIGURE 11-15 Flexor group of the leg, posterior view.

Procedure

Identify the following **leg muscles**, using a *leg torso model* or *prosected cadaver specimen*. The origin, insertion, and action of the muscles that move the leg are presented in *Table 11-7* and should be reviewed and referred to as you proceed in your observation. Observe the leg muscles first on the torso before proceeding to the cadaver. Use *Color Plates 39 to 40, 49 to 54*, and *Figures 11-15* and *11-22* for reference.

To Be Observed On The Anterior Surface Of The Thigh

With the cadaver in the supine position, the extensors of the knee can be observed. The four knee extensor muscles collectively known as the **quadriceps femoris** make up the bulk of the anterior thigh. These muscles show up as distinct masses superior to the **patella** as the muscles collectively merge to insert into the **patellar ligament**. Observe the **rectus femoris muscle** as it runs vertically in the center of the thigh. The majority of the muscle can be traced from the anterior inferior iliac spine to insert on the tibial tuberosity via the patellar ligament. Medially to the rectus, identify the **vastus medialis muscle,** which arises on the entire length of the linea aspera of the femur and inserts in the same manner as the rectus femoris. Lateral to the rectus, now identify the **vastus lateralis muscle**, which arises inferior to the greater trochanter along the linea aspera and inserts the same as the rectus femoris. Deep to the rectus femoris lies the fourth extensor of the leg, the **vastus intermedius muscle**. Reflect the rectus femoris and observe the intermedius between the vastus medialis and lateralis muscles. Identify the long belt-like **sartorius muscle**, which runs inferiorly at an oblique angle from the anterior superior iliac spine, medially to the medial surface of the tibia. The sartorius is a flexor and lateral rotator of the thigh and not an extensor muscle.

TABLE 11-7 Muscles That Move the Leg

Muscle	Origin	Insertion	Action	Innervation
Flexors of the Leg				
Biceps femoris	Ischial tuberosity and linea aspera of femur	Head of fibula, lateral condyle of tibia	Flexes leg, extends and laterally rotates thigh	Sciatic nerve; tibial portion (to long head) and common fibular branch (to short head)
Semimembranosus	Ischial tuberosity	Posterior surface of medial condyle of tibia	Flexes and medially rotates leg, extends thigh	Sciatic nerve (tibial portion)
Semitendinosus	Ischial tuberosity	Proximal, medial surface of tibia near insertion of gracilis	As above	As above
Sartorius	Anterior superior iliac spine	Medial surface of tibia near tibial tuberosity	Flexes leg, flexes and laterally rotates thigh	Femoral nerve
Popliteus	Lateral condyle of femur	Posterior surface of proximal tibial shaft	Medially rotates tibia (or laterally rotates femur)	Tibial nerve
Extensors of the Leg				
Rectus femoris	Anterior inferior iliac spine and superior acetabular rim of ilium	Tibial tuberosity via patellar ligament	Extends leg, flexes thigh	Femoral nerve
Vastus intermedius	Anterolateral surface of femur and linea aspera (distal half)	As above	Extends leg	As above
Vastus lateralis	Anterior and inferior to greater trochanter of femur and along linea aspera (proximal half)	As above	As above	As above
Vastus medialis	Entire length of linea aspera of femur	As above	As above	As above

To Be Observed On The Posterior Surface Of The Thigh

With the cadaver in the prone position, the muscles that flex the knee can be examined. Use *Color Plates 50 to 54* and *Figure 11-17* for reference. The *semimembranosus, semitendinosus,* and *biceps femoris* collectively make up the **hamstring muscles**. The hamstrings are discernible as separate flexor muscles on the posterior and medial surfaces of the thigh. These muscles simultaneously extend the hip and flex the knee, actions that occur together in walking. Collectively, the hamstrings originate on the ischial tuberosity and linea aspera of the femur, and insert around the tibia and fibula. By inserting around the knee, these muscles form a tripod and act as a knee stabilizer. Identify first the wide flat **semimembranosus muscle** as it descends on the posteromedial surface of the thigh. From the adductor magnus trace the semimembranosus as it descends to the medial condyle of the tibia. Now observe the cylindrical **semitendinosus muscle** as it lies over the semimembranosus muscle and descends in the middle of the posterior thigh. Reflect the gluteus maximus to observe the superior end of this muscle. Trace the semitendinosus inferiorly and note how its tendon inserts on the posteromedial surface of the tibia. The **biceps femoris muscle** lies somewhat parallel to these muscles as it descends on the posterolateral surface of the thigh. This muscle has two origins, a *long head* and a *short head*. The tendon of biceps femoris passes over the knee to insert laterally and anteriorly on the head of the fibula.

Superior to the popliteal fossa, the *sciatic nerve* divides into a *tibial nerve* and a *common fibular nerve*. [*Figures 14-8* and *14-9 pp211* and *213*; ∞ p372]. The tibial nerve can easily be observed deep to and between the biceps femoris and semitendinosus muscles. The femoral artery and vein continue into the leg respectively as the *popliteal artery* and *popliteal vein* and are easily observed at this location. [*Color Plates 88* and *90*; ∞ Figure 14-14, p372, 599] Parallel and slightly deep to the tibial nerve lies the popliteal vein; medial and parallel to the vein lies the popliteal artery. These nerves and blood vessels pass through the popliteal fossa into the leg.

To understand the relation between muscles, blood vessels, and nerves it is helpful to view these structures in transverse section. If a transverse section of the thigh [∞ Figure 11-12 p309] is available for viewing, these relations can easily be observed.

See: Color Plates 50 to 54, Figs. 11-16, 11-17

∞ *Figs.*
11-13, p310
11-14, p311-312
11-15, p314
14-14, p372

LOCATE

Extensors of the Leg (Anterior View)

__ Quadriceps femoris {
Rectus femoris
Vastus intermedius
Vastus lateralis
Vastus medialis
}

__ Sartorius (flexor)

Flexors of the Leg (Posterior View)

__ Biceps femoris   __ Sartorius

__ Semimembranosus   __ Popliteus

__ Semitendinosus

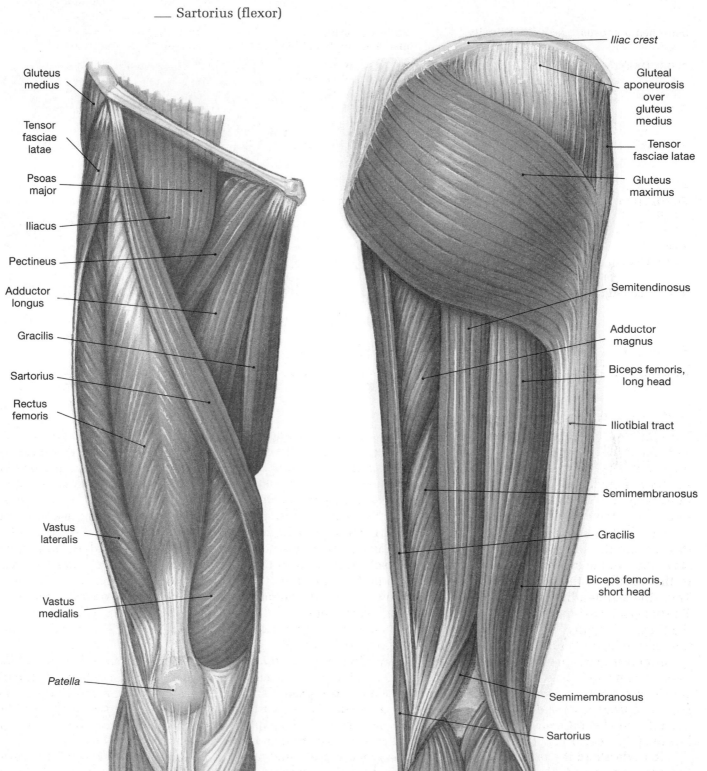

FIGURE 11-16 The quadriceps group, anterior view.

FIGURE 11-17 Thigh muscles, superficial, posterior view.

Muscles that Move the Foot and Toes

The muscles that move the foot and toes take part in the act of walking. [∞ p313] The muscles of the calf, *gastrocnemius* and *soleus*, allow for plantar flexion. These muscles share a single tendon, the *calcaneal tendon* or *Achilles tendon*. The muscles that move the toes originate on the tibia, fibula, or on both. The tendons of these muscles cross the ankle joint deep to stabilize the connective sheath bands, the *superior extensor retinaculum*, *inferior extensor retinaculum*, and *flexor retinaculum*. This arrangement of muscles is similar to those that move the hand.

Procedure

✓ Quick Check

Before you begin to examine the muscles that move the foot and toes, review the bone markings of the distal ends of the tibia and fibula (*Figures 7-16* and *7-17 pp83-84*), and ankle and bones of the foot. (*Figure 7-18 p85*) Keep an articulated skeleton nearby to aid you in identifying bony landmarks and points of muscle attachment as you proceed.

Identify the following **leg muscles, skeletal elements,** and **associated structures**, using a *leg torso model* or *prosected cadaver specimen*. The origin, insertion, and action of the muscles that move the foot and toes is presented in *Table 11-8*, and should be reviewed and referred to as you proceed in your observation. Observe these muscles first on the model before proceeding to the cadaver. Use *Color Plates 50 to 55* for reference.

TABLE 11-8 Extrinsic Muscles That Move the Foot and Toes

Muscle	Origin	Insertion	Action	Innervation
PRIMARY ACTION AT THE ANKLE **Dorsiflexor**				
Tibias anterior	Lateral condyle and proximal shaft of tibia	Base of 1st metatarsal bone and medial cuneiform bone	Dorsiflexes and inverts foot	Deep fibular nerve
Plantar flexors				
Gastrocnemius	Femoral condyles (medial and lateral)	Calcaneus via calcaneal tendon	Plantar flexes, inverts, and adducts foot; flexes leg	Tibial nerve
Fibularis brevis	Midlateral margin of fibula	Base of 5th metatarsal bone	Everts and plantar flexes foot	Superficial fibular nerve
longus	Lateral condyle of tibia, head and proximal shaft of fibula	Base of 1st metatarsal bone and medial cuneiform bone	Everts and plantar flexes foot; supports longitudinal arch	As above
Plantaris	Lateral supracondylar ridge	Posterior portion of calcaneus	Plantar flexes foot, flexes leg	Tibial nerve
Soleus	Head and proximal shaft of fibula, and adjacent posteromedial shaft of tibia	Calcaneus via calcaneal tendon (with gastrocnemius)	Plantar flexes, inverts, and adducts foot	Sciatic nerve, tibial branch
Tibialis posterior	Interosseous membrane and adjacent shafts of tibia and fibula	Tarsals and metatarsal bones	Adducts, inverts, and plantar flexes foot	As above
PRIMARY ACTION AT THE TOES **Flexors**				
Flexor digitorum longus	Posteromedial surface of tibia	Inferior surfaces of distal phalanges, toes 2–5	Plantar flexes toes 2–5	As above
Flexor hallucis longus	Posterior surface of fibula	Inferior surface, distal phalanx of great toe	Plantar flexes great toe	As above
Extensors				
Extensor digitorum longus	Lateral condyle of tibia, anterior surface of fibula	Superior surfaces of phalanges, toes 2–5	Extends toes 2–5	Deep fibular nerve
Extensor hallucis longus	Anterior surface of fibula	Superior surface, distal phalanx of great toe	Extends great toe	As above

To Be Observed On The Posterior Leg (Calf)

With the cadaver in the prone position, the muscles that move the foot and toes can be examined. [∞ p313] The superficial leg muscles, the gastrocnemius, soleus, and popliteus will be observed first. Use *Color Plates 50 to 53* for reference. Return to the popliteal fossa and at the inferior border of this area identify the **medial** and **lateral heads** of the **gastrocnemius muscle**. The heads originate respectively from just superior to the medial and lateral condyles of the femur. Trace the gastrocnemius inferiorly and observe how its fibers insert into the **calcaneal tendon** or **Achilles tendon**. The massive calcaneal tendon inserts on the posterior surface of the calcaneus bone. Deep to the gastrocnemius lies the thin, broad flat **soleus muscle**. The soleus arises from the head of the fibula and the posteromedial shaft of the tibia and is totally covered by the gastrocnemius, only its medial and lateral borders are exposed. Reflect the medial and lateral heads of the gastrocnemius (cut at their origins) and observe the soleus. Notice the soleus and gastrocnemius share a common tendon, the calcaneal tendon. The soleus muscle and both heads of the gastrocnemius form the muscular mass of the calf. On the medial surface of the gastrocnemius, the *saphenous nerve* [∞ p371] and *great saphenous vein* [*Color Plate 90* ∞ p599] may be observed.

The small slightly triangular **popliteus muscle** arises from the lateral femoral condyle within the fibrous capsule of the knee joint, crosses the knee joint at an oblique angle, from lateral to medial, to insert onto the proximal posterior surface of the tibia. Observe the popliteus deep to the lateral head of the gastrocnemius (reflected) within the popliteal fossa. The action of the popliteus is to rotate medially the tibia. The knee joint is locked during extension simply because there is more travel on the medial condyle than on the lateral condyle, which accounts for medial rotation of the femur on the tibia.

To Be Observed On The Medial Surface of The Leg

The deep calf muscles are primarily concerned with plantar flexion of the foot and flexion of the toes and send their tendons across the ankle into the foot. These muscles are best observed by reflecting the gastrocnemius and soleus. Use *Color Plates 51, 52,* and *Figures 11-20* and *11-22.* Keep these muscles reflected as you observe the deep muscles from medial to lateral sides of the leg. Begin on the medial side and identify the **flexor digitorum longus**. Trace its tendon as it passes just posterior to the medial malleolus at the ankle and under the **extensor retinaculum**, inserting on the inferior surfaces of toes 2–5. Now identify the **tibialis posterior**, which lies immediately lateral and runs parallel to the flexor digitorum longus. Trace its tendon and note that it runs a course similar to the flexor, but inserts at numerous points on the plantar surface of the foot. The *tibial nerve* [*Figures 14-8* and *14-9 pp211, 213;* ∞ pp371-372] may be observed on the surface of the tibialis posterior. Identify the **flexor hallucis longus** as it lies on the posterior surface of the fibula, just lateral to the tibialis posterior. Trace the tendon of the flexor hallucis longus and note that it runs posterior to the other muscles, inserting on the inferior surface of the distal phalanx of the great toe.

To Be Observed On The Anterior And Lateral Surfaces Of The Leg

With the cadaver in the supine position the muscles that move the foot and toes can be examined [∞ p313]. The anterior of the leg has very little musculature. Use *Color Plates 49, 51, 53* and *Figures 11-20* to *11-22* for reference. Tendons of the muscles that move the foot and toes must cross the ankle and pass deep to the retinacula. Identify on the anterior surface of these tendons, just superior to the ankle, the connective tissue fibers of the **superior extensor retinaculum**. Just inferior to the ankle identify the **inferior extensor retinaculum** as it arises from the lateral surface of the calcaneus and passes superiorly to overlie the tendons of the extensor digitorum longus and extensor hallucis longus.

On the anterolateral side of the leg identify the large **tibialis anterior muscle** as it runs inferiorly and parallel to the lateral surface of the tibial shaft. Trace the muscle and observe how it arises from the shaft and interosseous membrane to form a large tendon that crosses the anterior surface of the ankle. The tendon passes medially deep to the superior and inferior extensor retinacula to insert on the base of the first metatarsal.

The superficial **extensor hallucis longus muscle** lies lateral and somewhat deep to the tibialis anterior. Reflect the tibialis anterior medially and observe the extensor as it arises from the anterior medial surface of the fibula, then trace its tendon inferiorly. The **extensor digitorum longus** is best identified just lateral to the tendon of the tibialis anterior. Trace the muscle superiorly to the lateral condyle of the tibia. Now trace the muscle inferiorly and observe its tendon. The extensor tendons cross the ankle to pass deep to the superior and inferior extensor retinacula. The tendon of the extensor digitorum longus divides into four tendons to insert on the dorsal surfaces of toes 2-5.

Identify on the lateral surface of the leg the superficial **fibularis longus muscle**. Trace this muscle from the lateral condyle of the tibia to its tendon, which crosses the ankle posterior to the lateral malleolus to then pass under the inferior peroneal retinaculum to the base of the first metatarsal. Superior to the ankle the bulk of the **fibularis brevis muscle** can be identified under the tendon of the longus. The tendon of the brevis follows the same course, but inserts on the base of the fifth metatarsal. The *common fibular nerve* [*Figures 14-8* and *14-9 pp211-213*; ∞ p372] can be observed between the peroneus muscles near the neck of the fibula.

To understand the relation between muscles, blood vessels, and nerves, it is helpful to view the leg structures in transverse section.

See: Color Plates 50 to 55, Figs. 11-18 to 11-22

∞ *Figs.*
11-15a,b, p314
11-15c,d, p315
11-16, p316
11-17, p317

LOCATE

Superficial Muscles (Posterior Surface of the Leg)

___ Gastrocnemius (medial and lateral heads)

___ Soleus

___ Calcaneal (Achilles) tendon

___ Popliteus

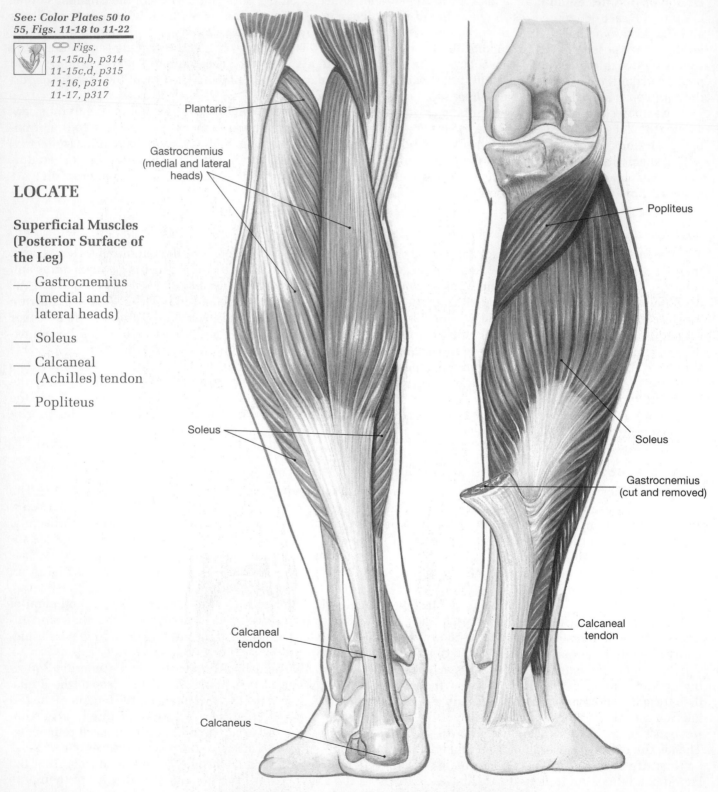

Plantaris

Gastrocnemius (medial and lateral heads)

Popliteus

Soleus

Soleus

Gastrocnemius (cut and removed)

Calcaneal tendon

Calcaneal tendon

Calcaneus

FIGURE 11-18 Superficial leg muscles that primarily perform plantar flexion, posterior view.

Deep Muscles (Posterior Surface of the Leg)

___ Fibularis longus

___ Fibularis brevis

___ Flexor hallucis longus

___ Tibialis posterior

___ Flexor digitorum longus

___ Tendon of fibularis brevis

___ Tendon of fibularis longus

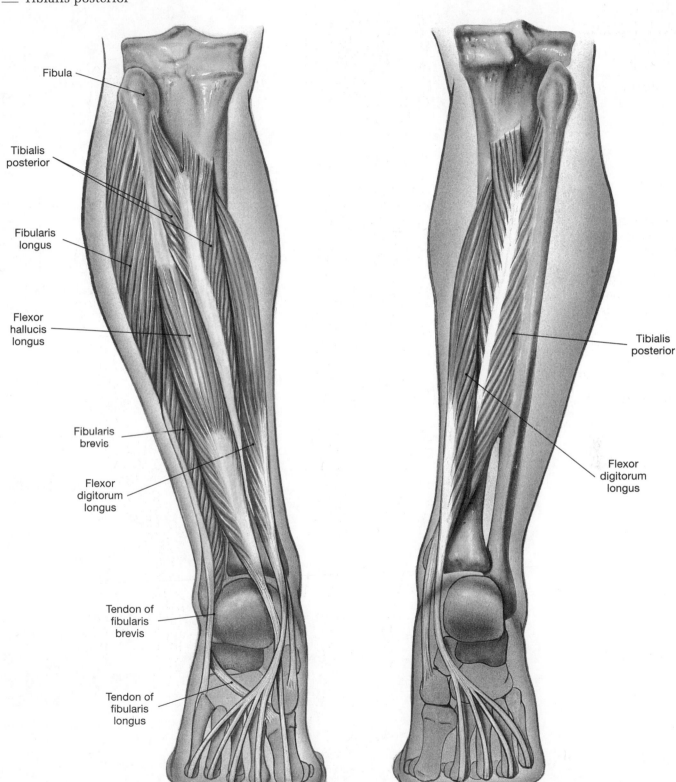

Fibula

Tibialis posterior

Fibularis longus

Flexor hallucis longus

Fibularis brevis

Flexor digitorum longus

Tendon of fibularis brevis

Tendon of fibularis longus

Tibialis posterior

Flexor digitorum longus

FIGURE 11-19 Deep leg muscles that primarily perform flexion of the foot and toes, posterior view.

Superficial Muscles of the Leg (Lateral View)

___ Tibialis anterior

___ Fibularis longus

___ Fibularis brevis

___ Extensor digitorum longus

___ Soleus

___ Lateral head of gastrocnemius

___ Calcaneal tendon

___ Inferior extensor retinaculum

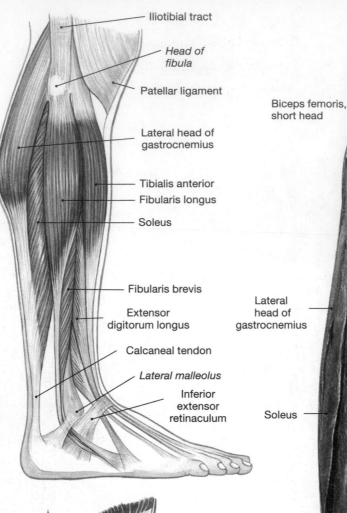

Iliotibial tract

Head of fibula

Patellar ligament

Lateral head of gastrocnemius

Tibialis anterior

Fibularis longus

Soleus

Fibularis brevis

Extensor digitorum longus

Calcaneal tendon

Lateral malleolus

Inferior extensor retinaculum

(a)

Superficial Muscles of the Leg (Medial View)

___ Patellar tendon (ligament)

___ Tibialis anterior

___ Soleus

___ Gastrocnemius

___ Calcaneal tendon

___ Tendon of tibialis anterior

Patella

Medial condyle of tibia

Patellar ligament

Medial surface of tibial shaft

Gastrocnemius

Tibialis anterior

Soleus

Calcaneal tendon

Medial malleolus

Tibialis anterior tendon

FIGURE 11-20 Leg muscles, (a) lateral view (b) medial view. **(b)**

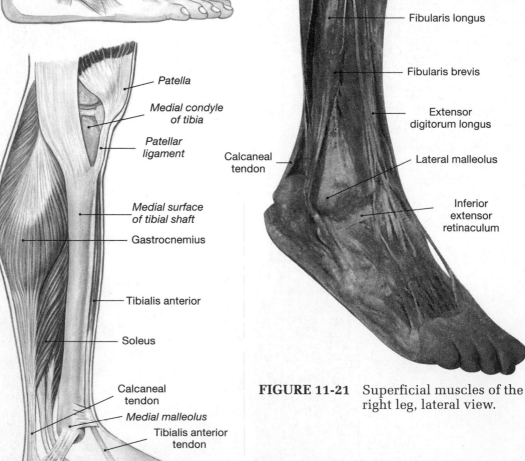

Iliotibial tract

Biceps femoris, short head

Patella

Head of fibula

Patellar ligament

Lateral head of gastrocnemius

Soleus

Tibialis anterior

Fibularis longus

Fibularis brevis

Extensor digitorum longus

Calcaneal tendon

Lateral malleolus

Inferior extensor retinaculum

FIGURE 11-21 Superficial muscles of the right leg, lateral view.

Anterior Surface (Lower Leg)

___ Patellar tendon (ligament)   ___ Extensor hallucis longus

___ Fibularis longus   ___ Superior extensor retinaculum

___ Tibialis anterior   ___ Inferior extensor retinaculum

___ Extensor digitorum longus

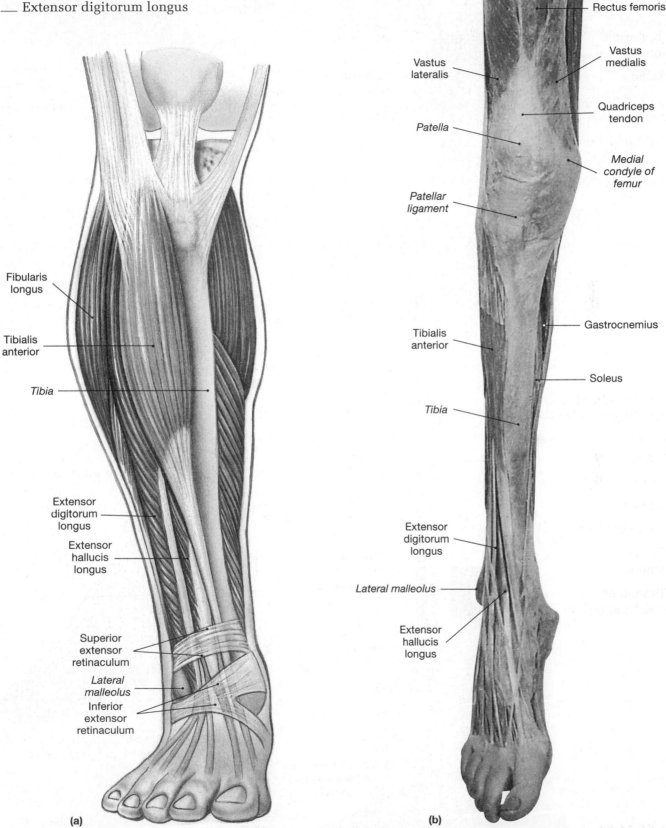

FIGURE 11-22 Anterior superficial leg muscles primarily concerned with dorsiflexion of the foot and extension of the toes (a) diagrammatic view (b) cadaver.

Muscles of the Foot

Procedure

✓ **Quick Check**

Before you begin to examine the muscles that move the foot and toes, review the bone markings of the distal end of the tibia and fibula (*Figures 7-16* and *7-17 pp83-84*), and ankle and foot. (*Figure 7-18 p85*) Keeping an articulated skeleton nearby will aid you in identifying bony landmarks and points of muscle attachment as you proceed.

Identify the following **muscles of the foot** and **associated structures**, using a *model* of the *foot* or *prosected cadaver specimen*. The origin, insertion, and action of the muscles of the foot are presented in *Table 11-9*, and should be reviewed and referred to as you proceed in your observation. Observe these muscles on the model first before proceeding to the cadaver. Use *Color Plates 53, 55,* and *Figures 11-23* and *11-24* for reference.

To Be Observed On The Dorsum Of The Foot

Review on the dorsum of the foot the inferior extensor retinaculum and the tendons of the extensor digitorum longus. Use *Figure 11-23* for reference. Lateral and deep to these tendons lies the thin flat **extensor digitorum brevis muscle**. This muscle arises from the calcaneus at about a 30 degree angle and forms four tendons, which insert on the dorsal surfaces of toes 1–4. The dorsal *interossei muscles* may be observed from the dorsum of the foot, between the tendons of the extensor digitorum brevis, but will be described and identified with the muscles of the sole of the foot.

To Be Observed On The Plantar Surface (Sole) Of The Foot

Before the muscles of the plantar surface of the foot can be observed, the dense **plantar aponeurosis** must first be removed. Use *Figure 11-24* for reference. If this has not been removed, check with your instructor before proceeding. The muscles that form the sole of the foot are categorized into four layers, one through four. Begin at the most superficial, the first layer of muscles, which lie deep to the plantar aponeurosis. Running down the center of the sole is the large **flexor digitorum brevis muscle**. This muscle arises from the calcaneus and plantar

---

TABLE 11-9   Intrinsic Muscles of the Foot

Muscle	Origin	Insertion	Action	Innervation
Extensor digitorum brevis	Calcaneus (superior and lateral surfaces)	Dorsal surfaces of toes 2-4	Extends proximal phalanges of toes 2-4	Deep fibular nerve
Abductor hallucis	Calcaneus (tuberosity on inferior surface)	Medial side of proximal phalanx of great toe	Abducts great toe	Medial plantar nerve
Flexor digitorum brevis	As above	Sides of middle phalanges, toes 2-5	Flexes the middle phalanx of toes 2-5	As above
Abductor digiti minimi	As above	Lateral side of proximal phalanx, little toe	Abducts the little toe	Lateral plantar nerve
Quadratus plantae	Calcaneus (medial surface and lateral border of inferior surface)	Tendon of flexor digitorum longus	Flexes toes 2-5	As above
Lumbricals (4)	Tendons of flexor digitorum longus	Insertions of extensor digitorum longus	Flexes proximal phalanges, extends middle phalanges, toes 2-5	Medial plantar nerve (1), lateral plantar nerve (2-4)
Flexor hallucis brevis	Cuboid and lateral cuneiform bones	Proximal phalanx of great toe	Flexes proximal phalanx of great toe	Medial plantar nerve
Adductor hallucis	Bases of metatarsals II-IV and plantar ligaments	Proximal phalanx of great toe	Adducts the great toe	Lateral plantar nerve
Flexor digiti minimi brevis	Base of 5th metatarsal bone	Lateral side of proximal phalanx of little (5th) toe	Flexes proximal phalanx of little toe	As above
Interossei dorsal (4)	Sides of metatarsal bones	Sides of toes 2-4	Abducts the toes	As above
plantar (3)	Bases and medial sides of metatarsal bones	Sides of toes 3-5	Adducts the toes	As above

aponeurosis and forms four tendons. Trace the tendons and observe how they insert on the sides of the middle phalanges of toes 1–4. Medial to this muscle identify the **abductor hallucis muscle**. Trace the muscle distally from the calcaneus and observe the insertion of its tendon on the proximal phalanx of the great toe. Return to the flexor digitorum brevis and proceed laterally to identify the **abductor digiti minimi muscle**. Observe how this muscle forms the lateral surface of the sole. Trace this muscle distally from the calcaneus and observe the insertion of the tendon on the proximal phalanx of the little toe.

Deep to the first muscle layer lies the second layer of muscles. The second layer is characterized by the long flexor tendons. To view layers two through four, keep retracted the cut portions of the previously identified flexor digitorum brevis, abductor hallucis, and abductor digiti minimi muscles. Locate the tendon of the flexor digitorum longus and observe deep to the tendon the **quadratus plantae muscle**. Trace the muscle from its origin, the plantar surface of the calcaneus, to its insertion into the lateral border of the tendon of the flexor digitorum longus. Return to the central tendon of the flexor digitorum longus and note how it divides into four tendons. Between these tendons observe four **lumbrical muscles**, arising from the tendons of the flexor digitorum longus.

Return to the medial side of the sole, proximal to the great toe. In the third layer the **flexor hallucis brevis muscle** can be identified as it arises from the cuboid and 3rd cuneiform bones. The muscle lies deep to and parallels the tendon of the flexor hallucis longus. Examine the divided belly of this muscle and observe a medial and a lateral part, inserting respectively on the medial and lateral surfaces of the proximal phalanx of the great toe. The **adductor hallucis muscle** consists of two small heads, arising from the bases of metatarsals 2–4 and plantar ligaments. The smaller head runs from lateral to medial at the bases of the metatarsals to join with the larger head, which arises from the bases of metatarsal bones 2–4. The heads join to insert on the lateral side of the proximal phalanx of the great toe.

Associated with the metatarsals are seven *interossei muscles*, which compose the fourth muscle layer. Using the tendon of the peroneus longus as a landmark, proceed proximally to the toes to view the interossei muscles. The four cigar-shaped **dorsal interossei muscles** (termed first dorsal interosseous muscle, etc.) arise from the sides of the metatarsals to insert on the sides of toes 2–4. The three **plantar interossei muscles** (termed first plantar interosseous muscle, etc.) arise from the bases and medial surface of metatarsals 3–5 to insert on sides of toes 3–5. The dorsal interossei abduct the toes and the plantar interossei muscles adduct the toes. Return to the dorsum of the foot and identify between the tendons of the extensor digitorum brevis, the dorsal interossei muscles.

See: Color Plates 53, 55, Figs. 11-23, 11-24

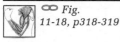

∞ *Fig. 11-18, p318-319*

LOCATE

Dorsal Surface

___ Superior/inferior extensor retinaculum

___ Extensor digitorum brevis

___ Abductor hallucis

___ Interossei dorsal (4)

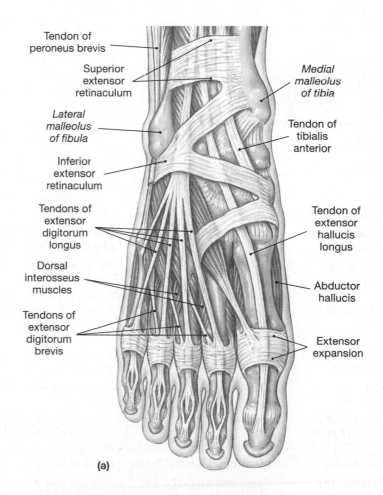

Tendon of peroneus brevis

Superior extensor retinaculum

Lateral malleolus of fibula

Inferior extensor retinaculum

Tendons of extensor digitorum longus

Dorsal interosseus muscles

Tendons of extensor digitorum brevis

Medial malleolus of tibia

Tendon of tibialis anterior

Tendon of extensor hallucis longus

Abductor hallucis

Extensor expansion

(a)

FIGURE 11-23 Muscles that move the foot and toes (a) dorsal view.

FIGURE 11-23 (Continued)   Muscles that
move the foot and toes (b) muscles,
tendons, and nerves, dorsal view.

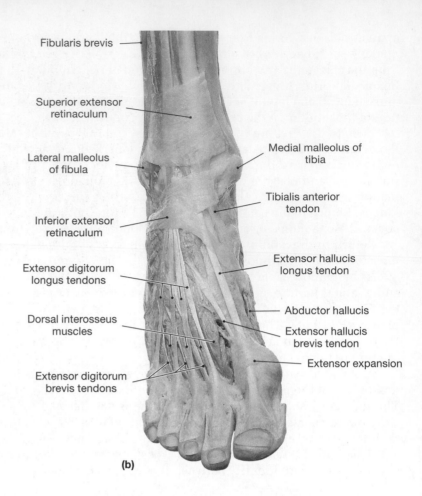

Fibularis brevis

Superior extensor
retinaculum

Lateral malleolus
of fibula

Medial malleolus of
tibia

Inferior extensor
retinaculum

Tibialis anterior
tendon

Extensor digitorum
longus tendons

Extensor hallucis
longus tendon

Dorsal interosseus
muscles

Abductor hallucis

Extensor hallucis
brevis tendon

Extensor expansion

Extensor digitorum
brevis tendons

(b)

Plantar View (Superficial and Deep Layers)

___ Plantar aponeurosis

___ Calcaneus

___ Abductor hallucis

___ Flexor digitorum brevis

___ Flexor digiti minimi

___ Abductor digiti minimi

___ Quadratus plantae

___ Lumbricals (4)

___ Flexor hallucis brevis

___ Adductor hallucis

___ Interossei, dorsal (4)

___ Interossei, plantar (3)

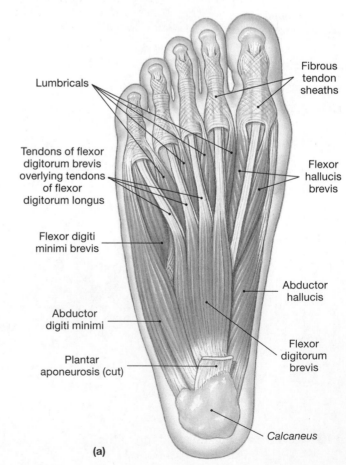

Lumbricals

Fibrous
tendon
sheaths

Tendons of flexor
digitorum brevis
overlying tendons
of flexor
digitorum longus

Flexor
hallucis
brevis

Flexor digiti
minimi brevis

Abductor
hallucis

Abductor
digiti minimi

Flexor
digitorum
brevis

Plantar
aponeurosis (cut)

Calcaneus

(a)

FIGURE 11-24   Muscles that move the foot and toes
(a) superficial layer, plantar view.

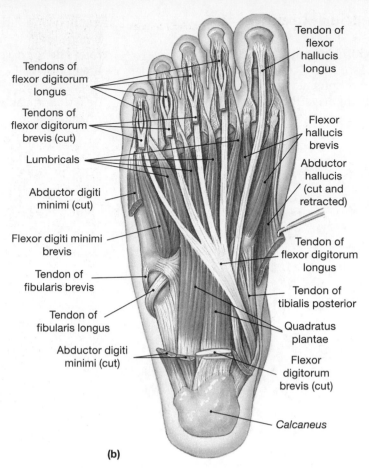

Tendon of flexor hallucis longus

Tendons of flexor digitorum longus

Tendons of flexor digitorum brevis (cut)

Lumbricals

Abductor digiti minimi (cut)

Flexor digiti minimi brevis

Tendon of fibularis brevis

Tendon of fibularis longus

Abductor digiti minimi (cut)

Flexor hallucis brevis

Abductor hallucis (cut and retracted)

Tendon of flexor digitorum longus

Tendon of tibialis posterior

Quadratus plantae

Flexor digitorum brevis (cut)

Calcaneus

(b)

FIGURE 11-24 (Continued) Muscles that move the foot and toes (b) deep layer, plantar view.

Anatomical Identification Review

1. Identify and label the following muscles: deltoid, latissimus dorsi, supraspinatus, teres major, sternocleidomastoid, rhomboideus major, trapezius, and levator scapulae. Additionally, describe the action of each of these muscles.

2. Identify and label the following muscles: biceps brachii, palmaris longus, pronator teres, flexor carpi ulnaris, brachialis, brachioradialis, pronator quadratus, and flexor carpi radialis. Additionally, describe the action of each of these muscles.

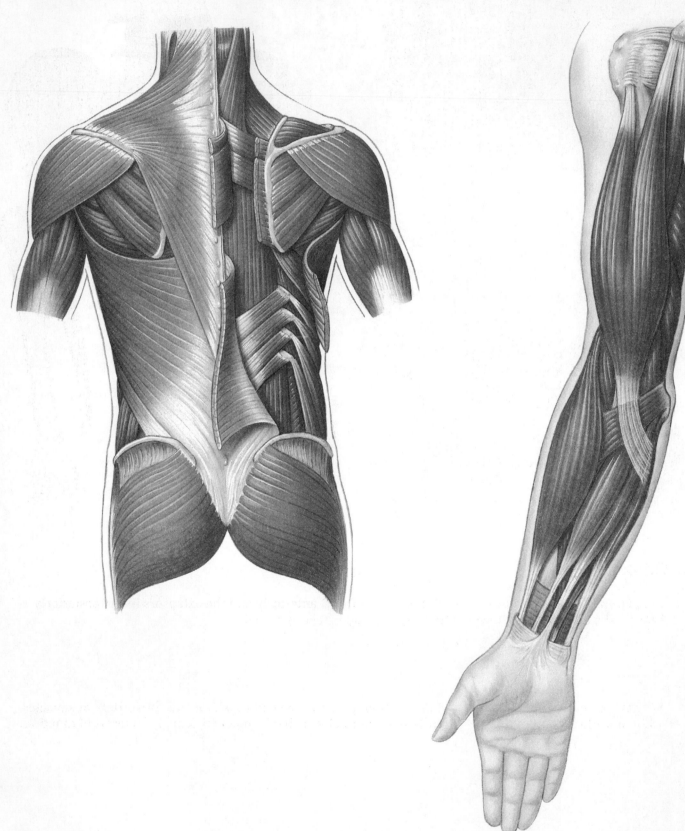

3. Identify and label the following muscles: gracilis, adductor brevis, adductor longus, pectineus, iliacus, adductor magnus, and psoas. Additionally, describe the action of each of these muscles.

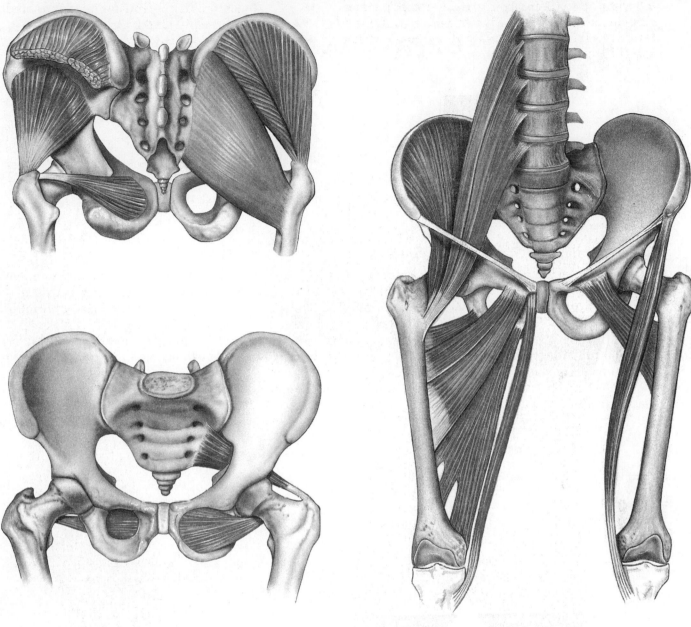

To Think About

1. In the upper limb, distal to each joint, the flexors insert anteriorly and the extensors insert posteriorly. What is the pattern in the lower extremity and why is it so arranged?

2. Why is the anatomical attachment of the pectoral girdle to the axial skeleton almost entirely by muscles while the attachment of the pelvic girdle is anchored by skeletal elements? Explain in terms of function.

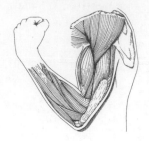

12

SURFACE ANATOMY

Objectives

1. Define surface anatomy and describe its importance both in the study of anatomy and in clinical settings.

2. Identify on a partner through visual observation and palpation the following surface anatomy regions: the head and neck, the thorax, the abdomen, the upper and lower limbs. Use the labeled photographs in this chapter for reference.

Regional Approach to Surface Anatomy

The study of anatomical landmarks as they appear on the exterior of the human body is called **surface anatomy**. Surface anatomy is best studied using an approach that divides the body into five regions: (1) *head and neck*, (2) *thorax*, (3) *abdomen*, (4) *upper limb*, and (5) *lower limb*. The photographs that appear in *Human Anatomy* follow the regional approach and serve as the basis for structure identification. [∞ p324]

Surface anatomy provides the means to relate anatomical structures to visible or palpable features. The structural and functional relationships between skeletal and muscular systems become obvious as you study surface anatomy. For example, in the anatomy laboratory, surface anatomy locates landmarks during the dissection process. In the clinical setting, surface anatomy is used to help pinpoint patient complaints and to perform non-invasive (e.g., CAT scans or MRIs) laboratory diagnostic procedures.

When comparing the body to the photographs in this chapter, it should be kept in mind that a layer of subcutaneous fat on the body may obscure the structures to be observed.

Observation of Head and Neck Surface Anatomy

Procedure

✓ Quick Check

Before you begin to examine the surface anatomy of the head and neck, review both the external bone markings of the adult skull and features of the axial skeleton. (*Figures 6-4* and *6-8, pp42, 45*). Keep a skull or articulated skeleton nearby during your observations to assist in the identification of bony landmarks.

Locate and identify on your laboratory partner, through **visual observation** and **palpation**, the **surface anatomy** of the **head and neck region**, using the labeled photographs in *Figures 12-1* and *12-2* for reference. For you to observe the surface anatomy in this region, your partner may be seated. Estimate the location of the structure, then locate it specifically with visual observation and palpation.

On the head, **supraorbital margins** are easily palpable on the frontal bone, and the **zygomatic bone, body of mandible, angle of mandible, ramus of mandible, mental protuberance, zygomatic arch**, and **mastoid process** are easily observed.

In the neck region, the **sternocleidomastoid** muscle forms a distinct diagonal muscular marking. In males the slope of the **trapezius** muscle in the shoulders is readily observed, as is the **hyoid bone**. A prominent **thyroid cartilage** (Adam's apple) and the **cricoid cartilage** in the neck can be easily observed during swallowing movements. The pulsing of the **carotid artery** is often observed and always palpable, at the superior edge of the sternum. When the breath is held, the **jugular veins** will also stand out visibly. The **suprasternal (jugular) notch** at the superior edge of the sternum is visible or palpable.

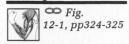
LOCATE

Head and Neck (from Anterior)

____ Supraorbital margin

____ Auricle (external ear)

____ Zygomatic bone

____ Body of mandible

____ Mental protuberance

____ Thyroid cartilage

____ Cricoid cartilage

____ Suprasternal notch

____ Sternum (manubrium)

____ Sternocleidomastoid muscle { Clavicular head / Sternal head

____ Trapezius muscle

____ Clavicle

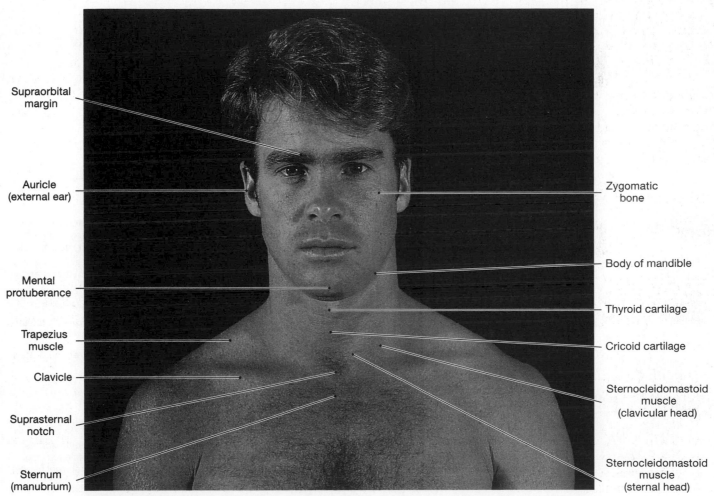

FIGURE 12-1 Head and neck, anterior view.

Head and Neck (Posterior and Anterior Triangles)

_____ Angle of mandible

_____ Mastoid process

_____ Site for palpation of submandibular gland and submandibular lymph nodes

_____ Hyoid bone

_____ Site for palpation of pulse of facial artery

_____ Trapezius muscle

_____ Thyroid cartilage

_____ Supraclavicular fossa

_____ Anterior triangle

_____ Suprasternal notch

_____ Site for palpation of carotid pulse

_____ External jugular vein beneath platysma muscle

_____ Posterior triangle

_____ Origin of brachial plexus

_____ Clavicle

_____ Sternocleidomastoid muscle { Clavicular head / Sternal head }

_____ Acromion process

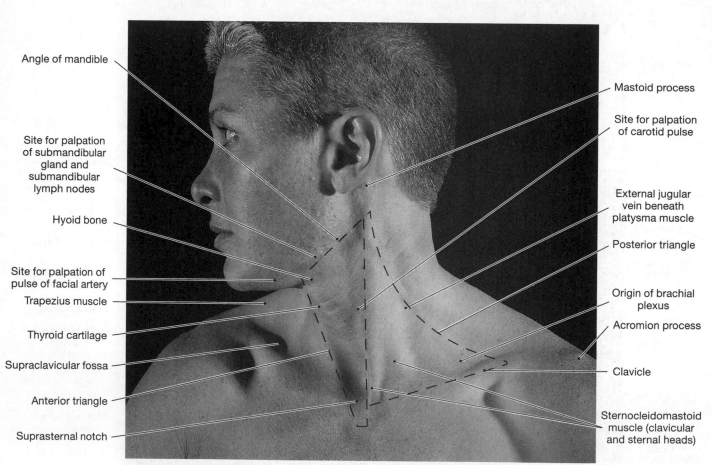

FIGURE 12-2 Head and neck, anterior and posterior triangles.

Observation of Thorax Surface Anatomy
Procedure

✓ **Quick Check**

Before you begin to examine the surface anatomy of the thorax, review the bone markings of the thorax, costal margin, and xiphoid process. (*Figure 6-41, p66*) Keep an articulated skeleton nearby during your observations to assist in the identification of bony landmarks and points of muscle attachment as you proceed.

Identify on your laboratory partner, through **visual observation** and **palpation**, the **surface anatomy** of the **thorax**, using the labeled photograph in *Figure 12-3* for reference. For you to observe the surface anatomy in this region, your partner must be standing with arms slightly abducted and extended. Estimate the location of the structure, then locate it specifically with visual observation and palpation.

The **xiphoid process** can be felt as a depression at the inferior edge of the sternum. The **clavicle** may be palpated and its curvature easily detected for all of its length. The clavicle can be palpated adjoining the **acromion process** as a projection at the most superior and lateral area of the shoulder. The margins of the **pectoralis major muscle** may be observed outlining a major part of the chest. With the subject momentarily holding a breath after a deep inspiration, the **costal margin** of the ribs is easily observed and palpable.

See: Fig. 12-3

∞ *Fig. 12-2, p326*

LOCATE

Thorax (from Anterior)

_____ Suprasternal notch

_____ Clavicle

_____ Acromion process

_____ Sternum { Manubrium / Body / Xiphoid process

_____ Sternocleidomastoid muscle (sternal and clavicular heads)

_____ Trapezius

_____ Xiphoid process

_____ Costal margin of ribs

_____ Deltoid muscle

_____ Pectoralis major muscle

_____ Areola and nipple

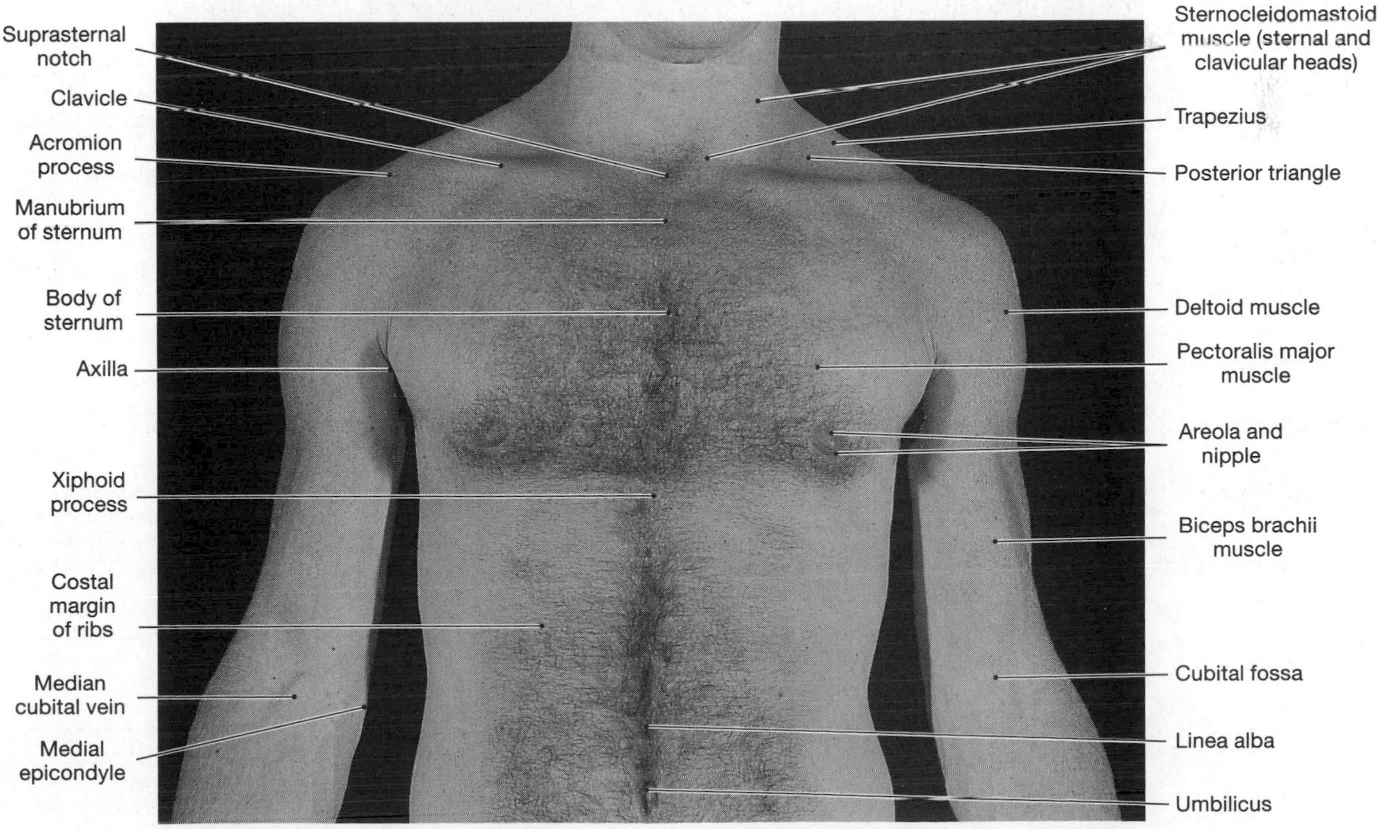

FIGURE 12-3 Thorax, anterior view.

Observation of Back and Shoulder Surface Anatomy
Procedure

✓ Quick Check

Before you begin to examine the surface anatomy of the back and shoulder, review both the bone markings of the scapula (*Figures 7-2 to 7-4, pp69-71*) and posterior features of the vertebral column. (*Figure 6-28, p58*) Keep an articulated skeleton nearby to assist you in the identification of bony landmarks and points of muscle attachment as you proceed.

Identify on your laboratory partner, through **visual observation** and **palpation,** the **surface anatomy** of the **back** and **shoulder**, using the labeled photograph in *Figure 12-4* for reference. For you to observe the surface anatomy in this region, your partner must be standing with arms slightly abducted and extended. Estimate the location of the structure, and then locate it specifically with visual observation and palpation.

On the posterior of the subject, the **scapular spine** is easily traced medially from the acromion toward the vertebral column, and both the **vertebral** (medial) and **lateral borders** of the scapula can be palpated. A **furrow** over the **spinous processes of thoracic vertebrae** can be observed to be bordered on either side by the *spinalis* group of muscles. On the upper back, an **infraspinatus muscle** may protrude inferior to the scapular spine or can be palpated as a muscle mass inferior to the spine. The rounded eminence capping the shoulder is the **deltoid muscle**, and its insertion can be observed as a depression if the arm is abducted. The **trapezius muscle** in the shoulder is more prominent in males.

See: Fig. 12-4

∞ *Fig. 12-2, p326*

LOCATE
Back and Shoulder Regions

_____ Vertebra prominens (C₇)

_____ Scapula { Superior angle / Inferior angle / Vertebral border / Lateral border }

_____ Acromion process

_____ Deltoid muscle

_____ Trapezius muscle

_____ Infraspinatus muscle

_____ Furrow over spinous processes of thoracic vertebrae

_____ Latissimus dorsi muscle

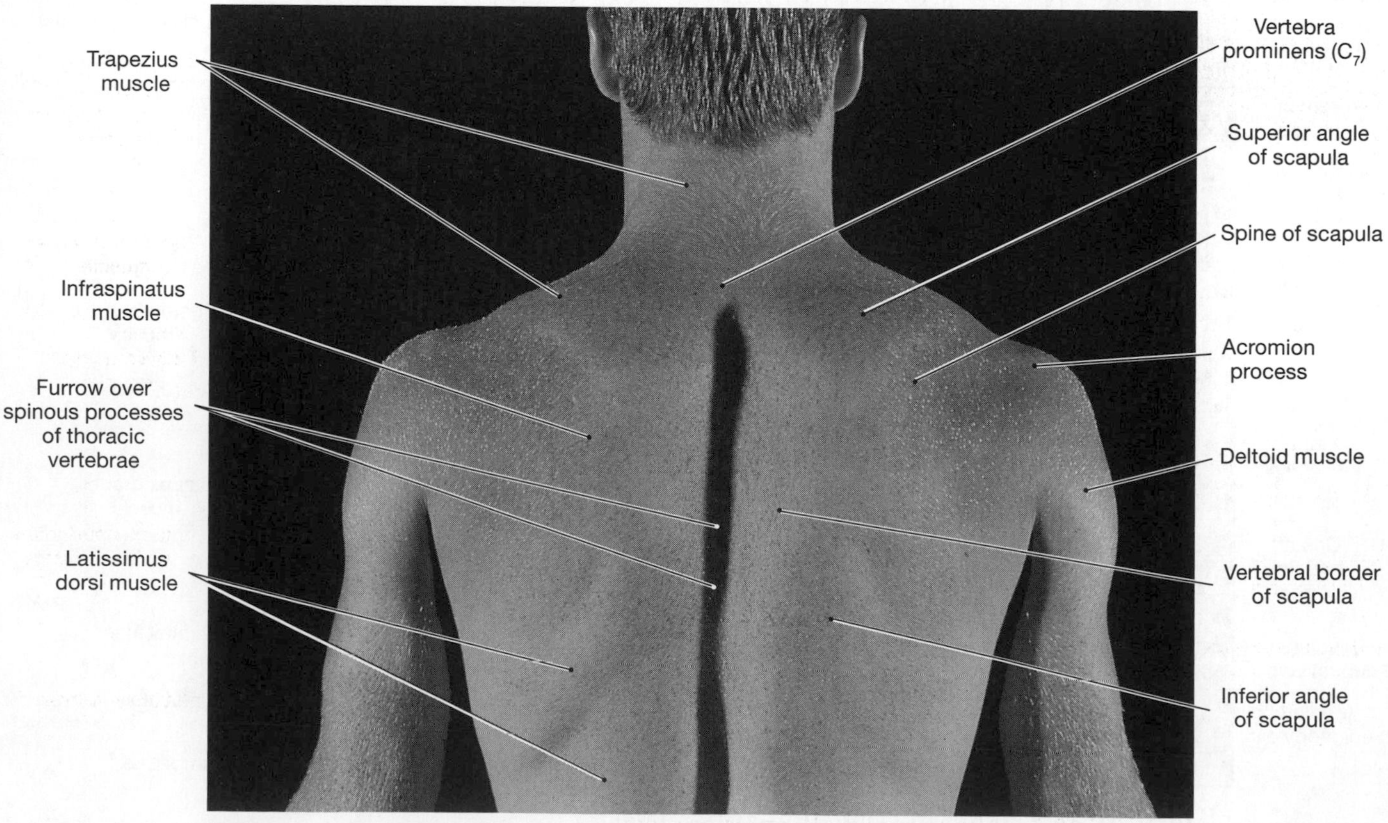

FIGURE 12-4   Back and shoulder regions.

Observation of Abdominal Wall Surface Anatomy

Procedure

✓ **Quick Check**

Before you begin to examine the surface anatomy of the abdominal wall, identify rib borders, costal margin, xiphoid process (*Figure 6-41, p66*), and bone markings of the superior surface of the pelvis. (*Figures 7-9* and *7-11, pp76* and *78*) Keep an articulated skeleton nearby during your observations to assist in the identification of bony landmarks and points of muscle attachment as you proceed.

Identify on your laboratory partner, through **visual observation** and **palpation**, the **surface anatomy** of the **abdominal wall**, using the labeled photographs in *Figures 12-5* and *12-6* for reference. For you to observe the surface anatomy in this region, your partner must be standing in a pose similar to that shown in the photographs that follow. Estimate the location of the structure, then locate it specifically with visual observation and palpation.

The **rectus abdominis muscle** resembles a "six pack" on the anterior surface of the abdomen, with the **linea alba** and the **umbilicus** as the central divider, and the **tendinous inscriptions** of the rectus muscle as the horizontal dividers. The sides of the abdominal wall are formed by the **external oblique muscles**. The **anterior superior iliac spines** are observed readily when the subject is either relatively thin or in the supine position. The anterior bony part of the pelvis is the **pubic symphysis**. The **inguinal ligament** draws an oblique line across from the **anterior superior iliac spine** to the pubic tubercle. The **iliac crest** is the widest and highest part of the hips, the part your belt rides above.

See: Figs. 12-5, 12-6

 ∞ *Fig. 12-3, p327*

LOCATE

Abdominal Wall (from Anterior)

_____ Xiphoid process

_____ Rectus abdominis muscle

_____ Tendinous inscriptions of rectus abdominis muscle

_____ Serratus anterior muscle

_____ Latissimus dorsi muscle

_____ External oblique muscle

_____ Linea alba

_____ Umbilicus

_____ Anterior superior iliac spine

_____ Inguinal ligament

_____ Inguinal canal

_____ Pubic symphysis

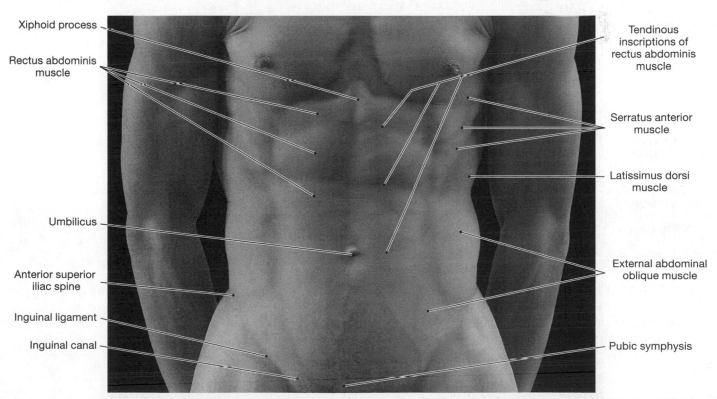

FIGURE 12-5 Abdominal wall, anterior view.

Abdominal Wall (from Lateral)

____ Xiphoid process	____ Rectus abdominis muscle
____ Costal margin	____ Iliac crest
____ Latissimus dorsi muscle	____ Anterior superior iliac spine
____ External oblique muscle	

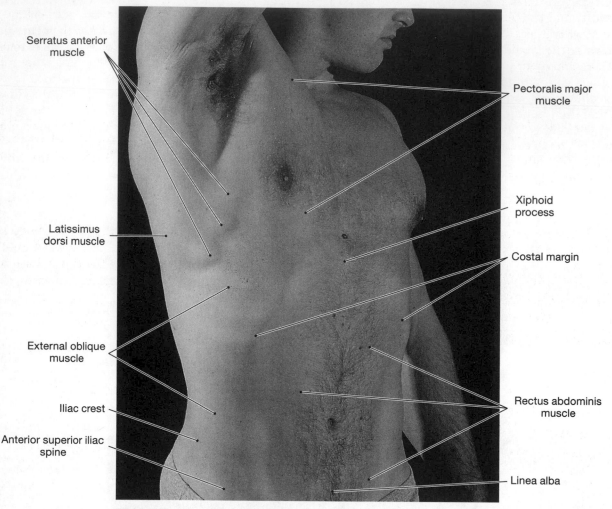

Serratus anterior muscle

Pectoralis major muscle

Latissimus dorsi muscle

Xiphoid process

Costal margin

External oblique muscle

Iliac crest

Rectus abdominis muscle

Anterior superior iliac spine

Linea alba

FIGURE 12-6   Abdominal wall, lateral view.

Observation of Upper Limb Surface Anatomy
Procedure

✓ Quick Check

Before you begin to examine the surface anatomy of the upper limb, review the bone markings of the scapula (*Figures 7-2 and 7-3, pp69-70*), clavicle (*Figure 7-1, p69*), arm (*Figures 7-4 to 7-7, pp70-74*), and hand (*Figure 7-8, p75*). Keep an articulated skeleton nearby during your observations to assist in the identification of bony landmarks and points of muscle attachment as you proceed.

Identify on your laboratory partner, through **visual observation** and **palpation**, the **surface anatomy** of the **upper limb**, using the labeled photographs in *Figures 12-7* to *12-9* for reference. For you to observe the surface anatomy in this region, your partner may sit or stand in a pose like that shown in the photographs that follow. Estimate the location of the structure, then locate it specifically with visual observation and palpation.

Arm

The **biceps brachii muscle** is clearly visible when your partner flexes the forearm. The separate heads of the **triceps muscle** may be visible with the arm in the extended position and slightly abducted. The **cubital fossa** is visible at the anterior surface of the elbow between the insertions of the biceps muscle on the radius and the *brachialis muscle* on the ulna. The **median cubital vein** crosses the cubital fossa and in the clinical setting is the vein of choice for obtaining blood samples. Just under the skin of the upper arm and over the biceps brachii muscle, the **cephalic vein** runs superiorly towards the shoulder. The **basilic vein** runs along the medial side of the arm, just underlying the skin, and disappears near the axilla. In muscular individuals, these veins are readily observed. The **medial** and **lateral epicondyles** of the humerus are easily palpated as the greatest width at the elbow. The **olecranon process** is easily observed. The **olecranon fossa** can be felt as the "funny bone" when the elbow is flexed. The **ulnar nerve** lies in the ulnar groove on the posterior of the medial epicondyle, medial to the olecranon fossa and can be palpated as it crosses this area.

Forearm

The **brachioradialis muscle** forms a bulge on the lateral side of the forearm along with tendons from most of the superficial forearm muscles (extensors and flexors). These muscles and their tendons are visible at the wrist when the wrist is flexed with tension. Tendons in the posterior side of the hand, when the fingers are fully extended, are those of the **extensor digitorum muscle**.

Wrist and Hand

The **pisiform bone** can be palpated at the proximal part of the medial side of the wrist, as can the **hamate bone** at the heel of the hand, and the **styloid processes** of the **radius** and of the **ulna** on either margin of the wrist joint. The **pollicis muscles** are visible at the base of the thumb, with the *opponens pollicis* forming the fleshy base of the thumb, the *abductor pollicis brevis* forming the flesh pad at the lateral margin of the hand, and the *adductor pollices* forming the webbing between the thumb and hand.

See: Figs. 12-7 to 12-9

∞ *Figs.*
12-4, p328
12-5, p329

LOCATE

Right Upper Limb (from Lateral)

____ Acromial end of clavicle

____ Deltoid muscle

____ Teres major muscle

____ Triceps brachii, { Long / muscle heads { Lateral

____ Lateral epicondyle of humerus

____ Olecranon process

____ Biceps brachii muscle

____ Brachialis muscle

____ Tendon of biceps brachii muscle

____ Brachioradialis muscle

____ Extensor carpi radialis longus muscle

____ Extensor carpi radialis brevis muscle

____ Extensor digitorum muscle

____ Styloid process of radius

____ Head of ulna

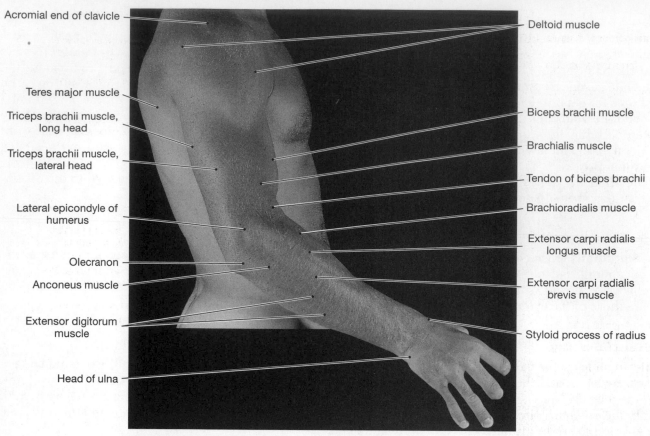

Acromial end of clavicle

Teres major muscle

Triceps brachii muscle, long head

Triceps brachii muscle, lateral head

Lateral epicondyle of humerus

Olecranon

Anconeus muscle

Extensor digitorum muscle

Head of ulna

Deltoid muscle

Biceps brachii muscle

Brachialis muscle

Tendon of biceps brachii

Brachioradialis muscle

Extensor carpi radialis longus muscle

Extensor carpi radialis brevis muscle

Styloid process of radius

FIGURE 12-7   Right upper limb, lateral view.

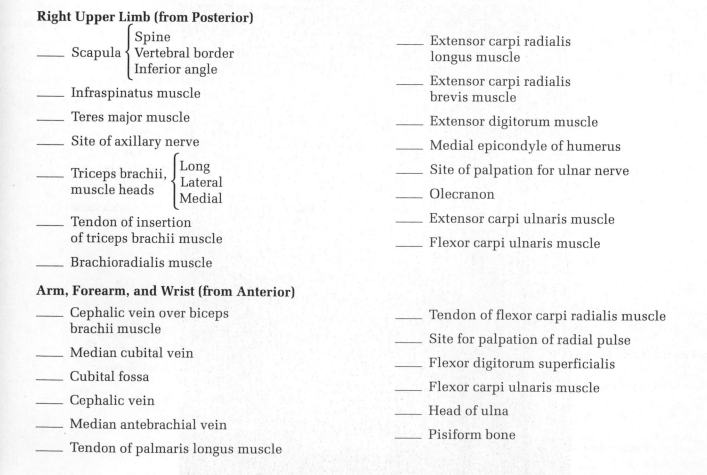

Right Upper Limb (from Posterior)

____ Scapula { Spine / Vertebral border / Inferior angle

____ Infraspinatus muscle

____ Teres major muscle

____ Site of axillary nerve

____ Triceps brachii, muscle heads { Long / Lateral / Medial

____ Tendon of insertion of triceps brachii muscle

____ Brachioradialis muscle

____ Extensor carpi radialis longus muscle

____ Extensor carpi radialis brevis muscle

____ Extensor digitorum muscle

____ Medial epicondyle of humerus

____ Site of palpation for ulnar nerve

____ Olecranon

____ Extensor carpi ulnaris muscle

____ Flexor carpi ulnaris muscle

Arm, Forearm, and Wrist (from Anterior)

____ Cephalic vein over biceps brachii muscle

____ Median cubital vein

____ Cubital fossa

____ Cephalic vein

____ Median antebrachial vein

____ Tendon of palmaris longus muscle

____ Tendon of flexor carpi radialis muscle

____ Site for palpation of radial pulse

____ Flexor digitorum superficialis

____ Flexor carpi ulnaris muscle

____ Head of ulna

____ Pisiform bone

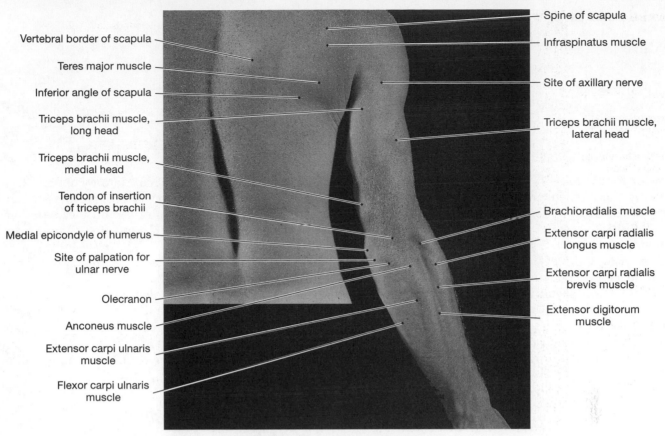

Spine of scapula

Infraspinatus muscle

Site of axillary nerve

Triceps brachii muscle, lateral head

Vertebral border of scapula

Teres major muscle

Inferior angle of scapula

Triceps brachii muscle, long head

Triceps brachii muscle, medial head

Tendon of insertion of triceps brachii

Brachioradialis muscle

Medial epicondyle of humerus

Extensor carpi radialis longus muscle

Site of palpation for ulnar nerve

Extensor carpi radialis brevis muscle

Olecranon

Extensor digitorum muscle

Anconeus muscle

Extensor carpi ulnaris muscle

Flexor carpi ulnaris muscle

FIGURE 12-8 Right upper limb, posterior view.

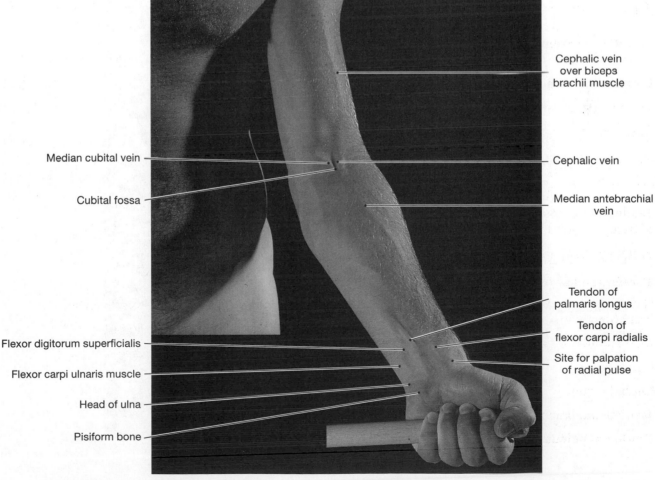

Cephalic vein over biceps brachii muscle

Median cubital vein

Cephalic vein

Cubital fossa

Median antebrachial vein

Tendon of palmaris longus

Tendon of flexor carpi radialis

Flexor digitorum superficialis

Site for palpation of radial pulse

Flexor carpi ulnaris muscle

Head of ulna

Pisiform bone

FIGURE 12-9 Arm, forearm and wrist, anterior view.

Observation of Lower Limb Surface Anatomy

Procedure

✓ **Quick Check**

Before you begin to examine the surface anatomy of the lower limb, review the bone markings of the pelvis (*Figures 7-11 and 7-12, pp78-79*), femur (*Figures 7-13 and 7-14, pp80-81*), and tibia/fibula (*Figures 7-16 and 7-17, pp83-84*). Keep an articulated skeleton nearby during your observations to assist in the identification of bony landmarks and points of muscle attachment as you proceed.

Identify on your laboratory partner, through **visual observation** and **palpation,** the **surface anatomy** of the **lower limb**, using the labeled photographs in *Figures 12-10* to *12-12* for reference. For you to observe the surface anatomy in this region, your partner must be standing. Applying tension to the thigh and leg muscles will give greater definition to some muscles, improving your ability to observe them. Estimate the location of the structure, then locate it specifically with visual observation and palpation.

The inferior width of the pelvis extends between the **greater trochanters** of the femurs. The lumbar vertebrae and **median sacral crest** of the sacrum are palpable just superior to the **fold of buttock**.

In the thigh, the **sartorius muscle** crosses the thigh diagonally from outside the hip (laterally) to the inner knee (medially), while the **tensor fasciae latae** forms a tight band, the **iliotibial tract**, on the lateral side of the thigh. The area of the **femoral triangle** is formed by three structures: (1) superiorly, by the **inguinal ligament**; (2) laterally, by the medial border of the sartorius muscle; and (3) medially, by the medial border of the **adductor longus muscle**. The femoral triangle is an important clinical site for obtaining blood samples or performing vascular

See: Figs. 12-10 to
12-12

∞ Fig.
12-6, p330

LOCATE

The Pelvis and Lower Limb (from Anterior)

____ Inguinal ligament

____ Area of femoral triangle, site for palpation of femoral artery/vein

____ Tensor fasciae latae muscle

____ Sartorius muscle

____ Rectus femoris muscle

____ Vastus lateralis muscle

____ Vastus medialis muscle

____ Adductor longus muscle

____ Gracilis muscle

____ Patella

____ Tibial tuberosity

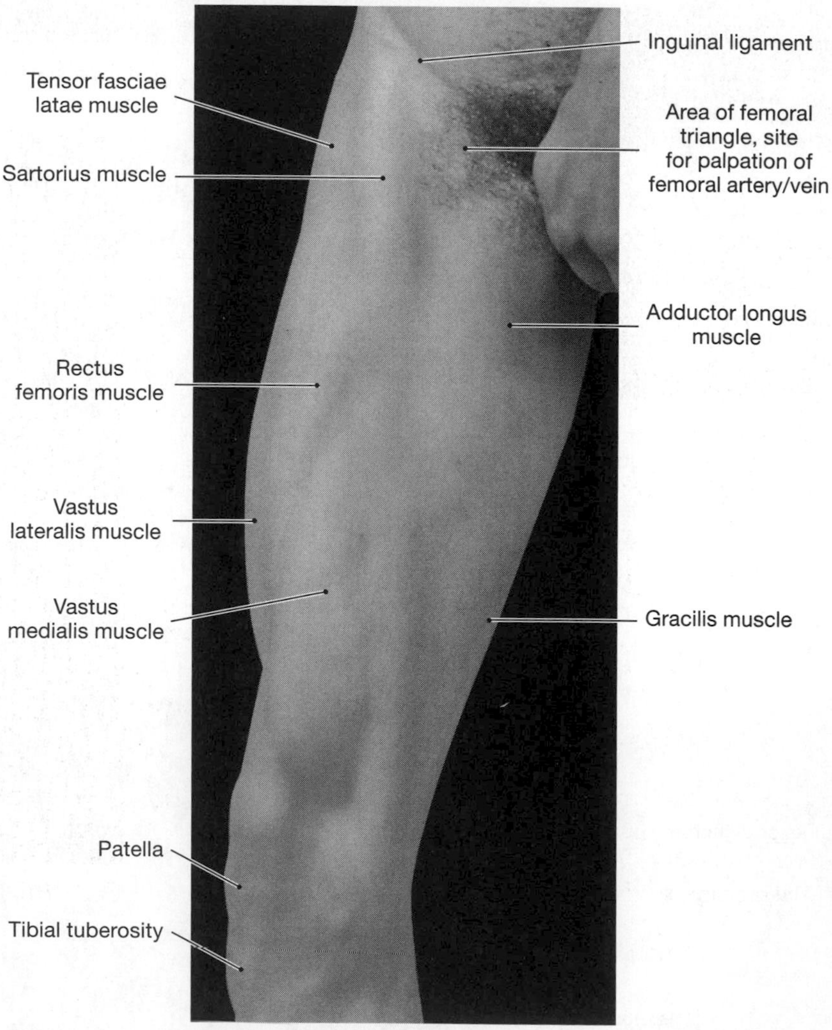

FIGURE 12-10   Pelvis and lower limb, anterior view.

Labels on figure:
Tensor fasciae latae muscle
Sartorius muscle
Rectus femoris muscle
Vastus lateralis muscle
Vastus medialis muscle
Patella
Tibial tuberosity
Inguinal ligament
Area of femoral triangle, site for palpation of femoral artery/vein
Adductor longus muscle
Gracilis muscle

catherization procedures. The **hamstring muscles** are discernible as separate muscles on the posterior of the thigh. The **vastus lateralis, rectus femoris**, and **vastus medialis muscles** show up as distinct masses superior to the **patella** as the muscles merge into the **patellar ligament**. As you view the lower extremity from the posterior, the concave pit formed at the posterior side of the knee, the **popliteal fossa**, is delineated superiorly by the **tendons of biceps femoris, semimembranosus**, and **semitendinosus muscles** (hamstring muscles) and inferiorly by the medial and lateral heads of the **gastrocnemius muscle**. The **gluteal muscles** show quite well with the **gluteus medius muscle** appearing superior to the greater trochanter of the femur and the inferior border of the **gluteus maximus muscle** appearing as the **fold of buttock**.

Typically, the muscles of the leg show quite well. As you view the leg from the anterior, the **anterior tibialis muscle** is visible immediately to the lateral side of the **tibial crest**, the **tendon of extensor digitorum longus** is immediately lateral to that, and the **fibularis longus muscle** is lateral to the extensor digitorum longus. On the anterior surface of the tibia, the **tibial tuberosity** can be palpated where the quadriceps inserts. On the posterior side of the lower leg, the calf exhibits the **medial** and **lateral heads of the gastrocnemius**, and the **soleus muscle** peeks out from both the lateral and medial side deep underlying the gastrocnemius. These muscles are best displayed with your partner standing with both feet on tiptoe.

The Pelvis and Lower Limb (from Lateral)

____ Tensor fasciae latae muscle

____ Gluteus muscles { Medius
                       Maximus

____ Iliotibial tract

____ Vastus lateralis muscle

____ Semitendinosus and semimembranosus muscles

____ Tendon of biceps femoris muscle

____ Popliteal fossa

____ Gastrocnemius muscle

____ Soleus muscle

____ Head of fibula

____ Patella

____ Patellar ligament

____ Tibial tuberosity

____ Fibularis longus muscle

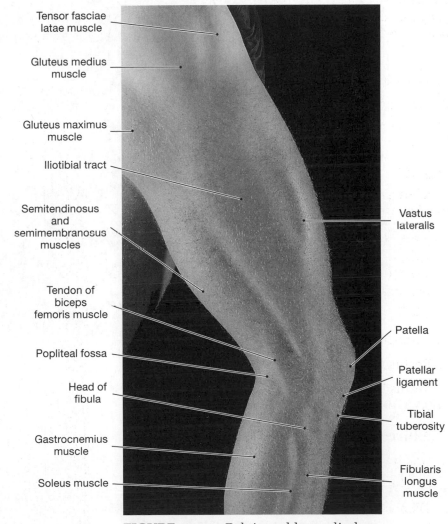

FIGURE 12-11 Pelvis and lower limb, lateral view.

The Pelvis and Lower Limb (from Posterior)

____ Iliac crest

____ Median sacral crest

____ Gluteus muscles { Medius / Maximus

____ Greater trochanter of femur

____ Location of sciatic nerve

____ Fold of buttock

____ Hamstring muscle group

____ Popliteal fossa

____ Tendon of biceps femoris muscle

____ Tendon of semitendinosus muscle

____ Site for palpation of popliteal artery

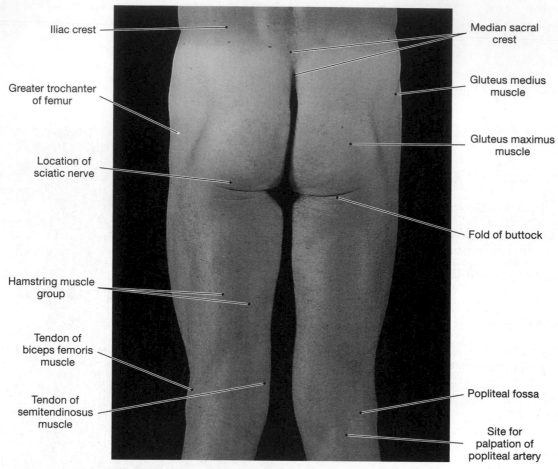

Iliac crest

Greater trochanter of femur

Location of sciatic nerve

Hamstring muscle group

Tendon of biceps femoris muscle

Tendon of semitendinosus muscle

Median sacral crest

Gluteus medius muscle

Gluteus maximus muscle

Fold of buttock

Popliteal fossa

Site for palpation of popliteal artery

FIGURE 12-12   Pelvis and lower limb, posterior view.

Observation of Ankle and Foot

Procedure

✓ Quick Check

Before you begin to examine the surface anatomy of the ankle and foot, review the bone markings of the distal ends of the tibia and fibula (*Figures 7-16* and *7-17, pp83-84*), ankle, and foot. (*Figure 7-18, p85*). Keep an articulated skeleton nearby during your observations to assist in the identification of bony landmarks and points of muscle attachment as you proceed.

Identify on your laboratory partner, through **visual observation** and **palpation**, the **surface anatomy** of the **ankle and foot**, using the labeled photographs in *Figure 12-13* and *12-14* for reference. The subject should be standing so that you can observe the surface anatomy in this region. Estimate the location of the structure, then locate it specifically with visual observation and palpation.

The **medial malleolus** and **lateral malleolus** form the visible ankle bones. The **extensor digitorum longus muscle** passes anterior to the lateral malleolus, with the **tendon of the tibialis anterior muscle** passing anterior to the medial malleolus. Between these two structures is the site for palpation of the **dorsal pedis artery**. The **calcaneal tendon** is prominent in all individuals and extends to the **calcaneus** bone. The **tendon of peroneus longus** passes immediately posterior to the lateral malleolus. Visible on the superior aspect of the foot when the toes are spread are the **tendon of extensor hallucis longus** (great toe) and **tendons of extensor digitorum longus** muscles. The **dorsal venous arch** with its branches passes over these tendons and is also visible on the anterior side of the foot.

See: Figs. 12-13, 12-14

 ∞ *Fig. 12-7, p331*

LOCATE

Knee, Leg, Ankle and Foot (from Anterior)

____ Patella

____ Tibial tuberosity

____ Gastrocnemius muscle

____ Soleus muscle

____ Fibularis longus muscle

____ Anterior border of tibia

____ Tibialis anterior muscle

____ Medial malleolus

____ Great saphenous vein

____ Lateral malleolus

____ Tendon of extensor hallucis longus muscle

____ Tendons of extensor digitorum longus muscle

____ Dorsal venous arch

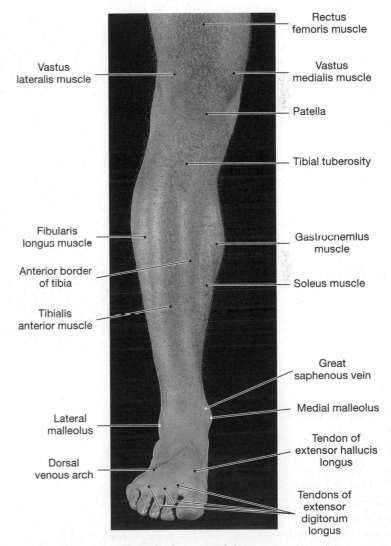

FIGURE 12-13 Right knee and foot, anterior view.

Knee, Leg, Ankle, and Foot (from Posterior)

_____ Site for palpation of popliteal artery

_____ Site for palpation of common peroneal nerve

_____ Gastrocnemius, { Lateral
muscle heads { Medial

_____ Soleus

_____ Calcaneal tendon

_____ Tendon of fibularis longus muscle

_____ Medial malleolus

_____ Lateral malleolus

_____ Site for palpation of posterior tibial artery

_____ Calcaneus

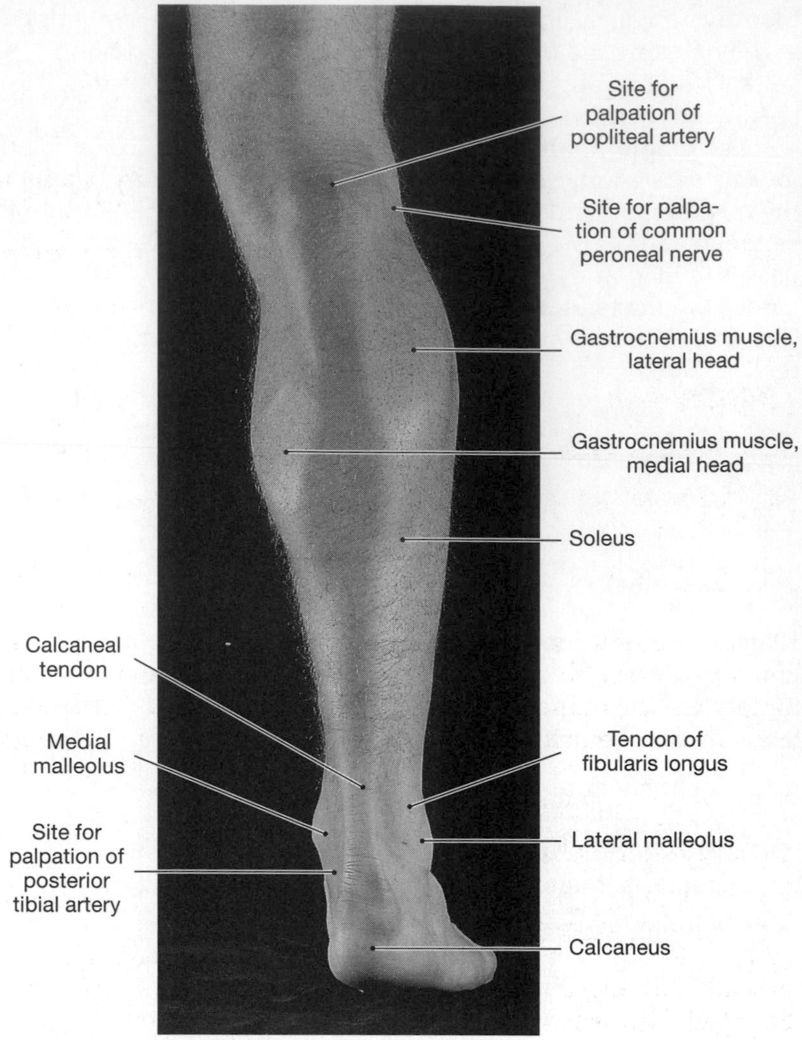

FIGURE 12-14   Right knee and foot, posterior view.

To Think About

1.   How does the process of palpation reveal anatomical structures?

2.   Why is a good working knowledge of surface anatomy essential for any person in a health-related field of study?

3.   How are the structures of the foot and leg affected by wearing high-heeled shoes on a regular basis for long periods of time? What surface anatomical structures would you expect to change when a person puts on this type of footwear?

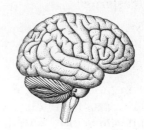

13
THE NERVOUS SYSTEM:
Neural Tissue

Objectives

1. Identify the structural parts of a motor neuron on a neuron model; describe the function of each part.

2. Using histology sections, describe motor neurons and their structural parts, and distinguish these cells from neuroglia.

3. Identify the structural components of a chemical synapse on a model of a synaptic knob.

The nervous system is one of two body systems that control and direct activities of the other body systems. The nervous system functions to: provide information about the external and internal environments, coordinate voluntary and involuntary motor activities, perceive and coordinate sensory information, and control and regulate body systems, organs, and tissues. The nervous system includes all of the **neural tissue** in the body. [∞ p335]

Description of Neural Tissue

The nervous system is composed of two distinct components: *neurons* and *neuroglia*. The **neurons** are responsible for the receipt, transfer, and processing of information. The **neuroglia (glial cells)** provide a framework of support for all nervous tissue. Some of these supporting cells may act as phagocytes. Structurally, neurons are classified as **unipolar, bipolar,** or **multipolar**, based on the number of processes that project from the cell body. Functionally, neurons are classified as *sensory neurons, motor neurons,* or *interneurons*. **Sensory neurons** form the afferent division of the peripheral nervous system (PNS) and deliver information to the central nervous system (CNS). **Motor neurons** stimulate or modify the activity of a tissue, organ, or organ system, and form the efferent division of the peripheral nervous system. **Interneurons (association neurons)** may be situated between sensory and motor neurons. They are located entirely within the central nervous system, where they analyze sensory inputs and coordinate motor outputs. [∞ p344]

Nerve impulses are conducted along **axons**, single cellular processes that lead away from the cell body, until the impulse reaches a terminal point, the **synapse**. Axons may branch along their length, and such branches are termed **collaterals**. Axons and collaterals end in a series of fine extensions called **telodendria**, which ultimately terminate in expanded endings called **synaptic knobs**. At synapses, information (nerve) impulses pass from neuron to neuron or neuron to another cell type, such as a muscle fiber. Synapses that pass the nerve impulse from neuron to another cell type are called *neuroeffector junctions*. There are two major classes of these junctions, *neuromuscular junctions* and *neuroglandular junctions*. At a neuromuscular junction, the neuron communicates with a muscle cell whereas, at a neuroglandular junction, a neuron controls or regulates the activity of a secretory cell. [∞ p346]

The following laboratory observations emphasize the structure of multipolar neurons and a chemical synapse.

Observation of a Multipolar Neuron

Locate and identify the following **neuronal structures**, using a *model* of a neuron. Use *Color Plate 26* for reference. Neurons are characterized by a **cell body** (**soma**) containing a large round **nucleus** that houses a conspicuous **nucleolus**. Extending from the cell body are numerous, relatively short **dendrites** and a single elongated **axon**. Dendritic processes serve as a receptive region and transmit nerve impulses toward the cell body. Axons are neuron processes that carry impulses away from the cell body. That portion of a neuron capable of producing a nerve impulse is the axon and the **axon hillock**. This is a conical region where the axon attaches to the soma. Cytoplasm around the nucleus of neurons is called the **perikaryon**. It contains a cytoskeleton of neurofibrils made up of **neurofilaments, microtubules**, and **microfilaments**. Other organelles of the perikaryon include numerous mitochondria and an extensive network of rough endoplasmic reticulum (ER), called **Nissl bodies**. [∞ p342]

In the CNS, the axons of many neurons are completely ensheathed by the extensions of oligodendrocytes. Collectively, many *oligodendrocytes* form the *myelin sheath* along the entire length of a *myelinated axon*. [∞ p338] The relatively large areas of the axon wrapped in myelin are termed **internodes**. Identify the small gaps in the myelin wrapping, called **nodes of Ranvier**, that exist along the length of the myelinated axon. **Schwann cells** produce a sheath around every peripheral axon, whether it is myelinated or unmyelinated. [∞ p340]

See: Color Plate 26

∞ *Fig.*
13-8, p341
13-9, p342

LOCATE

_____ Cell body

_____ Perikaryon

_____ Nucleus

_____ Nucleolus

_____ Nissl bodies/endoplasmic reticulum

_____ Mitochondria

_____ Dendrite

_____ Axon hillock

_____ Axon

_____ Telodendria

_____ Synaptic knob

_____ Myelinated internode

_____ Node of Ranvier

_____ Schwann cell

Microscopic Observation of Neural Tissue

✓ **Quick Check**

After observing the structure of a motor neuron, return to *Chapter 9, p104* in this laboratory manual and re-examine the microscope slide entitled "Motor End Plate" to observe again a neuromuscular junction. (*Figure 9-2, p105*)

Identify the structure of **motor neurons** using the *microscope slide* provided, first by viewing under low power magnification and then under high power magnification. Use *Color Plate 22* and *Figure 13-1* for reference. Motor neurons are large cells with multiple branches, whereas the neuroglia, which are very abundant in the background, are tiny branched cells. Typically, in most smear preparations, motor neurons stain violet and neuroglia pink. Within the perikaryon note the large round nucleus and abundant Nissl bodies. A nucleolus may be visible within the nucleus. Extending from the soma are two types of branching processes, multiple dendrites and a single axon. Since a neuron has only one axon, it is very rare to see one. The axon is always much thinner than the dendrites. Dendrites are often misidentified as axons. The axon hillock may be observed as a lighter staining triangular area containing very few neurotubules and Nissl bodies. There are numerous neurofibrils streaming into the axon hillock and axon; you just don't see them with this type of stain. Below the photo draw and label the motor neurons that you observe under the high power microscope.

Slide # _____ Motor Neurons

See: Color Plate 22,
Fig. 13-1

∞ *Fig.*
13-9a, p342

LOCATE

____ Motor neurons

____ Dendrites

____ Axon

____ Axon hillock

____ Perikaryon

____ Nucleus

____ Nissl bodies

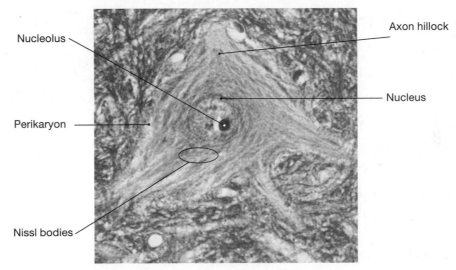

Nucleolus

Axon hillock

Nucleus

Perikaryon

Nissl bodies

FIGURE 13-1 Motor neuron from the
spinal cord × 1600

Anatomical Identification Review

Label the structures of a typical multipolar neuron.

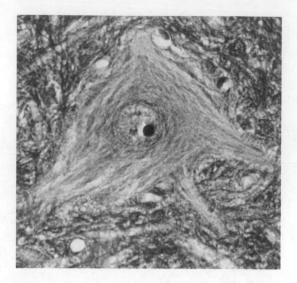

To Think About

Trace the route of a nerve impulse traveling from your fingertip to the brain, and then back to those skeletal muscles that flex the fingers.

14

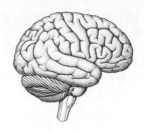

THE NERVOUS SYSTEM:
The Spinal Cord and Spinal Nerves

Objectives

1. Identify and name the gross structures of the spinal cord and describe the general functions of this organ.

2. Observe and identify the histological structure of the spinal cord on a microscope slide.

3. Locate, identify, and name the spinal meninges (dura mater, arachnoid, and pia mater) and describe both their structure and purpose.

4. Locate, identify, and name the spinal nerves, nerve plexuses, and the anatomical regions or structures innervated by these nerves.

Gross Anatomy of the Spinal Cord, Spinal Nerves, and Spinal Meninges

The central nervous system (CNS) is divided into the *brain* and *spinal cord*. [∞ p356] The spinal cord is encased within the vertebral canal. It is further protected by both a shock-absorbing fluid and its ensheathment by three connective tissue membranes. The spinal cord is one pathway between body and brain and contains neurons for carrying both sensory information to the brain and motor impulses away from the brain. Thirty-one pairs of spinal nerves convey information to and from the spinal cord and the body's tissues.

Procedure

✓ **Quick Check**

Before you begin to examine the spinal cord and spinal nerves, review the regions of the vertebral column (*Figure 6-27, p57*), and bone markings characteristic of cervical (*Figures 6-29 to 6-33, pp59-61*), thoracic (*Figures 6-34 and 6-35, pp62*), and lumbar vertebrae. (*Figures 6-36 and 6-37, p63*) Keep an articulated skeleton nearby to aid you in identifying the vertebrae and their characteristic markings.

Locate and identify structures of the **spinal cord** and **emerging spinal nerves**, using a *model* or *specimen* of the *spinal cord/vertebral column*. Observe the spinal cord and spinal nerves on the torso model first. Use *Color Plate 56* to *58* and *Figure 14-1* for reference. With the cadaver in the prone position, the spinal cord and meninges can be examined. Observation of the meninges is described later under a separate heading.

The spinal cord contains all of the **ascending** (carry sensory information to the brain) and **descending** (convey motor impulses away from the brain) **tracts** of myelinated axons as well as the cell bodies for many neurons. [∞ p362] Observe how the **spinal cord** emerges from the foramen magnum lying in the vertebral canal (*Figure 8-1, p88*) where it is protected by the lamina and pedicels of the vertebrae (*Figure 6-36, p63*).

As it descends caudally, observe how the spinal cord diameter varies, showing two enlargements along its length. Enlargements correspond to the regions containing additional gray matter and cell bodies for controlling the limbs. [∞ p356] The **cervical enlargement** contributes nerve roots that form the *brachial plexus* [∞ p367] to control the upper extremity. The **lumbar enlargement** contributes nerve roots that form the *lumbar plexus* and the *sacral plexus* to control pelvic structures and the lower extremity. [∞ p370] The spinal cord tapers and terminates in the cone-shaped **conus medullaris** at about L_2. The most caudal spinal nerve roots form the **cauda equina** and continue the length of the vertebral column before exiting. Caudally, the conus medullaris portion of the spinal cord is anchored to the coccyx by means of a strand of connective tissue, the **filum terminale**.

The 31 pairs of *spinal nerves* are arranged in *segments*. [∞ p362] Segments of the spinal cord correspond to vertebrae and are named in the same manner, except that there are 31 **spinal cord segments** and 30 vertebrae. In the cervical region, there are 8 pairs of spinal nerves. The first spinal nerve pair is located superior to the C_1 vertebra. However, beginning with the thoracic region and continuing in a caudal direction, spinal nerve pairs are named for the vertebra immediately superior to the nerve. [∞ p362] This difference is because the spinal cord is shorter than the vertebral column. Each spinal nerve pair is numbered and named according to its adjacent vertebra. Spinal nerves are *mixed nerves*, so termed because each nerve contains both *sensory* and *motor fibers*; [∞ p344] thus each nerve has the ability to transmit both incoming and outgoing messages.

Projecting laterally from the lateral surfaces of the spinal cord and within the dural-arachnoid membrane at intervertebral foramina, observe fine nerve filaments merging to form **dorsal** and **ventral nerve roots** of the spinal nerves. At intervertebral foramina, dorsal roots enter the **dorsal root ganglia**. These are the collected cell bodies of the axons of the entering sensory neurons. In contrast, the ventral roots contain the axons of motor neurons that carry impulses away from the CNS. The roots come together after the dorsal root ganglion and emerge from the intervertebral foramen to form a single **spinal nerve**.

Each spinal nerve divides into a **dorsal ramus** and a **ventral ramus**. [∞ p363] The dorsal rami reflect to the back and innervate the skin and muscles of the back. The ventral rami follow the curvature of the body walls anteriorly to supply the structures in the body wall and extremities.

See: Color Plates 56-58, Fig. 14-1

 ∞ *Figs. 14-1, p357 14-3, p359*

LOCATE

____ Spinal cord emerging from foramen magnum of occipital bone

____ The 31 pairs of spinal nerves

____ Cervical spinal nerves (C_1–C_8)

____ Cervical enlargement

____ Thoracic spinal nerves (T_1–T_{12})

____ Lumbar spinal nerves (L_1–L_5)

____ Lumbar enlargement

____ Sacral spinal nerves (S_1–S_5) emerging from sacral foramina

____ Cauda equina

____ Filum terminale

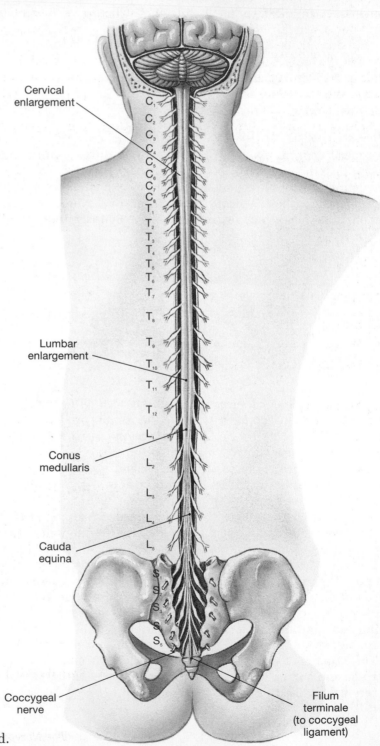

FIGURE 14-1 Gross anatomy of the spinal cord.

Observation of the Spinal Cord, Spinal Nerves, and Meninges

The brain and spinal cord are ensheathed by three distinct connective tissue layers called **meninges**. From outside to inside, they include: (1) *dura mater*, (2) *arachnoid*, and (3) *pia mater*. [∞ p356] The meninges protect the CNS and provide support for blood vessels, nerves and cerebrospinal fluid. **Cerebrospinal fluid (CSF)** acts as a fluid shock absorber and a medium by which nutrients, wastes, and other metabolic products can be transported. [∞ p359]

Procedure

Locate and identify the following **structures** of the **spinal cord**, using either a *model* of the *spinal cord* or *cadaver specimen*. The cadaver must be in the prone position to examine these structures. To view the spinal cord, spinal nerves, and meninges, a "window" within the posterior portion of the vertebrae must be prepared. Observe these structures on the torso first. Use *Color Plates 56* to *58* for reference. The laminae, spinous processes, and ligaments of selected vertebrae [∞ p223] have been removed, along with portions of erector spinae muscles (mostly the spinalis muscles). (*Figure 10-8 p121*) If this "window" has not been prepared, check with your instructor before proceeding. For the cadaver, use *Color Plates 57* and *58* for reference.

The **dura mater** can be identified as the tough, pearl-white outer meningeal layer that ensheathes the spinal cord and the brain, and divides the space surrounding them into **epidural** and **subdural spaces**. Deep to the dura mater is the thin **arachnoid**, which forms the middle layer and in life encloses CSF in the **subarachnoid space**. Deep to the arachnoid is the innermost layer, the **pia mater**, which follows and adheres tightly to the surface of the spinal cord and brain. Do not attempt to separate this layer from the cord. Arising from the sides of the spinal cord along its entire length, observe connective tissue filaments, called **denticulate ligaments**, which originate from the pia mater. These ligaments pierce through arachnoid-dura mater membranes to prevent rotation of the spinal cord within the **vertebral canal**. All three membranes extend out through the intervertebral foramina, ensheathing each spinal nerve to become continuous with the **epineurium** of nerves. [∞ p362]

See: Color Plates 56-58, Fig. 14-2

∞ Fig. 14-2, p358

LOCATE

____ Spinal cord

____ Meninges { Dura mater
Arachnoid
Pia mater

____ Spinal nerves

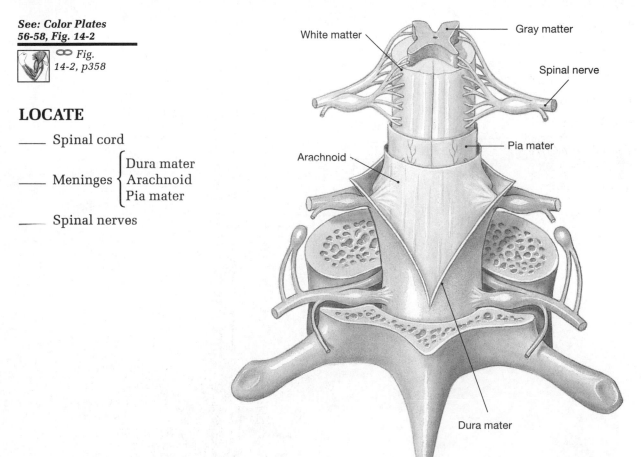

FIGURE 14-2 Spinal cord and spinal meninges, posterior view.

Sectional Anatomy of the Spinal Cord: Microscopic Identification

Procedure

Locate and identify the following **structures**, using the *microscope slide* provided, first by viewing with a dissecting microscope. The general organizational plan of the spinal cord is best observed with the dissecting microscope at various magnification levels (low to high). Use *Figure 14-3* for reference. Then use your compound microscope, viewing first with low power magnification and then with high magnification to reveal the details of these structures.

Identify the areas of gray and white matter. The **gray matter** resembles outstretched butterfly wings. Observe the oval **central canal** within the middle of the spinal cord for the flow of CSF. Large motor neurons (*Figure 13-1 p197*) can be observed in the large **anterior (ventral) gray horns** of the gray matter. The **posterior (dorsal) gray horns** are smaller, and are the more pointed portions of the wings. **White matter** surrounds the gray matter and is made up of *anterior, posterior*, and *lateral columns* containing fibers of *ascending sensory* and *descending motor tracts*. [Figures 16-2 and 16-4 ∞ pp430, 433] The spinal cord is scored into two halves by a deep **anterior median fissure** and a shallow **posterior median sulcus**. Identify the wide anterior median fissure within the white matter between the anterior gray horns. Also, locate on the posterior side the narrow posterior median sulcus within the white matter, but between the posterior gray horns.

The meninges may be visible on your slide specimen, but this depends upon the manner in which the specimen was prepared. Most often, only the pia mater containing blood vessels is visible. **Ventral roots** [∞ p354] may be visible emanating from the anterior and lateral surfaces of the spinal cord, as seen in *Figure 14-4*.

Slide # _____ Spinal Cord

See: Fig. 14-3

∞ *Fig. 14-5a, p361*

LOCATE

_____ Spinal cord

_____ Dura mater

_____ Arachnoid

_____ Pia mater

_____ Ventral root

_____ Gray matter

_____ White matter

_____ Anterior median fissure

_____ Posterior median sulcus

_____ Central canal

_____ Dorsal root ganglion

_____ Dorsal root

_____ Gray horns { Posterior / Anterior / Lateral }

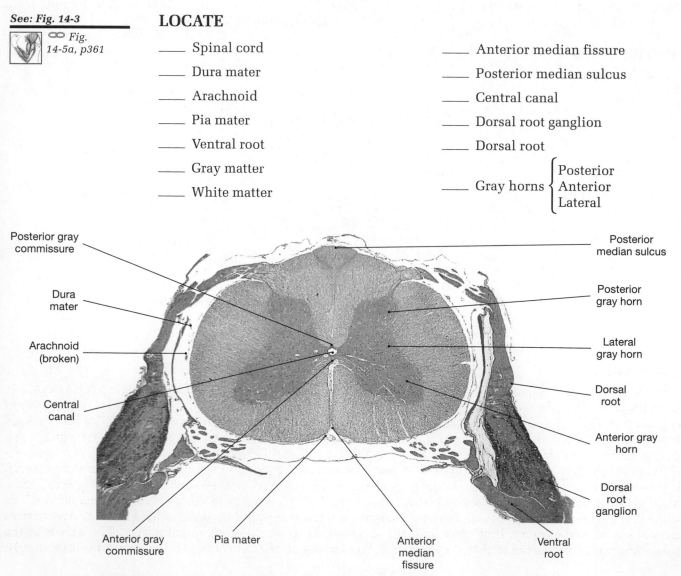

FIGURE 14-3 Spinal cord, transverse section.

Gross Sectional Anatomy of the Spinal Cord and Meninges

Procedure

Locate and identify the following **structures** of the **spinal cord** as they appear in section, using a *sectional model* or *sectioned specimen* of the spinal *cord* with vertebra. Return to the previous observation titled "Microscopic Identification" and use that description for this identification of the gross observation of the spinal cord in section, using a model and or specimen. Use *Color Plates 57-58* and *Figure 14-4* for reference.

See: Color Plates 57, 58, Fig. 14-4

∞ Fig. 14-2, p358

LOCATE

___ Vertebral body	___ Arachnoid	___ Dorsal root ganglion
___ Vertebral foramen	___ Subarachnoid space	___ Dorsal root
___ Spinal cord	___ Pia mater	___ Dorsal ramus
___ Epidural space	___ Denticulate ligaments	___ Ventral ramus
___ Dura mater	___ Ventral root	

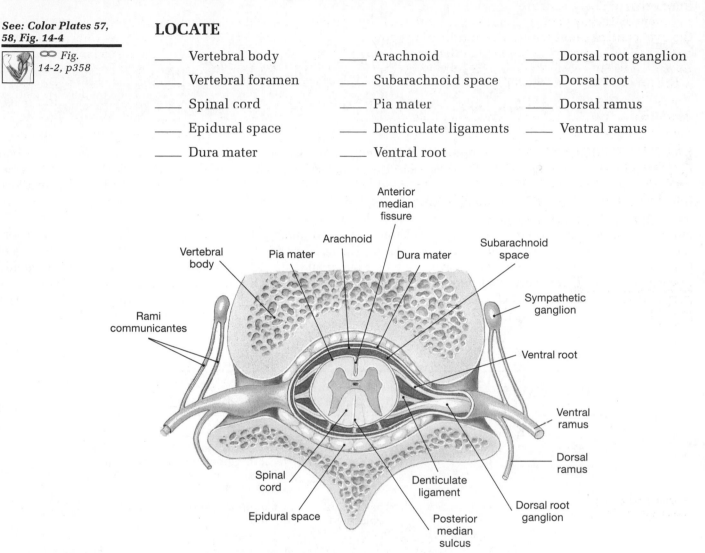

FIGURE 14-4 Spinal cord and spinal meninges, sectional view.

Nerve Plexuses

As nerves emerge from the spinal cord, they merge into *trunks*, divide into *divisions*, and then merge again to form *cords* or trunks. This pattern of merging and dividing forms nerve **plexuses** associated with the segments of the spinal cord that control neck structures and the upper and the lower extremities. [∞ p363] The plexuses are the **cervical, brachial**, **lumbar**, and **sacral**. The term plexus means "braid," which is an accurate description and visual picture of the anatomical arrangement of these spinal nerves. The ventral rami of most spinal nerves, except thoracic, are braided together to convey sensations to the CNS and to provide motor control to the limbs. The nerves that emerge from the plexus contain a mixture of fibers from those nerves entering the plexus. Thus, a spinal nerve entering the plexus has its fibers braided or mixed together with other entering spinal nerves. The result is that the emerging nerve branches are different from the entering. This arrangement helps to ensure that, in case of spinal nerve injury, limb areas are still supplied but from different spinal nerves. The nerves of the plexuses are named according to their location (e.g., bone or region they are closest to) or structures they supply.

Cervical Plexus

The **cervical plexus** is formed by the ventral rami of spinal nerves C_1–C_4 and some fibers from C_5. [∞ p364] Branches from the cervical plexus innervate most of the muscles of the neck, extrinsic laryngeal muscles, and the diaphragm. Other cutaneous branches from the cervical plexus supply the skin of the upper chest, shoulder, neck, and some portions of the head.

Procedure

✓ Quick Check

Before you begin to examine the nerves that form the cervical plexus, review the cervical region of the vertebral column (*Figure 6-27, p57*), bone markings characteristic of cervical vertebrae (*Figures 6-29 to 6-33, pp59-61*), and structures of the posterior triangle. (*Color Plate 59* and *Figure 10-7, p118*) Keep an articulated skeleton nearby to aid you both in associating the nerve roots with their characteristic cervical vertebrae and in visualizing the course taken by the nerve branches.

Locate and identify the **nerve roots** of the **cervical plexus**, using a *model* or *cadaver specimen*. Use *Color Plates 58, 59* and *Figure 14-5* for reference. The spinal segment and the distribution of nerves of the cervical plexus is presented in *Table 14-1* and should be reviewed and referred to as you proceed in your observation. With the cadaver in the supine position, adjust the head so the chin is lateral. This will provide you with a better view of the posterior triangle region on the side you will be observing. Reflect the cut sternocleidomastoid muscle to observe the cervical plexus. Deep to the sternocleidomastoid, about 2 inches inferior to its insertion, the first four cervical ventral rami of the cervical plexus can be observed on the anterior and lateral surfaces of the scalenus muscles. The **phrenic nerve** is a major nerve emerging from this plexus. It is easy to observe and identify as it passes inferiorly descending along the ventral surface of the anterior scalenus muscle. The phrenic nerve arises from C_3, C_4, and C_5 nerves and typically enters the thorax deep to the *subclavian vein* (*Color Plate 89*) to innervate the diaphragm.

TABLE 14-1	The Cervical Plexus	
Spinal Segment	*Nerves*	*Distribution*
C_1–C_4	Ansa cervicalis (superior and inferior branches)	Five of the extrinsic laryngeal muscles (sternothyroid, sternohyoid, omohyoid; geniohyoid and thyrohyoid via N XII) and only C_1 fibers
C_2–C_3	Lesser occipital, transverse cervical, supraclavicular, and greater auricular nerves	Skin of upper chest, shoulder, neck, and ear
C_3–C_5	Phrenic nerve	Diaphragm
C_1–C_5	Cervical nerves	Levator scapulae, scalenes, sternocleidomastoid, and trapezius (with N XI)

See: Color Plates 58, 59, 89, Fig. 14-5

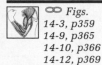

∞ Figs.
14-3, p359
14-9, p365
14-10, p366
14-12, p369

LOCATE

____ Cervical spinal nerves (C_1–C_8)

____ Nerve roots of Cervical plexus

C_1–C_4: Ansa cervicalis complex

C_2–C_3: lesser occipital, transverse cervical, supraclavicular, and greater auricular

C_3–C_5: phrenic nerve

C_1–C_5: cervical nerves

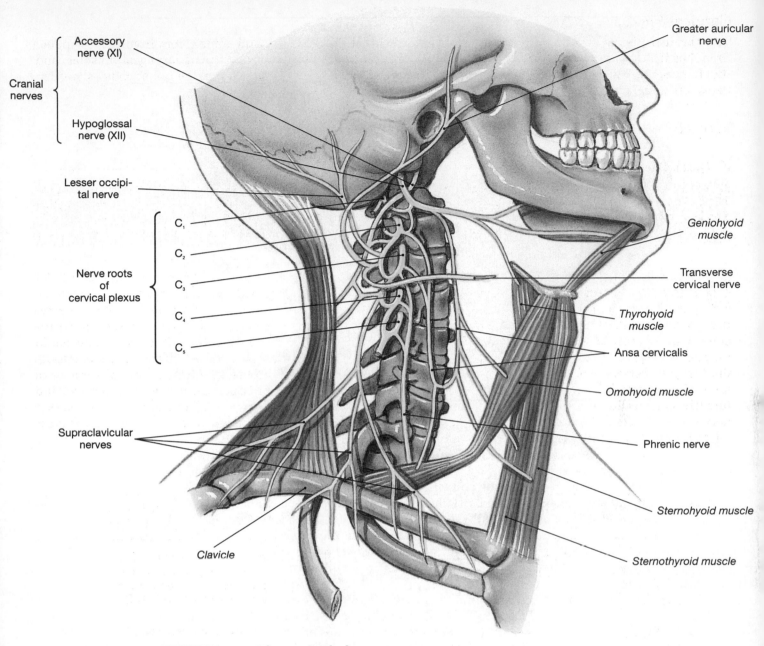

FIGURE 14-5 The cervical plexus.

Brachial Plexus

The organization of ventral rami of spinal nerves C_5 to T_1 forms the **brachial plexus**, which serves to supply innervation to most of the structures of the superficial back, shoulder, and upper limb, along with skin receptors in these regions. The spinal segment and the distribution of nerves of the brachial plexus is presented in *Table 14-2* and should be reviewed and referred to as you proceed in your observation. [∞ p367]

Procedure

✓ **Quick Check**

Before you begin to examine the nerves that form the brachial plexus, review the cervical and thoracic regions of the vertebral column (*Figure 6-27, p57*), and bone markings characteristic of cervical (*Figures 6-29 to 6-33, pp59-61*) and thoracic (*Figures 6-34 and 6-35, p62*) vertebrae. Also review the position of the bones that compose the pectoral girdle (*Figures 7-1 to 7-3, pp69-70*) and the bone markings characteristic of the humerus. (*Figures 7-4 and 7-5, pp71-72*) Keep an articulated skeleton nearby to aid you both in associating the nerve roots with their characteristic vertebrae and in visualizing the course taken by the nerve branches along the bones of the pectoral girdle and upper limb. Review the previously identified prosected muscles of the upper extremity (See *Chapter 11, p144*).

TABLE 14-2 The Brachial Plexus

Spinal Segment	Nerves	Distribution
C$_5$, C$_6$	Axillary nerve	Deltoid and teres minor muscles
		Skin of shoulder
C$_5$–T$_1$	Radial nerve	Extensor muscles on the arm and forearm (triceps brachii, brachioradialis, extensor carpi radialis, and extensor carpi ulnaris)
		Digital extensors and abductor pollicis
		Skin over the posterolateral surface of the arm
C$_5$–C$_7$	Musculocutaneous nerve	Flexor muscles on arm (biceps brachii, brachialis, coracobrachialis)
		Skin over lateral surface of forearm
C$_6$–T$_1$	Median nerve	Flexor muscles on forearm (flexor carpi radialis, palmaris longus)
		Pronators (p. quadratus and p. teres)
		Digital flexors
		Skin over anterolateral surface of hand
C$_8$, T$_1$	Ulnar nerve	Flexor muscle on forearm (flexor carpi ulnaris)
		Adductor pollicis and small digital muscles
		Skin over medial surface of hand

Locate and identify the **nerve roots** and **cords** of the **brachial plexus**, using a *model* or *cadaver specimen*. Examine the brachial plexus with the cadaver in supine position, the upper limb to be observed abducted to about 45 degrees, and the head adjusted so the chin is lateral. This will provide you with a better view of the posterior triangle region on the side you will be observing. (*Color Plates 35* and *36*)

Reflect the cut sternocleidomastoid muscle and observe deep to the clavicular head (*Color Plate 59*) the origins of the **brachial plexus**. If the middle section of the clavicle has been removed, you will be able to observe the brachial plexus in greater detail. For landmarks, locate the *internal jugular vein* and omohyoid muscle. (*Color Plate 37*) Deep to these structures and inferior to the clavicle, the ventral rami of spinal nerves C$_5$ to T$_1$ merge to form trunks that divide again to form cords. Three cords form the brachial plexus: *lateral cord, medial cord*, and *posterior cord*.

The **lateral cord** contains fibers of spinal nerves C$_5$, C$_6$, and C$_7$. Identify the lateral cord lateral to the *axillary artery*. (*Figure 14-6* and *Color Plate 59*) Trace the lateral cord distally to its first branch, the **musculocutaneous nerve**, a small branch that innervates the upper arm flexor (biceps brachii, brachialis, and coracobrachialis) muscles. The larger branch, the **median nerve**, lies anterior and lateral to the axillary artery and continues distally to the wrist. The median nerve, composed of fibers from both medial and lateral cords, innervates the flexor muscles of the forearm and some hand muscles.

The **medial cord** contains fibers of spinal nerves C$_8$ and T$_1$. The medial cord gives off five branches, but we will focus on only two branches. The short **medial root of the median nerve** joins a branch of the lateral cord at an angle to form the median nerve. It can be identified as it passes over the anterior surface of the axillary artery. Immediately distal to the median root of the medial nerve, identify the other, larger branch as the **ulnar nerve**, which lies on the medial side of the humerus (*Color Plates 44* and *84*) and supplies some flexor muscles in the forearm, and most of the intrinsic hand muscles.

The **posterior cord** contains fibers of all the spinal nerves (C$_5$ to T$_1$) that form the brachial plexus. The posterior cord can be identified running posterior to the axillary artery and lying on the anterior surface of the subscapularis muscle. (*Figure 11-2, p148* and *Color Plate 44*) The posterior cord gives off five branches, but we will focus only on the two terminal branches. Identify the terminal branch that is passing deep posteriorly onto the subscapularis and teres major muscles [*Color Plate 59*] as the **axillary nerve**. This nerve passes closest to the armpit and innervates the deltoid and teres minor muscles. Also, identify the second terminal branch, the **radial nerve**, which passes along the radial groove (lateral margin) of the humerus (*Figure 7-5, pp71-72*) to the radial side of the arm, innervating all the arm and forearm extensor muscles (posterior and lateral).

See: Color Plates 35,
36, 44, 59, 84, Figs.
14-b, 14-7

∞ Figs.
14-3, p359
14-11, pp367-368
14-12, p369
22-12c, p585

LOCATE

_____ Cervical spinal nerves (C_5–C_8)

_____ Thoracic spinal nerves (T_1–T_2)

_____ Nerve roots C_5–T_1 of Brachial plexus

_____ Brachial plexus {
  Superior trunk
  Middle trunk
  Inferior trunk
}

_____ Lateral cord {
  Musculocutaneous nerve
  Lateral root of median nerve
}

_____ Medial cord {
  Medial root of median nerve
  Ulnar nerve
}

_____ Posterior cord {
  Axillary nerve
  Radial nerve
}

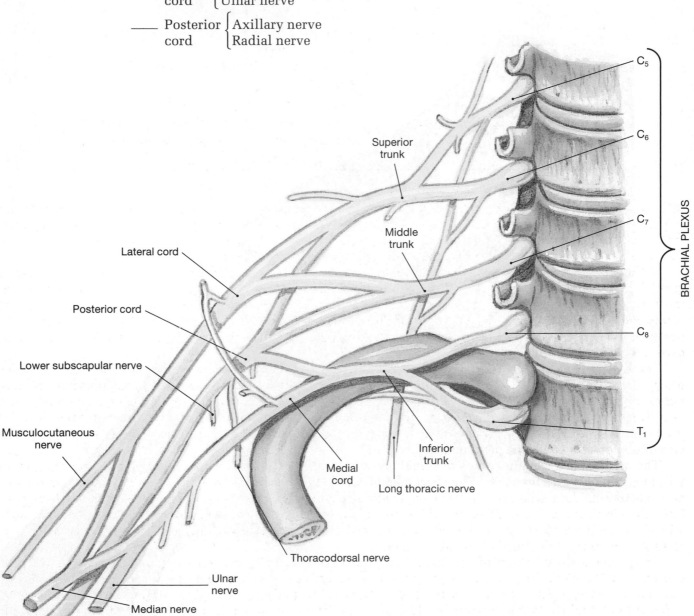

FIGURE 14-6 Brachial plexus, anterior view.

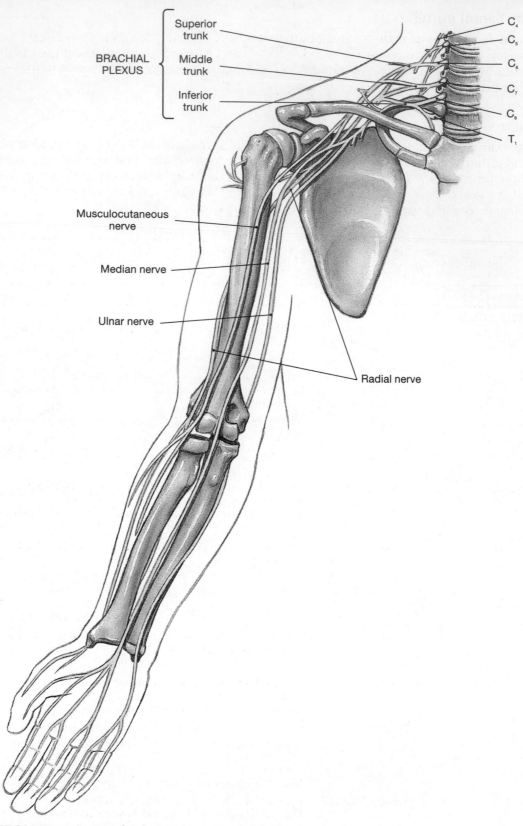

Superior
trunk

BRACHIAL
PLEXUS

Middle
trunk

Inferior
trunk

C$_4$
C$_5$
C$_6$
C$_7$
C$_8$
T$_1$

Musculocutaneous
nerve

Median nerve

Ulnar nerve

Radial nerve

FIGURE 14-7 Brachial plexus, anterior view.

The Lumbar Sacral and Plexus

The **lumbar** and **sacral plexus** is the source of innervation for all muscles of the lower limb. Additionally, they supply the sensory receptors to regions of the skin. [∞ p370] The spinal segment and the distribution of spinal nerves of the lumbar and sacral plexus are presented in *Table 14-3* and should be reviewed and referred to as you proceed in your observation.

Lumbar Plexus

The **lumbar plexus** is formed anteriorly from nerve roots of spinal nerves T_{12}–L_4. The lumbar plexus innervates the medial and anterior areas of the thigh and lower abdominal regions. The name of each nerve of the plexus specifies the region innervated. The lumbar plexus is observed deep within the abdominopelvic cavity. To observe the lumbar plexus, the abdominal and pelvic viscera must be retracted or removed (*Color Plate 120*). *Check with your instructor before proceeding with this observation.*

TABLE 14-3	The Lumbar and Sacral Plexuses	

Spinal Segment	Nerves	Distribution
THE LUMBAR PLEXUS		
T_{12}, L_1	Iliohypogastric nerve	Abdominal muscles (external and internal obliques, transversus abdominis) Skin over lower abdomen and buttocks
L_1	Ilioinguinal nerve	Abdominal muscles (with iliohypogastric) Skin over medial upper thigh and portions of external genitalia
L_1, L_2	Genitofemoral nerve	Skin over anteromedial surface of thigh and portions of external genitalia
L_2, L_3	Lateral femoral cutaneous nerve	Skin over anterior, lateral, and posterior surfaces of thigh
L_2–L_4	Femoral nerve	Anterior muscles of thigh (sartorius and quadriceps) Adductors of thigh (pectineus and iliopsoas) Skin over anteromedial surface of thigh, medial surface of leg and foot
L_2–L_4	Obturator nerve	Adductors of thigh (adductor magnus, brevis, longus) Gracilis muscle Skin over medial surface of thigh
L_2–L_4	Saphenous nerve	Skin over medial surface of leg
THE SACRAL PLEXUS		

L_4–S_2	Gluteal nerves:
	  Superior
	
	  Inferior
		
		Abductors of thigh (gluteus minimus, gluteus medius, and tensor fasciae latae)
		Extensor of thigh (gluteus maximus)

L_4–S_3	Sciatic nerve:	Two of the hamstrings (semimembranosus, semitendinosus) Adductor magnus (with obturator nerve)
	Tibial nerve	Flexors of leg and plantar flexors of foot (popliteus, gastrocnemius, soleus, tibialis posterior, biceps femoris (long head)) Flexors of toes Skin over posterior surface of leg, plantar surface of foot
	Peroneal nerve	Biceps femoris of hamstrings (short head) Fibularis (brevis and longus) and tibialis anterior Extensors of toes Skin over anterior surface of leg and dorsal surface of foot
S_2–S_4	Pudendal nerve	Skin and muscles of perineum, including urogenital diaphragm and external anal and urethral sphincters Skin of external genitalia and related skeletal muscles (bulbospongiosus and ischiocavernosus)

Procedure

✓ **Quick Check**

Before you begin to examine the nerves that form the lumbar plexus, review the lumbar region of the vertebral column (*Figure 6-27, p57*), and bone markings characteristic of thoracic (*Figures 6-34* and *6-35, p62*) and lumbar vertebrae. (*Figures 6-36* and *6-37, p63*) Keep an articulated skeleton nearby to aid you both in associating nerve roots with their characteristic vertebrae and in visualizing the course taken by the nerve branches. Review the previously identified muscles of the posterior abdominal wall (psoas, iliacus, and quadratus lumborum muscles) (See *Color Plate 121*).

Locate and identify the **nerve roots** of the **lumbar plexus**, using a *model* or *cadaver specimen*. The lumbar plexus can only be observed with the abdominopelvic organs retracted. *Check with your instructor for specific directions regarding the retracting of these organs.* With the cadaver in the supine position and the abdominopelvic organs retracted or removed, locate the cone-shaped psoas major muscle. (*Figure 11-14, p164*)

The lumbar plexus forms anteriorly from nerve roots of spinal nerves T_{12}–L_4 and penetrates through either the lateral or medial borders of the psoas major muscle. Identify the small **genitofemoral nerve** as it descends on the anterior surface of the psoas muscle, just superior to the iliac crest. The *genital nerve* continues descending inferiorly and exits the abdomen through the inguinal ring to supply the genital area. To identify the large **femoral nerve**, observe its descent between the psoas and iliacus muscles and exit of the abdomen with these muscles as they enter the femoral triangle. The femoral nerve lies immediately lateral to the femoral artery. (*Figure 22-7, p324* and *Color Plate 51*) Now return to the origin of the femoral nerve. Just superior to it and inferior to the anterior superior iliac spine, the **lateral femoral cutaneous nerve** passes over the anterior surface of the quadratus lumborum (*Figure 10-8, p121*), exiting the abdomen deep to the lateral end of the inguinal ligament.

Superior to the lateral femoral cutaneous nerve, lying on the anterior surface of the iliacus muscle at the level of L_2–L_3, identify both the **iliohypogastric** and **ilioinguinal nerves**. These nerves run laterally and inferiorly at an oblique angle to enter either the scrotum or labial folds.

The **saphenous nerve** is a branch of the femoral nerve and is easily identified on the medial surface of the medial head of the gastrocnemius muscle. It is a subcutaneous nerve and continues inferiorly to the foot.

The large **obturator nerve** can be identified at the level of L_4 as it passes over the medial edge of the psoas muscle into the obturator foramen (*Figure 7-11, p78*) and exits the pelvis into the thigh, deep to the pectineus muscle. This nerve innervates the adductor muscles of the thigh.

Sacral Plexus

The **sacral plexus** is formed posteriorly in the pelvis, from nerve roots of spinal nerves L_4–S_4 and extends inferiorly on the posterior of the thigh. The majority of the nerve branches enter the muscles of the buttock. The sacral plexus supplies the posterior portions of the buttock and the thigh, all the muscles inferior to the knee, and muscles of the pelvic diaphragm and genital area.

Procedure

✓ **Quick Check**

Before you begin to examine the nerves that form the sacral plexus, review the lumbar and sacral regions of the vertebral column (*Figure 6-27, p57*), bone markings characteristic of lumbar vertebrae (*Figures 6-36* and *6-37, p63*), sacrum (*Figures 6-38* and *6-39, p64*), and pelvis. (*Figures 7-11* and *7-12, pp78-79*) Keep an articulated skeleton nearby to aid you both in associating nerve roots with their characteristic vertebrae and in visualizing the course taken by the nerve branches. Review the previously identified gluteal muscles (*Figures 11-12* and *11-13, pp163-164*).

See: Fig. 14-8

  ∞ *Figs.*
14-3, p359
14-13, p371
14-14, p372

LOCATE

____ Thoracic spinal nerves (T_{10}–T_{12})

____ Lumbar spinal nerves (L_1–L_5)

____ Nerve roots T_{12}–L_4 of lumbar plexus

____ Iliohypogastric nerve (L_1)

____ Ilioinguinal nerve (L_1)

____ Genitofemoral nerve (L_1, L_2)

____ Lateral femoral cutaneous nerve (L_2–L_3)

____ Femoral nerve (L_2–L_4)

____ Obturator nerve (L_2–L_4)

____ Saphenous nerve (L_2–L_4)

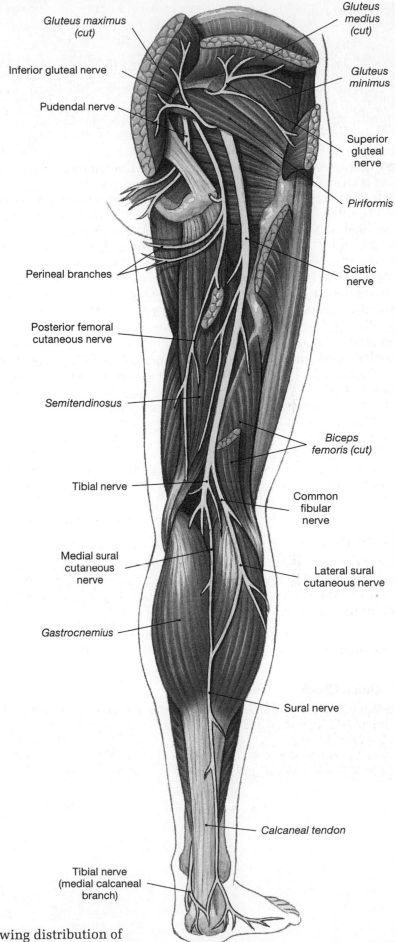

FIGURE 14-8 Lumbar and sacral plexuses showing distribution of peripheral nerves of the lower extremity, posterior view.

Locate and identify the **nerve roots** and nerve branches of the **sacral plexus**, using a *model* or *cadaver specimen*. Use *Color Plates 52, 54,* and *Figures 14-8* and *14-9* for reference. With the cadaver in the prone position, first reflect both the gluteus maximus and the gluteus medius muscles. Retract the piriformis muscle superiorly. Deep to the piriformis, identify the largest nerve of the sacral plexus, the **sciatic nerve**, which is also the largest nerve in the body. It is about the width of your thumb, and it represents two nerves, *tibial* and *common fibular*, which are bound together by connective tissue. The sciatic nerve innervates the posterior section of the thigh. The path of the sciatic nerve can be traced distally as it extends posteriorly to the adductor magnus muscle and deep to the semitendinosus and biceps femoris (long head) muscles.

Superior to the popliteal fossa, the sciatic nerve divides into a **tibial nerve** and a **common fibular nerve**. The tibial nerve can easily be identified deep to and between biceps femoris and semitendinosus muscles. At the popliteal fossa, the tibial nerve gives off a branch into the **medial sural cutaneous nerve**, which passes inferiorly and superficially on the surface between the heads of the gastrocnemius muscle. The tibial nerve and its branches innervate the flexor muscles (posterior muscles) of the leg and the plantar muscles of the foot.

Return to the common fibular nerve and observe how it extends laterally beyond the lateral head of the gastrocnemius muscle to branch into the **lateral sural cutaneous nerve**. On the posterior surface of the gastrocnemius, superior to the calcaneal tendon, the medial sural cutaneous nerve is joined by a branch of the lateral sural to form the **sural nerve**. The common fibular nerve and its branches innervate the extensor muscles (anterior muscles) of the leg and dorsum of the foot.

The gluteal branches, **superior gluteal nerve** and **inferior gluteal nerve**, innervate the gluteal muscles. Identify the superior gluteal nerve superior to the piriformis muscle, and note the fibers passing to the gluteus medius and minimus muscles. Identify the inferior gluteal nerve inferior to the piriformis and observe its fibers supplying the gluteus maximus muscle.

The **pudendal nerve** does not innervate any muscles or structures of the buttock, but passes through the buttock to supply the genital region and muscles of the perineum. Identify the pudendal nerve deep to the piriformis muscle as it lies medial to the sciatic nerve.

See: Color Plates 52, 54, Fig. 14-9

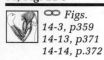 ∞ *Figs.*
14-3, p359
14-13, p371
14-14, p.372

LOCATE

_____ Lumbar spinal nerves (L_1–L_5)

_____ Sacral spinal nerves (S_1–S_5) emerging from sacral foramina

_____ Nerve roots L_4–S_4 of Sacral plexus

_____ Gluteal nerves $\Big\{$ Superior
(L_4–S_2)   Inferior

_____ Sciatic nerve $\Big\{$ Tibial branch
(L_4–S_3)   Common fibular branch

_____ Pudendal nerve (S_2–S_4)

_____ Lateral sural cutaneous nerve

_____ Medial sural cutaneous nerve

_____ Sural nerve

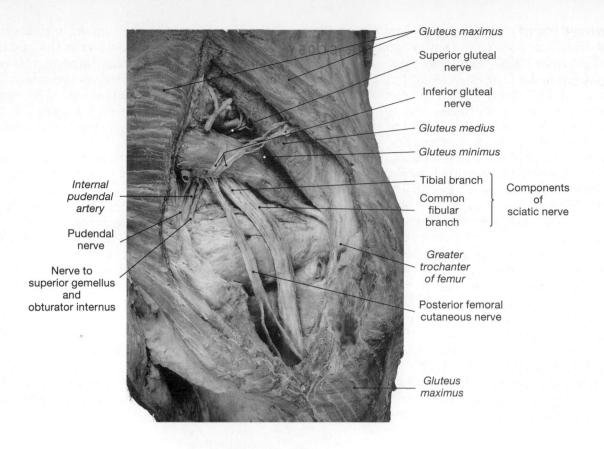

Gluteus maximus

Superior gluteal nerve

Inferior gluteal nerve

Gluteus medius

Gluteus minimus

Tibial branch

Common fibular branch

Components of sciatic nerve

Internal pudendal artery

Pudendal nerve

Nerve to superior gemellus and obturator internus

Greater trochanter of femur

Posterior femoral cutaneous nerve

Gluteus maximus

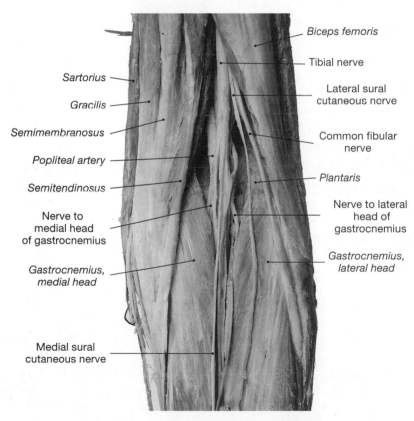

Biceps femoris

Tibial nerve

Sartorius

Gracilis

Semimembranosus

Popliteal artery

Semitendinosus

Nerve to medial head of gastrocnemius

Gastrocnemius, medial head

Medial sural cutaneous nerve

Lateral sural cutaneous nerve

Common fibular nerve

Plantaris

Nerve to lateral head of gastrocnemius

Gastrocnemius, lateral head

FIGURE 14-9 Lumbar and sacral plexuses, posterior view of right gluteal and popliteal fossa region.

Anatomical Identification Review

On the accompanying illustration, label the structures of the spinal cord, including the protective meningeal layers and bony landmarks.

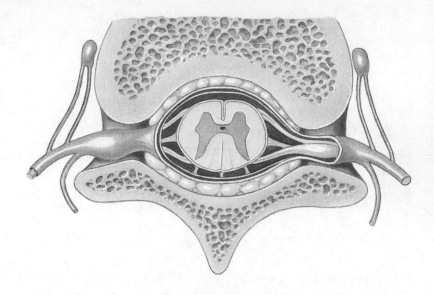

To Think About

1. A man who has a vasectomy subsequently suffers loss of the ability to maintain an erect penis. What is the likely cause of this problem?

2. A runner who has increased her training regimen dramatically in preparation for a marathon has begun to experience tingling and numbness down the rear of the right leg and into the right foot. What would you predict to be the cause of these problems?

3. Where is the spinal cord injury in a quadriplegic who has difficulty breathing?

4. Where would be the most likely location for damage to the spinal cord of a person who has become a paraplegic?

5. A nine-year-old boy falls from his tree fort, but manages to slow his descent somewhat by grabbing onto limbs he passes on the way to the ground. When he finally lands, he experiences some soreness on the buttocks, but otherwise does not think he is hurt until he discovers that he cannot move his fingers easily or accurately. What structures have been damaged?

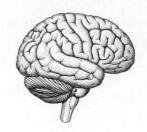

15

THE NERVOUS SYSTEM:
The Brain and Cranial Nerves

Objectives

1. Name and identify the major regions of the brain and describe their functions.

2. Develop a three-dimensional perspective of the organization of the brain (control center) and cranial nerves (connections).

3. Identify and describe the structures that cover, support, and protect the brain and relate these to their functions.

4. Identify both the choroid plexus and structures associated with cerebrospinal fluid (CSF) circulation, and describe their function(s).

5. Name and identify the structures that compose the following regions and describe their functions: telencephalon, diencephalon, mesencephalon, metencephalon, and myelencephalon.

6. Identify the 12 pairs of cranial nerves (some of the connections between brain and body) and correlate with their functions.

Organization of the Brain: Major Regions and Landmarks

The human brain contains at least 35 billion neurons, holds 98% of the neuronal tissue of the body and weighs about three pounds. The adult brain is organized into five divisions: (1) the *telencephalon*, (2) the *diencephalon*, (3) the *mesencephalon*, (4) the *metencephalon*, and (5) the *myelencephalon*. [∞ p383] These divisions are based upon embryological and fetal development. Due to the fragile nature of the brain, it is well protected by the cranial bones, cerebrospinal fluid, and the cranial meninges.

Procedure

✓ **Quick Check**

Before you begin to examine the human brain, remove the calvaria (the "skullcap" formed by the occipital, parietal, and frontal bones) on a dry human skull specimen (or model) and review the bones of the cranium (*Figures 6-3 to 6-5, 6-8 to 6-10, pp41-42, 45-47*), the bony landmarks of the cranial bones, and the cranial fossae (anterior, middle, and posterior) (*Color Plate 33*). Form a three-dimensional image of the space occupied by the brain. It fits into this space as does a piece into the middle of a puzzle.

Identify the following **major anatomical regions, divisions**, and **landmarks** in the adult brain, using a *model, specimen* of the human brain, or a *mammalian brain specimen.* Use *Color Plates 60* and *61* for reference. The **cerebrum** is the largest, most obvious region of the brain. Observe the surface of the cerebrum and note the folds, called **gyri**, which are separated by shallow depressions, termed **sulci**, or in selected areas by deep grooves, called **fissures**. Identify the **longitudinal fissure**, separating the cerebrum into left and right *cerebral hemispheres.* [∞ p392–393] The two hemispheres are held together at the inferior edge of the longitudinal fissure by nerve fibers. Use *Color Plates 60* and *63* for reference.

Identify the five divisions of the brain: (1) the **telencephalon** (*cerebrum*), (2) the **diencephalon** (*thalamus, hypothalamus*), (3) the **mesencephalon** (*midbrain*), (4) the **metencephalon** (*cerebellum, pons*) and (5) the **myelencephalon** (*medulla oblongata*), using *Color Plates 60, 61, 66, 67* and *Figure 15-1* for reference. The mesencephalon, metencephalon, and myelencephalon are collectively termed the **brain stem**. Observe the divisions of the brain with the intact brain and then in mid-sagittal section.

Your instructor may also provide you with a sheep brain to examine and dissect. The descriptions that follow for the examination of the human brain also apply to the sheep brain. But keep in mind that the references for directions (e.g., superior) are for humans and not for quadruped animals.

See: Color Plates 60, 61, 66, 67, Figs. 15-1

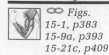

∞ *Figs.*
15-1, p383
15-9a, p393
15-21c, p409

LOCATE

Overview of the Brain

____ Cerebrum (right/left cerebral hemispheres)

____ Longitudinal fissure

____ Cerebellum

____ Pons

____ Medulla oblongata

____ Telencephalon (cerebrum)

____ Diencephalon (hypothalamus and thalamus)

____ Mesencephalon (midbrain)

____ Metencephalon (cerebellum & pons)

____ Myelencephalon (medulla oblongata)

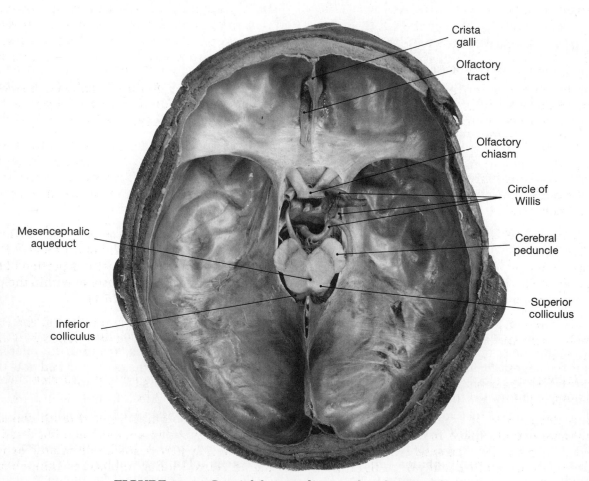

FIGURE 15-1 Cranial fossae, dissected and exposed.

Observation of the Meninges

The brain and spinal cord are encased by a connective tissue sheet called **meninges**, which form three distinct layers, observing from outside to inside: (1) *dura mater*, (2) *arachnoid*, and (3) *pia mater*. [∞ p384] The meninges protect the CNS and provide support for blood vessels, nerves, and cerebrospinal fluid. **Cerebrospinal fluid (CSF)** acts as a shock absorber and a medium by which nutrients, waste, and other metabolic products can be transported. [∞ p389] The cranial meninges are continuous with the spinal meninges, which are described and identified in Chapter 13.

Procedure

✓ Quick Check

Before you begin to examine the cranial meninges, remove the calvaria on a dry human skull specimen (or model) and review the bony landmarks of the cranial bones: crista galli, petrous portion of temporal bone, internal occipital crest, sella turcica, foramen magnum, and cranial fossae. (*Color Plate 33* and *Figures 6-9* and *6-10, pp46-47*) Also, review the description of the meninges described in *Chapter 14, p201*.

Identify the **cranial meninges** and **associated structures** that encase and protect the brain, along with the **cranial bones** that provide support and further protection. Use a *brain meninges model* or *specimen* of the *human brain* and *cranium* of prosected cadaver for this identification. Use Figures *15-2* and *15-3* for reference.

The **dura mater** can be identified as the tough, pearl-white outer connective tissue meningeal layer that encases the brain and spinal cord, dividing the space surrounding them into **epidural** and **subdural spaces**. The dura mater is actually two layers laminated together. Identify the outer *endosteal* layer, which is fused to the cranial bones and the smooth glistening inner meningeal layer. The *middle meningeal artery*, the blood vessel that supplies the cranial bones and dura mater, may still be visible in the endosteal layer of your specimen. On the deep surface of the calvaria, the grooves for this artery and its branches can easily be observed. Between outer and inner layers, pockets of space are visible that contain large veins, collectively called **dural sinuses**. [∞ p386] The veins of the brain open into the dural sinuses, which return venous blood to the internal jugular vein of the neck. (see *Color Plate 89*)

Within the cranial cavity, the inner dura mater is connected at four points to various bony landmarks. These attachment points serve to further stabilize the brain within the meninges. Within the cadaver cranium (calvaria removed) or on a model with meninges (or demonstration specimen) identify the following.

1. The **falx cerebri** is a crescent-shaped portion of the dura mater that lies within the longitudinal fissure and between left and right cerebral hemispheres (*Figure 15-2*), attaching anteriorly to the crista galli and posteriorly to the tentorium cerebelli and internal occipital crest. [∞ p386] The superior border attaches in the midline to the cranial bones. The **superior** and **inferior sagittal sinuses** can be identified within the falx. (*Figure 15-2*)

2. The **tentorium cerebelli** is a horizontal fold of dura mater that physically separates or partitions the cerebral hemispheres from those of the cerebellum. The *occipital lobes* of the cerebral hemispheres rest on the tentorium. It extends across the cranium, at almost a right angle, to attach to the petrous portion of the temporal bone. The falx cerebri attaches posteriorly to the tentorium cerebelli. Observe within the posterior area of the tentorium cerebelli the **right** and **left transverse sinuses**. (*Figure 15-2*)

3. The **falx cerebelli** lies in the midline inferior to the tentorium cerebelli and separates the left and right cerebellar hemispheres. (*Figure 15-2*) Its posterior border attaches to the internal occipital crest of the occipital bone.

4. The **diaphragma sellae** is that portion of the dura mater that forms the roof of the sella turcica of the sphenoid bone to encase the pituitary gland inferior to the brain. (*Color Plate 33*)

Identify deep to the dura mater the thin **arachnoid**, which forms the middle layer and in life contains CSF in the **subarachnoid space**. In the cadaver the subarachnoid space is collapsed. Keep in mind that once you reflect the dura mater, the **subdural space** that separates dura mater from arachnoid is now removed. A traumatic blow to the skull may result in the rupture of bridging veins, leading to bleeding into the subdural space and the subsequent subdural hemorrhage (hematoma).

Deep to the arachnoid identify the innermost cranial meninx, the delicate **pia mater**. Note how the pia mater follows and adheres tightly to the winding surface and fissures of the brain. The pia mater acts as a supporting floor to carry the large cerebral blood vessels as they branch and invade the neural tissue of the brain.

See: Figs. 15-2 to 15-4

∞ *Figs.*
15-3, p387
15-4, p388
15-6, p390

LOCATE

____ Cranial meninges
{ Dura mater
Arachnoid
Pia mater

____ Cerebral hemispheres

____ Cerebellum/medulla oblongata/spinal cord

____ Dura mater (connects at four locations in the cranial cavity)
{ Falx cerebri
Tentorium cerebelli
Falx cerebelli
Diaphragma sellae

____ Sinuses within the dura mater
{ Superior sagittal sinus
Inferior sagittal sinus
Transverse sagittal sinus
Straight sinus

____ Structures and spaces of the arachnoid
{ Arachnoid trabeculae
Subdural space
Subarachnoid space

____ Pia mater

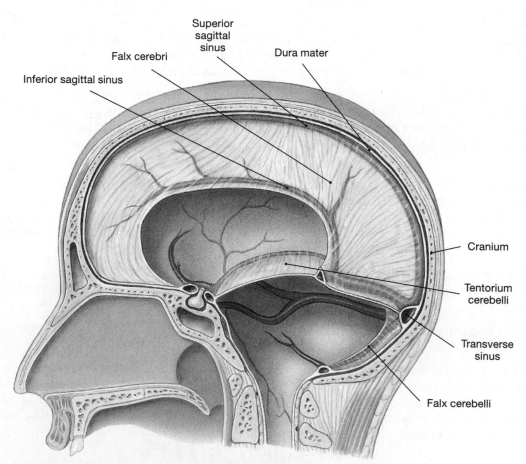

Superior sagittal sinus

Falx cerebri

Dura mater

Inferior sagittal sinus

Cranium

Tentorium cerebelli

Transverse sinus

Falx cerebelli

FIGURE 15-2a View of cranial cavity with brain removed showing the orientation and extent of the falx cerebri and tentorium cerebelli.

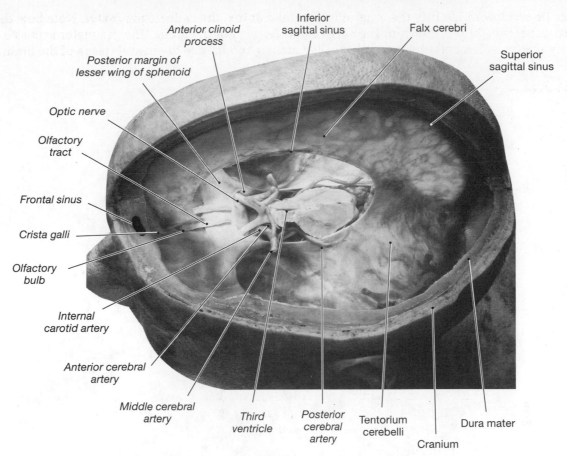

FIGURE 15-2b Superior view of the open cranial cavity with the
cerebrum and diencephalon removed.

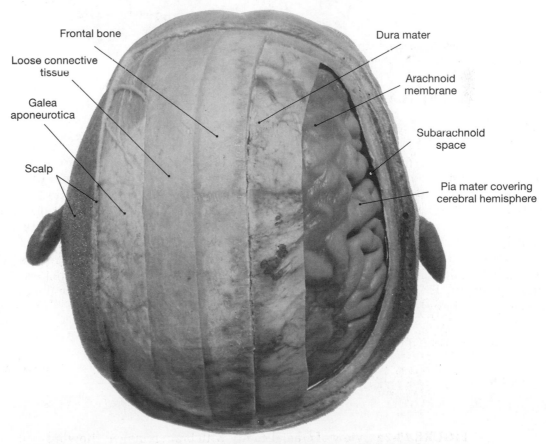

FIGURE 15-3 Meninges as seen in a step dissection of the scalp.

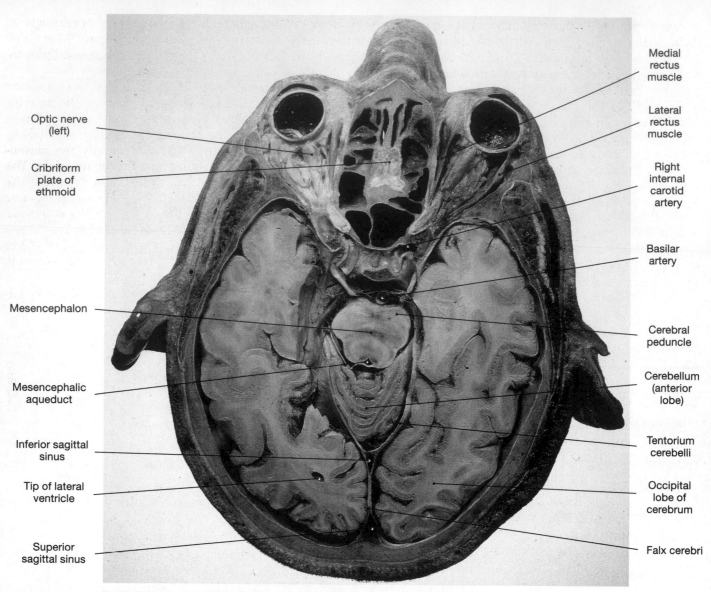

Optic nerve (left)

Cribriform plate of ethmoid

Mesencephalon

Mesencephalic aqueduct

Inferior sagittal sinus

Tip of lateral ventricle

Superior sagittal sinus

Medial rectus muscle

Lateral rectus muscle

Right internal carotid artery

Basilar artery

Cerebral peduncle

Cerebellum (anterior lobe)

Tentorium cerebelli

Occipital lobe of cerebrum

Falx cerebri

FIGURE 15-4   Brain, horizontal section.

Observation of the Ventricles, Choroid Plexus, and Circulation of Cerebrospinal Fluid

The brain and spinal cord develop in a unique way that allows the formation within the brain of hollow, fluid-filled cavities, termed **ventricles**. [∞ p337] The ventricles are interconnected and they communicate with the central canal of the spinal cord. Vascular folds in the ventricle walls are formed by specialized ependymal cells lining portions of the ventricles and the underlying vascular pia mater. This vascular coat, the *choroid plexus*, produces cerebrospinal fluid (CSF) by the selective absorption of blood components that are then secreted into the ventricular cavity. [∞ p389] CSF circulates and cascades through the ventricles, into the central canal of the spinal cord and the subarachnoid space that surrounds the brain and spinal cord. It serves as the medium for the transport of nutrients, chemical messengers, and waste products, and it has the crucial function of floating the brain. Thus, the brain floats within the meninges. The buoyancy provided by the CSF is enough to prevent damage to the brain by its own weight.

Procedure

✓ **Quick Check**

Before you begin, review the location of the superior sagittal sinus. [*Figure 15-2, p218*; ∞ Figure 15-3, p383]

Locate and identify the **ventricles** of the **brain** and their **associated channels** for distribution of **cerebrospinal fluid**, using a *brain model* or specially prepared *specimens* of *coronal sections* of a *human brain*. Use *Color Plates 62 to 64, Figures 15-4, 15-5 and 15-6* for reference. Additionally, identify the area of the **choroid plexus**, first using a model of the human brain, and then a *demonstration specimen*.

Identify the **lateral ventricles** (first and second) in the cerebral hemispheres, which lie lateral to either side of the midline, but are separated by a thin curtain of tissue, the **septum pellucidum**. Observe how the anterior and inferior portions of the two lateral ventricles communicate with a **third ventricle** by means of a delicate drain tube, the **interventricular foramen (foramen of Monro)**. The third ventricle lies in the midline of the diencephalon directly inferior and medial to the lateral ventricles. From the third ventricle, identify the **mesencephalic aqueduct (cerebral aqueduct)** and trace it inferiorly as it leads to a chamber, the **fourth ventricle**. The fourth ventricle lies between the cerebellum and the pons/medulla oblongata. (see *Color Plate 64*) From the fourth ventricle, CSF drains into the central canal of the spinal cord and into the subarachnoid space. It must be noted that very little CSF actually reaches the central canal. Re-observe the specimen and note the choroid plexus within the medial walls of the lateral ventricles, and on the roof of the third and fourth ventricle. Trace the circulation of cerebrospinal fluid on the model or specimen provided. Begin your tracing at the lateral (first and second) ventricles. As you trace the circulation of CSF, identify each structure involved. The vein receiving the reabsorbed CSF is the superior sagittal sinus

If available, view the ventricles of the brain in frontal (coronal) and horizontal section demonstration specimens.

See: Color Plates 62 to 64, Figs. 15-4 to 15-6

∞ *Figs.*
15-2, p385
15-6, p390
15-11, p396
15-19, p407

LOCATE

____ Lateral ventricles (1st and 2nd)

____ Septum pellucidum

____ Interventricular foramen (foramen of Monro)

____ Third ventricle

____ Mesencephalic aqueduct (cerebral aqueduct)

____ Fourth ventricle

____ Choroid plexus of lateral (1st, 2nd), and 3rd ventricles

____ Choroid plexus of 4th ventricle

____ Lateral/median apertures of 4th ventricle

____ Central canal of spinal cord

____ Subarachnoid space

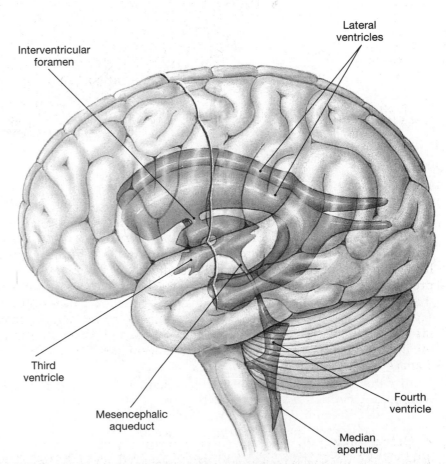

FIGURE 15-5 Ventricular cavities of the brain.

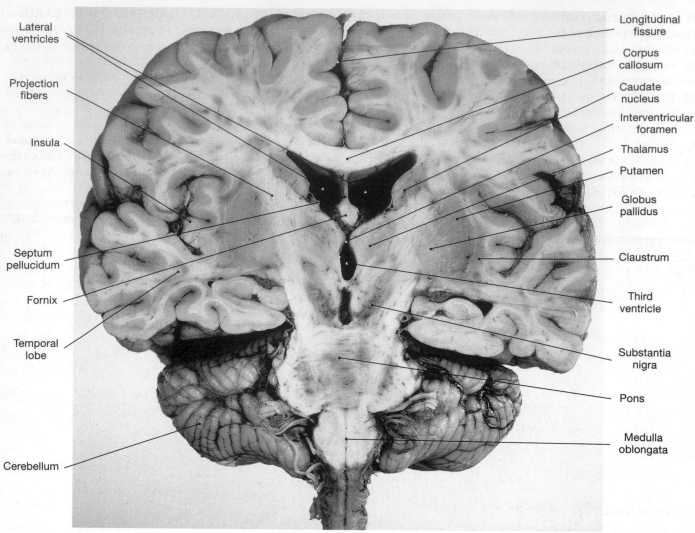

Lateral ventricles

Projection fibers

Insula

Septum pellucidum

Fornix

Temporal lobe

Cerebellum

Longitudinal fissure

Corpus callosum

Caudate nucleus

Interventricular foramen

Thalamus

Putamen

Globus pallidus

Claustrum

Third ventricle

Substantia nigra

Pons

Medulla oblongata

FIGURE 15-6   Frontal section of the brain.

Observation of the Telencephalon (Cerebrum)

From your previous examination of the brain, you observed that the cerebrum is clearly the largest region of the brain. Conscious thought, all intellectual processes (e.g., imagination), the processing of somatic and motor information (e.g., interpretation and appreciation of sensations) all occur in the cerebral hemispheres. [∞ p391]

Procedure

✓ **Quick Check**

Have a dry human skull available for immediate comparison and reference. When attempting to identify regions in the telencephalon, refer to the skull and make correlations between the appropriate structures and the space occupied within the skull.

Identify the following **structures** of the **cerebrum**, first using a *model* and then a *specimen* of the *human brain*. Hold the intact brain model in your hand or position it on a stand so that both lateral and superior surfaces are facing you. Use *Color Plates 60, 61, 66,* and *67* for reference. Identify left and right cerebral hemispheres and the longitudinal fissure. Use these structures as landmarks in your continuing exploration. Begin your identification at the midpoint of the superior surface of the cerebrum along the edge of the longitudinal fissure. From this point, proceed laterally and inferiorly on the surface of each hemisphere. Locate the deep **central sulcus**, and follow it to the deep **lateral sulcus**. The lateral sulcus runs horizontally at the level of the squamosal suture (refer to the human skull for reference point) (see *Figures 6-6* and *6-10, pp43, 47*). Return to the longitudinal fissure. In the area of the cerebrum that would be covered by the posterior border of the parietal bones [refer to the human skull], locate the **parieto-occipital sulcus**, which runs laterally from the fissure. [∞ p393]

Return to the central sulcus and locate, just anterior to it, the slightly raised cerebral fold known as the **precentral gyrus** and, just posterior to the sulcus, the slightly raised cerebral fold known as the **postcentral gyrus**. The gyri provide a way to increase the number of neurons in the cortical regions by increasing the surface area of the cerebral hemispheres. [∞ p393]

The cerebrum is divided functionally into four lobes, each named for the cranial bone that covers it [refer to the human skull]. The **frontal lobe** is anterior to the central sulcus. Inferior to the lateral sulcus is the **temporal lobe**. The cerebral area most posterior and inferior to the parieto-occipital sulcus is the **occipital lobe**. The cerebral area between the central sulcus, parieto-occipital sulcus, and lateral sulcus is the **parietal lobe**. Keep in mind that all lobes have both left and right parts.

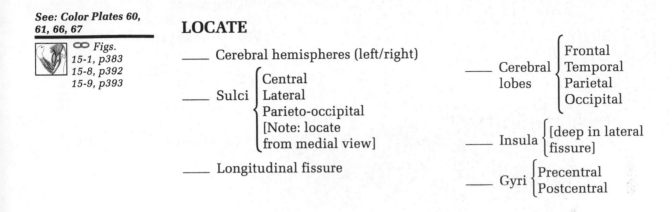

See: Color Plates 60, 61, 66, 67

∞ *Figs. 15-1, p383 15-8, p392 15-9, p393*

LOCATE

____ Cerebral hemispheres (left/right)

____ Sulci {
Central
Lateral
Parieto-occipital
[Note: locate
from medial view]
}

____ Longitudinal fissure

____ Cerebral lobes {
Frontal
Temporal
Parietal
Occipital
}

____ Insula {
[deep in lateral fissure]
}

____ Gyri {
Precentral
Postcentral
}

Observation of the Cerebral Cortex

The outer surface of the cerebral hemispheres is a relatively thin layer of gray matter (primarily neuron cell bodies) covering white matter (formed of myelinated nerve fibers). The **cerebral cortex** is the layer of gray matter that forms the surface of gyri and sulci. Functionally, the cerebrum is divided into lobes, with the left and right hemispheres being different. The right side of the brain processes information received from the left side of the body and vice versa. As you identify anatomical structures, keep in mind that the boundaries of functional areas do overlap and cross over anatomical structures. Functional areas of the brain include the *motor cortex, sensory cortex, association areas*, and *integrative areas*. [∞ p394]

Procedure

Identify the **motor** and the **sensory regions** of the **cortex** and their connections, the **association areas**, using a *model* of the *human brain* and/or specimen and *Figure 15-7* for reference. Locate first the sulci and the gyri of the cerebrum, then locate frontal, temporal, parietal, and occipital lobes.

Specific functional activities are associated with the separate lobes. The voluntary control of skeletal muscles occurs in the **primary motor cortex** area, located in the precentral gyrus region of the frontal lobe. The conscious awareness of sensations (e.g., temperature) occurs in the **primary sensory cortex**, located in the postcentral gyrus region of the parietal lobe. The special senses have their own regions of the cerebral cortex; the occipital lobe for vision, the temporal lobe for hearing and smell and the parietal lobe for taste. Surrounding most primary sensory cortex areas are association areas that serve to integrate and process sensory data (e.g., color). Information from several or many **association areas** is received, integrated, and processed within **integrative areas**. For example, the initiating of complex motor activities like the act of writing are performed by integrative areas of the parietal lobe.

Use a coronal (frontal) section of a brain specimen to identify *gyri, sulci, gray matter,* and *white matter* of the *cerebral cortex*. This type of section shows clearly the relationships of the structure and areas of the cerebral cortex. Use *Color Plate 62* and *Figure 15-6* for reference.

See: Fig. 15-7

 *Fig.*
15-9b, p393

LOCATE

____ Lobes
- Frontal
- Parietal
- Temporal
- Occipital

____ Sulci
- Central
- Lateral
- Parieto-occipital

____ Cortex
- Primary motor (precentral gyrus)
- Visual
- Auditory
- Olfactory
- Primary sensory (postcentral gyrus)

____ Association areas
- Somatic motor (premotor cortex)
- Visual association
- Auditory association
- Sensory association

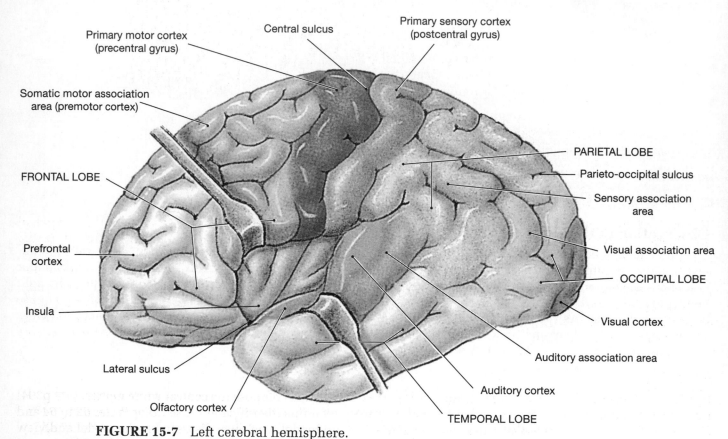

FIGURE 15-7 Left cerebral hemisphere.

The Neural Cortex

The neurons of the primary motor cortex direct voluntary movements by controlling somatic motor neurons in the brain stem and spinal cord. These neurons are called **pyramidal cells**. [∞ p394]

Procedure

Identify **pyramidal cells** using the *microscope slide* provided, first by viewing under low power magnification and then under high power magnification. Use *Figure 15-8* for reference. Pyramidal cells are located within the gray-matter portion of the neural cortex. Look for these cells in the middle of the neural cortex. These cells can be easily distinguished from other cell types under high power magnification. Their characteristic triangular shape, ink-black color, and numerous twig-like branches projecting from the cell body conveniently distinguishes pyramidal cells from other cell types.

Slide # _____ Cerebral Cortex, sec.

LOCATE

_____ Pyramidal cells within neural cortex

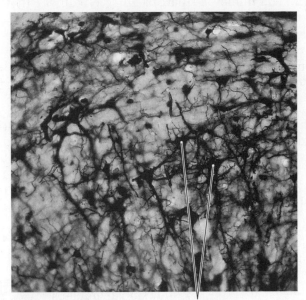

Pyramidal cells
within cerebral
gray matter
(neural cortex)

FIGURE 15-8 Pyramidal cells × 1000.

Observation of the Central White Matter

The bulk of the cerebral hemispheres consists of white matter, or simply nerve fibers. There are three types of myelinated fibers (axons) in the central white matter that link regions of the brain to each other and/or to the spinal cord. _Association, commissural,_ and _projection fibers_ provide the links. [∞ p394] Within a cerebral hemisphere, **association fibers** connect different gyrus regions. Large bundles of **commissural fibers** of the _corpus callosum_ link left and right hemispheres. The _anterior commissure_ is another prominent bundle of commissural fibers linking the hemispheres. **Projection fibers** relay impulses from the cerebral cortex to the brain stem, cerebellum, and spinal cord. Collectively, these projection fibers compose the fan-shaped **internal capsule**, which can be observed lateral to the _thalamus._

Procedure

Identify the location of the **three major groups** of fibers (axons), which compose the **central white matter**. [∞ p394] Use a _model_ of the human _brain_ and a demonstration _specimen_ for this identification. Use _Color Plates 62_ to _64_ and _Figure 15-6_ for reference. To observe the white matter on a brain model, you must disassemble the model and view the cerebral hemispheres in coronal and or horizontal sections. Identify the white matter on frontal and horizontal sections of brain demonstration specimens.

See: Color Plates 62 to 64, Fig. 15-6

∞ _Figs. 15-10, p395 15-11, pp396-397 15-12b, p398_

LOCATE

_____ Central white matter
  { Association fibers
    Commissural fibers
    Projection fibers

_____ Commissural fibers
  { Corpus callosum
    Anterior commissure

_____ Association fibers
  { Arcuate fibers
    Longitudinal fasciculi

_____ Projection fibers
  { Internal capsule

Observation of the Basal Nuclei

The **basal nuclei** or **basal ganglia** are collections of cell bodies that lie in the central white matter close to the lateral ventricles within each cerebral hemisphere. [∞ p394] The names of the nuclei are based upon their shapes, positions, or other attributes. They are intermediate centers between the cerebral cortex and other control regions of the brain. The nuclei are part of a nerve pathway that controls muscle tone and coordinates learned movement patterns and other somatic motor activities.

Procedure

Identify the **basal nuclei** or **basal ganglia**, using a *model* and or *specimen* of the *human brain*. The basal nuclei are masses of gray matter (nerve cell bodies) that are located inferiolaterally to the floor and lateral walls of the lateral ventricles. To observe the basal nuclei on a brain model, you must disassemble the model and view the cerebral hemispheres in coronal and or horizontal sections. Identify within the frontal lobe the **caudate nucleus**, which is the large comma-shaped mass of gray matter that runs from anterior to posterior along the lateral wall of each lateral ventricle. At the tail of the caudate nucleus (posterior portion) lies the almond-shaped **amygdaloid body**. Functionally, it belongs to the *limbic system*. Identify the largest mass of gray matter, the **lentiform nucleus** (composed of the *putamen* and *globus pallidus*). Lateral to the lentiform nucleus is the **claustrum**. The **corpus striatum** (striated body) encompasses the *caudate* and *lentiform* nuclei.

View a brain demonstration specimen in coronal (frontal) and horizontal section to identify the cerebral nuclei. Use *Color Plates 62* and *63* and *Figures 15-4* and *15-6* for reference.

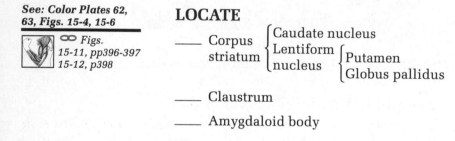

See: Color Plates 62, 63, Figs. 15-4, 15-6

∞ Figs. 15-11, pp396-397 15-12, p398

LOCATE

____ Corpus striatum ⎰ Caudate nucleus
⎱ Lentiform nucleus ⎰ Putamen
⎱ Globus pallidus

____ Claustrum

____ Amygdaloid body

Observation of the Limbic System

The **limbic system** includes the gyri nuclei and tracts along the border of the cerebrum and diencephalon. [∞ p395] It establishes emotional states and related behavioral drives, consolidates learning, and links our emotions to rational thinking.

Procedure

Identify the following **gyri nuclei** and **tracts** along the border between the cerebrum and diencephalon, which are collectively termed the **limbic system**, using a *model* and or *specimen* of the *human brain*. To observe the structures of the limbic system on a brain model, disassemble the model and view the brain in midsagittal section. Use *Color Plates 64* and *66* and *Figure 15-9* for reference.

Identify the **limbic lobe** as the deepest portion of the cerebrum, lying superior to the corpus callosum, and follow its curve to the medial surface of the temporal lobe. The limbic lobe consists of two gyri, the *cingulate gyrus* and the *parahippocampal gyrus*. The **cingulate gyrus** can be identified as lying deep within the longitudinal fissure, but just superior to the corpus callosum. The **parahippocampal gyrus** and the **hippocampus** lie just medial to the lateral ventricles. The hippocampus and *hypothalamus*, a small region in the diencephalon, which is critical for many life functions [∞ p400], are connected together by a tract of white matter, the **fornix**. Many of these white matter fibers end in the **mamillary bodies**, which are paired structures located in the floor of the hypothalamus. The mamillary bodies control reflex movements associated with eating (i.e., chewing and swallowing).

View a brain demonstration specimen in midsagittal section to identify the above structures of the limbic system.

See: Color Plates 64, 66, Fig. 15-9

∞ Fig.
15-12, p398

LOCATE

____ Limbic lobe {
Cingulate gyrus
Parahippocampal
gyrus
}

____ Temporal lobe

____ Corpus callosum

____ Hippocampus

____ Fornix

____ Mamillary body

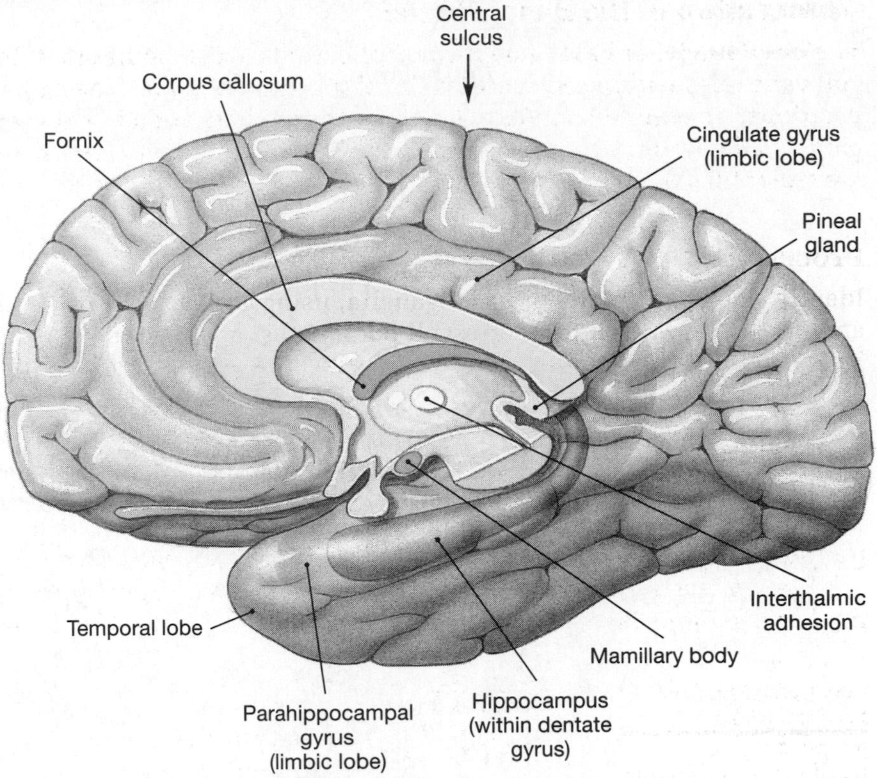

FIGURE 15-9 The limbic system, sagittal section.

Observation of the Diencephalon (Thalamus and Hypothalamus)

The **diencephalon** or **forebrain** is a small region that serves as a relay and integrative center for much of the stimuli received by the body and the responses consequent to these stimuli. Most of the neural tissue of the diencephalon is concentrated into two structures, the thalamus and the hypothalamus. [∞ p400]

Procedure

Identify the location and structure of the **diencephalon** and its relationship to other landmarks of the brain, using a *model* and or *specimen* of the *human brain*. To observe the structures of the diencephalon on a brain model, disassemble the model and view the brain in midsagittal section. After identifying the structures of the diencephalon in midsagittal section, if your model can be disassembled to isolate and show the diencephalon, then view these structures from this portion of the model. Use *Color Plate 64* and *Figure 15-10* for reference.

The diencephalon can be located immediately inferior to the lateral ventricles and completely encases the third ventricle. The roof of the diencephalon, called the **epithalamus**, has an extensive area of choroid plexus and contains the **pineal gland**, which is an endocrine structure that secretes the hormone *melatonin*. Melatonin is critical in setting and maintaining our internal clock for day and night cycles.

Identify the **thalamus** portion of the diencephalon as the structure that extends from the anterior commissure to the pineal gland. The thalamus consists of two lobes that are connected medially by a round projection of gray matter, called the **interthalmic adhesion** (seen in midsagittal section). Each lobe of the thalamus contains groups of **thalamic nuclei**, which act as relay and processing centers for sensory information.

Now identify the **hypothalamus** portion of the diencephalon as the region that extends from the optic chiasma to the mamillary bodies. The hypothalamus contains centers for controlling emotions, autonomic functions (e.g. regulating heart rate), and production of hormones. The *pituitary gland* [∞ p510] is connected to the floor of the hypothalamus by the **infundibulum**. This gland is located in the sella turcica and is held in position by the diaphragma sellae. Locate the infundibulum, a thin stalk that connects the pituitary to the hypothalamus, immediately posterior to the optic chiasma with the pituitary gland attached. Between the optic chiasma and the mamillary body, identify the **tuber cinereum** and a raised area of the tuber called the **median eminence**.

View a brain demonstration specimen in midsagittal section to identify the structures of the diencephalon and use the same color plates for reference.

See: Color Plate 64,
Fig. 15-10

∞ *Figs.*
15-8, p392
15-13a,b, p399
15-14, p401
15-16, p404

LOCATE

____ Epithalamus	____ Optic chiasma
____ Pineal gland	____ Infundibulum
____ Thalamus (walls of the diencephalon surrounding the 3rd ventricle)	____ Hypothalamus (floor of the 3rd ventricle) { Tuber cinereum / Paraventricular nucleus / Supraoptic nucleus / Preoptic area
____ Interthalmic adhesion	

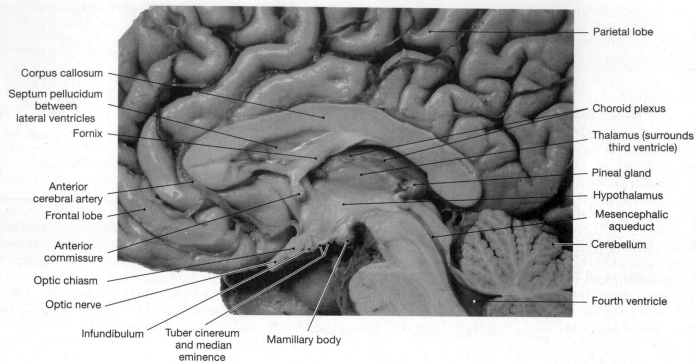

Labels on image: Corpus callosum, Septum pellucidum between lateral ventricles, Fornix, Anterior cerebral artery, Frontal lobe, Anterior commissure, Optic chiasm, Optic nerve, Infundibulum, Tuber cinereum and median eminence, Mamillary body, Parietal lobe, Choroid plexus, Thalamus (surrounds third ventricle), Pineal gland, Hypothalamus, Mesencephalic aqueduct, Cerebellum, Fourth ventricle

FIGURE 15-10 Midsagittal section through the diencephalon.

Observation of the Mesencephalon (Midbrain)

The **mesencephalon** is centrally located at the superior end of the brain stem. From this vantage point, it has structures that process visual and auditory information, initiate involuntary somatic motor responses, and maintain our conscious awareness. [∞ p403]

Procedure

Identify the location and structure of the **mesencephalon**, and its relationship to other landmarks of the brain, using a *model* and or *specimen* of the *human brain*. To observe the structures of the mesencephalon on a brain model, disassemble the model and view the brain in midsagittal section. After identifying the structures of the mesencephalon in midsagittal section, if your model can be disassembled to isolate and show only the brain stem, then view these structures from this portion of the model. Use *Color Plate 64* and *Figure 15-10* for reference.

Return to the thalamus and immediately inferior to it identify the **mesencephalon** portion of the brain. Its roof (dorsal side) is the **tectum**, which contains two pairs of sensory nuclei, the **superior colliculi** and the **inferior colliculi**. Collectively, they are called the **corpora quadrigemina**. Identify on the ventrolateral surfaces of the mesencephalon the **cerebral peduncles**. The axons in the peduncles connect the motor cortex with the motor neurons in the brain stem and spinal cord. The peduncles also contain ascending fibers that carry sensory information to the thalamus. Additionally, the mesencephalon contains the **red nucleus**, which integrates information from the cerebrum and cerebellum to maintain muscle tone and posture, and the **substantia nigra**, which helps regulate motor output of the cerebral nuclei.

View a brain demonstration specimen in midsagittal section to identify the structures of the mesencephalon and use the same color plates and Figure 15-10 for reference.

See: Color Plate 64, Fig. 15-10

∞ Figs.
15-13, p399
15-15, p402
15-16, p404

LOCATE

____ Thalamus

____ Tectum { Superior colliculus
(roof) { Inferior colliculus

____ Corpora quadrigemina

____ Wall and { Red nucleus
floor { Substantia nigra
structures

____ Cerebral peduncles

Observation of the Metencephalon (Cerebellum and Pons)

The **metencephalon** is made up of two portions, the *cerebellum* and the *pons*. [∞ p405] The primary function of the cerebellum is to adjust activity of skeletal muscles of the body (relaxed versus contracted) for the appropriate patterns of contractions to accomplish the desired voluntary or involuntary movements. The pons acts as a bridge between the cerebellar hemispheres and the mesencephalon, diencephalon, cerebrum, and spinal cord.

Procedure

Identify the location of the **metencephalon**, its two components, the cerebellum and pons, and its relationship to other landmarks of the brain, using a *model* and or *specimen* of the *human brain*. Observe the structures of the metencephalon on the intact brain model, then disassemble the model and view it in midsagittal section. Use *Color Plates 60, 61, 64, 66*, and *Figures 15-11, 15-12* for reference. After identifying the structures of the metencephalon in midsagittal section, if your model can be disassembled to show the brain stem, then also view the structures of the metencephalon from this portion of the model.

Identify the **cerebellum** on the intact brain model immediately inferior to and posterior to the cerebral hemispheres. [see *Color Plates 61* and *66*] In life, the two cerebral hemispheres and cerebellum are separated by the tentorium cerebelli. [see *Figure 15-3b*] Observe the subtle folds, called **folia**, on the surface of the cerebellum. View from a superior and posterior vantage to identify the **anterior** and **posterior lobes** of the cerebellum. Notice that a narrow fissure (very subtle or absent in some models), called the **primary fissure**, separates the two lobes. The cerebellum, like the cerebrum, consists of left and right hemispheres, termed **cerebellar hemispheres**. Identify these from both a superior and posterior view. Inferior and anterior to these hemispheres, identify the small **flocculonodular lobe**. View the brain in midsagittal section to identify the **vermis**, a narrow band of cerebellar cortex that lies in the midline of the cerebellum. Observe the pattern of white matter in the cerebellar hemispheres that resembles a tree, termed **arbor vitae**.

Just inferior to the flocculonodular lobe, in the lateral walls of the metencephalon (which encase the fourth ventricle) lie the **superior, middle**, and **inferior cerebellar peduncles**. These are tracts that link the cerebellum to the midbrain, pons, and medulla. Observe on the ventral surface of the brain an obvious bulge just inferior to the mesencephalon that appears to connect left and right cerebellar hemispheres. Identify this structure as the **pons**. The fiber tracts of the pons link together the cerebellar hemispheres with the cerebrum, diencephalon, mesencephalon, and spinal cord.

Identify the structures of the metencephalon on an intact brain demonstration specimen, then view a midsagittal sectioned specimen. Use the same color plates and *Figures 15-11* and *15-12* for reference.

See: Color Plates 60, 61, 64, 66, Figs. 15-11, 15-12

∞ Figs.
15-8, p392
15-12, p398
15-19, p407

LOCATE

Cerebellum

____ Folia

____ Lobes { Anterior / Posterior / Flocculonodular

____ Primary fissure

____ Vermis

____ Cerebellar hemispheres

____ Cerebellar cortex

____ Arbor vitae

____ Cerebellar peduncles { Superior / Middle / Inferior

____ Medulla oblongata

____ Choroid plexus of 4th ventricle

Pons

____ Gray matter

____ White matter

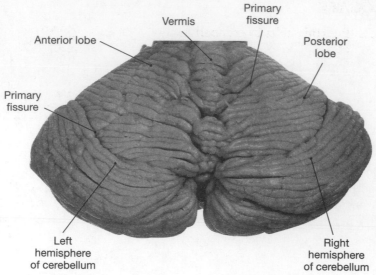

FIGURE 15-11   Superior view of the cerebellum.

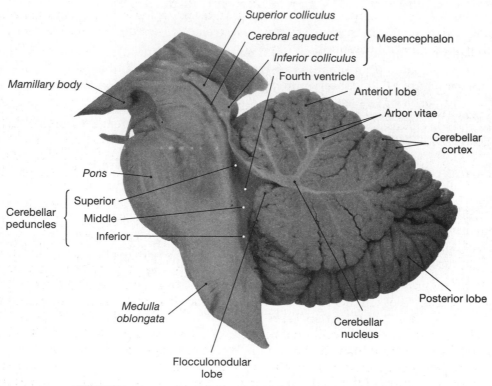

FIGURE 15-12   Sagittal view of the cerebellum.

Observation of Cerebellar Cortex

Identify **Purkinje cells**, using the *microscope slide* provided, first by viewing the specimen under low power magnification and then under high power magnification to see the details of these cells. Locate these large cells under high power magnification in the middle portion of the gray matter (outermost layer) of the cortex. Depending upon the stain used for the specimen on your slide, the color of Purkinje cells may be either violet or black. Recognize Purkinje cells by their pear shape. You may be able to observe prominent dendrites as they project outward into the gray matter of the cerebellar cortex. Less prominent are axons, which project toward the white matter of the cerebellum (arbor vitae) through the granular layer of the cerebellar cortex. These may be difficult to observe in some slide preparations.

Slide # _____ Cerebellar Cortex, sec.

See: Fig. 15-13

∞ Fig.
15-19b, p407

LOCATE

_____ Purkinje cells in cerebellar
cortex

Dendrites projecting
into the gray matter
of the cerebellum

Soma of
Purkinje
cell

Axons of Purk-
inje cells project-
ing into the white
matter of the
cerebellum

FIGURE 15-13   Purkinje cells × 120.

Observation of the Myelencephalon (Medulla Oblongata)

The **medulla oblongata** is contained within the cranium and is continuous with the spinal cord, thus providing the link between brain and spinal cord. All communication between these two occurs via ascending and descending fiber tracts, which must pass through the medulla to reach their destination. The medulla also contains centers for regulating heart rate and breathing. [∞ p408]

Procedure

Using a _model_ and or _specimen_ of the _human brain_, identify the location and structure of the **medulla oblongata**, and relationship to the **pons** and **spinal cord**. Observe the medulla oblongata from the inferior surface on the assembled brain model, then disassemble the model and view it in midsagittal section. Use _Color Plates 64, 66_, and _Figure 15-14_ for reference. After identifying the structures of the medulla in midsagittal section, if your model can be disassembled to show the brain stem, then also view the structures of the medulla from this portion of the model.

On the intact brain model, identify the **medulla oblongata**, which is located immediately inferior to the pons. Notice how there is no sharp distinction between the spinal cord and medulla. On the ventral surface of the medulla oblongata are two longitudinal ridges called the **pyramids** formed by the large pyramidal (corticospinal) tracts descending from the motor cortex. On the dorsal surface of the medulla, the _nucleus gracilis_ and _nucleus cuneatus_ act as relay stations for sensory information that passes between the spinal cord and the brain. Observe the prominent bulges, called **olives**, on the ventral and lateral surfaces of the medulla. The olives are formed by the _olivary nuclei._ These nuclei act as relay stations for sensory and motor information that passes between the brain, the spinal cord, and the cerebellum. Typically, you will not be able to observe these nuclei because they are not depicted on standard brain models. Four cranial nerves (IX, X, XI, and XII) have their origins in the nuclei of the medulla. We will identify the cranial nerves in a separate observation.

View first an intact brain demonstration specimen (human mammalian) and then a midsagittal sectioned specimen to identify the structures of the medulla oblongata. Use the same color plates and _Figure 15-14_ for reference.

See: Color Plates 64, 66, Figs. 15-14

∞ Figs.
15-13 p399
15-16 p404
15-18, p406

LOCATE

_____ Pons

_____ Medulla oblongata

_____ Spinal cord

_____ Olives

_____ Olivary nuclei

_____ Nucleus gracilis

_____ Nucleus cuneatus

_____ Nuclei of cranial nerves { N VIII, N IX N X, N XI, N XII

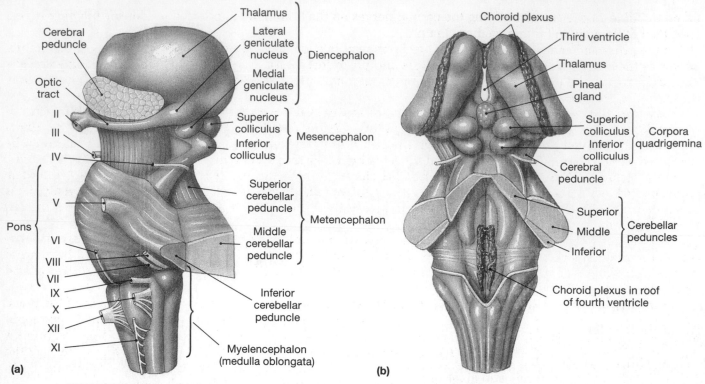

FIGURE 15-14   The brain stem (a) lateral view (b) posterior view.

Observation of the Cranial Nerves

Twelve pairs of nerves originate on the ventrolateral and dorsal surfaces of the brain to exit the cranial cavity. Collectively they are called the **cranial nerves**. [∞ p409] These nerves are part of the peripheral nervous system that connects to the brain. Typically, the cell bodies of neurons for the cranial nerves are located within the central nervous system or in cranial ganglia. Cranial nerves are sensory, motor, or mixed. Cranial nerves are named according to their appearance or function and are identified with Roman numerals. Numbering begins anteriorly at the cerebrum, continues posteriorly and inferiorly. It is common practice for anatomists to identify the cranial nerves with the prefix N followed by its Roman numeral. This system of identification clearly distinguishes cranial nerves from spinal nerves.

Procedure

✓ Quick Check

Before you begin to examine the cranial nerves, review on a dry human skull specimen the bony landmarks of the orbit, (*Figure 6-21, p54*) the cranial fossae regions, (*Color Plate 33*) and the foramina of the skull as seen from both internal and external views. (*Figure 6-8, p45*) Keep a skull specimen nearby to aid you in associating the cranial nerves with their foramina.

Identify the **twelve pairs** of **cranial nerves, associated structures**, and the **specific regions** of the **brain** from which they arise, using first a *model* and then a *human brain* or a mammalian *brain specimen*. Hold an intact brain model in your hand or position it on a stand so the inferior surface is facing you and the posterior surface is directed towards you. Use *Color Plates 33, 66* to *68* and *Figure 15-15* for reference. Identify all of the paired cranial nerves on the ventrolateral surface of the brain, except for the trochlear nerve (N IV), which originates on the dorsal surface. Begin your identification with cranial nerve N I located at the anterior of the cerebrum and work posteriorly to cranial nerve N XII, associating functions with each nerve. After identifying the cranial nerves on the intact model, if your model can be taken apart so you can work with just the brain stem portion of the model, then do so. By using just the brain stem portion of the model you will be able to observe clearly the origins of the nerve rootlets of the cranial nerves. A summary of the basic distribution and function of each cranial nerve is detailed in *Table 15-1*. After identifying the cranial nerves on the brain model, proceed to identify the nerves on a specimen of the human (or mammalian) brain.

In the prosected cadaver, the brain has been removed in such a way as to preserve the stumps of the cranial nerves as they exit the cranial (anterior, middle and posterior) fossae. (see *Figure 15-1* and *Color Plate 86*) The view from the cranial fossae of the cadaver permits an examination of the cut edges of each cranial nerve as they pass into their respective foramina.

Olfactory Nerve (N I) *SENSORY* Identify the **olfactory bulbs**, which lie on the anterior-inferior surface of the frontal lobes and on either side of the longitudinal fissure. (see *Color Plate 66*) The **olfactory nerve fibers** arise from the olfactory bulbs to pass through the *cribriform plate* of the ethmoid bone. These nerves cannot be identified on your brain model. Identify and trace the **olfactory tracts** from the olfactory bulbs to the cerebrum.

Optic Nerve (N II) *SENSORY* Identify just anterior to the sella turcica (see *Color Plate 33* and *67*) the **optic chiasma**. Proceed posteriorly and identify the left and right **optic tracts**, which connect to the lateral geniculates of the thalamus. From the optic chiasma, proceed anteriorly and identify the **optic nerves** that will pass through the left and right *optic canals* to enter the eyes.

Oculomotor Nerve (N III) *MOTOR* On either side of the midline, just anterior to the pons, observe the **oculomotor nerves**. (see *Color Plates 66* and *67*) Identify these small diameter nerves as they arise from the ventral surface of the mesencephalon to pass through the *superior orbital fissures* of the sphenoid to enter the left and right orbits.

Trochlear Nerve (N IV) *MOTOR* Identify on the ventrolateral surfaces of the anterior portion of the pons, the **trochlear nerves**. (see *Color Plates 66* and *67*) These are the smallest in diameter of all the cranial nerves and are the only cranial nerves to arise from the dorsal surface of the brain stem. They arise dorsally from the inferior colliculus of the corpora quadrigemina (see *Color Plate 64*) and wind laterally and ventrally to pass through the superior *orbital fissures* of the sphenoid. If your model can be taken apart so you can work with just the brain stem portion of the model, than do so in order to observe the dorsal origin of the trochlear nerve.

Trigeminal Nerve (N V) *MIXED* Identify on the lateral surfaces of the pons, the **trigeminal nerves**, (see *Color Plates 66* and *67*) which are the largest in diameter of all the cranial nerves. From your point of observation, the trigeminal nerve will lie over and lateral to the trochlear nerve. As the nerve passes over the superior border of the petrous portion of the temporal bone, (see *Color Plate 33*) its sensory and motor roots form the *trigeminal (semilunar) ganglion* from which emanate three nerve branches. Check with your instructor before proceeding to identify these branches, because they may only be observed on specially prepared demonstration models and specimens.

1. **Ophthalmic branch**, passes through the *superior orbital fissure*; sensory for orbital structures, the nasal cavity and sinuses, and skin of the forehead.

2. **Maxillary branch**, passes through the *foramen rotundum*; sensory for upper jaw structures.

3. **Mandibular branch**, passes through the *foramen ovale*; motor and sensory for innervating structures of the mandible.

Abducens Nerve (N VI) *MOTOR* Identify on either side of the midline, at the anterior portion of the medulla and just posterior to the pons, the **abducens nerves**. (see *Color Plates 66* and *67*) These nerves pass through the superior *orbital fissures* to enter the orbits.

Facial Nerve (N VII) *MIXED* Proceed laterally from the abducens nerves to identify the **facial nerves** as they also arise from the anterior portion of the medulla. (see *Color Plates 66* and *67*) The facial nerves enter the petrous portion of the temporal bone to pass through the *internal acoustic meatus* to exit at the *stylomastoid foramen*. (see *Color Plate 33*)

Vestibulocochlear Nerve (N VIII) *SENSORY* Identify the **vestibulocochlear nerve** at its origin just posterior to the facial nerve. The vestibulocochlear nerve runs both lateral to the facial nerve and with it into the *internal acoustic meatus*. (see *Color Plates 66* and *67*) This nerve contains two distinct sensory branches. Check with your instructor before proceeding to identify these branches, because they typically may only be observed on specially prepared demonstration models and specimens.

1. **Vestibular nerve** serves the inner ear for balancing sensations.

2. **Cochlear nerve** serves the inner ear for the sense of hearing.

Glossopharyngeal Nerve (N IX) *MIXED* Observe posterior to the olive (raised portion) on the ventrolateral surface of the medulla oblongata the rootlets of three nerves. The nerve rootlets appear in order as superior, middle and inferior rootlets. Identify the most superior rootlets as the **glossopharyngeal nerve**. (see *Color Plates 66* and *67*) This nerve passes through the *jugular foramen*. The middle and inferior rootlets are cranial nerves N X and N XI.

TABLE 15-1 The Cranial Nerves

Cranial Nerve (No.)	Sensory Ganglion	Branch	Primary Function	Foramen	Innervation
Olfactory (I)			Special sensory	Cribiform plate of ethmoid bone	Olfactory epithelium
Optic (II)			Special sensory	Optic foramen	Retina of eye
Oculomotor (III)			Motor	Superior orbital fissure	Inferior, medial, superior rectus, inferior oblique, and levator palpebrae muscles; intrinsic muscles of eye
Trochlear (IV)			Motor	Superior orbital fissure	Superior oblique muscle
Trigeminal (V)	Semilunar		Mixed		Areas associated with the jaws
		Ophthalmic	Sensory	Superior orbital fissure	Orbital structures, nasal cavity, skin of forehead, upper eyelid, eyebrows, nose (part)
		Maxillary	Sensory	Foramen rotundum	Lower eyelid; upper lip, gums, and teeth; cheek, nose (part), palate and pharynx (part)
		Mandibular	Mixed	Foramen ovale	*Sensory* to lower gums, teeth, lips; palate (part) and tongue (part); *motor* to muscles of mastication
Abducens (VI)			Motor	Superior orbital fissure	Lateral rectus muscle
Facial (VII)	Geniculate		Mixed	Internal acoustic meatus to facial canal; exits at stylomastoid foramen	*Sensory* to taste receptors on anterior 2/3 of tongue; *motor* to muscles of facial expression, lacrimal gland, submandibular salivary gland, sublingual salivary glands
Vestibulocochlear (Acoustic) (VIII)		Cochlear	Special sensory	Internal acoustic meatus	Cochlea (receptors for hearing)
		Vestibular	Special sensory	As above	Vestibule (receptors for motion and balance)
Glossopharyngeal (IX)	Superior and inferior (petrosal)		Mixed	Jugular foramen	*Sensory* from posterior 1/3 of tongue; pharynx and palate (part); carotid body (monitors blood pressure, pH, and levels of respiratory gases); *motor* to pharyngeal muscles, parotid salivary gland
Vagus (X)	Jugular and nodose		Mixed	Jugular foramen	*Sensory* from pharynx; pinna and external meatus; diaphragm; visceral organs in thoracic and abdominopelvic cavities; *motor* to palatal and pharyngeal muscles, and visceral organs in thoracic and abdominopelvic cavities
Accessory (XI)		Medullary (cranial) portion	Motor	Jugular foramen	Voluntary muscles of palate, pharynx and larynx (with branches of the vagus nerve)
		Spinal portion			Sternocleidomastoid and trapezius muscles
Hypoglossal (XII)			Motor	Hypoglossal canal	Tongue musculature

Vagus Nerve (N X) *MIXED* Using the glossopharyngeal nerve (N IX) as a landmark, locate the middle rootlets of the nerve group. Identify the middle rootlet as the **vagus nerve**. (see *Color Plates 66* and *67*) Fibers from several other cranial nerves contribute to the formation of the vagus nerve rootlets. The vagus nerve passes through the *jugular foramen*. Also passing through this foramen are the internal jugular vein along with cranial nerves N IX and N XI. Only on specially prepared models and demonstration specimens may the *superior (jugular)* and *inferior ganglia* be identified. Observation of branches of the vagus are described with the cardiovascular and digestive systems.

Accessory Nerve (N XI) *MOTOR* Using the vagus nerve (N X) as a landmark, locate the most inferior rootlets of the group, as the **accessory nerve**. (see *Color Plates 66* and *67*) This nerve, along with the vagus (N X), passes through the *jugular foramen*. The accessory nerve consists of two branches: (1) *cranial (medullary)*, which joins with nerve fibers of N X to exit the cranial cavity and (2) *spinal*, which receives nerve fibers that originate from the lateral gray horns of spinal segments C_1 to C_5, which pass through the foramen magnum to enter the cranium. Check with your instructor before proceeding to identify these branches, because they may be observed only on specially prepared demonstration models and specimens.

Hypoglossal Nerve (N XII) *MOTOR* Identify anterior to the olive of the medulla oblongata the **hypoglossal nerve**. (see *Color Plates 66* and *67*) The nerve exits the cranial cavity by passing through the hypoglossal *canal*, passing inferior to the occipital condyles (see *Figure 6-8, p45*) and exits the skull lateral to the internal carotid artery (see *Figure 22-4, p320* and *Color Plate 86*) to enter the anterior triangle of the neck.

See: Fig. 15-15

∞ *Figs.*
15-21 to 15-30
pp409-418

LOCATE

____ Cerebrum

____ Olfactory nerve (N I)
(*Passes through cribriform plate*)
{ Olfactory nerve fibers
Olfactory bulb
Olfactory tract

____ Diencephalon

____ Optic Nerve (N II)
(*Passes through optic canal*)
{ Optic tracts
Optic chiasma

____ Mesencephalon

____ Oculomotor nerve (N III) {Ciliary ganglion
(*Passes through superior orbital fissure*)

____ Trochlear nerve (N IV)
(*Passes through superior orbital fissure*)

____ Pons

____ Trigeminal nerve (N V) and branches
{ Branches
Semilunar (trigeminal) ganglion
Ophthalmic (*Passes through superior orbital fissure*)
Maxillary (*Passes through foramen rotundum*)
Mandibular (*Passes through foramen ovale*)

_____ Abducens nerve (N VI) |
(*Passes through superior orbital fissure*)

_____ Facial nerve (N VII) { Geniculate ganglion
and associated { Petropalatine ganglion
ganglia (*Passes* { Submandibular ganglion
through internal
acoustic meatus to stylomastoid foramen)

_____ Vestibulocochlear nerve { <u>Branches</u>
(N VIII) { Vestibular nerve
(*Passes through internal* { Cochlear nerve
acoustic meatus)

_____ Glossopharyngeal nerve { <u>Ganglia</u>
(N IX) { Superior ganglion
(*Passes through* { Inferior (petrosal)
jugular foramen) { ganglion

_____ Vagus nerve (N X) { <u>Ganglia</u>
(*Passes through* { Superior (jugular) ganglion
jugular foramen) { Inferior (nodose) ganglion
{ <u>Nerve Branches</u>
{ Superior laryngeal
{ Internal laryngeal
{ External laryngeal
{ Recurrent laryngeal nerve
{ Cardiac nerves

_____ Accessory nerve (N XI) { <u>Branches</u>
(*Passes through jugular* { Medullary (cranial)
foramen) { Spinal

_____ Hypoglossal nerve (N XII)
(*Passes through hypoglossal canal*)

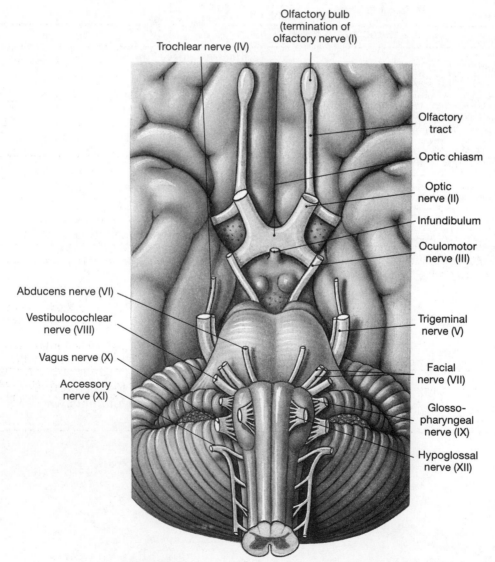

FIGURE 15-15   Origins of the cranial nerves.

Anatomical Identification Review

Fill in the spaces in the table to identify the appropriate region, structure, or function of the brain. Use this table to identify the levels of brain organization and interconnected substructures.

MAJOR ANATOMIC REGIONS OF THE BRAIN		
Region	*Structure*	*Function*
I. Cerebral Hemispheres		
A. _____ (forebrain)	_____ - (paired hemispheres) -motor, sensory, and association regions arranged in lobes -_____ -_____ - TEMPORAL -_____ limbic system	conscious thought processes, intellectual functions, memory storage
II. Brain Stem		
A. Diencephalon (forebrain)	_____ . hypothalamus	sensory information process center for _____ _____ _____
B. _____ (midbrain)	superior/inferior _____ reticular formation	process _____ / _____ information maintenance of consciousness
C. Metencephalon (hindbrain)	_____ _____	connects _____ to _____ _____: some somatic and visceral motor control adjusts voluntary and involuntary motor activities
D. _____	_____	sensory information relay centers for regulating _____

To Think About

1. What are the corpora quadrigemina and what are their functions?

2. Recently, in the news it was reported that a young woman had a javelin point thrust through her temple. Fortunately, she was not seriously injured because the downward aimed point missed critical anatomic structures. If it had entered and passed upward, which brain structures would probably have been damaged?

3. A young physical therapy student wakes up one morning and discovers that his face is paralyzed on the left side and he has no sensation of taste from the anterior two-thirds of the tongue on the same side. What is the cause of these symptoms, and what can be done to assist the situation?

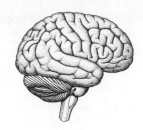

16

THE NERVOUS SYSTEM:
Pathways And Higher-Order Functions

Objectives

1. Locate, identify, and name the principal ascending (sensory) tracts and the principal descending (motor) tracts in the spinal cord.

2. Compare, identify, and name the structures and functions of the corticospinal pathway (pyramidal system) and medial and lateral pathways.

3. Locate, identify, and name the major structural specializations of the cerebral cortex and briefly describe their general functions.

Observation of Sensory and Motor Pathways

The nervous system works nonstop, day and night, to monitor, control, adjust, and sustain body activities and internal physiological environment. All of this nervous system activity requires constant and instantaneous communication between the central nervous system (CNS) and the peripheral nervous system (PNS). Sensory tracts extend from the body's sensory receptors (e.g., touch, temperature, pain, etc.) to the spinal cord. Here, they ascend the spinal cord to specific sensory areas in the brain. [∞ p428]

Motor pathways convey information in the form of motor impulses to both skeletal muscles and visceral effectors, such as smooth muscles, cardiac muscles, and glands. Motor impulses are routed via two different pathways, the somatic and the autonomic. The autonomic, more precisely called the *autonomic nervous system* (ANS), is described and observed in Chapter 17. [∞ p447]

Observation of Sensory Pathways

Procedure

✓ **Quick Check**

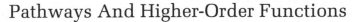

Before you begin to identify the sensory and motor tracts of the spinal cord, review the transverse (cross section) gross sectional anatomy of the spinal cord. (See *Figure 14-4, p203*) Prepare for your observations by reviewing a sectional model of the spinal cord, specifically the regions of gray and white matter, the gray horns, the anterior median fissure, the posterior median sulcus, and the locations of the anterior, posterior, and lateral columns within the white matter.

Locate and identify the following **principal ascending (sensory) tracts** in the **spinal cord section**, using a *cross sectional model* of the *human spinal cord*. [∞ p429] Sensory pathways (tracts) conduct sensory information from their points of origin towards the cerebral cortex. Sensory information undergoes very little processing until it reaches the spinal cord or brain stem. The name of the sensory tract provides us with exact information about its origin and destination. These tracts are not easy to display or identify in actual dissection. Use *Figure 14-3, p202* and *Figure 16-1* for reference. Three major ascending (sensory) tracts are identified in the spinal cord:

1. The **posterior column pathway** consists of two tracts, both of which are located within the white matter between the posterior gray horns. Locate and identify the *fasciculus gracilis* on either side of the posterior median sulcus. Also, locate and identify the *fasciculus cuneatus* between the fasciculus gracilis and the medial border of the posterior gray horn. The posterior column pathway transmits localized information about fine touch, pressure, vibration, and proprioceptive sensations to the medulla oblongata and then on to the thalamus. From the thalamus, these sensations are then projected to the specific sensory area of the cerebral cortex for interpretation.

2. The **spinothalamic pathway** consists of two tracts, both of which are located adjacent to the anterior gray horns. Identify the *lateral spinothalamic tracts* at the lateral edges of the widest part of the anterior gray horns. The *anterior spinothalamic tracts* are located anterior to the most anterior portion of the anterior gray horns, between the anterior gray horns and the edge of the spinal cord on either side of the anterior median fissure. This pathway transmits information about touch, temperature, pain, and pressure sensations to the thalamus. From the thalamus, these sensations are projected to the specific sensory area of the cerebral cortex for interpretation.

3. The **spinocerebellar pathway** also consists of two tracts. Identify the *posterior spinocerebellar tracts*, located directly lateral to the posterior gray horns at the lateral edge of the spinal cord. The *anterior spinocerebellar tract* can be identified lateral to the widest portion of the anterior gray horn at the edge of the spinal cord. The spinocerebellar pathway carries proprioceptive information about the status of the body's muscles from the spinal cord to the cerebellum.

It may be convenient to remember that if the name of a tract begins with "spino" then the tract has its origin in the spinal cord. On the other hand, if the name of a tract ends in "spinal", then its destination is the spinal cord.

See: Fig. 16-1

∞ *Figs.*
16-1, p429
16-2, pp430-431
16-3, p432

LOCATE

____ Posterior column pathway (fasciculus gracilis and fasciculus cuneatus)

____ Spinothalamic pathway (lateral and anterior spinothalamic tracts)

____ Spinocerebellar pathway (posterior and anterior spinocerebellar tracts)

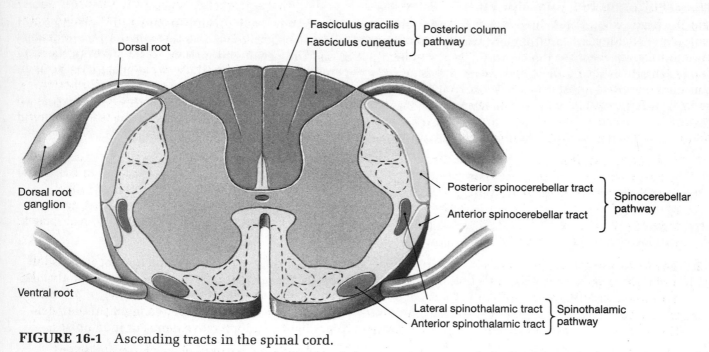

FIGURE 16-1 Ascending tracts in the spinal cord.

Motor Pathways: Corticospinal (Pyramidal System), Medial and Lateral Pathways

Procedure

Use the same *model* that you used for the identification of the ascending tracts to now locate and identify the following **principal descending (motor) tracts** in the **spinal cord**. [∞ p433] Use *Figure 16-2* for reference. There are two principal descending or motor pathways, the *corticospinal pathway* and the *medial and lateral pathways*. The **corticospinal pathway** (or *pyramidal system*) provides for the voluntary control of muscles, such as muscles for eye movements, facial muscles, tongue muscles, and superficial muscles of the neck and back that are innervated by the cranial nerves (N III, N IV, N V, N VI, N VII, N X, N XI, and N XII). This system is divided into two pairs of descending (motor) tracts:

1. The **lateral corticospinal tract** consists of a left and a right tract located within the white matter. These tracts can be identified lateral to the lateral surfaces of the posterior gray horns. They lie between the gray horns and the posterior spinocerebellar tracts.

2. The **anterior corticospinal tract** consists of a left and a right tract located within the white matter. These tracts can be identified in the white matter on each side immediately lateral to the anterior median fissure.

The corticospinal tracts pass through the brain stem to synapse in the gray horns of the spinal cord. These tracts may be identified externally, as the pyramids along the ventral surface of the medulla oblongata on either the intact brain model or specimen.

Procedure

✓ **Quick Check**

Before you begin to locate and identify the medial and lateral pathways on a model, review the basal nuclei, which are components of these pathways. The basal nuclei are described and identified in Chapter 15 under the heading "Observation of the Basal Nuclei." (*p226*)

The **medial** and **lateral pathways** consist of neurons (second order) that originate at synapses in the basal nuclei and the brain stem, and extend into the spinal cord to an appropriate level within the gray horns. These pathways carry unconscious and involuntary motor commands. The most important components of these pathways are the basal nuclei of the cerebrum. These nuclei adjust the motor commands issued in the other processing centers (tracts of the corticospinal and medial and lateral pathways), as in the pattern of swinging your arms in conjunction with the leg movements associated with walking. To produce graceful and precise motor activities, as in walking, requires monitoring and regulating the corticospinal, medial and lateral pathways. The precise control of voluntary and involuntary motor activities is performed by the cerebellum. The medial and lateral pathways are divided into four major tracts:

1. The **vestibulospinal tract** begins with the vestibular nuclei (posterolateral surface of superior end of the medulla and inferior of the pons) and descends as left and right tracts within the white matter of the spinal cord. The tracts can be identified anterior to the most anterior portion of the anterior gray horns of the spinal cord. They lie in the white matter between the anterior gray horns and the anterior border of the spinal cord. The vestibulospinal tract carries motor commands to alter the position of the eyes, head, neck, and limbs in order to maintain posture and balance.

2. The **tectospinal tract** originates in the superior colliculi or tectum (roof) of midbrain and descends as left and right tracts within the white matter of the spinal cord. These tracts can be identified immediately adjacent to the anterior median fissure. They lie within the white matter just antero-lateral to the anterior corticospinal tracts. The tectospinal tract carries motor commands for the involuntary changing of the position of the eyes, head, neck, and arms in response to bright lights, sudden movements, or loud noises.

3. The **reticulospinal tract** originates with the reticular formation (network of nuclei in brain stem) and descends as left and right tracts within the white matter of the spinal cord. These tracts can be identified anterior to the anterior gray horns and between anterior corticospinal and vestibulospinal tracts. The reticulospinal tract carries motor commands that regulate involuntary reflex activity and autonomic functions.

4. The **rubrospinal tract** begins in the red nucleus (within midbrain) and descends as left and right tracts within the white matter of the spinal cord. These tracts co-mingle with the corticospinal tracts and can be identified immediately anterior to the lateral corticospinal tracts and lateral to the posterior gray horns. The rubrospinal tract facilitates motor commands for flexor motor neurons of proximal limb muscles, mostly upper limb muscles.

LOCATE

_____ Corticospinal pathway (Lateral and Anterior Corticospinal tracts)

_____ Medial and Lateral pathways (Vestibulospinal tract, Tectospinal tract, Rubrospinal tract, and Reticulospinal tract)

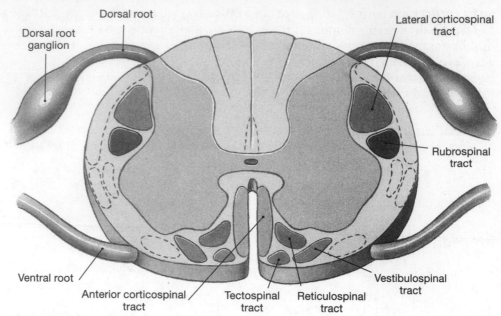

FIGURE 16-2 Corticospinal, medial and lateral pathways.

Hemispheric Specialization: The Functional Areas of the Cerebral Cortex

Higher-order functions are performed by the cerebral cortex and involve the complex interaction between conscious and unconscious information processing. These functions consist of intricate interactions between areas completely within the cortex and between the cerebral cortex and other areas of the brain. Higher-order functions have four characteristics: (1) they are performed by the cerebral cortex; (2) they involve complex interactions either between areas of the cerebral cortex or between the cortex and other areas of the brain; (3) they involve conscious and unconscious information processing; and (4) they are subject to modification and adjustment over time. The higher-order integrative centers include: the *prefrontal cortex*, the *general interpretive area*, and the *speech center (Broca's area)*. [∞ p438]

Higher-order centers in the left and right cerebral hemispheres have different but complementary functions. Among other functions, the left hemisphere is specialized to contain the general interpretive and speech centers and is responsible for language-based skills. The right hemisphere is usually concerned with spatial relationships and analyses. [∞ p438]

Procedure

✓ **Quick Check**

Before you begin to identify the functional areas of the cerebral cortex, review the fissures and sulci of the cerebral hemispheres, the lobes of the cerebrum, the motor and sensory regions of the cerebral cortex, and the association areas. These landmarks, regions and areas are described and identified in Chapter 15 under the headings "Observation of the Telencephalon (Cerebrum)" and "Observation of the Cerebral Cortex." (*p222-223*)

Locate the **centers responsible** for **higher-order functions** (e.g., speech center), using a painted *brain model* showing the *functional areas* of the *cerebral cortex*. Hold an intact brain model or position it on a stand so that lateral and superior surfaces are facing you. Use *Color Plates 60* and *61* for reference. Begin by reviewing the landmarks of the brain, and the regions and the association areas noted in Quick Check. Identify the **prefrontal cortex** located in the frontal lobes. The prefrontal cortex coordinates information from the higher order association areas of the entire cortex and performs abstract intellectual functions.

Locate and identify the **general interpretive area** posterior to the lateral sulcus within the parietal and temporal lobes, by observing the left cerebral hemisphere and using *Figure 16-3* for reference. This area receives auditory input and information from all the sensory association areas and is present in only one hemisphere, usually the left. Identify both the **primary visual cortex** (occipital lobe) and the **primary auditory cortex** (temporal lobe), which are involved in the conscious perception of visual and auditory stimuli, respectively. These primary cortices are connected to **association areas** (visual and auditory, respectively) that interpret incoming data or coordinate a motor response. Also, locate the premotor cortex and identify the **speech center (Broca's area)** that lies along the antero-lateral edge of the premotor cortex. This center regulates the patterns of breathing and vocalization needed for normal speech. Again, the centers responsible for higher-order functions are not found in both cerebral hemispheres. The left cerebral hemisphere usually contains the general interpretive area (for language and mathematical calculation) and the speech center.

See: Fig. 16-3

 ∞ *Figs.*
15-9, p393
16-7, p439
16-8, p440

LOCATE

Cerebral Hemispheres

____ Lobes $\begin{cases} \text{Frontal} \\ \text{Parietal} \\ \text{Temporal} \\ \text{Occipital} \end{cases}$

____ Primary motor cortex

____ Somatic motor association area

____ Primary somatic sensory cortex

____ Somatic sensory association area

____ Primary visual cortex

____ Visual association area

____ Primary auditory cortex

____ Auditory association area

____ Prefrontal cortex

Left Cerebral Hemisphere

____ General interpretive area (for language and mathematical calculation)

____ Speech center (Broca's area)

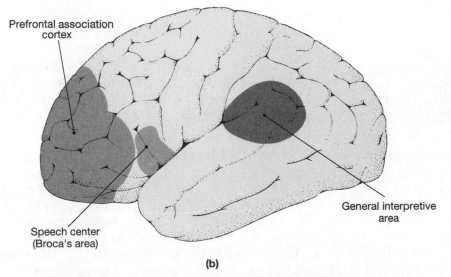

FIGURE 16-3 Functional areas of the cerebral hemisphere showing association areas (a) Centers contained in both areas (b) Centers contained in only the left hemisphere.

Anatomical Identification Review

1. Identify, by coloring and labeling, the principal ascending (sensory) and descending (motor) tracts on the following illustration of a cross section of the spinal cord. Use one group of colors to identify the ascending tracts and a second group of colors to identify the descending tracts. Label each tract.

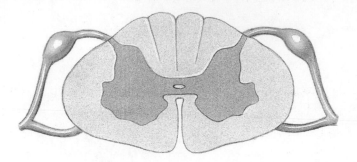

2. Identify, by coloring and labeling, the motor and sensory association areas on the following illustration of the left cerebral hemisphere. Include the lobes and the fissures in your labeling.

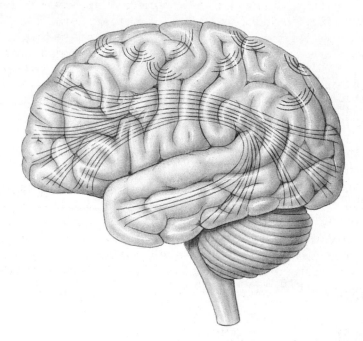

To Think About

1. What condition would be expected in an older individual who displays forgetfulness, inability to control emotions, and decreased ability to perform motor skills?

2. How is the role of the corpus callosum of the brain understood in terms of the disconnection syndrome?

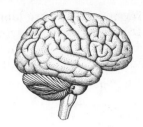

17

THE NERVOUS SYSTEM:
Autonomic Division

Objectives

1. Compare the anatomical and functional differences between sympathetic (thoracolumbar) and parasympathetic (craniosacral) divisions of the autonomic nervous system.

2. Identify the structures of the sympathetic (thoracolumbar) division and the parasympathetic (craniosacral) division.

The Autonomic Nervous System

The physiological needs of the body change continuously as the body attempts to maintain homeostasis. Homeostasis refers to the existence of a stable internal environment. Homeostatic regulation refers to the adjustments in physiological systems that are responsible for the preservation of homeostasis. For example, physiological demands to maintain homeostasis may be dramatically less during sleep than during exertion in a stressful endeavor. The *autonomic nervous system* monitors and adjusts body systems to maintain homeostasis. It is abbreviated ANS, and we will follow this custom and use that abbreviation throughout this laboratory manual. The ANS coordinates activities of the respiratory, cardiovascular, digestive, excretory, and reproductive systems of the body, and it regulates body temperature. [∞ p448] The ANS differs from the somatic nervous system by always involving two motor neurons, which occur in series with a synapse point between the spinal cord and a peripheral effector (such as smooth muscle, cardiac muscle or a gland).

Observation of the Sympathetic and Parasympathetic Divisions of the Autonomic Nervous System

The ANS is organized into two divisions, a *sympathetic or thoracolumbar division* and a *parasympathetic or craniosacral division.* The sympathetic division is distributed throughout the body and usually stimulates tissue metabolism, causes an increase in heart beat, and elevates mental alertness as the body begins to cope with an emergency. Sympathetic division activation prepares the body for "fight" or "flight" action. In contrast, the parasympathetic division conserves body energy by promoting both minimum physical activity and sedentary activities, such as "rest and recuperation" functions. It is not as widely distributed throughout the body, and is limited to the head, neck, and selected viscera.

Procedure

✓ **Quick Check**

Before you begin to identify structures of the autonomic nervous system, review the origin and distribution of cranial nerves N III, N VII, and N IX, the origin and distribution of the vagus nerve (N X) (*Color Plates 66, 67* and *Table 15-1, p234*), the cervical spinal nerves (C_1–C_8), the thoracic spinal nerves (T_1–T_{12}), the lumbar spinal nerves (L_1–L_5), and sacral spinal nerves (S_1–S_5) emerging from the sacral foramina. (*Color Plates 87, 106, 107,* and *Table 14-3, p209*)

Locate and identify the **spinal cord segments** that compose both divisions of the ANS, using *laboratory charts illustrating the CNS and ANS* and/or a *model of the nervous system.* The **sympathetic (thoracolumbar) division** consists of all the preganglionic neurons that send their fibers through the ventral roots of the thoracic and the upper lumbar nerves to synapse in ganglia adjacent to the spinal cord. It is called thoracolumbar because the cell bodies for all of the first neurons in the series (*preganglionic*) reside in spinal cord lateral gray horns in either the thoracic or the lumbar region. Axons of ganglionic neurons are called *postganglionic* fibers because they transmit impulses away from the ganglion. Locate and identify on the laboratory chart provided to you and/or on the nervous system model, the spinal nerves T_1 to T_{12} and L_1 to L_2. Use *Color Plates 56, 66, 67,* and *Figure 17-1* for reference.

The cell bodies for the **parasympathetic (craniosacral) division** are located either in the brain stem (associated with cranial nerves N III, N VII, N IX, and N X) or in the spinal cord gray horns in the sacral region. The cranial nerves involved are motor nerves, and these control both the secretions of tear and salivary glands and the structures associated with focusing the eyes. The pelvic splanchnic nerves (formed from spinal cord segments S_2, S_3, and S_4) innervate the large intestine, kidneys, urinary bladder, and reproductive organs. Now identify on the chart and or model the origins of cranial nerves (N III, N VII, N IX, and N X) and spinal nerves S_2 to S_4.

See: Color Plates 56, 66, 67, Fig. 17-1

∞ *Figs. 14-3, p359 17-1, p449*

LOCATE

Sympathetic Division
(Thoracolumbar) T_1–L_2
Spinal segment: Thoracic

_____ Thoracic nerves T_1–T_{12}

Spinal segment: Lumbar

_____ Lumbar nerves L_1, L_2

Parasympathetic Division
(Craniosacral) Brainstem
as origin
Spinal segment: Sacral

_____ Spinal nerves only
S_2–S_4

Observation of Sympathetic Chain (Paravertebral) and Ganglia in the Thorax

The ventral roots of spinal segments T_1 to L_2 contain, in part, sympathetic preganglionic fibers. After passing through the intervertebral foramen, each ventral ramus of a spinal nerve gives rise to a white ramus, which carries preganglionic fibers into a nearby *sympathetic chain ganglion, (or paravertebral ganglion)* where a synapse with the second neuron in the ANS series could occur. [∞ p452] Some preganglionic fibers extend between sympathetic chain ganglia to interconnect them, thus making the chain resemble a string

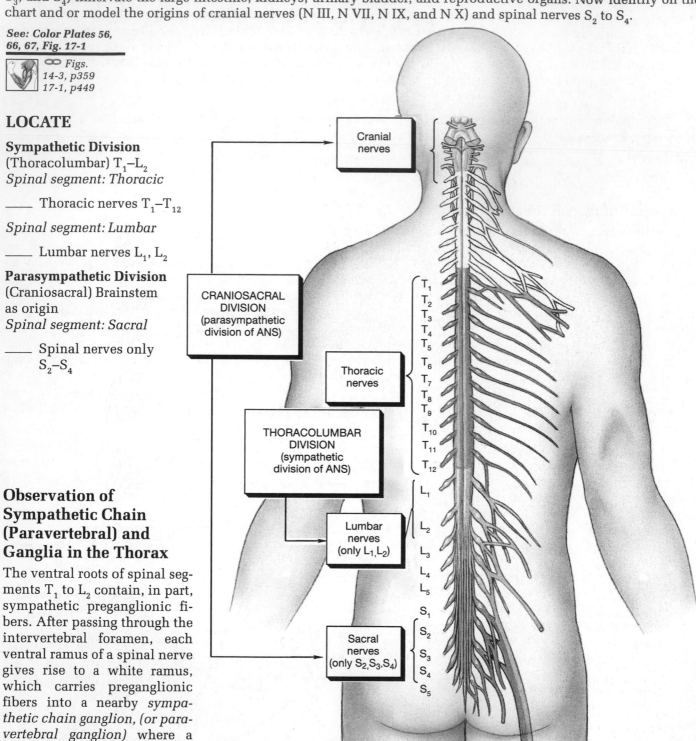

FIGURE 17-1 Anatomical divisions of the autonomic nervous system.

of widely separated beads. If a preganglionic fiber synapses in the sympathetic chain with the second neuron in the series, the postganglionic fiber extending from this neuron may either form the gray ramus and return to the spinal nerve for distribution or form an autonomic nerve and proceed directly to its target. Each ganglion in the chain provides sympathetic motor innervation via spinal nerves to a particular body segment or group of segments.

Identify the **sympathetic ganglia** in the **sympathetic chain** within the thorax, using *laboratory charts illustrating the CNS and ANS* and/or a *model of the nervous system.* If using a torso model, then you must remove the heart and lungs to view these structures. Use *Color Plates 71* and *87* for reference. In the thoracic cavity, locate the two sympathetic chains, one on each side of the vertebral column. The chains look like flat, narrow ribbons of tissue rather than tubular structures. Each sympathetic ganglion appears as one of many small swellings along the sympathetic chain. Locate the ganglia in the chain on the inferior border of the ribs lateral to the thoracic bodies. Each ganglion lies adjacent to the intercostal *artery* of each rib. (*Color Plate 87*)

See: Color Plates 71, 87, Fig. 17-2

∞ *Fig. 17-4, p453*

LOCATE

_____ Sympathetic ganglia in sympathetic chain

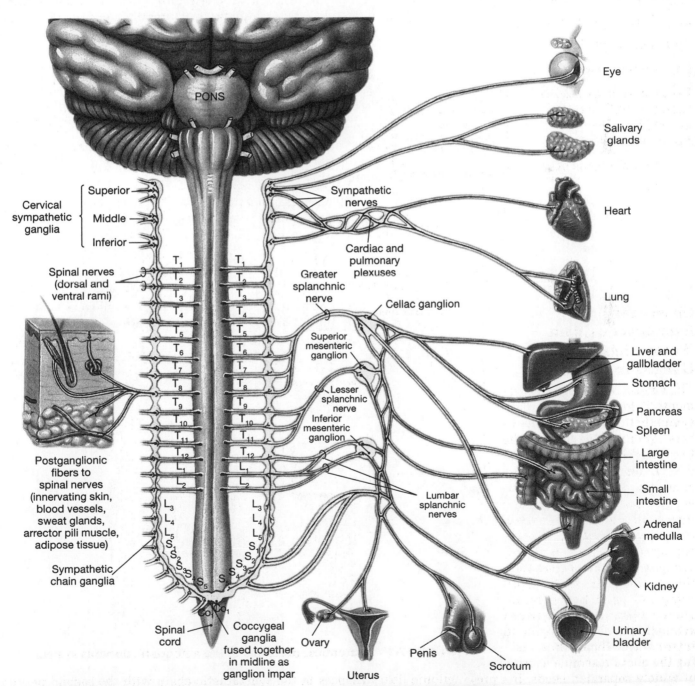

FIGURE 17-2 Anatomical distribution of sympathetic postganglionic fibers.

If a sagittal section of the trunk of a *cadaver specimen* is available for viewing, you can identify in the thoracic cavity the sympathetic chain and ganglia. The chain and ganglia are visible in the thoracic cavity when viewed from the anterior. The thoracic viscera must be reflected and a portion of the parietal pleura lateral to the vertebral column must be removed. The chain and ganglia lie deep to the pleural membrane. Before proceeding with any cadaveric observation, check with your instructor.

Observation of Collateral (Prevertebral) Ganglia of the Sympathetic Chain

The *collateral ganglia* are parts of the sympathetic division that innervate with postganglionic fibers some visceral effectors in tissues and organs in the abdominopelvic cavity. [∞ p452] These tissues and organs are recipients of sympathetic innervation via postganglionic fibers that originate from synapses with preganglionic fibers that pass through the sympathetic chain without synapsing and then proceed to a collateral ganglion. Preganglionic fibers that innervate the collateral ganglia form the **splanchnic nerves** in the dorsal wall of the abdominopelvic cavity. These nerves may be observed on both the left and right sides of the dorsal body wall. They converge on and innervate the collateral ganglia, including the *celiac ganglion*, the *superior mesenteric ganglion*, and the *inferior mesenteric ganglion*.

Identify the **collateral ganglia** of the sympathetic chain, using *laboratory charts illustrating the CNS and ANS* and or a *model of the nervous system*. Most collateral ganglia are single structures and very small. The collateral ganglia can be located anterior and sometimes lateral to the bodies of the lumbar vertebrae in the abdominopelvic cavity.

1. The **celiac ganglion** is the largest of the collateral ganglia and can be located on the lateral surface of the aorta at the base of the celiac artery (*Color Plates 87* and *106*) and at the level of the crura (L_2–L_3) of the diaphragm. (*Figure 10-11, p125*) Postganglionic fibers from the celiac ganglion innervate some of the digestive organs and spleen.

2. The small **superior mesenteric ganglion** is located near the *superior mesenteric artery.* (*Color Plate 87*) One or more ganglia may be present. Postganglionic fibers from the superior mesenteric ganglion innervate the small intestine and the first part of the large intestine.

3. The **inferior mesenteric ganglion** is extremely small (almost impossible to find in the cadaver specimen) and is located along the stump of the *inferior mesenteric artery.* (*Color Plate 87*) Postganglionic fibers from the inferior mesenteric ganglion innervate the terminal portion of the large intestine, the kidneys, urinary bladder, and reproductive organs.

If a sagittal section of the trunk of a *cadaver specimen* is available for viewing, you may be able to identify some of the collateral ganglia. It is not always possible to display or locate the ganglia for viewing in the prosected cadaver because the dissection is tedious and because these structures are so delicate they do not lend themselves to examination. Before proceeding with any cadaveric observation, check with your instructor.

See: Color Plates 87, 106, Fig. 17-3

∞ *Figs.*
17-4, p453
17-9, p460

LOCATE

___ Celiac ganglion

___ Superior mesenteric ganglion

___ Inferior mesenteric ganglion

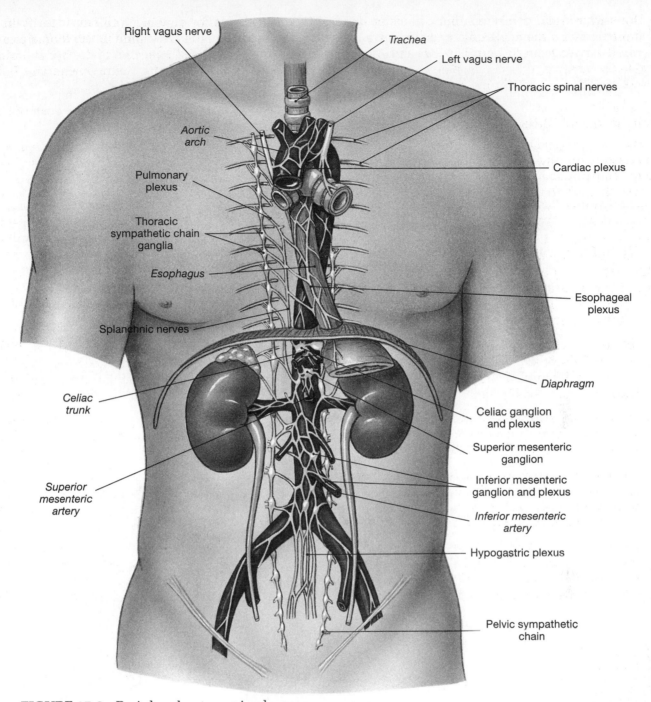

Right vagus nerve

Trachea

Left vagus nerve

Thoracic spinal nerves

Aortic arch

Cardiac plexus

Pulmonary plexus

Thoracic sympathetic chain ganglia

Esophagus

Esophageal plexus

Splanchnic nerves

Diaphragm

Celiac trunk

Celiac ganglion and plexus

Superior mesenteric ganglion

Inferior mesenteric ganglion and plexus

Inferior mesenteric artery

Superior mesenteric artery

Hypogastric plexus

Pelvic sympathetic chain

FIGURE 17-3   Peripheral autonomic plexuses.

Observation of the Autonomic Nerve Plexuses

In the thoracic and abdominopelvic cavities, the sympathetic postganglionic fibers mingle with parasympathetic preganglionic fibers at a series of plexuses. The nerves that exit these plexuses travel with the blood vessels and lymphatics that supply the thoracic, abdominal, and pelvic visceral organs. Locate and identify the following **plexuses**, using a *torso model, laboratory charts illustrating the CNS and ANS*, and/or *cadaver specimen*, and *Figure 17-3* for reference.

1. Sympathetic and parasympathetic fibers co-mingle in the thoracic cavity on the anterior surfaces of the aortic arch, pulmonary trunk, and trachea to form the **cardiac plexus**. The autonomic fibers of this plexus innervate the heart. (*Color Plates 69* and *73*)

2. Autonomic fibers from the cardiac plexus branch and extend to form the **pulmonary plexus**, which supplies sympathetic and parasympathetic autonomic fibers to the lungs. These fibers control the smooth muscle responsible for constriction of the bronchioles in the lungs. Locate this plexus on the posterior surface of the trachea where it splits into right and left primary bronchi. (*Color Plates 70* and *71*)

3. The **esophageal plexus** is formed by descending branches of both the vagus nerve and splanchnic nerves exiting the sympathetic chain on either side. This plexus is located on the esophagus in the thoracic cavity, just superior to the diaphragm.

4. Shortly after the aorta passes through the diaphragm, the **celiac (solar) plexus** is formed on the outer aorta wall near the branch point of the celiac and superior mesenteric arteries. (*Color Plate 87*) It forms by the co-mingling of parasympathetic preganglionic fibers from the vagus nerve (which entered the abdomino-pelvic cavity with the esophagus) and sympathetic postganglionic fibers. These fibers innervate the stomach, duodenum, pancreas, liver, gallbladder, small intestine, ascending colon, two thirds of transverse colon, and spleen from the celiac plexus. The **inferior mesenteric plexus** is a small plexus located on the aorta near the branch point of the inferior mesenteric artery. It receives its fibers from the celiac plexus and, in turn, innervates the large intestine.

5. The **hypogastric plexus** contains the parasympathetic outflow of the pelvic splanchnic nerves and sympathetic postganglionic fibers from both the inferior mesenteric ganglion and splanchnic nerves from the sacral sympathetic chain. This plexus supplies nerve fibers to all pelvic organs (colon, urinary bladder, and reproductive organs). Locate the hypogastric plexus just inferior to where the abdominal aorta splits into left and right vessels (*common iliac arteries*) to supply the legs. (*Color Plates 87, 121* and *Figure 22-6, p323*)

 If a sagittal section of the trunk of a *cadaver specimen* is available for viewing, you may be able to identify the above described plexuses. To view the cardiac plexus, the pericardial sac must be removed or opened to expose the anterior of the heart. It is not always possible to display the plexuses for viewing in the pro-sected cadaver because the dissection is again intricate and these structures are fragile to examine. So before proceeding with any cadaveric observation, check with your instructor.

See: Color Plates
69-71, 73, 87, 121,
Fig. 17-3

∞ *Fig.*
17-9, p460

LOCATE

____ Cardiac plexus

____ Pulmonary plexus

____ Esophageal plexus

____ Celiac (solar) plexus

____ Inferior mesenteric plexus

____ Hypogastric plexus

Anatomical Identification Review

Using the illustration shown, label the spinal cord segments that compose the sympathetic (thoraco-lumbar) division and the parasympathetic (cranio-sacral) division.

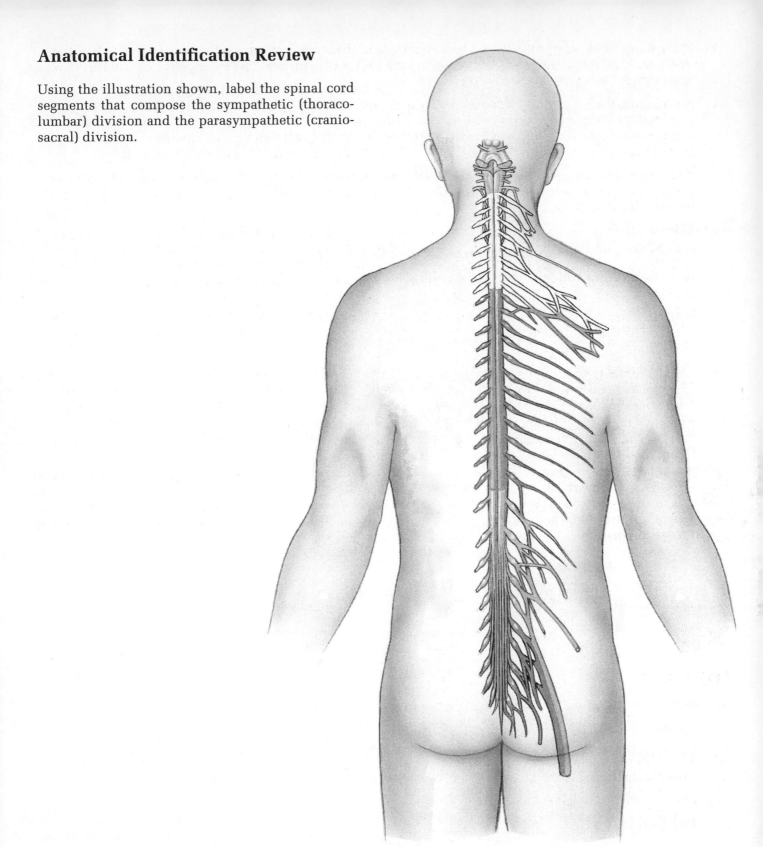

To Think About

1. What factors contribute to the more widespread effects the sympathetic division of the ANS has, in comparison to the parasympathetic division of the ANS?

2. Why is the pain from a visceral organ (e.g. gallbladder) typically not felt in that particular organ?

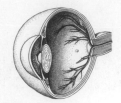

18

THE NERVOUS SYSTEM:
General and Special Senses

Objectives

1. Distinguish between the general and special senses.

2. Identify and describe the location of general sense receptors found in the skin, and briefly describe their function.

3. Identify, locate receptors, describe neural pathways/cranial nerve involvement and/or structures of the following: sense of smell (olfaction), sense of taste (gustation), senses of hearing and equilibrium, and sense of sight (vision).

The body contains vast numbers of specialized nerve cells, termed sensory **receptors**, that monitor conditions either in the body or in the external environment (outside the body) and transmit this information as nerve impulses via nerves to the CNS. [∞ p373] Sensory receptors are very specific and respond to only one type of stimulus. When sensory receptors are stimulated, they transmit information about this stimulus in the form of a sensory code. The sensory code contains information such as the strength of the stimulus and its duration. Once sensory information is received by the CNS, it must be interpreted.

The **general senses** convey information about *temperature, pressure, touch, pain, vibration*, and *proprioception*. [∞ p469] The general sense receptors transmit their information via the spinal nerves and spinal pathways to the somatosensory cortex of the brain. General senses can be categorized in two ways. The first is based upon the source of the stimulus: (1) **exteroceptors** are those receptors that monitor the environment outside the body and (2) **interoceptors** are those receptors that monitor and provide information about the internal workings of the body. The general senses can also be classified a second way into four major categories based upon the type of stimulus that excites them:

1. **Nociceptors** respond to pain due to injury or inflammation and are located in the skin and most deep tissues. [∞ p469]

2. **Thermoreceptors** respond to temperature changes and are limited to the dermis of the skin, hypothalamus, and liver. [∞ p469]

3. **Mechanoreceptors** respond to touch or pressure and are subdivided into three types: (1) *tactile receptors* respond to various degrees of touch; (2) *baroreceptors* monitor and respond to changes in blood pressure, lung expansion, urinary bladder expansion, and digestive tract expansion; and (3) *proprioceptors* monitor and respond to joint movements and the state of contraction or relaxation of the body's skeletal muscles. [∞ p470]

4. **Chemoreceptors** monitor the chemical composition of specific body fluids (e.g., concentration of carbon dioxide, ions, or pH of the blood) and the concentration of specific chemicals in solution for taste and smell. [∞ p472]

The **special senses** are *olfaction* (smell), *gustation* (taste), *equilibrium* (balance), *hearing*, and *vision*. [p470] The receptors for the special senses are located in the specialized **sense organs**, the nose, tongue, ear, and eye. From these sense organs, the special sense receptors transmit their individual sensory coded information along individual cranial nerves to specific areas of the cerebral cortex for interpretation.

Observation of the General Senses of the Skin

Procedure

✓ **Quick Check**

Before you begin to identify the sensory receptors of the skin, review the skin model (*Color Plate 27*) for layers of the skin (epidermis, dermis, and subcutaneous), and accessory structures of the skin (hair and glands) (*Color Plate 28*)

Identify the general senses' **sensory receptors** located in the skin, using a *model* of the *skin*. Use *Color Plate 27* and *Figure 18-1* for reference. Begin your observation by locating on the model the layers of the skin and accessory structures noted in Quick Check. Six different types of tactile receptors may be identified on the skin model.

1. **Free nerve endings**, which are classified as nociceptors, thermoreceptors, and mechanoreceptors may be identified in the outer region of the dermis. The free nerve endings are dendrites of sensory neurons and appear as dark branching threads within the dermal layer.

2. **Tactile (Meissner's) corpuscles** detect very light touch and can be identified as dark-colored, oval-to-spiral-striped corpuscle structures. These corpuscles are best observed in the dermal papillae.

3. **Lamellated (Pacinian) corpuscles** are stimulated by direct pressure, and can be identified by their oval-to-round concentric ring structure. Locate oval-to-round "bulls eye" or "sliced onion-like" structures in the subcutaneous region.

4. **Merkel cells** are stimulated by fine pressure and touch, and can be identified as small groups of darkly colored cells located within the stratum germinativum layer. A dendritic branch emerges from the dermis to penetrate into the germinativum layer to the Merkel cell.

5. **Free nerve endings of root hair plexus** detect hair follicle movements associated with movements of hair and can be identified as fine branching fibers around the connective tissue sheath of hair follicles. Locate hair follicles in the dermis and look for these nerve endings.

6. **Ruffini corpuscles** are stimulated by distortions of the skin, such as stretching, and may be identified as spindle-shaped structures within the upper dermis.

See: *Color Plate 27, Fig. 18-1*

 ∞ *Figs. 4-2, p89 18-3, p471*

LOCATE

____ Free nerve endings

____ Merkel cells

____ Pacinian corpuscle

____ Meissner's corpuscle

____ Ruffini corpuscle

____ Free nerve endings of root hair plexus

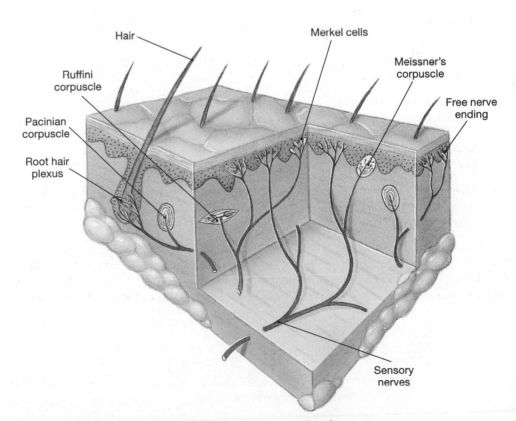

FIGURE 18-1 Tactile receptors in the skin.

Microscopic Observation of Skin Sensory Receptors

Procedure

✓ **Quick Check**

Before you begin to identify microscopically the sensory receptors of the skin, review a microscope slide of the skin for layers of the skin (epidermis, dermis, and subcutaneous) and accessory structures of the skin (hair and glands). See *Figure 4-1, p25* and *Color Plates 23* and *28*.

 Identify the following **receptors** in the **skin**, using the *microscope slide* provided and *Figure 18-2* for reference, first by viewing under low power magnification and then under high power magnification to see the details of these receptors. Your microscope slide will contain a section of skin specifically selected and stained for these sensory receptors. (NOTE: Your instructor may provide you with a set of slides identified for each receptor, as opposed to one slide containing all types.) Begin your observation under low power magnification by locating the layers of the skin and accessory structures noted in Quick Check. Under high magnification, locate **free nerve endings** at the upper region of the dermis near the epidermis border. The free nerve endings are dendrites of sensory neurons and appear as minute, darkly-stained branching threads. Return to low power magnification and observe the dermal papillae for dark-staining oval structures. Under high magnification, identify a dark-stained, spirally striated, oval structure as a **tactile receptor** known as **Meissner's corpuscle**. These corpuscles detect very light touch. Return to low magnification and scan the subcutaneous region for oval-to-round "bulls-eye" structures. Under high magnification, identify a dark-staining, oval-to-round concentric ring structure as a **Lamellated (Pacinian) corpuscle**. These corpuscles are also tactile receptors, but respond to direct pressure and are also located in the mesenteries and wall of the urinary bladder.

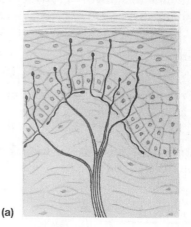
(a)

Slide # _____ Skin

See: Fig. 18-2

∞ *Fig.*
18-3, p471

LOCATE

_____ Free nerve ending

_____ Lamellated (Pacinian) corpuscle

_____ Tactile (Meissner's) corpuscle

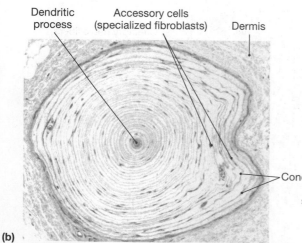
(b)

Dendritic process — Accessory cells (specialized fibroblasts) — Dermis

Concentric layers (lamellae) of collagen fibers separated by fluids

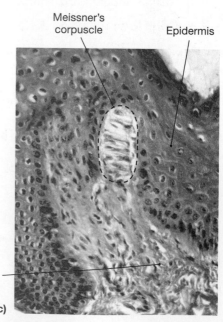
Meissner's corpuscle — Epidermis

Dermis

(c)

FIGURE 18-2   Tactile receptors in the skin (a) free nerve endings (b) Pacinian corpuscle × 125 (c) Meissner's corpuscle × 550.

Observation of the Sense Organ for Olfaction (Smell)

The nose is the special sense organ for **olfaction** or smell. The sensory receptors for olfaction are located in the paired **olfactory organs**, which are located in the nasal cavity of the nose on either side of the nasal septum. The olfactory organs consist of *olfactory epithelium* and *olfactory (Bowman's) glands*. The receptors in our nose can distinguish over 50 primary smells. [∞ p472] Keep in mind that all of the special senses supply sensory information that is used to assist in the interpretation of a specific sense. For example, think of the last time you had a cold. Not only could you not smell, but what you ate had very little taste.

Procedure

✓ **Quick Check**

Before you begin to examine the sensory receptors for smell, review the following: on a dry human skull specimen, the nasal conchae, (*Figure 6-20, p53*) the cribriform plate and crista galli, (*Figure 6-9, p46*) and the anterior cranial fossae region; (*Color Plate 33*) and on a brain model, cranial nerve N I (Olfactory Nerve).

Identify the olfactory organs, olfactory pathway, and associated structures, using a *model* of the *nose* in sagittal section, *Color Plate 92*, and *Figure 18-3* for reference. Identify on a model of the nose or a dry human skull the nasal conchae (turbinates), cribriform plate and crista galli as noted in Quick Check. Identify the **olfactory epithelium** covering the roof of the nasal cavity, the superior nasal conchae, and superior portion of the nasal septum. Located in the olfactory epithelium are the *olfactory receptors*, which are specialized receptors for smell. The **olfactory bulbs** can be identified as they lie on the inferior surface of the frontal lobes of the brain towards the anterior and on either side of the longitudinal fissure and crista galli. [See *Color Plate 66*] Identify the **olfactory nerve fibers**, which arise from the olfactory bulbs to pass through the *cribriform plate* of the ethmoid bone. [∞ Figure 15-21, p405] Now locate and trace the **olfactory tracts** from the olfactory bulbs to the cerebrum. Recall that the olfactory nerves are the only cranial nerves attached directly to the cerebrum.

See: Color Plates 66, 92, Fig. 18-3

 ∞ *Fig. 18-6, p474*

LOCATE

____ Nasal conchae (turbinates)

____ Cribriform plate

____ Crista galli

____ Olfactory epithelium

____ Olfactory bulb

____ Olfactory tract

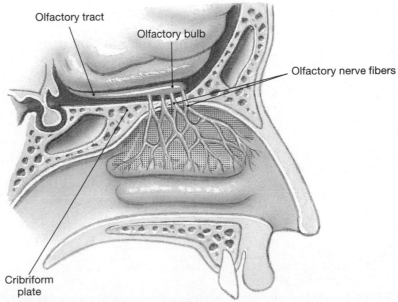

FIGURE 18-3   Olfactory organs.

Microscopic Observation of Olfactory Receptors

Procedure

✓ **Quick Check**

Before you begin to identify the olfactory epithelium, review from Chapter 4 microscope slides of simple cuboidal epithelium, [*Color Plate 3*] simple columnar epithelium, [*Color Plate 6*] and pseudostratified columnar epithelium. [*Color Plate 8*]

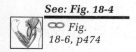 Identify the **olfactory epithelium** using the *microscope slide* provided and *Figure 18-4* for reference, first by viewing under low power magnification and then under high power magnification. Under low power magnification, scan your microscope slide to determine the outer free surface of the olfactory epithelium. Do this by locating the basement membrane and lamina propria portion of the connective tissue. Locate the free epithelial surface. The air we breathe comes in contact with this surface. Under high power magnification, the olfactory epithelium can be identified as the tall pseudostratified columnar epithelial cells. The olfactory epithelium is composed of three types of cells, but typically only two types can be distinguished. The tall slender cells with oval-shaped nuclei located at the surface may be identified as **supporting cells**. The torpedo-shaped cells with round nuclei may be identified as the **olfactory receptor cells**. These cells lie interior to the supporting cells and are highly modified neurons that respond to stimulation by dissolved chemical compounds in the nasal mucus. [∞ p472]

Under low power, locate the connective tissue that supports the epithelium and identify the numerous ducts of the **olfactory (Bowman's) glands**. [∞ p472] Observe that the walls of the ducts vary in shape (oval or round) and are formed of simple cuboidal epithelium. The olfactory glands produce a thick mucus secretion that is transported via the ducts to the surface of the respiratory epithelium. To stimulate the olfactory receptor cells, water and lipid-soluble secretions must diffuse through this mucus secretion.

Slide # _____ Olfactory Epithelium

See: Fig. 18-4
∞ *Fig. 18-6, p474*

LOCATE

_____ Olfactory epithelium

_____ Olfactory receptor cells

_____ Supporting cells

_____ Olfactory (Bowman's) gland

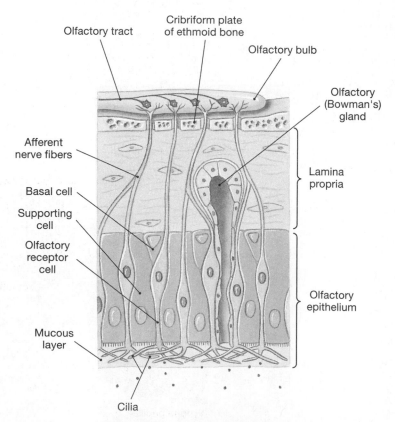

FIGURE 18-4 Olfactory epithelium.

Observation of the Tongue

The *tongue* houses the special sense organs for **gustation**. The most important gustatory receptors are the *taste buds* of the tongue. [∞ p475] Most of the sensory receptors for gustation are located on the dorsal surface of the tongue. To a minor extent taste receptors are also located in the pharynx and larynx. As you recall from study of the muscular system, the tongue consists of extrinsic and intrinsic skeletal muscles [∞ p274] wrapped in a blanket of epithelium. Our sense of taste is strongly related to seeing and smelling of foods. The muscular tongue serves not only as the special sense organ for taste, but also assists in the processes of chewing and swallowing, and in speech.

Procedure

✓ **Quick Check** ──

Before you begin to identify the surface features of the tongue, identify the location and attachment of the tongue in the oral cavity. [See *Color Plate 38*]

Identify the following surface features of the tongue, using a *model* or *specimen* of the *tongue* and *Figure 18-5* for reference. Locate the **dorsal surface (dorsum)** of the **tongue** and lateral margins. Determine the position of epithelial projections called **papillae** on the dorsal and lateral surfaces of the tongue. Use a magnifying glass and mirror to observe and identify the papillae on your tongue. Three types of papillae can be recognized on the human tongue. At the anterior tip of the tongue, identify the cone-shaped **filiform papillae**. Along the lateral margin of the tongue, identify the mushroom-shaped **fungiform papillae**. At the posterior portion of the dorsal surface, identify the large crater-like **circumvallate papillae**. The *taste buds* of the tongue lie along the lateral walls of the papillae. The tongue also contains sensory receptors for touch and temperature. These senses are important for protecting the tongue during the manipulation of hot or cold foods and liquids.

See: Fig. 18-5

∞ *Fig.*
18-7, p475

LOCATE

____ Dorsal surface (dorsum) of tongue

____ Distribution of papillae { Filiform Fungiform Circumvallate

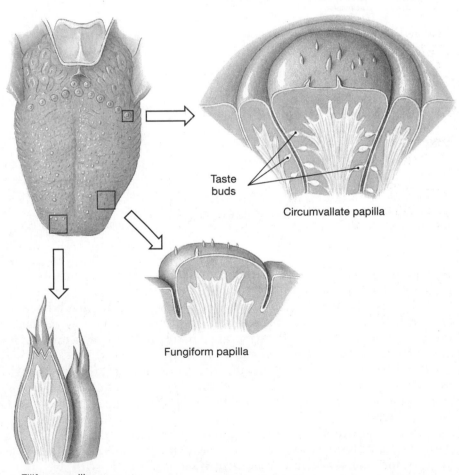

Taste buds

Circumvallate papilla

Fungiform papilla

Filiform papillae

FIGURE 18-5 Papillae of the tongue.

Observation of Cranial Nerves of Gustation

Procedure

✓ Quick Check ───

Before you begin to identify the cranial nerves of gustation, review: on a dry human skull specimen the cranial fossae regions (*Color Plate 33*) and the foramina of the skull, as seen from both internal and external views; (*Figures 6-8* and *6-9, pp45-46*) the cranial nerves (*Table 15-1, p234*) as described under the heading "Observation of the Cranial Nerves" [*Chapter 15, p232*] focusing on N VII, N IX and N X.

Identify the three cranial nerves that carry gustatory information to the primary sensory cortex of the cerebrum, using *models* and or *specimens* of the *human brain* and the *tongue*. Use *Color Plates 66, 67,* and *86* for reference. As identified in Quick Check, review on a brain model the cranial nerves (N VII, N IX, and N X) responsible for transmitting the sensory codes for gustation to the brain. Identify the **facial nerve** (N VII) as it arises from the anterior portion of the pons [∞ Figure 15-26, p411] to pass through the internal acoustic meatus to exit at the stylomastoid foramen. Taste on the anterior two-thirds of the tongue is monitored by this nerve. Identify the **glossopharyngeal nerve** (N IX) [∞ Figure 15-28, p412] on the ventrolateral surface of the medulla oblongata, just posterior to the olive. This nerve passes through the jugular foramen. Identify this nerve as it enters the posterior of the tongue deep to the styloglossus and hyoglossus muscles. Taste on the posterior third of the tongue is monitored by N IX. Identify the rootlets of the **vagus nerve** (N X) [∞ Figure 15-29, p413], which also pass through the jugular foramen. Locate a small sensory branch, the superior laryngeal nerve, at the posterior base of the tongue and the epiglottis. [*Color Plates 38* and *92*] This nerve branch of the vagus supplies the sensory fibers for monitoring taste in this region.

See: Color Plates 66, 67, 86

∞ Figs.
15-21, p409
18-8, p476

LOCATE

_____ Facial (N VII)

_____ Glossopharyngeal (N IX)

_____ Vagus (N X)

Microscopic Observation of Gustatory (Taste Buds) Receptors

Procedure

✓ Quick Check ───

Before you begin to identify the gustatory receptors of the tongue, review the microscopic appearance of stratified squamous epithelium [*Color Plate 2*] as described under the heading "Microscopic Observation of Stratified Squamous Epithelium." [*Chapter 3 p13*]

Identify the following structures associated with gustation, using the *microscope slide* provided and *Figure 18-6* for reference, first by viewing under low power magnification and then under high power magnification. The slide provided to you is a longitudinal section of the tongue. The tongue consists of an outer layer of stratified squamous epithelium connected to an underlying region of dense irregular connective tissue. Scan the slide under low magnification and distinguish between the violet-stained stratified squamous epithelium and the pink-stained dense irregular connective tissue.

Papillae can be observed on your slide, but on most slides only filiform and fungiform can be distinguished. Identify the cone-shaped **filiform papillae** and the small mushroom-shaped **fungiform papillae**. Large crater-shaped **circumvallate papillae** may be recognized on only those sections that are obtained from the posterior region of the tongue.

Observe clusters of onion-shaped structures embedded within the stratified squamous epithelial wall of papillae. Identify these structures under high power magnification as **taste buds**. Each taste bud contains about 40 **gustatory cells**, which can be identified as the cells with dark purple-staining nuclei and light violet cytoplasm. Between gustatory cells are narrow torpedo-shaped **supporting cells**. Locate the opening of each taste bud into the surrounding fluid as the **taste pore**. Several **microvilli**, or **taste hairs** of gustatory cells, can be observed extending from the taste pore. The microvilli of gustatory cells are stimulated by dissolved chemicals within the surrounding solution of the oral cavity. If possible, observe the above cells and structures with the oil immersion objective.

Slide # _____ Tongue, l.s.

See: Fig. 18-6

∞ *Fig. 18-7, p475*

LOCATE

_____ Papillae $\begin{cases} \text{Filiform} \\ \text{Fungiform} \\ \text{Circumvallate} \end{cases}$

_____ Taste buds

_____ Gustatory cell

_____ Supporting cell

_____ Taste pore

_____ Microvilli (taste hairs)

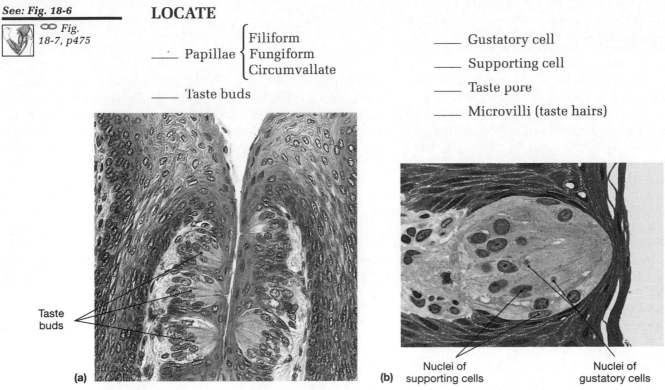

FIGURE 18-6   Gustatory reception (a) tastebuds in a circumvallate papilla × 280 (b) detail of a taste bud × 650.

Observation of Ear Anatomy

Hearing and equilibrium functions are located in the ear and are anatomically associated with each other. The ear is divided into three anatomical regions; outer, middle, and inner ear. [∞ p477] The visible portion of the ear is the *external* or *outer ear* and is responsible for collecting and directing sound waves to the eardrum. The *middle ear* is an air chamber located within the petrous portion of the temporal bone. It contains the auditory ossicles, which amplify and transmit sound waves to the inner ear. The *internal* or *inner ear* is also located within the petrous portion of the temporal bone and contains the sensory organs for hearing and equilibrium. The structures associated with each ear region are described and identified by region.

Procedure

✓ **Quick Check**

Before you begin to identify the structures of the ear, review on a dry skull the petrous portion of the temporal bone and the internal acoustic canal; (*Color Plate 33* and *Figure 6-10, p47*) cranial nerve N VIII (vestibulocochlear nerve) [∞ p405] under the heading "Observation of The Cranial Nerves" [*Chapter 15, p233*] and its two distinct nerve divisions: the cochlear nerve and the vestibular nerve.

The External Ear

Use a *model* of the ear to identify the **divisions** and **structures** of the **ear**. Identify on your model the divisions of the ear and the following structures associated with each division, using *Figure 18-7* for reference. Identify the **external**, the **internal**, and the **middle ear** divisions. Observe the fleshy skin flap over cartilage, which is the outer ear, **auricle**. Think about sound waves being collected by the auricle and directed toward the **external acoustic meatus** to strike the **tympanic membrane** or eardrum (tympanum). Identify these structures. The division between the external and middle ears is the tympanic membrane. External to this membrane is the external ear; internal to the membrane is the middle ear.

In the middle ear, identify the location of three small ear bones, the **auditory ossicles**, and observe their respective positions within the **petrous portion** of the **temporal bone**. Notice how the ossicles are suspended by their attachment points within the middle ear. Locate the floor (inferior portion) of the middle ear and identify the tubular structure known as the **auditory** (*pharyngotympanic* or Eustachian) **tube** that leads towards the throat. This tube contains air and is the means for equalizing the pressure within the middle ear so the tympanic membrane does not rupture. Normally the tube is closed, but swallowing, chewing, and yawning opens the tube to equalize the pressure on either side of the eardrum.

To observe the structures of the inner ear, a section of the petrous portion of temporal bone must be removed from your model. Remove this portion of the temporal bone and set it aside. Observe within the temporal bone the snail-shaped structure with attached semicircular rings. Identify this structure as the **bony labyrinth of** the **inner ear**. Remove the bony labyrinth from the model and observe its structure and shape. The bony labyrinth is a shell of dense bone that encases structures for hearing and balance. Identify the snail-shaped **cochlea**, which contains the receptor structures for hearing. Identify the **vestibular complex** as the combination of vestibule and semicircular canals which are continuous with the cochlea. The vestibular complex consists of those structures associated with position and balance. The structures of the vestibular complex are described and identified in a separate observation under the heading "Inner Ear." Find the large **vestibulocochlear nerve (N VIII)** as it emerges from the internal acoustic meatus. Trace this nerve internally from the meatus towards the vestibular complex and cochlea to observe how this cranial nerve is actually formed by two nerve branches. Observe the nerve branch emanating from the cochlea as the **cochlear nerve**, which serves for the sense of hearing. Now identify the nerve branch emanating from the vestibular complex as the **vestibular nerve**, which serves for balancing and positioning sensations.

Additional middle ear and inner ear structures are described and identified under separate headings in the observations that follow.

See: Fig. 18-7

 ∞ *Fig. 18-9, p477*

LOCATE

_____ External, middle, and inner ear divisions

_____ Auricle

_____ External acoustic meatus

_____ Tympanic membrane (tympanum)

_____ Auditory ossicles

_____ Temporal bone (petrous portion)

_____ Auditory tube (pharyngotympanic tube or Eustachian tube)

_____ Bony labyrinth of inner ear

_____ Vestibular complex

_____ Cochlea

_____ Vestibulocochlear (N VIII) nerve { Cochlear division / Vestibular division

The Middle Ear

The middle ear is a very small air-filled chamber containing the ear (auditory) ossicles. The auditory ossicles are vital to the conduction of sound waves from the external ear to the inner ear. Because the middle ear is an air-filled chamber, but separated from the external acoustic canal, there must be a mechanism to equalize the pres-

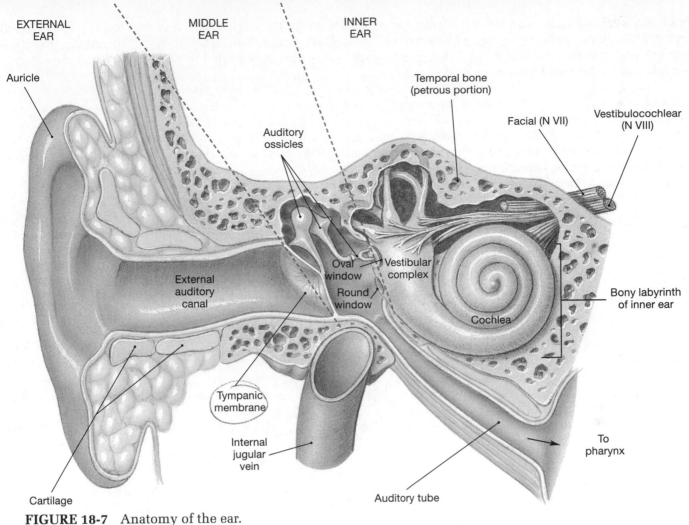

EXTERNAL
EAR

MIDDLE
EAR

INNER
EAR

Auricle

Auditory
ossicles

Temporal bone
(petrous portion)

Facial (N VII)

Vestibulocochlear
(N VIII)

External
auditory
canal

Oval
window

Vestibular
complex

Round
window

Cochlea

Bony labyrinth
of inner ear

Tympanic
membrane

Internal
jugular
vein

Cartilage

Auditory tube

To
pharynx

FIGURE 18-7   Anatomy of the ear.

sure to prevent the tympanic membrane from rupturing. The auditory (pharyngotympanic or Eustachian) tube serves to equalize the pressure to prevent damage to the eardrum. [∞ p477]

Procedure

Identify the following **middle ear** structures, using an *ear model*. Use *Figures 18-7* and *18-8* for reference. Observe the position of the tympanic membrane within the most medial portion of the external acoustic meatus. View the internal surface of the tympanic membrane and identify the **auditory ossicles** that occupy the air chamber of the middle ear. Examine and identify the three auditory ossicles: beginning with the outermost ossicle, the **malleus**, which is attached to the inner surface of the tympanic membrane, the middle ossicle, the **incus**, and the innermost ossicle, the **stapes**. The **base (footplate) of** the **stapes** attaches to the *oval window* of the inner ear. In life, the three bones are linked together by synovial joints to transfer sound waves from the tympanic membrane to the inner ear. The bones are linked in such a way to act as levers that amplify the sound waves from the tympanic membrane to the receptors of the inner ear. Movement of the footplate of the stapes transfers sound waves to the fluid contents of the inner ear.

The delicate receptors of the inner ear are protected by two small ribbons of muscle in the middle ear (Note: On some models these muscles are not present). Identify on the "handle" portion of the malleus the **tensor tympani muscle**. This ribbon of muscle has its origin on the petrous portion of the temporal bone and when contracted it stiffens the tympanic membrane, thereby reducing the amount of possible movement. Identify the **stapedius muscle** attached to the stapes. When this muscle contracts, the stapes is pulled away from the oval window, which reduces its movement at the oval window. Both muscles contract in a reflex manner during loud noises to prevent excessive vibrations from damaging delicate inner ear structures (hair cells). The muscles do not protect the eardrum or ossicles.

See: Figs. 18-7, 18-8

⚙ Fig.
18-10, p478

LOCATE

____ External acoustic meatus

____ Auditory ossicles { Malleus
Incus
Stapes

____ Base of stapes in oval window

____ Oval window

____ Tensor tympani muscle

____ Stapedius muscle

____ Auditory tube

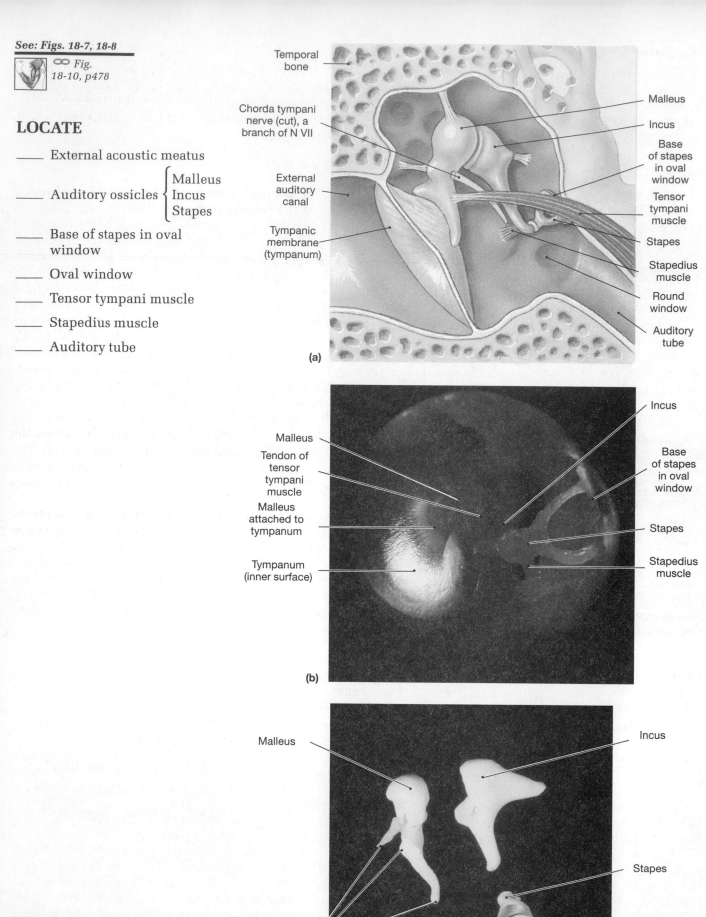

FIGURE 18-8 Middle ear (a) structures (b) tympanic membrane and auditory ossicles as seen by fiberoptic tube (c) isolated auditory ossicles.

The Inner Ear

Of all the regions of the ear, the inner ear is the most complex. It is composed of two components: the cochlea is associated with hearing and the vestibular complex is associated with balance and position. Both components are embedded in a bony labyrinth, which is lined by a membranous labyrinth filled with fluid. The sensory receptors both for hearing and for equilibrium and position respond to the movement of the fluid contained within the structures that compose the membranous labyrinth. [∞ p479]

Procedure

Identify the following **inner ear** structures, using an *ear model* or *labyrinth model* and *Figure 18-9* for reference. If you are using an ear model, remove the detachable section of the petrous portion of the temporal bone and then remove the bony labyrinth from the model. If you are using a labyrinth model, separate the cochlea portion. Observe the shell of dense bone that forms the bony labyrinth and encases the structures for hearing and balance. The bony labyrinth is lined by a collection of fluid-filled tubes and chambers collectively called the **membranous labyrinth**. Distinguish between the bony and membranous labyrinths. The fluid that flows between the bony and membranous labyrinths, called *perilymph*, resembles CSF. The fluid within the membranous labyrinth is termed *endolymph*. (Note that these fluids are not shown in models.) Differentiate between the snail-shaped cochlea and the vestibular complex. Now identify the three semicircular rings of the vestibular complex as the **semicircular canals** and note that they are positioned at right angles to each other. View the semicircular canals in the anatomical position and identify the **anterior, posterior**, and **lateral** semicircular canals. Each canal contains an endolymph-filled membranous **semicircular duct** having the same name as the canal. Identify the expanded regions of each duct, called the **ampullae**, which contain the sensory hair cell receptors that respond to rotational movements of the head. (Hair cells, cristae, and cupulae are not depicted on models.)

Examine the vestibule portion of the model and identify the **utricle**, the membranous sac that communicates with the semicircular canals. Inferior to the utricle identify a second membranous sac, the **saccule**. Locate the **endolymphatic duct** that is continuous with the slender passageway that connects the utricle and saccule. The **endolymphatic sac** can be identified as the terminal end of the endolymphatic duct. In life, the endolymphatic sac projects through the dura mater into the subdural space. The receptor hair cells in the utricle and saccule are clustered in the **maculae** and provide us with sensations of gravity and linear acceleration.

Return to the separated portion of the cochlea and examine it. Notice how the entire complex makes 2½ turns around a central bony core from which emerges the cochlear nerve. Hold the separated portion of the cochlea so the cochlear nerve points down. Now observe the compartments contained in each turn. Select the bottom turn compartment and observe three tubes. Identify the top tube within the compartment as the **vestibular duct (scala vestibuli)**, the bottom tube within the compartment as the **tympanic duct (scala tympani)**, and the middle tube as the **cochlear duct (scala media)**. Observe on the exterior at the base of the cochlear spiral two distinct membranous areas, the **round window** and the **oval window**. The footplate portion of the stapes connects to the oval window to conduct sound waves from middle ear to inner ear.

See: Fig. 18-9

∞ *Fig. 18-12, p480*

LOCATE

____ Cochlea

____ Vestibular complex

____ Bony labyrinth

____ Membranous labyrinth

____ Semicircular canals { Anterior, Posterior, Lateral }

____ Ampullae (one for each duct)

____ Utricle

____ Saccule

____ Endolymphatic duct

____ Endolymphatic sac

____ Macula in saccule

____ Macula in utricle

____ Vestibular duct (scala vestibuli)

____ Cochlear duct (scala media)

____ Tympanic duct (scala tympani)

____ Round window

____ Oval window

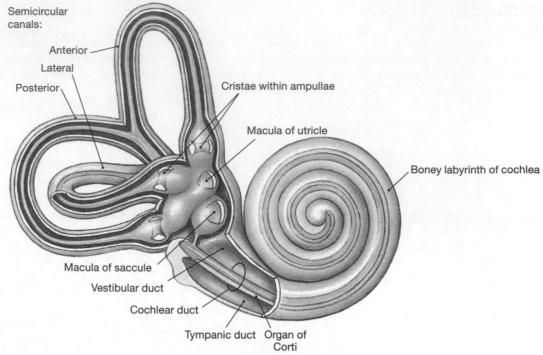

Semicircular canals:

Anterior

Lateral

Posterior

Cristae within ampullae

Macula of utricle

Boney labyrinth of cochlea

Macula of saccule

Vestibular duct

Cochlear duct

Tympanic duct Organ of Corti

FIGURE 18-9 Anterior view of the bony labyrinth showing the enclosed membranous labyrinth.

Microscopic Observation of Cochlea and Organ of Corti

Procedure

✓ **Quick Check**

Before you begin to identify the structures of the cochlea and the organ of Corti, review on a model the bony labyrinth of the inner ear as it appears in section. Identify the vestibular duct, cochlear duct, tympanic duct, and basilar membrane. [⚭ Figure 18-16, pp484-485]

Identify the following **structures**, using the *microscope slide* provided and *Figures 18-10* and *18-11* for reference, first by viewing under low power magnification and then with high power magnification. The slide provided to you is a longitudinal section of the cochlea. Under low scanning magnification (or dissecting microscope), identify the general structure of the cochlea as it appears within a portion of the temporal bone. The color of the *bony cochlear wall* is typically royal blue in this type of slide. For orientation purposes, identify the apical and the basal turns. Observe how the turns coil around a central hub of bone, the *modiolus*. A portion of the **cochlear nerve** can be identified within the modiolus. This nerve is a branch of the **vestibulocochlear nerve** (N VIII).

Now select a compartment in which you can observe all three tubes of the membranous labyrinth and view this compartment under low power. Identify the top tube within the compartment as the **vestibular duct (scala vestibuli)**, the bottom tube as the **tympanic duct (scala tympani)**, and the middle tube as the **cochlear duct (scala media)**. Identify the **basilar membrane**, which separates the cochlear duct from the tympanic duct. Position the slide so that both the cochlear duct and basilar membrane are in your field of vision. Now switch to high dry magnification, and identify the **organ of Corti** and the ribbon-like **tectorial membrane** that lies over it. Observe the **hair cells** that make up the organ of Corti and note their position in relation to the tectorial membrane. The hair cells sit on the basilar membrane with the hairs just barely touching the tectorial membrane. Keep the basilar membrane in the field of vision and trace it to the modiolus to identify the **spiral ganglion cells** that form the cochlear nerve.

Slide # _____ Cochlea

See: Figs. 18-10, 18-11

∞ *Fig.*
18-16, pp484-485

LOCATE

____ Vestibular duct (scala vestibuli)

____ Bony cochlear wall

____ Cochlear duct (scala media)

____ Tectorial membrane

____ Organ of Corti ⎰ Basilar membrane
⎱ Hair cells

____ Tympanic duct (scala tympani)

____ Spiral ganglion of cochlear nerve

____ Cochlear nerve

____ Vestibulocochlear nerve (N VIII)

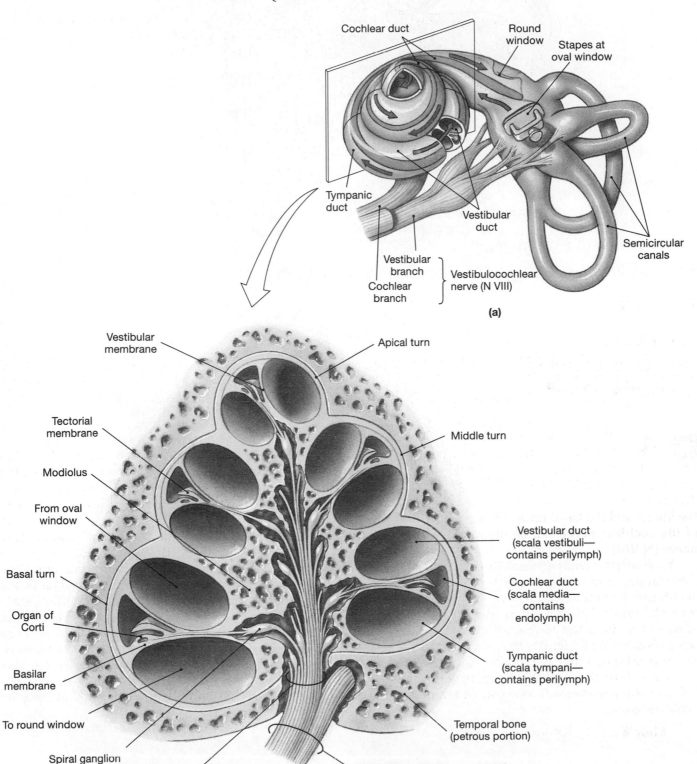

FIGURE 18-10 Cochlea and organ of Corti (a) structures of the cochlea (b) cochlea in section.

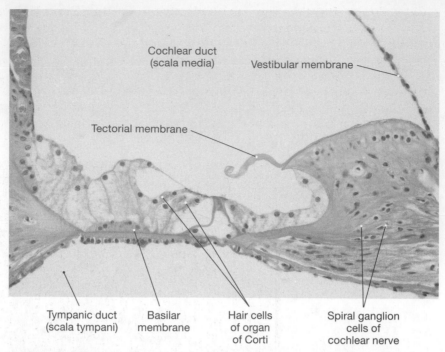

Cochlear duct
(scala media)

Vestibular membrane

Tectorial membrane

Tympanic duct
(scala tympani)

Basilar
membrane

Hair cells
of organ
of Corti

Spiral ganglion
cells of
cochlear nerve

FIGURE 18-11 Organ of Corti, histologic section × 650.

Observation of Eye Anatomy

Of all the senses, we rely on vision the most. The eye, along with its accessory structures, is responsible for vision. [∞ p487] The sensory receptors for vision are located within the eyes, and enable us to both detect light and create detailed visual images. Each eye is protected within a bony orbit, and is supported and cushioned by a mass of orbital fat. The eyeball is hollow and is formed by three layers called *tunics*. An assortment of structures for focusing images, and adjusting light intensities, along with the sensory receptors for vision, compose the tunics. The structures for each tunic are described and identified separately.

Observation of Accessory Structures of the Eye

The *accessory structures* are responsible for lubricating, protecting, and supporting the eyes. The accessory structures of the eye include the *eyelids*, and the structures associated with the production, secretion, and removal of tears. [∞ p487]

Superficial Anatomy of the Accessory Structures of the Eye

Procedure

Observe your *laboratory partner's eyes* for the identification of the superficial anatomy of the **accessory structures** of the **eye**. Use *Figure 18-12* for reference. On your laboratory partner, identify the upper and lower eyelids, termed **palpebrae**. These structures of skin and muscle protect and cover the eyes and spread tear fluid, which cleans and moistens the surface of the eyeball. The eyelids are supported and strengthened by broad sheets of connective tissue collectively called the *tarsal plate*. The muscle fibers of the *orbicularis oculi* and the *levator palpebrae superioris* lie between the tarsal plate and the skin. See *Color Plate 34*. The inner surface of the eyelids and the outer surface of the eye are covered by a mucous membrane termed the **conjunctiva**. The *palpebral conjunctiva* covers the inner surface of the eyelids, and the *ocular* (bulbar) *conjunctiva* covers the anterior surface of the eyeball.

Examine the palpebral margins and locate the **eyelashes**. When the eyes are open, notice how the free margins of the upper and lower eyelids are separated by a space, called the **palpebral fissure**. Identify the **medial canthus** as the point where the two lids are connected medially and the **lateral canthus** as the point where the lids join laterally. Just lateral to the medial canthus, locate an orange triangular structure known as the **lacrimal caruncle**. This structure contains the passageways to drain tear fluid from the eye.

On your partner observe the "white of the eye," anatomically termed the **sclera**. The sclera is part of the outermost tunic of the eyeball and consists of dense fibrous connective tissue that covers most of the ocular surface. Observe the small blood vessels that are embedded within the sclera. These vessels penetrate deep into the sclera to supply internal structures. The clear anterior window can be identified as the **cornea**. Examine the center of the *iris*, or colored portion of the eye, and identify its black central opening as the **pupil**. The reason the pupil looks like a black hole is because you are looking into a dark black chamber. Light rays must pass through the cornea and pupil to strike the sensory receptors for vision that are located within the eye. Have your partner cover his or her eyes with their hands for a minute, then remove his or her hands and look at direct light. Observe your partner's pupil change sizes according to light conditions.

See: Fig. 18-12

∞ *Fig.*
18-18a, p488

LOCATE

____ Palpebrae
(Eyelids)

____ Orbicularis oculi muscle

____ Conjunctiva

____ Eyelashes

____ Palpebral fissure

____ Medial canthus

____ Lateral canthus

____ Lacrimal caruncle

____ Sclera

____ Pupil

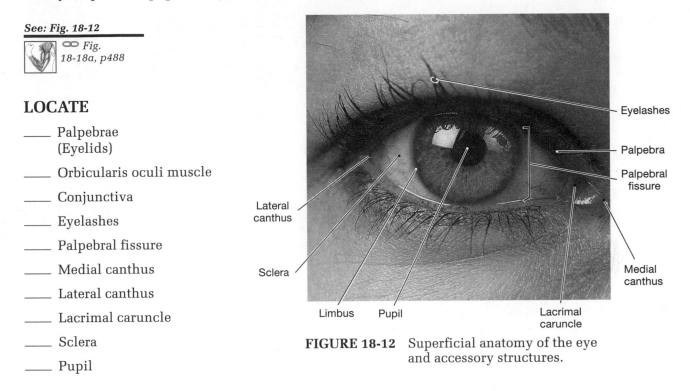

FIGURE 18-12 Superficial anatomy of the eye and accessory structures.

Anatomy of the Accessory Structures: The Lacrimal Apparatus

Procedure

✓ **Quick Check**

Before you begin to identify the accessory structures of the eye, review on a dry human skull specimen the bones and bony landmarks of the orbital complex. (*Figure 6-21, p54*)

Identify the following **accessory structures** of the **eye**, using a *dissectable eye in orbit model* or *specimen* of the *eye*. Use *Figure 18-13* for reference. Review on the eye model the following bone markings of the orbital complex: superior/inferior orbital margins, nasolacrimal fossa/groove, and supraorbital/infraorbital foramina. A continuous supply of tear fluid washes over the surface of the eyeball to keep the conjunctiva moist and clean. The production, distribution, and removal of tear fluid is performed by a group of four structures collectively called the **lacrimal apparatus**. Locate on the lateral and superior surface of the eyeball the **lacrimal gland** (tear gland). The **lacrimal canaliculi (lacrimal gland ducts)** can be identified emerging from the gland. These ducts lead tear fluid from the lacrimal gland to be spread across the corneal surface of the eyeball by the blinking action of the eyelids. Tear fluid flows continuously to the region of the medial canthus for removal. Identify two small openings on the margins of each eyelid in the area of the medial canthus. These are the **lacrimal puncta**. The puncta are the openings of the **superior** and **inferior lacrimal canals**. The function of these ducts is to drain tear fluid from the medial canthus area. Identify and trace these ducts to the **lacrimal sac**, located within the nasolacrimal fossa. Trace the lacrimal sac to a constriction that forms the **nasolacrimal duct**, located in the nasolacrimal groove. The duct drains tear fluid into the nasal cavity at a location just inferior to the inferior nasal conchae.

In the prosected cadaver, the lacrimal gland, lacrimal sac, and duct may be observed. These structures are best observed superior to the orbit. Typically the roof of the orbit is dissected open so as to create a "window" through which these structures may be viewed undisturbed. (see *Figure 18-14*) Check with your instructor before viewing as these structures are very delicate.

See: Fig. 18-13, 18-14

∞ *Fig. 18-18b p488*

LOCATE

_____ Upper/lower eyelids

_____ Lacrimal gland

_____ Lacrimal puncta

_____ Lacrimal canaliculi (ducts)

_____ Superior lacrimal canal

_____ Inferior lacrimal canal

_____ Lacrimal sac

_____ Nasolacrimal duct

_____ Opening of nasolacrimal duct

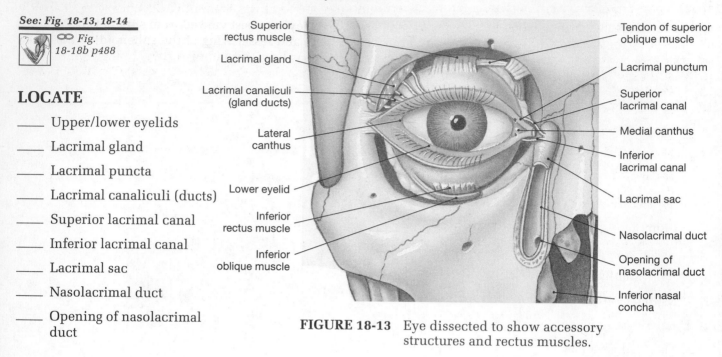

FIGURE 18-13 Eye dissected to show accessory structures and rectus muscles.

Observation of the Extraocular Eye Muscles

The *extraocular eye muscles* are outside the eyeball and rotate the eye within the bony orbit. (see *Figure 10-3, p112*) The contractions of the six extraocular eye muscles produce all the eye movements associated with the eyeball. For a detailed summary of the functions of the extraocular eye muscles, see *Table 10-2, p112*.

Procedure

✓ **Quick Check**

Before you begin to identify the accessory structures of the eye, review (1) on a dry human skull specimen, the bones and bony landmarks of the orbital complex, (*Figure 6-21, p54*) and (2) on a brain model the origins of the oculomotor (N III), trochlear (N VI), and (3) abducens nerves (N IV), which collectively pass through the superior orbital fissure of the sphenoid. [see *Color Plates 66, 67,* and *86*]

Identify the six **extraocular eye muscles** and review the associated cranial nerves, using a *dissectable eye in orbit model* or *specimen of the eye*. Use *Figures 18-13* and *18-14* for reference. Begin your observation of the extraocular eye muscles on the model from the lateral view perspective. The insertion end of each muscle is attached by a tendon to the sclera; the origin end is attached by a tendon to the bony orbit. The muscles are easily identified in partner pairs. First locate and identify on the superior surface of the eyeball the **superior rectus** muscle, then locate on the inferior surface the **inferior rectus** muscle. When these muscles are innervated by N III, the superior rectus causes the eye to look up while the inferior rectus causes the eye to look down. On the lateral surface of the eyeball identify the **lateral rectus** muscle, then on the medial surface locate the **medial rectus** muscle. When the lateral rectus is innervated by N VI, it causes abduction of the eye (rotates laterally), and when the medial is innervated by N III, it causes adduction of the eye (rotates medially).

Return to the superior surface of the eyeball and locate the superior rectus muscle. Immediately medial to it identify the **superior oblique** muscle. Notice how the tendon of the superior oblique passes inferior to the superior rectus muscle. When the superior oblique is innervated by N IV, the eyeball rolls, looking downward and to the side. Locate the lateral rectus muscle and observe inferior to it the tendon of the **inferior oblique** muscle, which courses medially from the superior surface of the maxilla. When the inferior oblique is innervated by N III the eyeball rolls, looking upward and to the side.

 After observing the extrinsic eye muscles on a model, proceed to the cadaver. In the prosected cadaver, the eyeball is often partially collapsed, but this will not prevent you from examining the muscles and associated structures. The muscles and nerves associated with the eye are best viewed from superior to the orbit. Typically the roof of the orbit (orbital portion of the frontal bone and lesser wing of the sphenoid) is dissected open so as to create a "window" through which these structures may be viewed undisturbed. Using *Figure 18-14*, which is a view of the eye from superior to the orbit, try to identify on the cadaver as many structures as possible.

See: Figs. 18-3, 18-14, 18-15

∞ *Figs.*
10-4, p270
10-5, p272
18-18b, p488
18-24, p495
18-25a, p496

LOCATE

____ Superior oblique muscle

____ Superior rectus muscle

____ Medial rectus muscle

____ Trochlear nerve (IV)

____ Oculomotor nerve (III)

____ Abducens nerve (VI)

____ Lateral rectus muscle

____ Inferior rectus muscle

____ Inferior oblique muscle

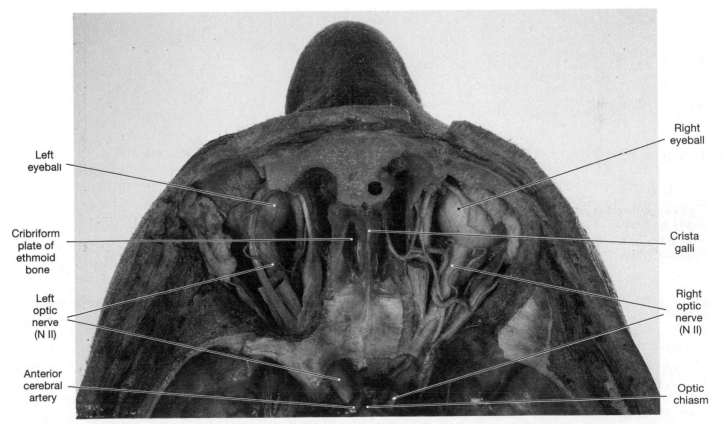

FIGURE 18-14 Horizontal section through human head at the level of the optic chiasm as seen from the superior view (frontal and ethmoid bones have been removed).

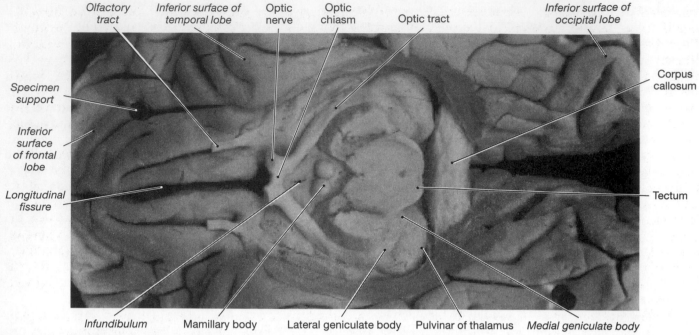

Olfactory tract · Inferior surface of temporal lobe · Optic nerve · Optic chiasm · Optic tract · Inferior surface of occipital lobe · Corpus callosum · Specimen support · Inferior surface of frontal lobe · Longitudinal fissure · Tectum · Infundibulum · Mamillary body · Lateral geniculate body · Pulvinar of thalamus · Medial geniculate body

FIGURE 18-15 Optic tract, inferior view of human brain dissected.

Sectional Anatomy of the Eye

When the eyeball is viewed in section, the tunics and associated structures may be easily observed and identified. Three tunics form the eyeball: an outer *fibrous tunic*, a middle *vascular tunic (uvea)*, and an inner *neural tunic*. A clear liquid, called the *aqueous humor*, and a gelatinous mass, termed the *vitreous (humor) body*, fill the hollow chambers of the eye and contribute to the shape of the eyeball.

Procedure

✓ **Quick Check**

Before you begin to examine the sectional anatomy of the eye, review on a brain model the **optic chiasma**, left and right **optic tracts**, and the **optic nerves (N II)**, which pass through either the left or right *optic canal* to enter the eye. (*Color Plates 66* and *86*)

Identify the following **layers** and **structures** of the **eye**, using a *model* and then a *specimen* of the *eye*. Use *Color Plate 65* and *Figures 18-16* and *18-17* as reference. View the eye in section and review the previously identified palpebral conjunctiva, which covers the inner surface of the eyelids, and the ocular (bulbar) conjunctiva, which covers the anterior surface of the eyeball. Locate and identify the space that is created where the conjunctiva of the eyelid connects with the conjunctiva of the eyeball as the **fornix**.

Fibrous Tunic Structures

A dense, opaque white fibrous connective tissue composes the outside of the eyeball proper and can be identified as the **sclera** portion of the **fibrous tunic**, in contrast to the anterior clear window, the **cornea**. The cornea is structurally continuous with the sclera. Identify the border between the two as the **limbus**.

Vascular Tunic (Uvea) Structures

The **vascular tunic** or **uvea** contains the blood vessel, lymphatics, and nerves to service the tissues of the eye. Structures of the vascular tunic regulate the amount of light entering the eye, control the shape of the lens for focusing, and circulate aqueous humor for transporting the nutrients and waste products of the eye. View from the front of the eyeball through the transparent cornea, and identify the colored (such as blue or brown) or pigmented part of the eye as the **iris**. The iris consists of three parts: an outer, visibly colored portion, a middle layer consisting of two layers of smooth muscle fibers, and an inner black pigmented portion. Muscles of the iris change the diameter of the central opening. Identify this opening as the **pupil** of the iris. One group of smooth-muscle fibers, termed the **pupillary constrictor muscles**, forms a series of concentric circles around the pupil. When these contract, the diameter of the pupil decreases. A second group, termed the **pupillary dilator muscles**, extends radially away from the edge of the pupil. Contraction of these enlarges the pupil. Identify the space between the iris and the cornea as the **anterior chamber**. The iris attaches to the **ciliary body**, which can be identified at the junction between the cornea and sclera. Most of the ciliary body consists of **ciliary muscle**. Follow and identify the ciliary muscle as it extends posteriorly to the serrated anterior edge of the neural retina, the **ora serrata**, which projects posteriorly. Observe how the epithelium of the ciliary body is thrown into numerous pleats or folds, termed the **ciliary process**.

Locate the **lens** and observe how it is attached to the ciliary process by fibers of connective tissue, collectively called the **suspensory ligaments**. Contraction and relaxation of the ciliary muscle causes movement of the ciliary body and suspensory ligaments. These movements change the tension in the suspensory ligaments causing the lens to change shape. The lens must change its shape for the eye to view objects that are either near or far. Observe the space between the iris and the suspensory ligaments. This is the **posterior chamber** of the eye. In life, the anterior and posterior chambers are filled with a clear fluid, the *aqueous humor*. Find the large **vitreous chamber** posterior to the lens and suspensory ligaments, which in life contains a clear gelatinous mass termed the **vitreous body (humor)**. Aqueous humor circulates between the anterior and posterior chambers and drains into the **canal of Schlemm**. It can be identified on the interior border of the cornea and sclera. Waste products contained within the aqueous humor are drained by the canal into veins of the eye.

The **choroid** portion of the vascular tunic can be identified by its extensive vascular supply and black pigment. Observe how the outermost portion of the choroid contacts the sclera, while the innermost portion attaches to the neural tunic.

The Neural Tunic

Identify the innermost layer of the eye as the **neural tunic**. The neural tunic consists of two layers and covers. The outer layer of the neural tunic that is adjacent to the vascul tunic is the *pigmented layer*. Proceed inward to identify the remaining layer as the **neural retina** only as far anterior as the ora serrata. The neural retina contains the photoreceptors and supporting neurons for responding to light and the initial processing of visual information. The neural retina contains two types of photoreceptors, *rods and cones*. (Rods and cones are typically not depicted on models and can only be observed with the microscope; their identification is described under the heading "Microscopic Observation of Retina" [see *p273*]). Rods and cones are not evenly distributed across the outer surface of the neural retina. At the posterior of the eyeball, observe a yellow spot on the retinal surface, called the **macula lutea**. Locate the central portion of the macula lutea, the **fovea** (fovea centralis). The fovea contains the highest concentration of cones.

Axons from ganglion cells of the neural retinal layer converge in an area, called the **optic disc**, on the posterior of the eye and proceed out of the eyeball as the optic nerve (N II). Locate the optic disc medial to the fovea. The optic disc is also referred to as the blind spot because there are no rod and cone cells at the disc. Inspect the optic disc and identify the *central retinal artery* (red color) and *central retinal vein* (blue color). These vessels pass through the center of the optic nerve and emerge on the surface of the optic disc to supply blood to the retina.

Your instructor may provide you with a mammalian eyeball to dissect. (Specimen eyeballs are often partially collapsed, but this will not prevent you from examining its structure.) Examine the external surface of the specimen eyeball. Typically an abundance of fat is present on these specimens, so you will need to excise this tissue to expose structures. Locate the cornea, optic nerve, sclera, and stumps of the extrinsic eye muscles. The internal structures of the eye must be viewed in section. Some structures are better revealed in coronal section and others in sagittal section. Check with your instructor for precise directions for the sectioning of your specimen and identification of specific structures.

See: Color Plate 65,
Figs. 18-16, 18-17

∞ *Figs.*
18-20, pp490-491
18-22, p494

LOCATE

Tunics (Layers) of the Eye

_____ Fibrous (sclera, cornea)

_____ Vascular (choroid, ciliary body, iris)

_____ Neural (retina)

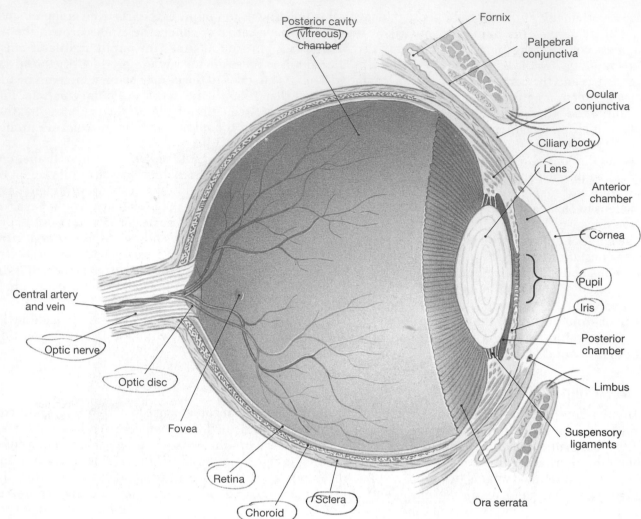

FIGURE 18-16 Eyeball, midsagittal view to show tunics and vitreous body.

Structures

_____ Conjunctiva {Ocular / Palpebral

_____ Fornix

_____ Sclera

_____ Cornea

_____ Limbus

_____ Anterior chamber

_____ Pupil

_____ Iris

_____ Ciliary body

_____ Ora serrata

_____ Lens

_____ Suspensory ligaments

_____ Posterior chamber

_____ Vitreous (humor) chamber/ body

_____ Choroid

_____ Pigmented layer

_____ Retina

_____ Fovea (fovea centralis)

_____ Optic disc

_____ Optic nerve

_____ Central retinal artery

_____ Central retinal vein

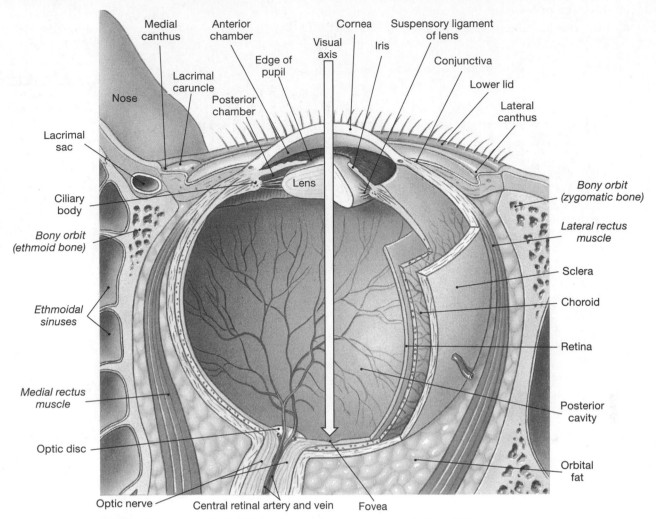

FIGURE 18-17 Eyeball, horizontal section through the right eye.

Microscopic Observation of Retina

Identify the following **structures**, using the *microscope slide* provided and *Color Plate 24, Figures 18-18 and 18-19* for reference, first by viewing under low power magnification. The slide provided to you is a sagittal section of the eyeball. View the slide under the dissecting microscope to identify the general structure of the eyeball. First, position the slide so the entire eyeball is in view. To see the entire eyeball, use the lowest magnification of the dissecting microscope. You will now be able to identify on your specimen most of the same structures that were identified on the eye model. Begin your identification with the cornea and then proceed posteriorly. It is best to identify the structures of the eye in this order: cornea, anterior chamber, pupil, iris, ciliary process, ciliary muscle, suspensory ligaments, lens, posterior chamber, vitreous chamber and body, optic disc, and optic nerve (typically visible on only mid-sagittal sections). Now locate and identify the layers of the eyeball. Begin on the external surface with the sclera and proceed internally. The choroid appears as a thin black line that adheres to the inner surface of the sclera. The retinal layer can be identified by a band of violet and pink that is located internally from the choroid.

To examine the retina, choroid, and sclera, place your slide on the compound microscope and locate the layers first under low magnification and then view with high power magnification. Position the slide so you are viewing the layers in the region of the fovea and examine with high magnification. Use *Color Plate 24* and *Figure 18-19* for reference as you identify the layers from externally to internally, beginning with the sclera. The sclera can be identified by its characteristic pink collagen fibers. Characteristic of the choroid are the numerous deeply pigmented melanocytes and blood vessels. The (outermost) layer of the retina lies adjacent to the choroid. This outermost layer is composed of cuboidal cells containing melanin pigment. Proceed internally to identify the narrow, pencil-thin **rods** and the bullet-shaped **cones**. The rods will appear light pink and the cones dark in color. Rods permit vision in dark or dimly lit situations. Cones provide us with color vision. The rods and cones synapse with **bipolar cells**, which in turn synapse with a layer of **ganglion cells**. The dark-staining cell bodies of the ganglion cells can be identified as those cells that face the vitreous chamber.

Using high magnification, the fovea and the optic disc may be observed on the posterior of the retina. Locate the area of the retina that appears indented. If this area contains an abundance of cones, then identify this as the fovea. Locate the optic disc by scanning the retina until you arrive at a region that is devoid of both rods and cones.

Slide # _____ Eye, Mid-Sagittal Section

See: Color Plate 24,
Fig. 18-18, 18-19

 Figs.
18-20d, p490
18-22a, p494

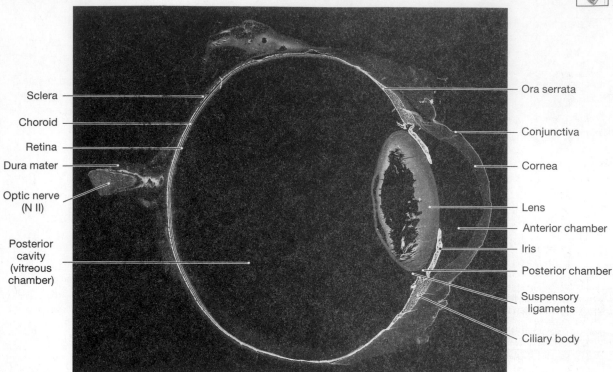

Sclera
Choroid
Retina
Dura mater
Optic nerve (N II)
Posterior cavity (vitreous chamber)

Ora serrata
Conjunctiva
Cornea
Lens
Anterior chamber
Iris
Posterior chamber
Suspensory ligaments
Ciliary body

FIGURE 18-18 Eyeball, sagittal section.

LOCATE

_____ Sclera

_____ Choroid

_____ Pigment layer of retina

_____ Photoreceptors { Rods Cones

_____ Bipolar cells

_____ Ganglion cells

_____ Fovea

_____ Optic disc

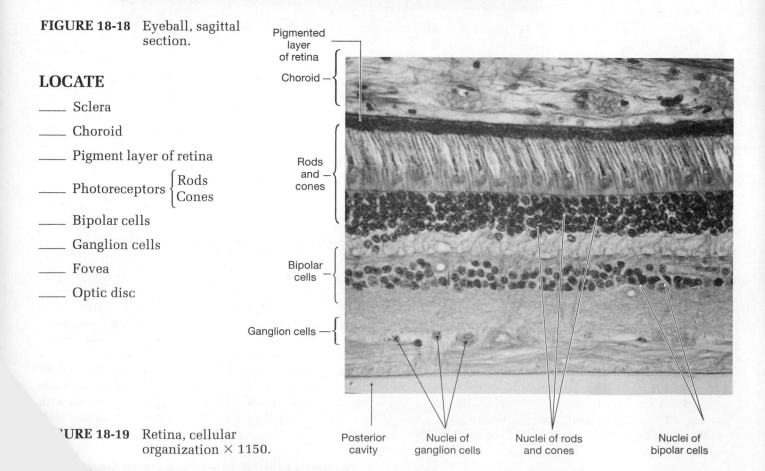

Pigmented layer of retina
Choroid
Rods and cones
Bipolar cells
Ganglion cells

Posterior cavity
Nuclei of ganglion cells
Nuclei of rods and cones
Nuclei of bipolar cells

'URE 18-19 Retina, cellular organization × 1150.

Anatomical Identification Review

Identify by coloring and labeling the structures of the eye that may be viewed in section.

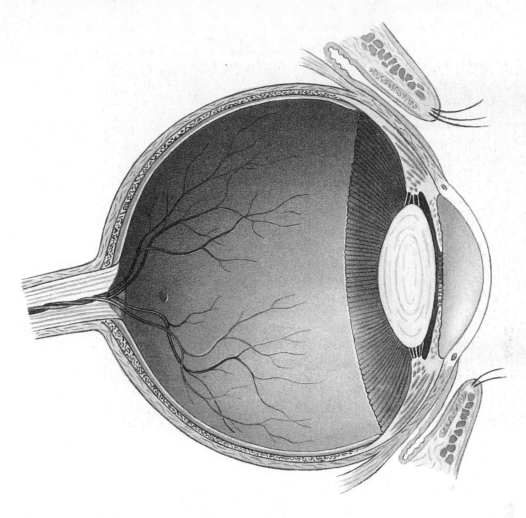

To Think About

1. How is temperature perceived by the body receptors?

2. What happens to reduce the effectiveness of your sense of taste when you have a cold?

3. What is otitis media?

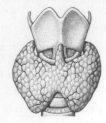

19

THE ENDOCRINE SYSTEM

Objectives

1. Locate and describe the hypophysis (pituitary), thyroid, parathyroid, adrenal glands, and endocrine pancreas, and identify their gross histology features.

2. Identify the hormones secreted by each endocrine gland/tissue and briefly note their target and function.

The nervous and endocrine systems cooperate to regulate body activities and processes. On its own, the endocrine system also monitors and adjusts body activities, but with extended effects. The **endocrine system** includes all of the endocrine cells and tissues of the body. Endocrine cells are glandular secretory cells (epithelia) that release chemical messengers that influence other cells. Those messengers that affect only adjacent cells are **local hormones** (e.g., cytokines, prostaglandins), whereas **general hormones** are released into and transported by the circulatory system to reach target cells in other tissues (e.g., growth hormone, thyroid hormone). Cells that are sensitive to the influence of specific hormones are known as **target cells**. [∞ p508] A hormone may alter simultaneously the metabolic activities of many cells, which may be widespread in various tissues and organs. Recall that all metabolic activities are controlled by enzymes. Hormones alter cellular activities by changing the types, activities, or quantities of key enzymes and structural proteins within cells.

Control centers within the hypothalamus integrate the coordination and regulation of nervous and endocrine system activities. The hypothalamus exerts three types of control over the endocrine organs: (1) direct neural control over the secretory activity of the *adrenal medulla*, (2) hypothalamic neuron secretion of *ADH* and *oxytocin* (hormones that produce specific responses in peripheral target organs), [∞ Table 19-2, p510] and (3) regulatory hormone released by the hypothalamic neurons for control of the secretory activity of the *anterior pituitary gland*. [∞ p512]

In this chapter, we describe and identify those endocrine organs that are typically observed: the pituitary gland (hypophysis), thyroid gland, parathyroid glands, adrenal (suprarenal) glands, and the endocrine pancreas.

The testes and ovaries are the male and female gonads, respectively. Each produces gametes and secretes hormones that influence gamete maturation, reproductive system maturation, and development of secondary sexual characteristics. These are described and identified in *Chapter 27*.

Observation of the Pituitary Gland (Hypophysis)

The **pituitary gland** or **hypophysis** is a small, pea-size gland that lies cradled within the sella turcica of the sphenoid bone (*Figure 6-9, p46*) and is held in this protected position by the diaphragma sellae. [∞ p510] The gland lies inferior to the hypothalamus, but is connected to it by a stalk, the *infundibulum*, which is encircled by the diaphragma sellae. Blood is supplied to the gland through the *inferior hypophyseal artery*. Anatomically, the pituitary gland is divided into *posterior* and *anterior* regions. It releases nine hormones (growth hormone, thyroid-stimulating hormone, adrenocorticotropic hormone, follicle-stimulating hormone, luteinizing hormone, prolactin, melanocyte-stimulating hormone, antidiuretic hormone, and oxytocin), many of which influence other endocrine glands. A summary of the pituitary hormones and their targets is presented in Table 19-2. [∞ p510].

Procedure

✓ **Quick Check**

Before you begin to identify the hypophysis (pituitary gland), review the sella turcica of the sphenoid bone in a dry skull specimen (*Figures 6-9* and *6-10, pp46-47*) and the landmarks of the diencephalon, both in sagittal section *Color Plate 92* and on the intact brain *Color Plate 66*.

Locate and identify the landmarks of the **diencephalon** (forebrain), using a *torso head, a brain model,* and *specimen* of the *midbrain.* Examine the landmarks of the diencephalon on the torso head model first. Observe the brain within the cranial cavity in sagittal section. Identify the optic chiasm, mamillary body, median eminence, and the sella turcica of the sphenoid bone. Now locate the optic chiasm and mamillary bodies on the inferior surface of an intact brain model. Between these two landmarks, identify the **infundibulum** and the **pituitary gland (hypophysis)**. The *anterior lobe* is that region of the pituitary gland anterior to the infundibulum, and the *posterior lobe* is the region that is connected directly to the infundibulum. In our next observation, we will use the microscope to identify cells in the major regions of these lobes. These regions can be observed only with the microscope. From the sagittal view, notice the location of the pituitary gland within the sella turcica of the sphenoid bone and its connection via the infundibulum to the hypothalamus.

 Observation of the pituitary gland in the cadaver specimen typically is not possible because of the manner in which the brain must be removed. What may be observed is its location within the sella turcica of the sphenoid bone and portions of the diaphragma sellae. [∞ pp146, 510]

See: Color Plates 66, 92, Fig. 19-1

 ∞ *Figs.*
15-3, p387
15-15, p402
15-21, p409

LOCATE

____ Sella turcica of sphenoid

____ Optic chiasm

____ Third ventricle

____ Median eminence

____ Mamillary body

____ Infundibulum

____ Pituitary gland

____ Posterior pituitary (Pars nervosa)

____ Anterior pituitary (Pars intermedia)

____ Anterior pituitary (Pars distalis)

____ Anterior pituitary (Pars tuberalis)

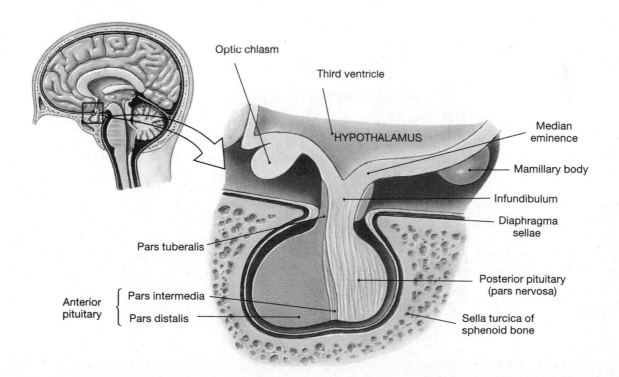

FIGURE 19-1 Anatomy and orientation of the pituitary gland.

Microscopic Identification of the Pituitary Gland (Hypophysis)

Procedure

 Identify the major **regions** of the **pituitary gland**, using the *microscope slide* provided, by viewing first with a dissection microscope, then with the compound microscope under low power magnification. Your microscope slide will contain a section of the entire gland. First identify the stalk that projects from the superior surface of the pituitary gland as the **infundibulum**. Use this as a landmark for determining the divisions. The pituitary gland is divided into four regions, but only three may be easily observed. Identify the largest round mass of brightly-stained cells as the *anterior pituitary*. The anterior pituitary is composed of three discrete regions of cells, the **pars distalis** (large, major portion of the gland), the **pars intermedia** (narrow band bordering posterior pituitary), and the **pars tuberalis** (wrapping around infundibulum). Identify the smaller rounded mass of lighter-stained cells as the *posterior pituitary* region. The cells that form the posterior pituitary are collectively termed the **pars nervosa**.

Under high magnification, the various types of cells that compose each region may be examined. Use *Figure 19-2* for reference. Your instructor may demonstrate these cells to you after you have located the regions of the pituitary, so check before proceeding to the next observation. After you have identified the regions of the pituitary on the slide, color the reference photograph using the same colors that you observed in the tissue under high power magnification.

Slide # _____ Hypophysis (Pituitary Gland) [Mallory Triple], sec.

See: Fig. 19-2

∞ *Fig. 19-4, p510*

LOCATE

____ Pars distalis (Anterior pituitary)

____ Pars intermedia (Anterior pituitary)

____ Pars nervosa (Posterior pituitary)

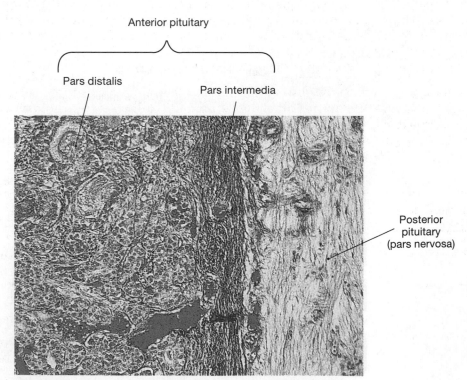

FIGURE 19-2 Pituitary gland tissue, anterior and posterior × 77.

Observation of Thyroid Gland

The *thyroid gland* lies on the anterior and lateral surfaces of the *trachea* at the inferior region of the larynx (*cricoid cartilage*) [∞ p513]. The lateral surfaces of the gland are partially covered by the sternocleidomastoid, omohyoid, sternothyroid, and sternohyoid muscles. *Thyroid follicles* are simple cuboidal epithelial structures that surround the colloid-filled space termed the follicular cavity. They manufacture, store, and secrete the thyroid hormones, T_3 (*triiodothyronine*) and T_4 (*thyroxin*). [∞ p513] Both speed up the rate of cellular metabolism and increase the use of oxygen by cells. The hormone *calcitonin* is produced and released by the *C cells* (*parafollicular cells*) in the thyroid and functions in the regulation of calcium ion concentration in body fluids. The thyroid removes iodine from the blood, then concentrates and stores it in the follicular cavity for later incorporation into thyroid hormones. An extensive blood supply to the gland provides quick access for these hormones to enter the bloodstream. A summary of the targets and effects of the thyroid hormones is presented in Table 19-3 [∞ p515].

Procedure

✓ **Quick Check**

Before you begin to examine the thyroid gland, review the anterior muscles of the neck: sternocleidomastoid, omohyoid, sternothyroid, and sternohyoid muscles. (*Color Plates 34, 36*, and *Figure 10-7, p118*).

Observe and identify the gross anatomy and surface features of the **thyroid gland**, using the *torso* and a *model* or *specimen* of the *thyroid gland*. Using *Color Plates 34, 36*, and *Figure 19-3* for reference, examine the thyroid gland located on the anterior and lateral surfaces of the *trachea*. Identify the superior border of the gland at the inferior portion of the *larynx* (*thyroid* and *cricoid cartilages*) and the inferior border of the gland at the second and third cartilage rings of the trachea. The thyroid gland has a "butterfly-like" appearance and consists of two lobes. Note both lobes of the **thyroid gland** as they curve around the cartilage rings. Observe that the two lobes are connected by a slender ribbon of tissue, the **isthmus**, at the level of the second or third tracheal rings. A capsule of connective tissue binds the gland to the trachea, but it is not present on models. Blood is supplied to the gland from two sources. Identify the *superior thyroid artery*, which is a branch of the *external carotid artery*, and the *inferior thyroid artery*, a branch of the *thyrocervical trunk*. Blood is drained from the gland by three veins. Locate and identify the *superior* and *middle thyroid veins*, which terminate in the internal *jugular vein*, and the *inferior thyroid veins*, which end at the *brachiocephalic veins*.

Use the above description also to observe the thyroid glands in the cadaver. Observation of the thyroid gland in the cadaver requires retraction of both the sternocleidomastoid and omohyoid muscles. Each lobe of the thyroid can be observed as a dark brown, wedge-shaped structure. Examine the posterior surface of each lobe and note the convex shape. This shape permits its close attachment to the trachea. It is not always possible to identify all of the arteries and veins that service the thyroid.

See: Color Plates 34, 36, Fig. 19-3

∞ *Fig. 19-7a, p514*

LOCATE

____ Hyoid bone

____ Larynx {Thyroid cartilage / Cricoid cartilage}

____ Trachea

____ Right and left lateral lobes of thyroid gland

____ Isthmus of thyroid gland

____ Internal jugular vein

____ Thyroid veins {Superior / Middle / Inferior}

____ Brachiocephalic vein

____ Common carotid artery

____ Thyroid arteries {Superior / Inferior}

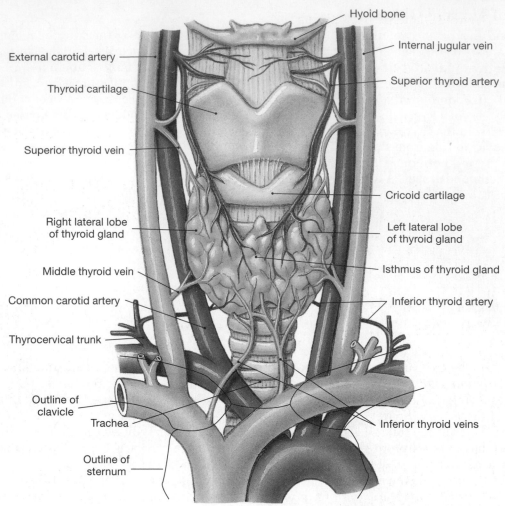

FIGURE 19-3 Location and anatomy of the thyroid gland.

Labels (clockwise from top): Hyoid bone · Internal jugular vein · Superior thyroid artery · Cricoid cartilage · Left lateral lobe of thyroid gland · Isthmus of thyroid gland · Inferior thyroid artery · Inferior thyroid veins · Outline of sternum · Trachea · Outline of clavicle · Thyrocervical trunk · Common carotid artery · Middle thyroid vein · Right lateral lobe of thyroid gland · Superior thyroid vein · Thyroid cartilage · External carotid artery

Microscopic Identification of the Thyroid Gland

Procedure

✓ **Quick Check**

Before you begin your microscopic observation of the thyroid, review the appearance of simple epithelial cuboidal cells. (*Color Plate 3*)

Locate and identify the **thyroid structures**, using the *microscope slide* provided, first by viewing under low power magnification and then under high power magnification to see the details of these structures/cells. Use *Figure 19-4* for reference. Observe, first under low magnification, the *capsule of connective tissue* that both surrounds the gland and penetrates into it. This connective tissue separates the lobes of the gland into **thyroid follicles**. Collectively, the thyroid follicles have been described as appearing like a stained-glass mosaic.

Identify under high magnification the simple cuboidal cells, termed **follicular cells**, which form each follicle. Follicular cells surround a **follicular cavity**, which is filled with **colloid** containing the stored protein **thyroglobulin**. Follicle cells manufacture thyroglobulin for storage in the colloid until it is used to produce thyroid hormones. The follicles remove iodine from the blood supplied to them and then concentrate the iodine to be incorporated into their hormones. Thus, follicles manufacture, store, and secrete the thyroid hormones.

Outside of and between individual follicles, locate and identify large irregular shaped, pale-staining **C cells (parafollicular cells)**. They produce the hormone **calcitonin**, which aids in the regulation of calcium ion concentration in body fluids. After you have identified these structures and cells, draw and color the tissue you observed in the space provided, using the same colors that you observed in the tissue under high power magnification. Label each region after coloring.

Slide # _____ Thyroid Gland (H/E stain), sec.

See: Fig. 19-4

∞ *Fig.*
19-7b,c p514

LOCATE

_____ Capsule of connective tissue

_____ C cells (parafollicular cells)

_____ Thyroid follicle

_____ Follicular cells

_____ Follicular cavity filled with colloid containing stored thyroglobulin

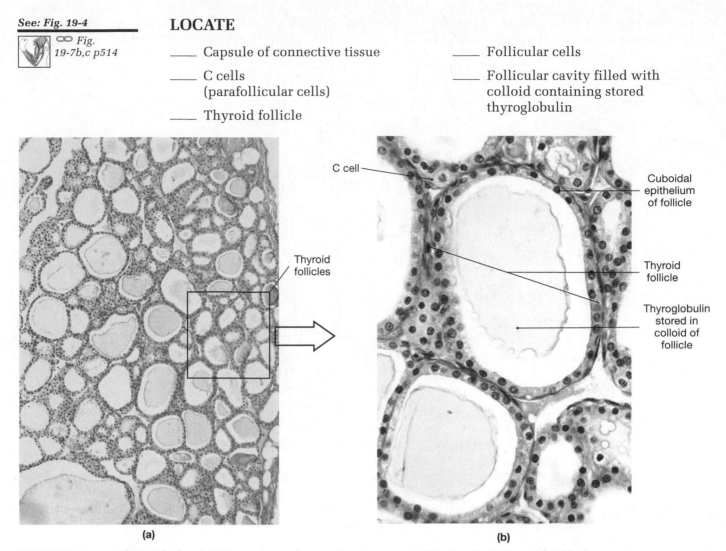

FIGURE 19-4 Thyroid gland (a) histological organization × 122 (b) detail of thyroid follicle × 450.

Observation of the Parathyroid Glands

The four nodules that compose the **parathyroid glands** are attached to the external, posterior surface of the thyroid gland by the thyroid capsule. [∞ p515] These glands secrete *parathormone (PTH)*, which regulates body fluid calcium ion concentration by opposing the effects of calcitonin. PTH is released when circulating calcium ion concentrations drop below normal. The level of blood calcium is increased by preventing its loss by kidney excretion, by removing it from storage in the bones, and by increasing its absorption across the digestive tract. A summary of the targets and effects of PTH are presented in Table 19-3. [∞ p515]

Procedure

Locate and identify the gross anatomy and surface features of the **parathyroid gland**, using the *torso* and a *model* or *specimen* of the *parathyroid glands*. Use *Figure 19-5* for reference to observe the four pea-sized (nodules) parathyroid glands on the posterior surface of the thyroid lobes. On each lobe, identify both a superior and an inferior parathyroid gland nodule. The superior glands are located at the level of the first tracheal ring and the inferior glands are located at the level of the third tracheal ring, within the inferior portion of each thyroid lobe. The connective tissue capsule of the thyroid binds the parathyroid glands to the thyroid gland, but does not surround them. This is not shown on models. Blood is supplied to the superior parathyroid glands by branches of the superior thyroid artery and to the inferior glands by branches of the inferior thyroid artery.

Use the above description to observe also the parathyroid glands in the cadaver. Examine the posterior surface of both the right and left lobes of the thyroid. The connective tissue capsule of the thyroid serves not only to connect the parathyroid glands to the thyroid, but also, in most individuals, it separates the two glands. Occasionally the parathyroid glands are embedded within the tissue of the thyroid. In such cases you will not be able to observe these glands.

See: Fig. 19-5

∞ *Fig. 19-9a, p516*

LOCATE

____ Thyroid gland

____ Connective tissue capsule of parathyroid gland

____ Parathyroid glands (total of 4; 2-superior and 2-inferior)

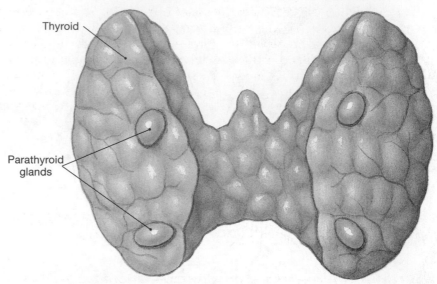

FIGURE 19-5 Parathyroid glands.

Microscopic Identification of the Parathyroid Glands

Locate and identify the following **parathyroid structures**, using the *microscope slide* provided first by viewing under low power magnification and then under high power magnification to see the details of these cells. Use *Figure 19-6* for reference. Your microscope slide typically contains small portions of both parathyroid and thyroid glands. Under low power magnification, first distinguish the thyroid portion. Identify the connective tissue capsule that separates the thyroid and parathyroid glands. Now observe only the parathyroid gland tissue and note its densely packed cells arranged in groups or rows. Under high power, identify the **principal (chief) cells** of the parathyroid gland as the pale-staining cells with large nuclei. These cells secrete PTH, which opposes the action of calcitonin to regulate calcium ion concentration in body fluids.

After you have identified the connective tissue capsule of the parathyroid gland and principal cells, draw in the space provided the tissue you observed under high power magnification.

Slide # ____ Parathyroid

See: Fig. 19-6

∞ *Fig. 19-9b, c, p516*

LOCATE

____ Connective tissue capsule of parathyroid gland

____ Principal (chief) cells

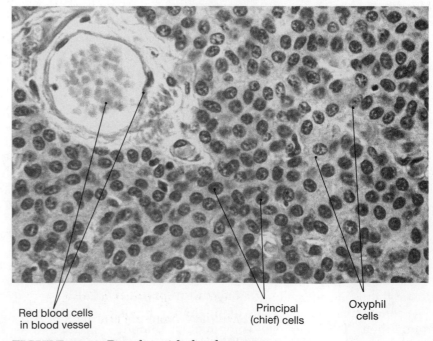

Red blood cells in blood vessel

Principal (chief) cells

Oxyphil cells

FIGURE 19-6 Parathyroid glands × 850.

Observation of the Adrenal (Suprarenal) Gland

The *adrenal (suprarenal) glands* lie on and cover the superior borders of the kidneys. (*Color Plates 120* and *123*) Structurally and functionally, the adrenal glands are divided into two regions: a superficial *adrenal cortex* and an internal *adrenal medulla*. [∞ p516] The cortex is subdivided into three regions or *zona*: (1) the outer *zona glomerulosa*, (2) the middle *zona fasciculata*, and (3) the inner *zona reticularis*. Each region produces different steroid hormones. The zona glomerulosa produces a collection of hormones termed the *mineralocorticoid group*, with *aldosterone* being the most significant of this group. The zona fasciculata produces a collection of hormones termed the *glucocorticoids*, with *cortisone* and *cortisol* being the most notable of this group. The zona reticularis is an additional source of sex hormones, both *estrogen* and *androgens*. [∞ p518]

The inner core of the gland is the adrenal medulla, which produces both *adrenaline (epinephrine)* and *noradrenalin (norepinephrine)*. The medulla is surrounded by and in contact with the zona reticularis of the cortex. Like the other endocrine glands, the adrenal glands are highly vascularized, with blood being supplied directly to the medulla first and then emanating out to the cortex. A summary of the targets and effects of the adrenal hormones is presented in Table 19-4. [∞ p518]

Procedure

✓ **Quick Check**

Before you begin to observe the adrenal glands, review the abdominopelvic regions and viscera. Focus your attention on the lumbar and umbilical regions. (*Figure 1-4, p3*)

Locate the following **adrenal gland structures**, using a *torso* and a *model* or *specimen* of the *adrenal gland*. Use *Color Plates 120, 121, 123*, and *Figures 19-7* and *19-8* for reference to observe the adrenal glands and associated structures. To observe the adrenal glands clearly, remove all of the abdominopelvic viscera from the torso. Examine both the right and left lumbar regions and identify the brown bean-shaped *kidneys*. [∞ p703] Identify the pyramid-shaped **adrenal (suprarenal) glands** that adhere to the superior and slightly medial surface of each kidney. Each gland is wrapped in a dense, fibrous connective tissue **capsule**, which binds the gland to the kidney and separates the two structures. Located medial to each adrenal gland and projecting along the posterior abdominal wall are the two major blood vessels, the *abdominal aorta* supplying blood to all of the abdominopelvic structures and lower extremities, and the *inferior vena cava* draining blood from these regions back to the heart. Identify the blood supply both to the kidneys and to the adrenal glands. The *renal arteries*, which are major branches of the aorta, supply blood to each kidney. Identify branches of the renal arteries, the **inferior suprarenal arteries**, which supply blood to the inferior portions of the adrenal glands. Superior to the renal arteries, identify the **middle suprarenal arteries** and the **superior suprarenal arteries**. Both of these branches supply blood to the remainder of the glands. Blood is drained from the adrenals by the **suprarenal veins**.

Observe the adrenal gland in sectional view to examine its regions. From the exterior inward, identify the connective tissue capsule that surrounds the gland. Immediately under the connective tissue layer, identify the thick **adrenal cortex**. Deep to the cortex, the highly vascularized inner core, the **adrenal medulla**, is completely surrounded by cortex.

Use the above description to observe the adrenal glands in the cadaver.

See: Color Plates 120, 121, 123, Figs. 19-7, 19-8

∞ *Fig. 19-10 a, b, p517*

LOCATE

_____ Kidneys

_____ Adrenal glands

_____ Cortex

_____ Medulla

_____ Renal arteries

_____ Renal veins

_____ Superior suprarenal arteries

_____ Middle suprarenal arteries

_____ Inferior suprarenal arteries

_____ Suprarenal veins

Left gastric artery

Common hepatic artery

Splenic artery

Celiac trunk

Celiac ganglion

Left suprarenal gland

Left suprarenal vein

Left renal vein

Left renal artery

Left kidney

Superior mesenteric artery

Left ureter

Left gonadal vein

Gonadal arteries

Abdominal aorta

A hepatic vein (stump)

Left renal vein

Right suprarenal gland

Inferior vena cava

Right renal vein

Right renal artery

Right kidney

Peritoneum

Right ureter

Right gonadal vein

Inferior mesenteric artery

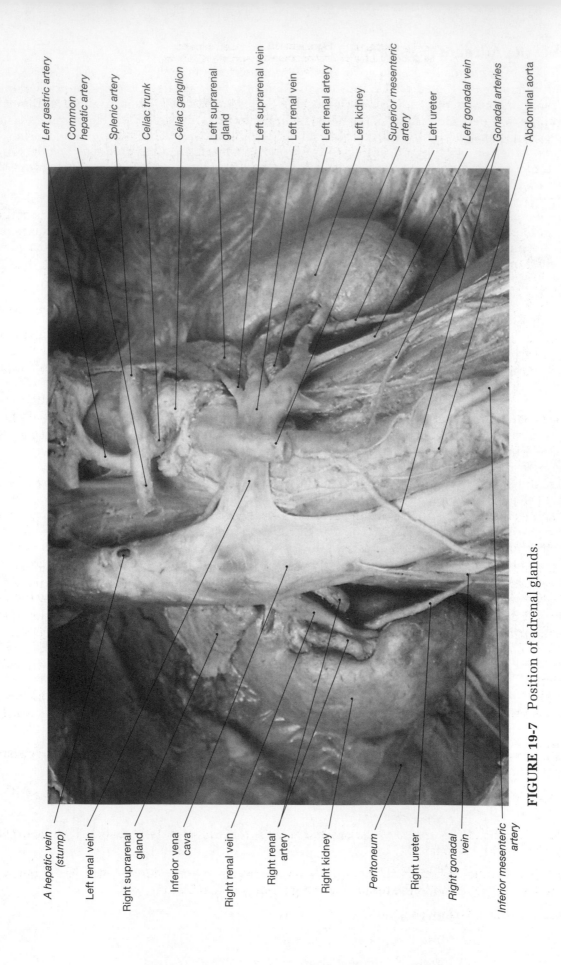

FIGURE 19-7 Position of adrenal glands.

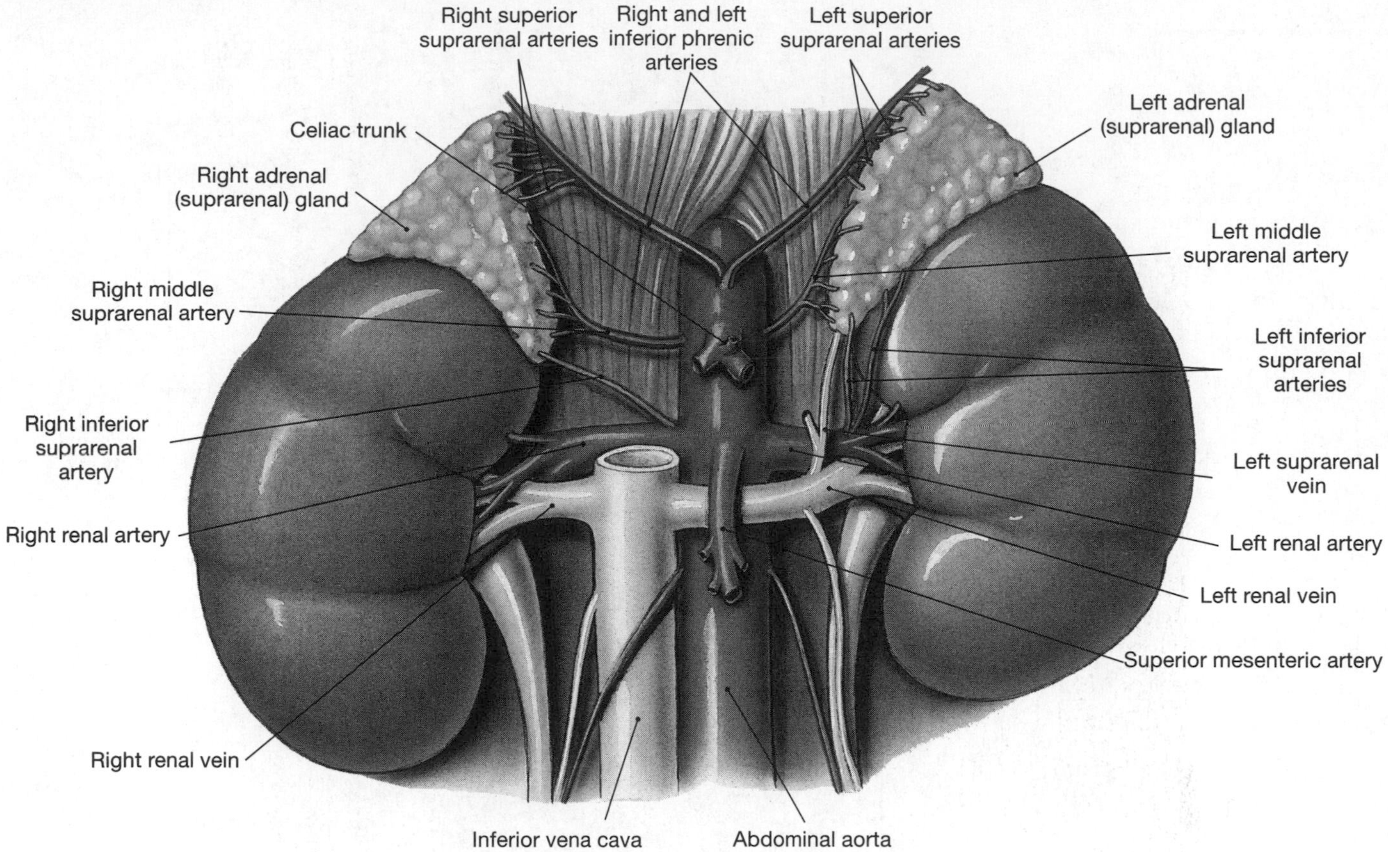

Right superior suprarenal arteries
Right and left inferior phrenic arteries
Left superior suprarenal arteries
Celiac trunk
Left adrenal (suprarenal) gland
Right adrenal (suprarenal) gland
Left middle suprarenal artery
Right middle suprarenal artery
Left inferior suprarenal arteries
Right inferior suprarenal artery
Left suprarenal vein
Right renal artery
Left renal artery
Left renal vein
Superior mesenteric artery
Right renal vein
Inferior vena cava
Abdominal aorta

FIGURE 19-8 Kidney and adrenals, superficial view.

Microscopic Identification of the Adrenal (Suprarenal) Gland

Procedure

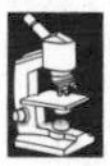

Identify the regions of the **adrenal gland**, using the *microscope slide* provided, first by viewing under low power magnification and then under high power magnification to observe the cells that compose these regions. Under low scanning magnification (or dissecting microscope), identify the connective tissue **capsule** and observe how the general appearance of the gland changes from the capsule toward the interior of the gland. Identify two regions, the **adrenal cortex** and **adrenal medulla**. The cortex is characterized by cells arranged in orderly rows. The outer groups of these cells stain lighter and the inner groups stain darker. The boundary between the cortex and medulla is not always easily distinguished, but numerous blood vessels and large, rounded cells are characteristic of the medulla.

Return to the outer layer of the cortex and observe it under high magnification. To identify the three regions of the cortex, it is best to begin viewing from the connective tissue capsule and work your way toward the interior. Identify:

1. The outer *zona glomerulosa* lies immediately interior to the connective tissue capsule and consists of lightly-stained short columnar cells arranged in groups or very short rows.
2. The middle *zona fasciculata* consists of larger cells with lipid vacuoles. These cells are arranged in long rows or columns.
3. The inner *zona reticularis* consists of darker-stained cells that are irregularly arranged. Some of these cells come in contact with the medulla.

After you have identified the medulla and the regions of the adrenal cortex, color the reference photograph using the same colors that you observed in the tissue under high power magnification.

Slide # _____ Adrenal Gland, sec.

See: Fig. 19-9

∞ *Fig.
19-10b, c, p517*

LOCATE

____ Adrenal gland

____ Cortex {
Zona glomerulosa
Zona fasciculata
Zona reticularis
}

____ Medulla

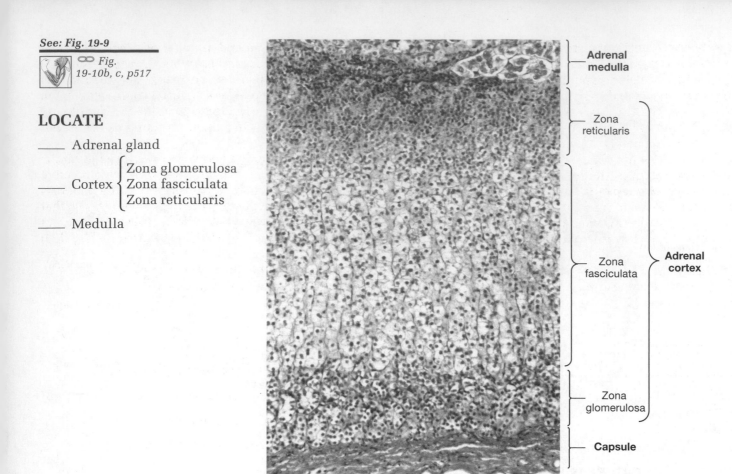

FIGURE 19-9   Adrenal gland regions × 173.

Observation of the Endocrine Pancreas

The *pancreas* lies within the abdominal cavity between the stomach and small intestine. (*Color Plates 103* and *109*) This slender organ has both exocrine (digestive) and endocrine activities. It is mostly an exocrine gland, secreting digestive enzymes into the digestive tract. For this reason, the anatomy of the pancreas is described in detail in Chapter 25. The endocrine portion of the pancreas consists of *pancreatic islets or islets of Langerhans*. [∞ p519] These are isolated cell clusters containing four cell types. Each islet produces secretions responsible for regulation of blood glucose levels, suppression of hormones that regulate blood glucose levels, and alteration of rates of food absorption and enzyme secretion.

1. **Alpha cells** make the hormone **glucagon**, which increases blood sugar levels by causing the liver to release stored glycogen.

2. **Beta cells** produce the hormone **insulin**, which reduces blood sugar levels by causing the cells to take up more glucose for use by the mitochondria.

3. **Delta cells** make the hormone **somatostatin**, which inhibits the production and secretion of glucagon and insulin, and slows the rate of nutrient absorption and enzyme secretion along the digestive tract.

A summary of the targets and effects of the hormones secreted by the pancreas are presented in Table 19-5. [∞ p521]

Procedure

✓ **Quick Check**

Before you begin to identify the pancreas, review the abdominopelvic regions and viscera, focusing on the epigastric region. (*Figure 1-4, p3*)

Locate and identify the structures of the **endocrine pancreas**, using the *torso* and a *model* or *specimen* of the pancreas. Use *Color Plates 103, 106, 109, Figures 19-10* and *19-11* for reference in observing the endocrine pancreas and associated structures. Use the torso to observe both the relationship of the pancreas to the surrounding digestive organs and the arterial blood supply to the pancreas. To observe the pancreas in the torso, remove the liver, large intestine, and spleen. Observe the relationship of the pancreas to the stomach and to the *duodenum* region of the small intestine. (*Color Plate 103*) If a model of the pancreas is available, its structure can be examined in greater detail.

Notice how the pancreas is positioned posterior to the stomach. The pancreas is divided into three portions; **tail**, **body** and **head**. It is a slender, pink-gray organ with a nodular consistency. A thin, transparent connective tissue capsule wraps the entire organ. Identify the broad head and bluntly rounded tail portions of the pancreas along the posterior region of the inferior surface of the stomach. Locate and identify the two major arteries that supply blood to the pancreas, the *pancreaticoduodenal artery* and the *pancreatic artery*. These arteries are branches of the *celiac trunk artery* and *superior mesenteric artery*. (*Color Plates 87, 106, and Figure 25-20, p385*)

 Check with your instructor before observing the pancreas in the cadaver. Some instructors often prefer to examine both the endocrine and exocrine portions of the pancreas at the same time.

See: Color Plates 87, 103, 106, 109, Fig. 19-10

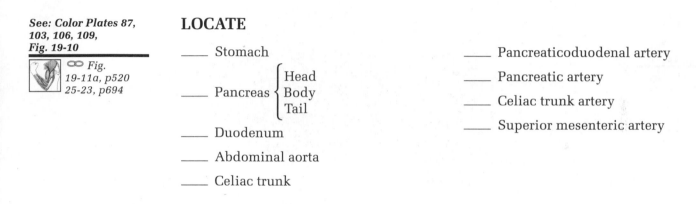

∞ Fig. 19-11a, p520 25-23, p694

LOCATE

____ Stomach

____ Pancreas { Head / Body / Tail

____ Duodenum

____ Abdominal aorta

____ Celiac trunk

____ Pancreaticoduodenal artery

____ Pancreatic artery

____ Celiac trunk artery

____ Superior mesenteric artery

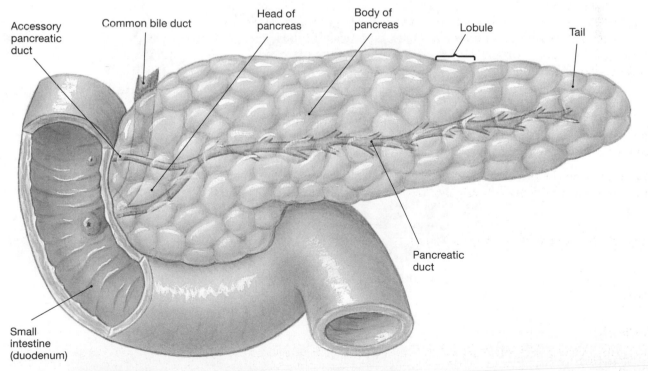

Accessory pancreatic duct

Common bile duct

Head of pancreas

Body of pancreas

Lobule

Tail

Pancreatic duct

Small intestine (duodenum)

FIGURE 19-10 Gross anatomy of the pancreas.

Microscopic Identification of the Endocrine Pancreas

Procedure

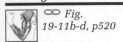 Locate and identify the structures of the **endocrine pancreas**, using the *microscope slide* provided, first by viewing under low power magnification and then under high power magnification. Use *Figure 19-11* for reference. Under low magnification observe the general appearance of the pancreas. The bulk of the pancreas contains the exocrine glandular tissue, which produces and secretes digestive enzymes. Locate and identify the exocrine portion of the gland as the darker stained **pancreatic acini (exocrine) cells**. The endocrine portion of the pancreas is scattered throughout the organ as the approximately one million **pancreatic islets (islets of Langerhans)**. Notice the pancreatic islets as the circular to oval masses of pale staining cells surrounded by the darker staining exocrine tissue regions. Now observe a pancreatic islet under high power. Each pancreatic islet contains four cell types: (1) alpha, (2) beta, and (3) delta. Ordinarily, it is not possible to differentiate between these cells, unless the slide provided to you has been specially stained for a particular cell type. Check with your instructor.

After you have identified exocrine and endocrine pancreatic structures, color the drawing provided using the same colors that you observed in the tissue under high power magnification. Label the structures and cells after coloring.

Slide # _____ Pancreas, sec.

See: Fig. 19-11

∞ *Fig. 19-11b-d, p520*

LOCATE

_____ Pancreatic acini (exocrine cells)

_____ Pancreatic islet (islet of Langerhans)

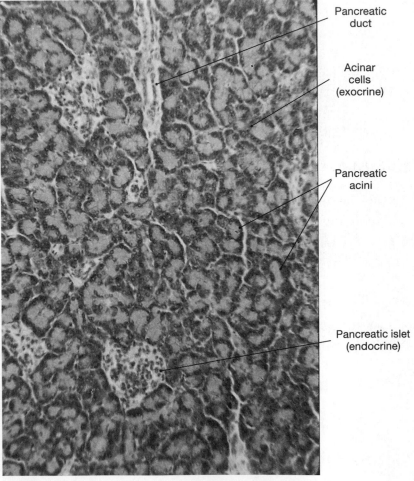

Pancreatic duct

Acinar cells (exocrine)

Pancreatic acini

Pancreatic islet (endocrine)

FIGURE 19-11 Pancreas, endocrine and exocrine tissues × 158.

Anatomical Identification Review

Label on the photo shown, the structure and cells of a thyroid follicle.

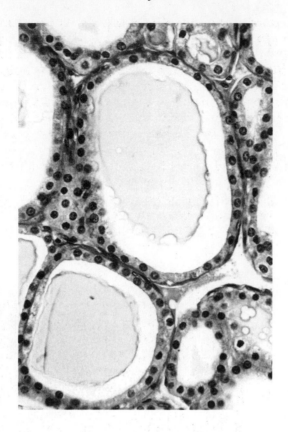

To Think About

1. How do the general responses of the endocrine system differ from those of the nervous system?

2. How are the nervous and endocrine systems regulated?

3. What is a portal system and how does it function?

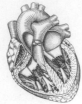

20

THE CARDIOVASCULAR SYSTEM:
Blood

Objectives

1. List and describe the functions of the blood.

2. Identify red blood cells (erythrocytes) and describe their function.

3. Identify and compare each type of white blood cell (leukocytes) and describe the function of each type.

4. Identify platelets (thrombocytes) and describe their function.

Observation of Blood Cells

The cardiovascular system links all parts of the body through a complex transport system. This system permits chemical exchange at specialized organs and tissues for the benefit of the entire body. Blood is the specialized connective tissue that flows through the cardiovascular system to distribute nutrients, oxygen, enzymes, hormones, and collect waste products throughout the body. Whole blood is composed of two components, the *plasma* (55%) and *formed elements* (45%). [∞ p533] Plasma is the liquid matrix of the blood. It contains water, dissolved gases, and dissolved organic (e.g. plasma proteins) and inorganic (e.g. electrolytes) molecules. Red blood cells, white blood cells, and platelets are the formed elements of the blood. The term formed elements is used instead of cell, because platelets are pieces or fragments of cells. All formed elements are derived from stem cells, which reside in the bone marrow.

Plasma serves as the medium to transport the formed elements. [∞ p534] It is made up of two major ingredients, water (92%) and plasma proteins (7%). Most of the plasma proteins are manufactured in the liver. Plasma proteins include: *albumins*, which are important in maintaining the osmotic pressure of the plasma; *globulins*, which are transport proteins and immunoglobulins; *fibrinogen*, a crucial ingredient in blood clotting; and *regulatory proteins*, which include enzymes, proenzymes, and hormones.

Red blood cells (RBCs), termed **erythrocytes**, are the most numerous of the blood cells and represent about half of the total blood volume. [∞ p535] Healthy adults differ in the number of erythrocytes contained in their blood. For example, one cubic millimeter (mm^3) of a man's whole blood contains about 5.4 million RBCs, while a woman's whole blood contains about 4.8 million RBCs. These unique cells are anatomically and physiologically designed to transport oxygen to body cells for use in metabolism and, to a lesser extent, carbon dioxide away from body cells for removal. Erythrocytes are thin, biconcave-shaped cells that lack a nucleus and other organelles (e.g., mitochondria). The lack of both a nucleus and organelles allows each RBC to contain the maximum amount of the molecule **hemoglobin**, which primarily carries oxygen. Each erythrocyte has up to 280 million hemoglobin molecules. The slender shape of the RBC permits it to flex and bend as it enters the smallest blood vessels, the capillaries.

White blood cells (WBCs), called **leukocytes**, are found both in the bloodstream and scattered throughout peripheral tissues. [∞ p540] WBCs help defend the body against pathogens, and they aid in removing toxins, wastes, and damaged or abnormal cells from the body. Leukocytes are less numerous then RBCs. A cubic millimeter (mm^3) of blood contains 6,000-9,000 WBCs. There are two classes of leukocytes, granular and agranular. *Granular leukocytes* have large, granular cytoplasmic inclusions while *agranular leukocytes* lack these granular inclusions. All blood cells originate from myeloid and lymphoid stem cells. [∞ Figure 20-8, p544].

Granular leukocytes include the *neutrophils, eosinophils*, and *basophils*. Neutrophils are phagocytic cells, which make up 60–70% of all white blood cells. Eosinophils (2–4%) are phagocytic cells, which are most active during chronic or parasitic diseases. Basophils (1%) manufacture the chemical *histamine* and then release it during inflammation. The agranular leukocytes are *monocytes* and *lymphocytes*. Monocytes (2–8%) are active in defense; they phagocytize cellular debris. Lymphocytes (20–30%) occur as one of three types, but they are identified only on a functional basis. Some lymphocytes manufacture antibodies (B-cells), others defend the body against foreign cells and tissues (T-cells), while others destroy abnormal tissue cells (NK cells).

Platelets, also called *thrombocytes*, are flattened, membrane-encased packets of cytoplasm. They are disc-shaped fragments or pieces of megakaryocyte cells from red bone marrow. Platelets are one of the critical participants in the vascular clotting system. [∞ p541]

Procedure

Locate and identify the **blood cells**, using the *microscope slide* provided. First focus under low power magnification and then view under high power magnification to see the details of the individual cells. Since the most commonly used stain for blood is Wright's blood stain, the description for observing blood cells is based upon this stain.

With low power magnification, scan the slide and observe that the majority of the cells (RBCs) are stained salmon color. Scattered among these cells are the darkly-stained WBCs. Identify **erythrocytes** now with high dry magnification. These are RBCs ($\times$ 430–450). Use *Figure 20-1* for reference. Note their characteristic biconcave disc shape. Observe how the central area of each cell is lightly stained. This is both because of the shape of the cell and the absence of a nucleus. At the same magnification, slowly scan the slide to observe purple staining fragments of cells, the **platelets**. Return to low power magnification and locate an area on the slide where a relatively high proportion of white blood cells is observed. Identify **leukocytes** under high dry magnification. Notice that leukocytes are larger than RBCs. They contain nuclei that appear to have different shapes. WBCs can be identified easily based on two characteristics, the shape of the nucleus and the presence of granules within the cytoplasm. Leukocytes are classified as granular or agranular based upon the presence or absence of cytoplasmic granules. It is best to view RBCs, WBCs, and platelets under the oil immersion lens. If you are unfamiliar with the use of this objective lens, check with your instructor for instructions on its use.

Granular Leukocytes

Granular leukocytes have a cytoplasm that appears grainy because it contains large granular inclusions. [∞ p541] These leukocytes are subdivided into three types based on their staining and nuclear characteristics and can be identified as:

1. **Neutrophils** or **polymorphonuclear leukocytes** are the most common granular leukocyte. They have a variable shaped nucleus and compose 60–70% of the WBC population. Locate neutrophils by identifying their multiple-lobed nucleus. It may have up to five lobes, but typically two to three are most commonly observed. The cytoplasm of neutrophils stains lilac or lavender and the nuclear lobes stain purple.

2. **Eosinophils** or **acidophils** are large cells that are sparse in number. They compose about 2–4% of the WBC population. Identify these cells by both their purple-stained, bilobed nucleus and large red-orange granules. The granules often obscure the nucleus. These cells will be more difficult for you to observe. Your instructor may have a demonstration slide prepared for your viewing of this type of WBC. Check with your instructor before proceeding.

3. **Basophils** are large cells that are the least numerous of all the leukocytes. Identify basophils on the basis of their large dark purple granules that obscure a pale staining nucleus. Since only about 1% of the white blood cells are basophils, this cell is the most difficult to identify. Before proceeding, check with your instructor, who may have a demonstration slide for you to view basophils.

Agranular Leukocytes

Agranular leukocytes have a cytoplasm that appears smooth or homogeneous because it lacks the large obvious granular inclusions. [∞ p541] These leukocytes are subdivided into two types based on their staining and nuclear characteristics and can be identified as:

1. **Lymphocytes** are the most common type of agranular leukocyte and range in size from small to large. They compose about 20–30% of the WBC population. Identify lymphocytes by their round, dark purple-stained nuclei, which may be so large as to obscure most of the cell's cytoplasm. Typically when the nucleus is large only a small rim of cytoplasm is visible.

2. **Monocytes** are fewer in number (about 2–8% of the population) and usually are much larger than lymphocytes. Identify monocytes by their kidney-bean, indented, or crescent-shaped purple-staining nucleus.

Slide # _____ Human Blood Smear (Wright's Blood Stain)

LOCATE

_____ Erythrocytes (RBCs)

_____ Neutrophil ⎫
_____ Eosinophil ⎪
_____ Basophil ⎬ WBCs
_____ Lymphocyte ⎪
_____ Monocyte ⎭

_____ Platelets (Thrombocytes)

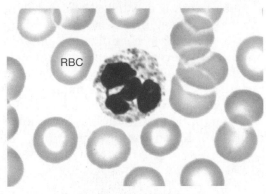

(a) Neutrophil

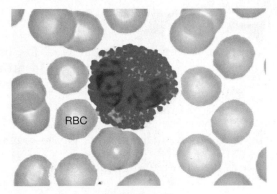

(b) Eosinophil

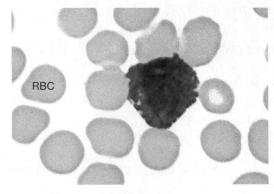

(c) Basophil

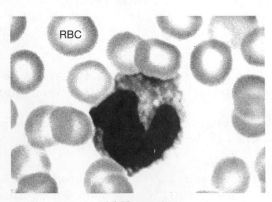

(d) Monocyte

(e) Lymphocyte

FIGURE 20-1 White blood cells × 1500.

Anatomical Identification Review

Label in the following light micrographs all of the formed elements.

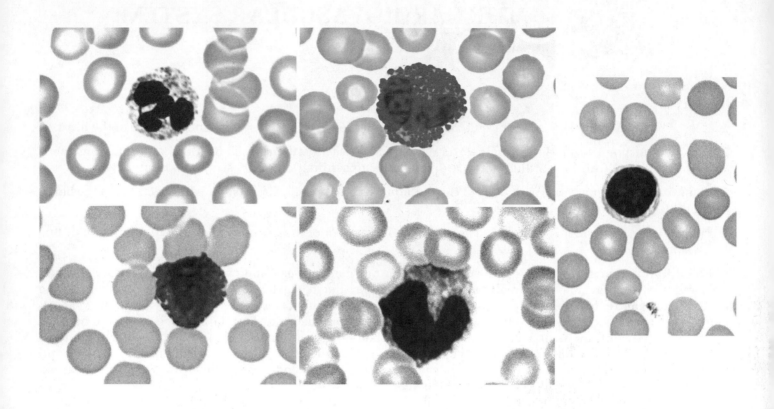

To Think About

1. What factors contribute to the short (approximately 120 days) life span of a red blood cell?

2. A college student who is constantly dieting decides to go to the infirmary as she has no energy, is experiencing muscular weakness, and sometimes becomes dizzy upon exertion. What is the likely diagnosis, and what is the cause of this condition?

3. Predict why athletes frequently move to elevations higher than those at which they will compete in athletic events several months prior to the start of the competition.

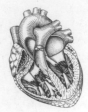

21

THE CARDIOVASCULAR SYSTEM:
The Heart

Objectives

1. Describe the basic design of the circulatory system, and the superficial anatomy and function of the heart.

2. Locate and identify the following heart structures and describe their specializations and functions: pericardium, heart wall (epicardium, myocardium, and endocardium), internal structures, and valves.

3. Identify the major arteries and veins of the pulmonary and systemic circuits, their connections to the heart, and locate the coronary blood vessels.

4. Identify the components of the electrical conduction system of the heart and describe the function of this system.

For blood to function as the body's transport system for nutrients, oxygen and carbon dioxide, hormones, vitamins, and wastes, it must be continually moving. Blood is circulated constantly through vessels by a pump. The **heart** is the organ responsible for pumping blood through the circulatory system. The heart has four compartments, two for receiving blood and two for pumping it away from the heart. Maintaining the motion of blood is vital to both the continuous delivery of nutrients to all body cells and the removal of cellular waste products from these cells. The **circulatory system** is a closed system of blood vessels that is divided into two separate and distinct patterns of circulation. Vessels that transport deoxygenated blood from the heart to the lungs for gas exchange and return newly oxygenated blood to the heart compose the **pulmonary circuit**. The vessels that transport blood in both directions between the heart and all body tissues compose the **systemic circuit**. [∞ p550]

Observation of the Heart in the Thoracic Cavity

The heart lies just to the left of the sternum within the thoracic cavity. It is a fist-sized pump in an adult male and actually lies in the *pericardial cavity* within the thoracic cavity. The pericardial cavity is located within the mediastinum between the pleural cavities and it is surrounded by a protective serous membrane, the *pericardium*. The anterior of the heart is posterior to the sternum. This position permits the heart sounds to be heard and monitored. Posterior to the heart surface lies the esophagus and the bodies of the thoracic vertebrae. [∞ Fig. 21-2d, p541] The heart contains four hollow muscular chambers. The two small superior chambers are termed *atria*, and receive blood returning to the heart from both the pulmonary and systemic circuits. Atria have relatively thin muscular walls that are capable of great distension. The two inferior chambers are called *ventricles*. These chambers, with thick muscular walls (3/8 to 1/2 inch thick) for the ejection of blood into the blood vessels, receive blood from the atria and pump it into the pulmonary and systemic circuits. [∞ p550]

Procedure

✓ **Quick Check**

Before you begin to identify the heart in the thoracic cavity, review the structure of the rib cage (*Figure 6-41, p66*) and diaphragm. (*Figure 10-11, p125*)

Identify the following **heart** and **thoracic structures**, using a *human torso model* or *cadaver specimen*. Use *Color Plates 68, 69, 72,* and *73* for reference. Remove the chest plate from the torso or cadaver to observe the heart in the thoracic cavity. Identify the location and position of the heart in the thoracic cavity. Observe the relationship among the heart, the diaphragm, right/left pleural cavities containing the lungs, and bodies of thoracic vertebrae. Observe that the mediastinum also contains the trachea, bronchi, and esophagus. Examine the blood vessels associated with the heart and note their position within the thoracic cavity and relationship to the thoracic viscera.

Identify the tough connective tissue membrane, the **pericardium**, that encases the heart within the mediastinum. (The pericardium may not be shown on your Torso model.) It functions to stabilize heart position and protect the heart from injury and overexpansion. Some fibers of the pericardium attach it to the dome of the diaphragm. The pericardium consists of two components: (1) the **visceral** layer of serous **pericardium** or *epicardium*, which covers the outer surface of the heart, and (2) the **parietal pericardium**, which forms the inner surface of the **pericardial sac** that surrounds the heart. The outer layer contains numerous collagen fibers and is called the fibrous pericardium. The pericardial sac surrounds the **pericardial cavity**, which contains the heart. In life, the pericardial cavity contains *pericardial fluid*, which is secreted by the pericardial membranes. This fluid acts as a lubricant to reduce friction between these membranes and allows free movement of the heart.

See: Color Plates 68, 69, 72, 73

∞ *Fig. 21-2, p551*

LOCATE

____ Heart

____ Right/left pleural cavities

____ Lungs

____ Diaphragm

____ Trachea

____ Esophagus

____ Bodies of thoracic vertebrae

____ Tissue of mediastinum

____ Pericardial sac

____ Fibrous attachment to diaphragm

____ Pericardial cavity

____ Pericardium { Outer layer: Parietal (fibrous)
Inner layer: Visceral (epicardium)

With the cadaver in the supine position, remove the chest plate to observe the heart in the thoracic cavity. Observe the relationship of the heart to the thoracic viscera. Use *Color Plates 72* and *73* for reference. As you explore the contents of the thoracic cavity, be careful to avoid the cut edges of the ribs—they are sharp. Examine the heart and pericardium in the same manner as described above. Observe the relationship of the pericardium to the heart. The following structures of the pericardium may be identified: pericardial sac, fibrous (parietal) pericardium, visceral pericardium (epicardium), and pericardial cavity containing the heart. The fibers that attach the pericardium to the diaphragm are often destroyed in the process of exposing the heart.

Observation of the Heart Wall

The bulk of the heart wall is thick cardiac muscle, which is covered on both sides by epithelium. The heart wall is composed of three layers: (1) an outer *epicardium* (or visceral pericardium), (2) a middle *myocardium*, and (3) an inner *endocardium*. [∞ p552]

Procedure

Identify the structures of the **heart wall**, using first a *human heart model*, then a *cadaver*, or *mammalian heart specimen,* and *Color Plates 73* and *75* for reference. Observe the outer surface of the heart on a model or prosected cadaver. Identify this outside portion of the heart wall as the **epicardium**. The epicardium is the visceral portion (layer) of the serous membrane. The exposed surface of this serous membrane produces pericardial fluid, and the underlying surface adheres to the cardiac muscle of the middle layer. Notice that adipose tissue lies deep to the epicardium.

Identify the thick muscular wall of the heart as the **myocardium**. This middle layer of the heart wall is composed of cardiac muscle cells and is thickest on the left side of the heart. All of the cardiac muscle cells are wrapped in a fascia layer network that both supports and separates groups of muscle fibers. It supports the twisting muscle fibers and holds them together while providing passageways for blood vessels as they invade the myocardium. Dense bands of collagen and elastic fibers encircle the heart valves and the bases of both the aorta and pulmonary trunk blood vessels to stabilize the positions of these structures. This arrangement of an internal connective tissue network is termed the *fibrous skeleton* of the heart.

Examine the inner surface of the heart, including the valves, and identify this epithelial lining as the **endocardium**.

See: Color Plates 73, 75

 ∞ *Fig.*
21-3a,b, p553

LOCATE

_____ Epicardium

_____ Myocardium

_____ Endocardium

Microscopic Observation of Cardiac Muscle

Identify **cardiac muscle tissue,** using the *microscope slide* provided, first by viewing under low power magnification and then under high power magnification. Use *Color Plate 21* and *Figure 21-1* for reference. Cardiac muscle is introduced and microscopically compared to smooth and skeletal muscle types in Chapter 3. (*Figures 3-21* to *3-23, pp21-22, Color Plates 19* and *20*) The specimen on your microscope slide is a small section of cardiac muscle. Cardiac muscle cells are termed **cardiocytes** and detailed observations must be conducted under high power magnification. Under low power magnification, cardiocytes can be identified by their large prominent dark-staining nuclei and subtle cross striations. Alignment of their sarcomeres gives cardiocytes a striated appearance. A unique characteristic of cardiac muscle is the *branching interconnections* that occur between cardiac muscle fibers. This branching design effectively creates one large muscle fiber for contraction, termed a *functional syncytium*. Another distinguishing characteristic of cardiocytes is their connection to each other by specialized junctional sites termed **intercalated discs**. Identify the darkest-staining lines as intercalated discs. These lines are orientated perpendicular to the long axis of the cardiocyte.

Slide # _____ Cardiac Muscle, sec.

See: Color Plate 21,
Fig. 21-1

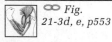 ∞ *Fig.*
21-3d, e, p553

LOCATE

_____ Cardiocytes (cardiac muscle fibers)

_____ Nucleus of cardiocyte

_____ Intercalated discs

_____ Branching interconnections between cardiac muscle fibers

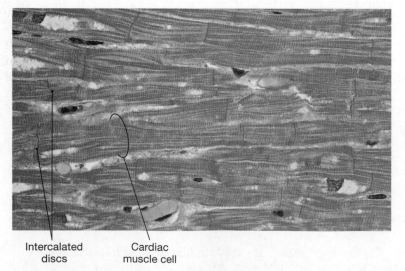

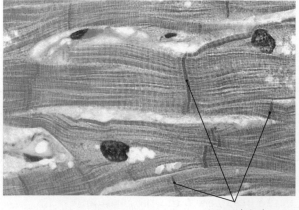

Intercalated discs   Cardiac muscle cell

Intercalated discs

FIGURE 21-1 Cardiac muscle tissue × 350, × 763.

Observation of the Superficial Anatomy of the Heart

The human heart is cone-shaped and does not resemble the valentine-type heart. The great blood vessels of the systemic and pulmonary circuits emerge from the superior base portion of the heart. The pointed, inferior **apex** of the heart is tipped slightly to the left and points toward the fifth intercostal space. [∞ p554]

Procedure

Identify the following **superficial heart structures**, using first a *human heart model,* then a *cadaver,* or *mammalian specimen.* Use *Color Plates 72* to *77* and *Figures 21-2 and 21-3* for reference. Identify the superior **base** portion of the heart to which the great veins and arteries connect, and the inferior pointed portion as the **apex** of the heart. The walls of the heart form four borders. Locate the **superior border** as that region that includes the bases of the great vessels and the two atria. The **right border** is formed by the right atrium and the **left border** is that formed by both the left ventricle and a portion of the left atrium. The left border extends in an inferior direction to the apex. Identify the wall of the right ventricle as the **inferior border**.

Observe and identify the three surfaces of the heart. These surfaces are formed by the external walls of the four chambers. The right atrium and right ventricle form the **anterior** or **sternocostal surface**. The posterior wall of the left ventricle, between the base and apex, forms the **diaphragmatic surface**. It rests upon the muscular diaphragm, hence its name. Identify that portion of the heart that comes in contact with the notched region of the left lung as the **pulmonary surface**.

Locate the two small upper chambers and the two large lower chambers of the heart. Observe that the **right atrium** receives blood from the systemic circuit. This blood enters the **right ventricle** from the right atrium, and the right ventricle ejects blood into the pulmonary circuit. Notice that the **left atrium** receives blood from the pulmonary circuit, and the **left ventricle** ejects blood into the systemic circuit. The right ventricle is smaller than the left ventricle. The difference in size is due to a relatively thin muscle wall (right) as compared to a thick muscle wall (left). (Anatomical differences between left and right ventricles can best be seen in sectional view and are described under the next heading.) The freely movable part of the atrium is a loose, expandable extension of the chamber called an **auricle**. The border between the atria and ventricles can be identified by a deep groove, called the **coronary sulcus**. Shallower depressions can be observed, called **anterior** and **posterior interventricular sulci**, crossing the anterior and posterior surfaces of the heart in a diagonal fashion. Within these sulci, identify the *coronary arteries* and *veins.* These blood vessels service the heart by bringing nutrients to the cardiocytes of the heart wall and removing their waste products.

Return to the right atrium and locate the two great veins that convey blood from the systemic circuit into the heart. Identify the vessel at the superior region of the right atrium as the **superior vena cava**. This vessel drains blood from the head, neck, upper extremities, and chest into the right atrium. At the inferior region of the right atrium, observe the **inferior vena cava**. This vessel returns blood into the right atrium from the trunk inferior to the heart, the viscera, and the lower extremities. Examine the anterior surface of the right ventricle and note the large blood vessel that emerges at a diagonal (on the cadaver or model inferior right to superior left as observed from ventral side) from the heart wall. Identify this vessel as the **pulmonary trunk**. Observe this vessel branching into **right** and **left pulmonary arteries**. Blood is ejected from the right ventricle into the pulmonary trunk to flow to the gas exchange surfaces in the lungs.

Locate the posterior wall of the left atrium, then examine the four blood vessels that convey blood from the lungs through the pulmonary circuit to the heart. Identify the two **left** and the two **right pulmonary veins** entering the posterior wall of the left atrium. Blood from the respiratory surfaces of the lungs is returned to the left atrium. This blood enters the left ventricle from the left atrium. It is pumped into the **aorta**, the major vessel of the systemic circuit. The aorta is divided into segments. Identify the segment that arises from the medial wall of the left ventricle as the **ascending aorta**. The next segment appears as an arch, termed the **aortic arch**. The arch continues posterior and inferior to the heart as a straight segment called the **descending thoracic aorta**. If the aorta were removed from the body along with all of its branches, the aorta would resemble a "cane."

See: Color Plates 72 to
77, Figs. 21-2, 21-3

 Fig.
21-4, p554
21-5, p555

LOCATE

____ Base

____ Apex

____ Borders { Superior
Inferior
Right
Left

____ Surfaces { Anterior
(Sternocostal)
Diaphragmatic
Pulmonary

____ Right atrium

____ Right ventricle

____ Left atrium

____ Left ventricle

____ Auricle of right atrium

____ Auricle of left atrium

____ Coronary sulcus

____ Interventricular sulci { Anterior
Posterior

____ Coronary vessels

____ Vena cavae { Superior
Inferior

____ Pulmonary trunk

____ Pulmonary arteries { Left
Right

____ Pulmonary veins { Left
Right

____ Aortic arch

____ Aorta { Ascending
Descending

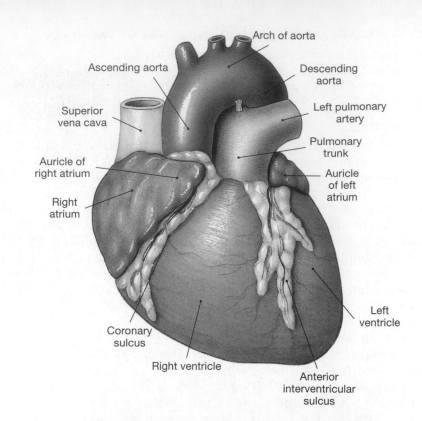

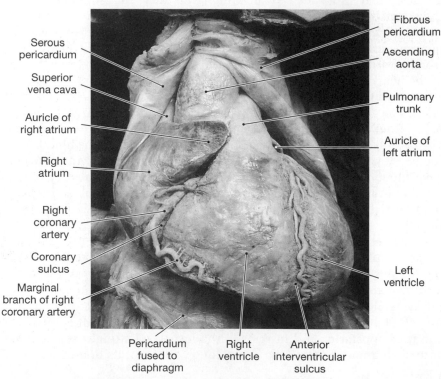

FIGURE 21-2 Superficial anatomy of the heart,
anterior (sternocostal) surface.

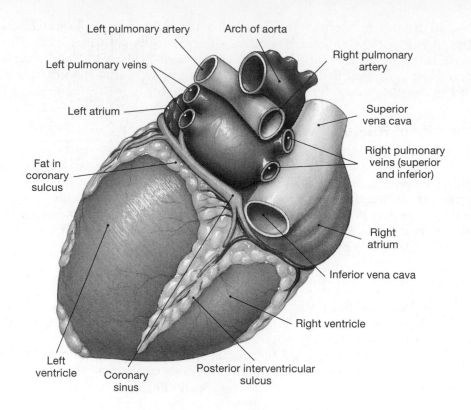

Left pulmonary artery

Arch of aorta

Left pulmonary veins

Right pulmonary artery

Left atrium

Superior vena cava

Fat in coronary sulcus

Right pulmonary veins (superior and inferior)

Right atrium

Inferior vena cava

Right ventricle

Left ventricle

Coronary sinus

Posterior interventricular sulcus

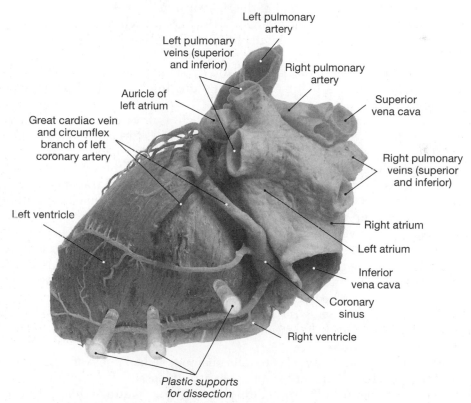

Left pulmonary artery

Left pulmonary veins (superior and inferior)

Right pulmonary artery

Auricle of left atrium

Superior vena cava

Great cardiac vein and circumflex branch of left coronary artery

Right pulmonary veins (superior and inferior)

Left ventricle

Right atrium

Left atrium

Inferior vena cava

Coronary sinus

Right ventricle

Plastic supports for dissection

FIGURE 21-3 Superficial anatomy of the heart, posterior (diaphragmatic) surface.

With the prosected cadaver in supine position, remove the chest plate to observe the heart in the thoracic cavity. You may be provided with a demonstration specimen of the human heart in lieu of examining the cadaver heart. Examine the heart in the same manner as described above. All of the surface structures of the heart previously described and identified on the model may be observed on the prosected cadaver. Use *Color Plates 72* to *77*, and *Figures 21-2* to *21-3* for reference.

Observation of Sectional Anatomy of the Heart

Keep in mind that the heart is two side-by-side pumps enclosed in the same container. The left and right side chambers are separated by the medial walls of the atria and ventricles. Structurally, the left and right ventricles are different. The left myocardium is more than twice as thick as the right myocardium because of the difference in pumping requirements. Separating atria from ventricles are the *atrioventricular valves*. [∞ p556] These valves prevent blood that was just pumped out from flowing back into the chamber. Valves are also located at the openings of the pulmonary trunk and the aorta. To observe and identify the internal anatomy of the heart and its associated blood vessels, it is necessary to observe the heart in section.

Procedure

Identify the following **internal heart structures**, using first a *human heart model*, then a *cadaver*, or *mammalian heart specimen*. Use *Color Plates 80, 81,* and *Figure 21-4* for reference. On the heart model, expose the internal structures for view in frontal section. Locate the structures associated with each chamber, beginning with the right atrium. Keep in mind that the internal surface of each chamber is the endocardium. Identify the auricle and observe it from both exterior and interior perspectives. Examine the interior of the right atrium and identify the prominent muscular ridges, termed **pectinate muscles**, that run along the inner surface of the auricle and across the adjacent anterior wall. Observe the openings of the superior and inferior vena cavae into the right atrium. Identify on the posterior wall of the right atrium, near the opening of the inferior vena cava, the opening of the **coronary sinus**. This is a large diameter vein that drains blood from the coronary veins into the right atrium. The right and left atria are separated by the **interatrial septum**. Identify the oval depression in the interatrial septum as the **fossa ovalis**. From the fifth week of embryonic development to birth, an opening termed the **foramen ovale** connects the right and left atria. This connection permits blood to flow from the right side directly into the left side of the heart, bypassing the non-functioning lungs. At birth, the foramen ovale closes and seals to form the fossa ovalis.

Examine the interior of the right ventricle and identify the **right atrioventricular (AV) valve**, or simply **tricuspid valve**. Blood flowing into this chamber from the right atrium passes through the tricuspid valve. Observe the three **cusps** or flaps of this valve. Note how each cusp is attached by connective tissue fibers, called **chordae tendineae,** to cone-shaped cylinders of muscle that project from the wall of the ventricle. Identify these muscular projections as **papillary muscles**. Observe the deep grooves and pleats, called **trabeculae carneae,** that compose the internal surface of the ventricle. Observe the thick **interventricular septum** that separates right and left ventricles. Locate the pulmonary trunk and identify three half-moon-shaped valve cusps at its origin. These form the **pulmonary semilunar valve**. This valve prevents the flow of pulmonary trunk blood back into the right ventricle. To help support the pulmonary trunk, the superior wall of the right ventricle tapers to a cone-shaped pouch, termed the **conus arteriosus**. Trace the pulmonary trunk to its branches, the left and right pulmonary arteries.

Identify the auricle of the left atrium. Examine the interior of the left atrium and observe the openings of the pulmonary veins on its posterior wall. The left atrium receives blood from the pulmonary veins. The walls of the left atrium are slightly thicker than the right atrium (may not be observable on your model). Identify the **left atrioventricular (AV) valve** or **bicuspid valve**. It contains only two cusps, hence the name bicuspid. In the clinical setting, it is referred to as the **mitral valve**.

Probe the interior of the left ventricle and note that its chamber is the largest and its walls the thickest of the heart chambers. Blood flows into this chamber from the left atrium passing through the bicuspid valve. Identify the **aortic semilunar valve** at the opening of the ascending aorta. The structure of this valve is the same as the pulmonary semilunar valve (both designed with three cusps). The design of the valve prevents the flow of aortic blood back into the left ventricle. Blood is ejected from the left ventricle through the aortic valve and into the aorta. From the aortic valve, trace the flow blood takes into the ascending aorta, aortic arch, and descending aorta.

Mammalian Heart Specimen Dissection

The heart specimen provided to you is beef, sheep, or pig. In these specimens, typically only short stumps of the great vessels will be present, and the pericardial sac may or may not be removed. If the pericardial sac has not been removed, then it must be removed to examine the heart and blood vessels. Remove the pericardial sac with your scissors. Examine the outer fibrous and inner serous portions (parietal layer) of the pericardium. As you examine the outer heart wall, you are viewing the smooth glistening epicardial surface (visceral pericardium) of the heart wall. In most specimens, you will need to remove the adipose tissue associated with the epicardium to observe the bases of the great vessels. The connective tissue of the epicardium, at both the coronary and interventricular sulci, contains substantial amounts of fat, which must be teased and removed to expose the underlying grooves containing the coronary blood vessels. Coronary vessels will appear as gray or purple threads.

Observe the surface anatomy of your specimen. Probe with a blunt instrument to identify each blood vessel. At this point in the dissection, it is usually not possible to identify with accuracy the superior/inferior vena cavae, aorta, and pulmonary vessels. Exact identification of these blood vessels can be made after sectioning the heart. The myocardium and endocardium can then be identified on the sectioned specimen. To identify the internal structures of the sheep heart, follow the directions for human heart dissection.

See: Color Plates 80, 81, Fig. 21-4

∞ *Fig. 21-6, p557*

LOCATE

____ Right atrium and auricle

____ Pectinate muscles

____ Vena cavae { Superior / Inferior

____ Coronary sinus

____ Interatrial septum

____ Foramen ovale (not present in the adult)

____ Fossa ovalis

____ Right ventricle

____ Cusps of right AV valve (tricuspid valve)

____ Chordae tendineae

____ Papillary muscles

____ Trabeculae carneae

____ Interventricular septum

____ Pulmonary trunk

____ Pulmonary semilunar valve

____ Conus arteriosus

____ Pulmonary arteries

____ Left atrium and auricle

____ Pulmonary veins

____ Cusps of left AV valve (bicuspid valve)

____ Left ventricle

____ Aortic semilunar valve

____ Ascending aorta

____ Aortic arch

____ Descending (thoracic) aorta

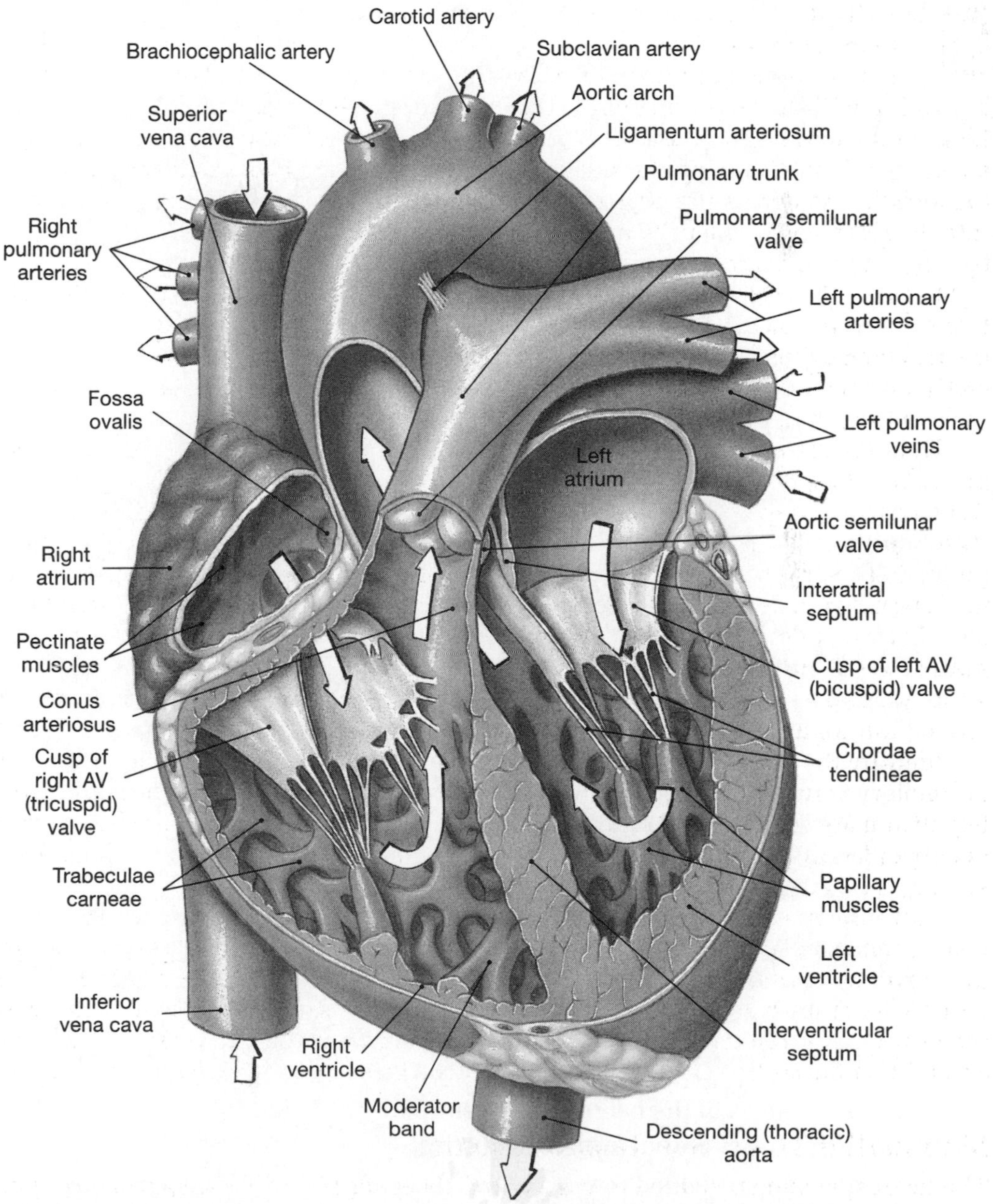

FIGURE 21-4 Frontal section through the heart.

The Cardiovascular System: The Heart **301**

Examine the heart of the prosected cadaver in the same manner as described above. You may be provided with a demonstration specimen of the human heart in lieu of examining the cadaver heart. All of the internal structures of the heart previously described and observed on the model may be identified on the cadaver specimen. Use *Color Plate 81* and *Figure 21-4* for reference.

Valves of the Heart Viewed in Transverse Section

Procedure

 Identify in a transverse section the **valves** of the **heart**, using a *human heart model, cadaver,* or *mammalian heart specimen*. The position and structure of the heart valves can be clearly observed in transverse and frontal sections. Observe both the tricuspid and mitral valves and their relationships to internal structures of the heart. Inspect the semilunar valves and note their relationship to internal structures of the heart. Note the cusps for each valve. Use *Color Plates 80, 81,* and *Figures 21-5* to *21-6* for reference.

See: Color Plates 80, 81, Figs. 21-5, 21-6

∞ *Fig. 21-7, p559*

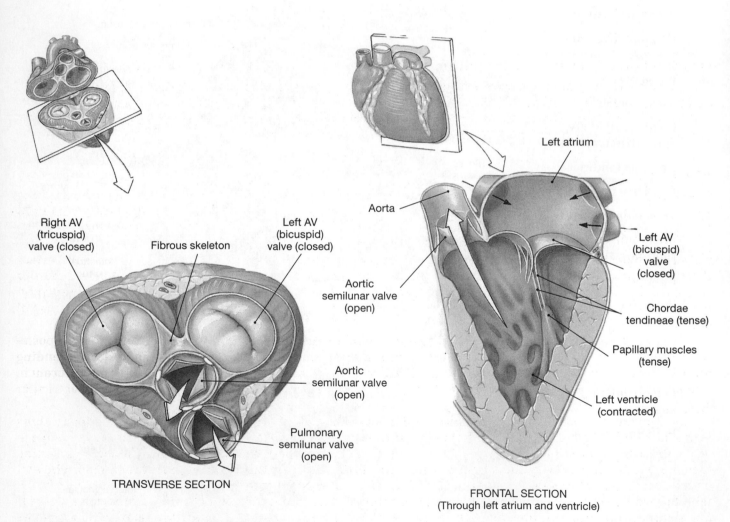

FIGURE 21-5 Valves of the heart during ventricular systole.

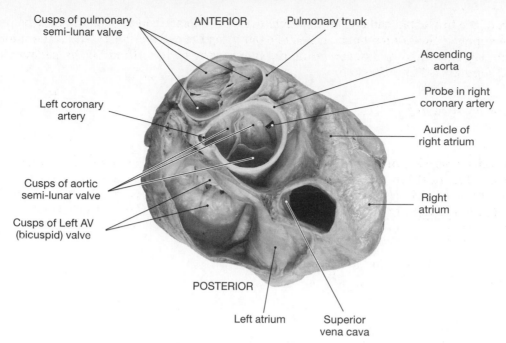

Cusps of pulmonary
semi-lunar valve

ANTERIOR

Pulmonary trunk

Ascending
aorta

Probe in right
coronary artery

Left coronary
artery

Auricle of
right atrium

Cusps of aortic
semi-lunar valve

Right
atrium

Cusps of Left AV
(bicuspid) valve

POSTERIOR

Left atrium

Superior
vena cava

FIGURE 21-6 Valves of the heart.

Observation of Coronary Circulation

The **coronary circulation** is a network of systemic blood vessels that exclusively supplies blood to the heart wall and returns it to the right atrium. For example, during strenuous vigorous exercise, the demand by the myocardium for oxygen and nutrients rises as much as nine times above resting levels. Coronary vessels supply blood to the heart even under periods of extreme physiological stress. [∞ p561]

Procedure

Identify the following **coronary vessels** distributed on the anterior and posterior surfaces of the heart, using a *human heart model, cadaver,* or *mammalian heart specimen.* Use *Color Plates 73, 78,* and *79.* The **right** and **left coronary arteries** can be identified on the anterior surface of the base of the ascending aorta. The coronary arteries can be quickly identified if you trace the right side separately from the left. Trace the right coronary artery into the atrioventricular sulcus. Near the right border of the heart, it gives rise to a **marginal artery** that extends across the ventricular surface. Posteriorly, the right coronary artery continues across the posterior surface of the heart into the posterior interventricular groove. Identify the **posterior interventricular branch** lying within this groove. The right coronary artery supplies blood to the wall of the right atrium and portions of both ventricles.

Return to the left coronary artery and trace it to where it branches. The artery splits into anterior and posterior branches. Identify the large anterior vessel as the **anterior interventricular branch (left anterior descending branch)**. Now trace the posterior branch to the left side of the heart and identify this as the **circumflex branch**. The left coronary artery supplies blood to the wall of the left atrium, left ventricle, and to the interventricular septum.

Identify the large sac-like **coronary sinus** on the posterior wall of the right atrium. Trace the coronary sinus to the left side of the heart where it is formed by smaller branch vessels. Locate the branch that projects superiorly along the posterior wall of the heart and identify this as the **posterior cardiac vein**. The large remaining branch comes to the coronary sinus from an anterior direction. Identify this vessel as the **great cardiac vein**, and observe that it lies alongside the anterior interventricular artery. Return to the coronary sinus and identify the **middle cardiac vein** as the branch that enters the coronary sinus from the middle surface of the heart where it lies alongside the posterior interventricular artery. From the coronary sinus, identify the **small cardiac vein** as the entering vessel that can be traced over the side of the right ventricle. The small cardiac vein originates anteriorly from anterior cardiac veins. This network of **coronary veins** carries blood from the myocardium and drains it into the coronary sinus, and then into the right atrium.

See: Color Plates 73, 78, 79

∞ *Fig. 21-9, p561*

LOCATE

_____ Aortic arch

_____ Coronary arteries { Right / Left

_____ Marginal branch of right coronary artery

_____ Posterior interventricular (descending branch)

_____ Anterior interventricular (descending branch)

_____ Circumflex branch of left coronary artery

_____ Posterior cardiac vein

_____ Cardiac veins { Great / Middle / Small

_____ Anterior cardiac veins

Observation of the Conduction System of the Heart

Two types of cardiac muscle cells are involved in a normal heartbeat. *Contractile cells* produce the powerful contractions that propel blood, while specialized muscle cells of the *conducting system* control and coordinate the activities of the contractile cells. Since the heart contracts as a unit, in an all-or-nothing manner, conducting system cardiac muscle fibers have their own intrinsic pace. Neural or hormonal stimuli can alter the basic rhythm of contraction. Each contraction follows a precise sequence. The atria contract first, followed by the ventricles. Contraction is termed **systole,** and it serves to eject blood from the chambers. The filling of the chambers with blood occurs in the relaxation phase, termed **diastole**. [∞ p563]

Procedure

Identify the following **structures** of the **conduction system** of the heart, using a *human heart model*. Use *Color Plate 80*. Cardiac contractions are coordinated by specialized cardiac muscle fibers that are incapable of contracting. *Nodal cells* are responsible for establishing the rate of cardiac contraction, and *conducting fibers* distribute and pass the contractile message to the myocardium of the heart. Disassemble the model to view the heart in frontal section. Identify the **sinoatrial (SA) node** within the interior of the right atrium, on the posterior wall at the opening of the superior vena cava. The SA node is referred to as the pacemaker and as such establishes the contractile rhythm of the heart. Within the floor of the right atrium near the opening of the coronary sinus, identify the **atrioventricular (AV) node**. The cells of the SA node communicate with those of the AV node.

Embedded within the myocardium of the interventricular septum, at its superior border, is a massive bundle of conducting fibers, known as the **AV bundle** or *bundle of His (HISS)*. Trace the AV bundle to where it divides into two branches. Identify a **right bundle branch** and a **left bundle branch**. The conducting fibers of the right bundle branch radiate into the myocardium of the right ventricle, and those of the left bundle branch into the myocardium of the left ventricle. Within the myocardium *Purkinje fibers* convey the stimulus for contraction.

See: Color Plate 80, Fig. 21-7

 ∞ *Fig. 21-12, p565*

LOCATE

_____ Sinoatrial (SA) node

_____ Atrioventricular (AV) node

_____ AV bundle

_____ Left/right bundle branches

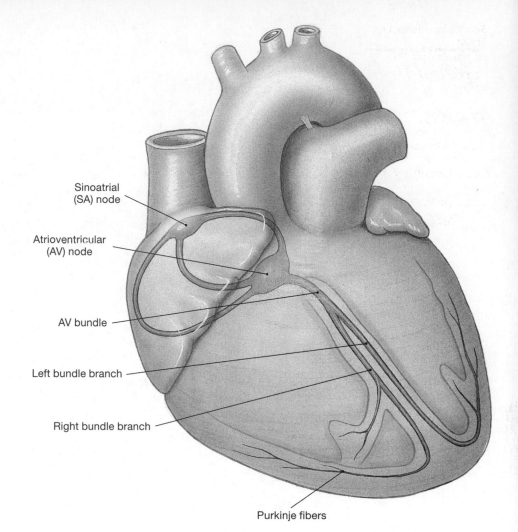

FIGURE 21-7 The conducting system of the heart.

Cat Dissection: Circulatory System

This exercise complements the study of the human circulation. To observe the heart and blood vessels, the thoracic and abdominal cavities must be opened. Expose the thoracic and abdominal organs of the cat by making a longitudinal midline incision through the muscles of the neck, thorax, and abdominal wall. Avoid cutting through the muscular diaphragm so you may identify the vessels and structures that pass between thoracic and abdominal cavities. Your instructor will provide you with specific instructions for exposing these cavities and for isolating the blood vessels. Occasionally clotted blood fills the thoracic and abdominal cavities. This must be removed. If you encounter clots, check with your instructor before proceeding.

Locate the large arteries leaving both the right and left ventricles, which direct blood away from the heart. Trace these arteries to smaller arteries, which direct blood to specific organs and tissues. Only those arteries that typically inject with latex are listed for identification. Keep in mind that the cat has more arteries than listed here and additional ones may be assigned by your instructor for you to identify.

Identification of Arteries

Arteries of the Head, Neck, and Thorax

1. Deoxygenated blood is pumped from the right ventricle toward the lungs through the **pulmonary trunk**, which divides into right and left branches of the **pulmonary artery** shortly after leaving the right ventricle; near the point of division, it is connected to the aorta by the *ligamentum arteriosum,* a strand of connective tissue representing the obliterated ductus arteriosus. The left branch of the pulmonary artery passes ventral to the *aorta* to reach the left lung; the right branch passes between the *aortic arch* and the heart to reach the right lung.

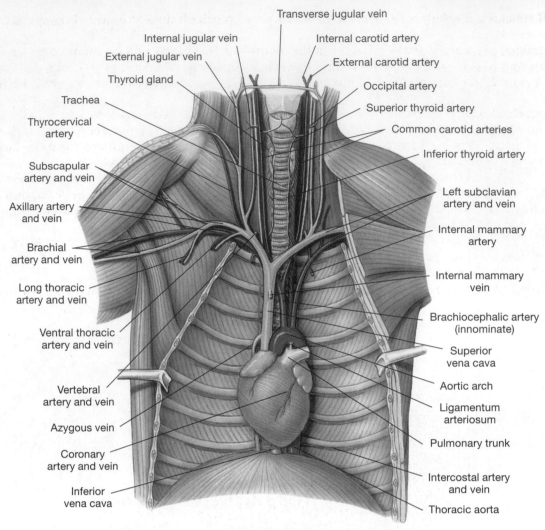

FIGURE 21-8 Arteries and veins of the cat head, neck, and thorax.

Labels in the figure:
- Transverse jugular vein
- Internal jugular vein
- External jugular vein
- Thyroid gland
- Trachea
- Thyrocervical artery
- Subscapular artery and vein
- Axillary artery and vein
- Brachial artery and vein
- Long thoracic artery and vein
- Ventral thoracic artery and vein
- Vertebral artery and vein
- Azygous vein
- Coronary artery and vein
- Inferior vena cava
- Internal carotid artery
- External carotid artery
- Occipital artery
- Superior thyroid artery
- Common carotid arteries
- Inferior thyroid artery
- Left subclavian artery and vein
- Internal mammary artery
- Internal mammary vein
- Brachiocephalic artery (innominate)
- Superior vena cava
- Aortic arch
- Ligamentum arteriosum
- Pulmonary trunk
- Intercostal artery and vein
- Thoracic aorta

2. The **aorta** is the large arterial trunk that conveys oxygenated blood from the left ventricle to the body. At its origin, it makes an abrupt curve to the left, the **aortic arch**, passing dorsal to the left pulmonary artery and continuing caudally along the left side of the vertebral column to the pelvis, where it divides into branches that supply the legs. The portion of the aorta anterior to the diaphragm is termed the **thoracic aorta**.

3. The first great vessel leaving the aortic arch is the **brachiocephalic** or **innominate**. It supplies the head and the right forelimbs.

4. Near the level of the second rib, the brachiocephalic divides into the **right subclavian** (which supplies the right forelimb) and the **right** and **left common carotids**.

5. The **superior thyroid artery** branches from the **common carotid artery** at the cranial end of the thyroid gland; it supplies the thyroid gland, superficial laryngeal muscles, and ventral neck muscles.

6. Near the point where the common carotid is crossed by the *hypoglossal nerve*, the common carotid artery gives off the **occipital** and the **internal carotid arteries**. Sometimes, they arise by a common vessel; the occipital artery supplies the deep muscles of the neck and extends toward the dorsal side of the tympanic bulla, continuing to the posterior side of the skull where it runs along the superior nuchal line. The internal carotid artery enters the skull via the foramen lacerum and joins the posterior cerebral artery.

7. After giving off the internal carotid artery, the common carotid continues as the **external carotid artery**.

8. The external carotid artery turns medially near the posterior margin of the masseter and continues as the **internal maxillary artery.** It gives off several branches, then ramifies to form the carotid plexus surrounding the maxillary branch of the trigeminal nerve near the foramen rotundum; the internal maxillary and the carotid plexus give various branches to brain, eye, and other deep structures of the head.

9. The **left subclavian artery** is the second great vessel to branch off the aortic arch. It supplies the left fore-limb.

10. The **internal mammary artery** arises from the ventral surface of the subclavian at about the level of the vertebral artery, and passes caudally to the ventral thoracic wall, giving off branches to the adjacent muscles, the pericardium, the mediastinum, and the diaphragm. It leaves the thorax and anastomoses with the inferior epigastric artery.

11. The **vertebral artery** arises from the dorsal surface of the subclavian and passes cranially through the transverse foramen of the cervical vertebrae. It gives off branches to the deep neck muscles and to the spinal cord near foramen magnum. Right and left branches enter the vertebral canal and unite to form the basilar artery, which lies along the ventral aspect of the medulla oblongata.

12. Distal to the vertebral artery, the **costocervical artery** arises from the dorsal surface of the subclavian artery. It sends branches to the deep muscles of the neck and shoulder and to the first two costal interspaces.

13. The **thyrocervical artery** arises from the cranial aspect of the subclavian (distal to the costocervical artery) and passes cranially and laterally, supplying the muscles of the neck and chest.

14. After the left subclavian artery branches from the aortic arch, paired **intercostal arteries** are given off by the thoracic aorta to the interspaces between the last eleven ribs.

15. Paired **bronchial arteries** arise from the **thoracic aorta** and supply the bronchi.

16. The **esophageal arteries** are several small vessels of varying origin along the thoracic aorta supplying the esophagus.

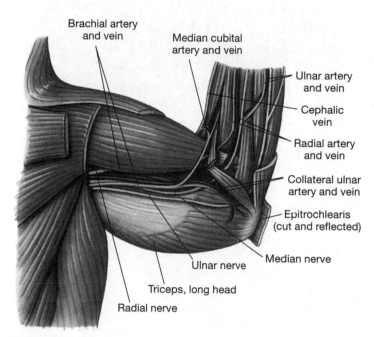

FIGURE 21-9 Arteries and veins of the cat forelimb (medial view).

Arteries of the Forelimb (Medial View)

17. Lateral to the first rib, the subclavian artery continues into the axilla as the **axillary artery**.

18. From the ventral surface of the axillary artery, just lateral to the first rib, the **ventral thoracic artery** arises and passes caudally to supply the medial ends of the pectoral muscles.

19. The **long thoracic artery** arises lateral to the ventral thoracic, passing caudally to the pectoral muscles and the latissimus dorsi.

20. The largest branch of the axillary artery is the **subscapular artery**. It passes laterally and dorsally between the long head of the triceps and the latissimus dorsi to supply the dorsal shoulder muscles. It gives off two branches, the thoracodorsal artery and posterior humeral circumflex artery.

21. Distal to the origin of the subscapular artery, the axillary artery continues as the **brachial artery**.

22. Distal to the elbow, the brachial artery gives rise to the superficial **radial artery** and **ulnar artery**.

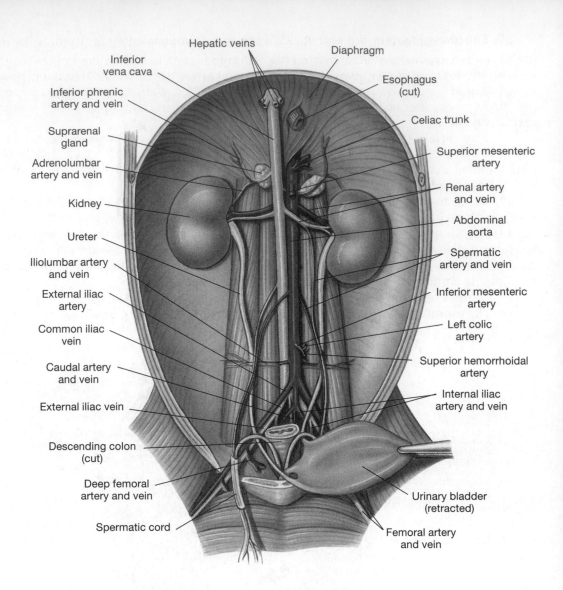

Hepatic veins

Inferior vena cava

Inferior phrenic artery and vein

Suprarenal gland

Adrenolumbar artery and vein

Kidney

Ureter

Iliolumbar artery and vein

External iliac artery

Common iliac vein

Caudal artery and vein

External iliac vein

Descending colon (cut)

Deep femoral artery and vein

Spermatic cord

Diaphragm

Esophagus (cut)

Celiac trunk

Superior mesenteric artery

Renal artery and vein

Abdominal aorta

Spermatic artery and vein

Inferior mesenteric artery

Left colic artery

Superior hemorrhoidal artery

Internal iliac artery and vein

Urinary bladder (retracted)

Femoral artery and vein

FIGURE 21-10 Arteries and veins of the cat abdominal cavity.

Arteries of the Abdominal Cavity

23. The portion of the aorta posterior to the diaphragm is called the **abdominal aorta** or **descending aorta** after passing through the diaphragm.

24. A single vessel, the **celiac trunk**, is the first arterial branch off the abdominal aorta. It divides into three branches — **hepatic, left gastric**, and **splenic arteries**.

25. Along the cranial border of the gastrosplenic part of the pancreas lies the **hepatic artery**. It turns cranially near the pylorus, lying in a fibrous sheath together with the portal vein and the common bile duct. Its branches include the **cystic artery** to the gallbladder and liver and the **gastroduodenal artery** near the pylorus, which further gives rise to **pyloric, anterior pancreaticoduodenal**, and **right gastroepiploic arteries**.

26. Along the lesser curvature of the stomach lies the **left gastric artery**, supplying many branches to both dorsal and ventral stomach walls.

27. The **splenic artery** is the largest branch of the celiac artery. It gives at least two branches to the dorsal surface of the stomach and divides into anterior and posterior branches to supply these portions of the spleen.

28. Just posterior to the celiac trunk, the abdominal aorta gives off the **superior mesenteric artery**. Its branches are the **posterior pancreaticoduodenal artery** (to caudal portions of pancreas and duodenum), the **middle colic artery** (to transverse and descending colon), and **ileocolic artery** (to caecum and ileum). The superior mesentic artery may also give off the separate **right colic artery** to the ascending colon. The superior mesenteric artery then divides into numerous intestinal branches that supply the small intestine.

29. The paired **adrenolumbar arteries** are just posterior to the superior mesenteric artery and supply the suprarenal glands. They pass laterally along the dorsal body wall and give rise to **phrenic** and **adrenal arteries**, then supply muscles of the dorsal body wall.

30. The **phrenic artery** is a branch off of the adrenolumbar artery. It supplies the diaphragm.

31. Paired arteries that emerge from the abdominal aorta to supply the kidneys are the **renal arteries.** In some specimens, the renal artery gives rise to the adrenal artery. Often double renal arteries supply each kidney.

32. Paired arteries (in the female) that arise from the aorta near the caudal ends of the kidneys are the **ovarian arteries.** They pass laterally in the broad ligament to supply the ovaries. Each artery gives a branch to the cranial end of the corresponding uterine horn.

33. In the male, **spermatic arteries** are paired internal arteries that arise from the abdominal aorta near the caudal ends of the kidneys. They lie on the surface of the iliopsoas muscle, passing caudally to the internal inguinal ring, and through the inguinal canal to the testes.

34. The **lumbar arteries** are seven pairs of arteries arising from the dorsal surface of the aorta, supplying the muscles of the dorsal abdominal wall.

35. At the level of the last lumbar vertebra, the **inferior mesenteric artery** arises from the abdominal aorta. Near its origin, it divides into the **left colic artery**, which passes anteriorly to supply the descending colon, and the **superior hemorrhoidal artery**, which passes posteriorly to supply the rectum.

36. Paired arteries, the **iliolumbar arteries,** arise near the inferior mesenteric artery and pass laterally across the iliopsoas muscle to supply the muscles of the dorsal abdominal wall.

37. Posterior to the iliolumbar arteries, near the sacrum, the abdominal aorta branches into right and left **external iliac arteries**, which lead toward the hindlimbs.

38. Posterior to branches of the external iliac arteries, the aorta gives rise to paired **internal iliac arteries**.

39. The first branch of the internal iliac artery is the **umbilical artery**, which arises near the origin of the internal iliac artery.

40. The **caudal (medial sacral) artery** is caudal to the point where the internal iliac arteries leave the aorta. This unpaired vessel continues into the ventral aspect of the tail as the caudal artery.

Arteries of the Hindlimb (Medial View)

41. Just prior to leaving the abdominal cavity, the external iliac artery gives off the **deep femoral artery,** which passes between the iliopsoas and the pectineus to supply the muscles of the thigh.

42. The external iliac artery continues outside of the abdominal cavity as the **femoral artery**, lying on the medial surface of the thigh.

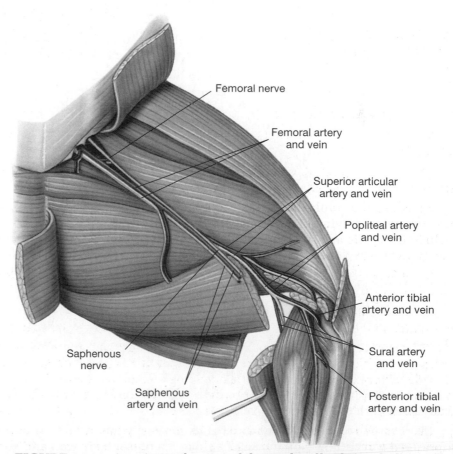

Femoral nerve

Femoral artery and vein

Superior articular artery and vein

Popliteal artery and vein

Anterior tibial artery and vein

Sural artery and vein

Posterior tibial artery and vein

Saphenous nerve

Saphenous artery and vein

FIGURE 21-11 Arteries and veins of the cat hindlimb (medial view).

Identification of Veins

Return to the heart and locate the vessels entering the atria. Trace these vessels back to smaller veins that direct blood flow back toward the heart. In general the veins follow the pattern of the arteries.

See: Fig. 21-8

Veins of the Head, Neck, and Thorax

1. To identify the **pulmonary veins**, gently pull the heart away from the lung on one side and examine the root of the lungs. From each lung lobe, a vein (usually uninjected) will be seen passing toward the dorsal side of the heart, to enter the left atrium.

2. The **superior vena cava** is the large vessel that drains blood from the head, neck and forelimbs to the right atrium. Its principal tributaries include: internal jugular veins (from brain and spinal cord); external jugular veins (from head and neck); subscapular veins (from the shoulder); and the axillary veins (from the forelimb).

3. The **inferior vena cava** is the large vessel that returns blood from the abdomen, hindlimbs, and pelvis to the right atrium. It drains blood from numerous vessels. The inferior vena cava pierces the diaphragm after it passes through the liver, within which it receives the hepatic veins. At this point it receives drainage from the inferior phrenic veins from the diaphragm. It then passes between the heart and the caudal lobe of the right lung to enter the right atrium.

4. Toward the anterior end of the superior vena cava, on its ventral side, is a common stem by which the paired **internal mammary veins** enter the superior vena cava. They lie on either side of the midline to drain the ventral chest wall.

5. If the heart is pushed to the left, the **azygous vein** can be seen arching over the root of the right lung and joining the superior vena cava near the right atrium. It lies along the right side of the vertebral column in the thorax and receives the **intercostal veins**.

6. The **brachiocephalic** or **innominate veins** are two short veins whose union forms the superior vena cava. They drain both the external jugular and subclavian veins.

7. A short vessel within the thorax joining the axillary vein to the brachiocephalic vein is the **subclavian vein**.

8. The **costocervical vein** is one of two veins (the other is the vertebral vein) that form a common stem (sometimes called **costovertebral vein**) and dorsally connect with the brachiocephalic vein. Its distribution corresponds in distribution to the artery of the same name.

9. The **vertebral vein** is one of two veins (the other is the costocervical vein) that form a common stem (sometimes called costovertebral vein) and dorsally connect with the brachiocephalic vein. Its distribution corresponds in distribution to the artery of the same name.

10. Left and right **external jugular veins** drain the head and neck. Together with the subclavian vein, each forms the brachiocephalic vein. Each is larger and more superficial than the internal jugular vein, and each is formed by the union of facial veins. The **anterior facial vein** is a superficial vein of the face. The **posterior facial vein** is a deep vein of the face. Its tributaries drain deep structures of the face.

11. Just superior to the hyoid bone, the **transverse jugular vein** connects the left and right external jugular veins.

12. The vessel that drains the brain and spinal cord and joins the external jugular vein near its union with the brachiocephalic vein is the **internal jugular vein**.

13. At the shoulder, the external jugular vein receives the large **transverse scapular vein** which drains the dorsal surface of the scapula.

See: Fig. 21-9

Veins of the Forelimb (Medial View)

14. The major vessel draining the limb in the axilla is the **axillary vein**. It forms from the brachial vein and is continuous with the subclavian vein.

15. A small vessel emptying into the axillary vein on its ventral surface is the **ventral thoracic vein**. Its location is near the point where the subscapular is found.

16. The **long thoracic vein** is another small vessel emptying into the axillary vein. It is distal to the ventral thoracic vein.

17. The **subscapular vein** is the largest tributary of the axillary vein. It drains vessels from the shoulder.

18. The **brachial vein** is the portion of the major vessel that drains the limb, and is distal to the axillary vein in the forelimb.

See: Fig. 21-10

Veins of the Abdominal Cavity

19. Blood circulating through the stomach, intestines, spleen, and pancreas returns to the liver via the portal vein. Within the liver, the portal vein ends in a system of capillaries termed *sinusoids* of the liver. Blood passes through the sinusoids of the liver and enters the inferior vena cava via the **hepatic veins**, which join the inferior vena cava within the liver.

20. The **hepatic portal vein** is formed by the union of the superior mesenteric and gastrosplenic veins. It is also joined by other veins from the gastrointestinal tract. The superior mesenteric vein is the largest branch of the hepatic portal vein.

21. The **adrenolumbar veins** are vessels that empty into the inferior vena cava and drain the adrenal glands.

22. Paired **renal veins** empty into the inferior vena cava and drain the kidneys. Often, double renal veins drain each kidney.

23. Posterior to the renal vein in males, the **internal spermatic veins** are vessels that drain the testes. Usually, the left internal spermatic vein empties into the renal vein. In females, the **ovarian veins** are posterior to the renal veins. These vessels empty into the inferior vena cava and drain the ovaries.

24. The **lumbar veins** are vessels that drain the dorsal abdominal wall and empty into the inferior vena cava.

25. Paired vessels that drain the muscles of the posterior dorsal abdominal wall and empty into the inferior vena cava are the **iliolumbar veins**.

26. The **common iliac veins** are paired veins that join together in the posterior abdominal region to form the inferior vena cava. The inferior vena cava lies dorsal to the aorta.

27. The vessel that drains the tail and empties into the inferior vena cava is the **caudal vein**.

28. The large **internal iliac veins** enter the common iliac vein in the pelvic cavity and drain the rectum, bladder, and internal reproductive organs.

29. Distal to the joining of the internal iliac vein to the common iliac vein, the **external iliac vein** is a continuation of the femoral vein from the hind limb.

See: Fig. 21-11

Veins of the Hindlimb (Medial View)

30. The **femoral vein** is a superficial vein on the anterior surface of the thigh. As it enters the abdominal cavity, it becomes the external iliac vein.

31. The **deep femoral vein** is a medial branch entering the femoral vein.

32. The medial and lateral branches of the **popliteal vein** drain the foot and calf region, uniting in the popliteal region to form the **saphenous vein**, which drains into the femoral vein. These vessels lie superficial to the muscles of the shank.

Anatomical Identification Review

1. Identify by coloring and labeling the structures and great vessels of the heart.

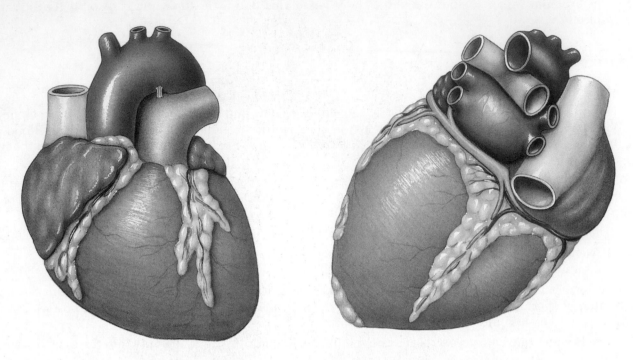

2. Identify the internal structures of the heart as seen in sectional view by coloring and labeling.

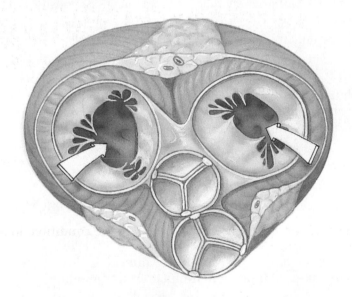

3. Identify the internal structures of the heart as seen in sectional view by coloring and labeling.

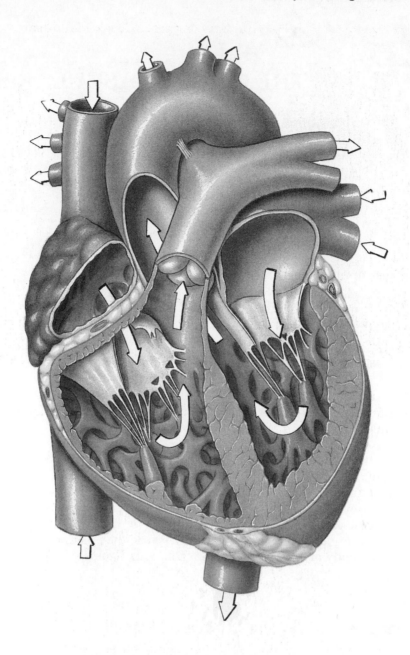

To Think About

1. If a person is diagnosed as having a heart murmur, what anatomic condition do they have, and how does it affect heart function?

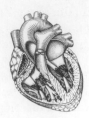

22

THE CARDIOVASCULAR SYSTEM:
Vessels and Circulation

Objectives

1. Describe the anatomical organization of the circulatory system and its relationship to the heart.

2. Recognize the types of blood vessels based on histological characteristics.

3. Identify the blood vessels of the pulmonary circuit, the major blood vessels of the systemic circuit, and the tissues and organs supplied by each vessel.

4. Locate the hepatic portal system and describe both its function and the blood vessels that compose it.

5. Describe the pattern of circulation before birth and the blood vessels of the newborn infant.

Blood travels through the circulatory system in vessels termed *arteries, veins*, and *capillaries*. Arteries transport blood away from the heart and occur in various sizes and types. Capillaries are the smallest of the body's blood vessels. The exchange of molecules between blood and tissues occurs only in capillaries. Blood flows very slowly through capillaries to permit this exchange of nutrients, waste products, and gases. The design of the capillary wall, together with other factors, facilitates the exchange. Veins also vary in size. They collect blood from the capillaries and transport it back to the heart. [∞ p577]

The circulatory system is divided into pulmonary and systemic circuits or patterns. The pulmonary circuit includes blood vessels that transport blood from the right side of the heart to the gas exchange surfaces in the lungs and back to the left side of the heart. (See *Chapter 21, p294*) The systemic circuit consists of the pattern of vessels transporting blood from the left side of the heart through capillary beds in all organs and back to the right side of the heart. The coronary vessels are part of the systemic circulation. [∞ p560] Blood vessels are named according to nearby landmarks (such as the skeletal system). Typically, arteries and veins share the same names. For example, the radial artery and vein are so named because they follow a course along the radius bone. The right and left sides of the body are nearly mirror images, with some slight asymmetry. Vessels are identified as right or left.

Microscopic Comparison of Typical Arteries and Veins

Blood leaving the heart flows through arteries. In sequential order from the heart, the types of arteries encountered are elastic, muscular, and arterioles. **Elastic arteries** are the large conducting arteries that transport large volumes of blood at high pressures away from the heart. The aorta and its branches, such as the common carotids, are examples of elastic arteries. **Medium-sized** or **muscular arteries** transport blood to organs (e.g., kidneys) and the skeletal muscles of the upper and lower extremities. These arteries are characterized by a thick middle muscle layer. Arteries of the smallest size are called **arterioles.** It is the smooth muscle in the wall of arterioles that plays the major role in controlling the amount of blood to a specific organ or tissue by the processes of *vasoconstriction* and *vasodilation*.[∞ p573]

Veins return blood to the heart. **Venules** collect blood from capillaries. From venules blood flows into **medium veins** and then into **large veins**. Medium veins are analogous to medium-sized arteries. Large veins are those that return blood directly to the heart: namely, the superior and inferior vena cavae. Most veins are characterized by internal valves. Valves located in the wall of venules and medium veins prevent the backflow of blood.[∞ p578]

Arteries and veins have three layers in their walls. Each layer is called a *tunic*. The innermost layer, called the **tunica interna** (*intima*), is composed of a single layer of endothelium (simple squamous epithelium), a basement membrane, and some elastic fibers. The middle layer, **tunica media,** is mostly sheets of smooth muscle and some elastic fibers. The outer layer, **tunica externa** (*adventitia*), is a thick supporting layer of connective tissue that helps bind blood vessels to the surrounding tissue. The internal diameter of certain blood vessels (not capillaries and venules) can be changed by contraction of the smooth muscle of the tunica media layer. The process of decreasing the diameter of the blood vessel is termed *vasoconstriction* and occurs when the muscle of the tunica media is stimulated. *Vasodilation* occurs when the smooth muscle relaxes, resulting in an increase in the internal diameter of the blood vessel.[∞ p573]

Arteries and veins can be distinguished easily from each other under the microscope. Arteries have thick walls and typically have a round profile when viewed in cross section. The thinner walled veins have an irregular or collapsed lumen (space where blood flows) when viewed in cross section. The walls of arteries are thick in order to transport large volumes of blood at very high pressures. Vein walls are thin in comparison to those of arteries, because veins transport blood at much lower blood pressures.

Procedure

 Identify and compare the **structure of arteries** and **veins** using the *microscope slide* provided, first by viewing under low power magnification and then under high power magnification. Use a dissecting microscope or the low power magnification of your compound microscope to distinguish between an artery and vein. Use *Figure 22-1* for reference. Identify arteries by their thick, round wall with a wrinkled inner layer. The wall of veins appears collapsed and more irregular in shape. Locate the tunica layers on both artery and vein under high power magnification. Observe the artery first. Begin at the tunica interna and work your way out to the tunica externa. Identify the tunica interna by the endothelium and wavy or scalloped appearance of the *internal elastic membrane*. The tunica media can be identified by both the smooth muscle fibers and the elastic fibers that form this layer. The tunica externa can be identified by the wavy collagenous and elastic fibers. The *external elastic membrane* forms the border between the media and externa layers. In small-to-medium-sized arteries, the externa layer is equal to or even thicker than the tunica media layer. In large-sized arteries, like the aorta, elastic tissue composes most of the tunica media, but the tunica adventitia is thinner in comparison.

Slide # _____ Artery and Vein

See: Fig. 22-1

∞ *Fig. 22-1, p573*

LOCATE

Artery

_____ Tunica interna (intima) $\begin{cases} \text{Endothelium} \\ \text{Internal elastic} \\ \text{membrane} \end{cases}$

_____ Tunica media $\begin{cases} \text{Smooth muscle fibers} \\ \text{External elastic} \\ \text{membrane} \end{cases}$

_____ Tunica externa (adventitia)

Vein

_____ Tunica interna (intima) $\{ \text{Endothelium}$

_____ Tunica media $\begin{cases} \text{Smooth muscle fibers} \\ \text{Elastic fibers (sparce)} \end{cases}$

_____ Tunica externa (adventitia)

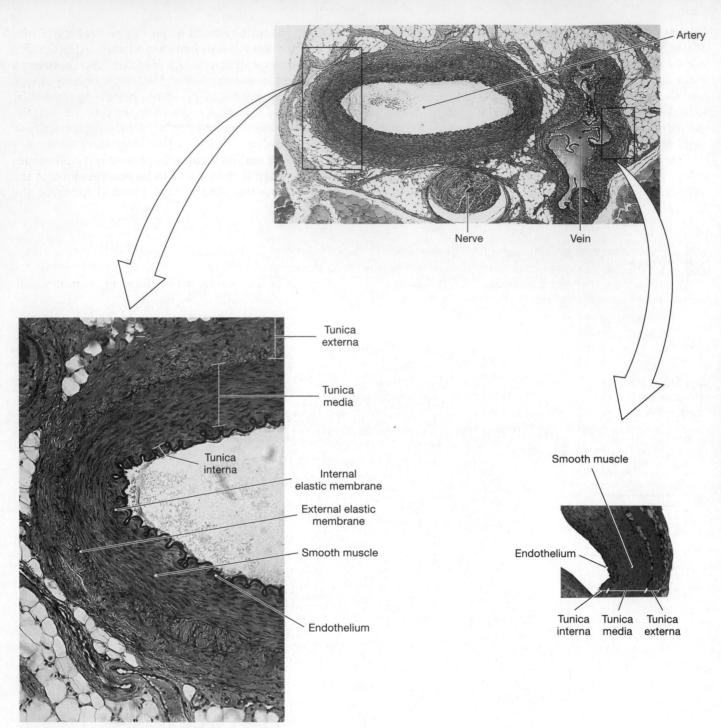

FIGURE 22-1 A comparison of typical arteries and veins × 95, × 35.

Observation of Pulmonary Circulation

The vessels of the pulmonary circuit transport blood between the heart and the gas exchange surfaces of the lungs. This circuit facilitates carbon dioxide excretion from the blood and the replenishment of oxygen to the blood. This blood is returned to the heart for distribution via the systemic circuit to the rest of the body. [∞ p580]

Procedure

✓ **Quick Check**

Before you begin to identify the vessels of the pulmonary circuit, review from Chapter 21 the external and internal anatomy of the heart. (*Figures 21-2* to *21-4, pp298-299, 301*)

Review the primary circulatory routes within the pulmonary and systemic circuits using *Figure 22-2* and *Color Plates 69 to 71* for reference. Identify those **arteries** and **veins** that form the **pulmonary circuit** using a *torso model* or *cadaver specimen*. Anatomical models and charts depict arteries and veins in color. The traditional color red indicates the oxygenated blood carried by systemic arteries, whereas blue indicates the deoxygenated blood carried by systemic veins. The pulmonary circuit is the exception to the oxygenation pattern indicated by these colors. The pulmonary arteries are shown in blue to indicate the transport of deoxygenated blood and the pulmonary veins are shown in red to indicate the transport of oxygenated blood. Again, color refers to oxygen content: oxygenated blood is shown in red and deoxygenated blood in blue.

Identify the origin of the pulmonary trunk from the right ventricle and its bifurcation into a left pulmonary artery and a right pulmonary artery. Deoxygenated blood travels through these vessels to be reoxygenated at the gas exchange surfaces of the lungs. Identify the four pulmonary veins that return oxygenated blood from the lungs to the left atrium.

See: Color Plates 69 to 71, Fig. 22-2

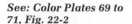

Figs.
22-8, p580
22-9, p581
22-10, p582

LOCATE

____ Heart and chambers

____ Pulmonary trunk

____ Right/left pulmonary arteries

____ Right/left pulmonary veins

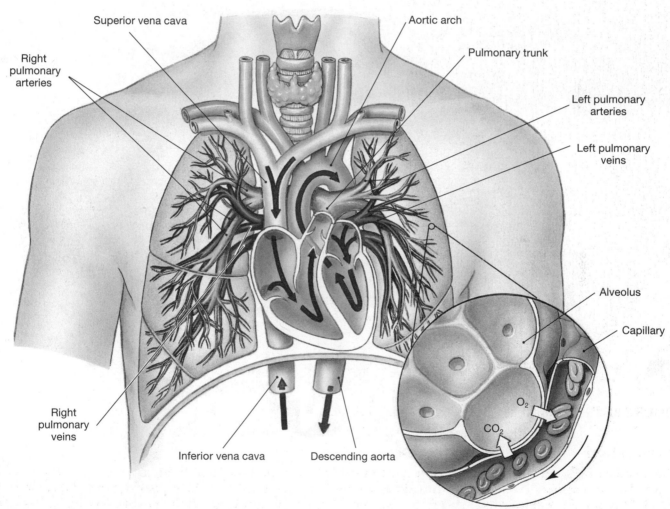

FIGURE 22-2 The pulmonary circuit.

With the prosected cadaver in the supine position, remove the chest plate to observe the heart in the thoracic cavity. You may be provided with a demonstration specimen of the human heart in lieu of examining the cadaver heart and lungs. Examine the pulmonary circuit in the same manner as described above. All of the vessels of the heart previously described and identified on the model may be identified on the prosected cadaver. Use *Figures 21-2, 21-3,* and *Color Plate 73* for reference.

Observation of the Systemic Circulation: Systemic Arteries

Procedure

Identify the relative locations of the **systemic arteries** using first a *torso model*, then *arm* and *leg models* and, if available, a *cadaver specimen*. The arrangement of the systemic arteries and their branches should be located on both right and left sides. Locate the segments of the aorta beginning with the ascending aorta. The aorta resembles a "cane." Keep in mind that the coronary arteries are part of the systemic circulation. [∞ p560] Remove all of the internal organs from the model but keep the diaphragm in position.

See: Color Plates 109–111, Fig. 22-3

∞ *Figs. 21-6, p557 22-10 to 22-20, pp582-594*

LOCATE

The Ascending Aorta

____ Aortic semilunar valve

____ Ascending aorta

____ Left/right coronary arteries

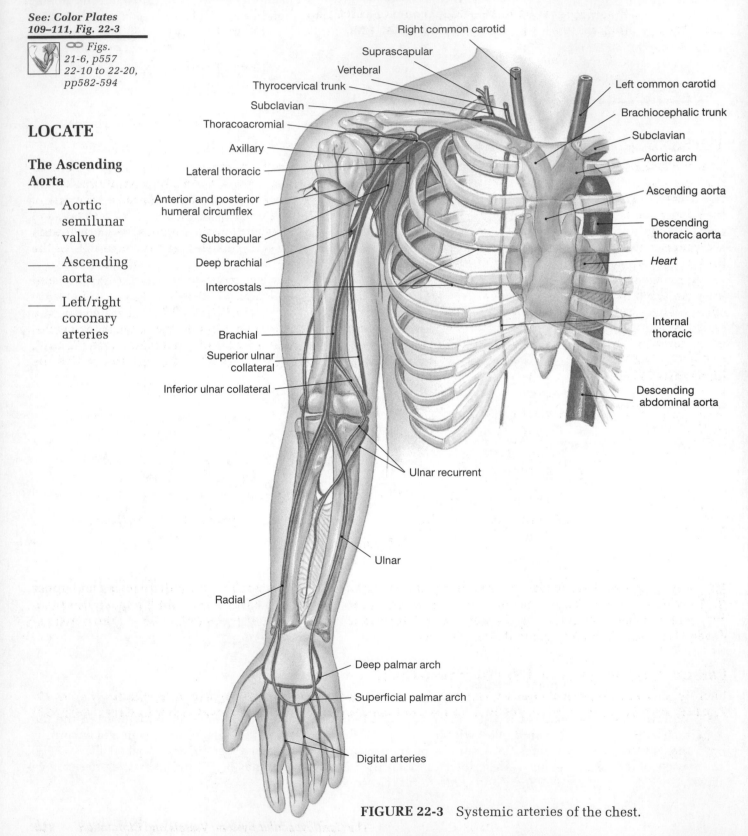

Right common carotid
Suprascapular
Vertebral
Thyrocervical trunk
Subclavian
Thoracoacromial
Axillary
Lateral thoracic
Anterior and posterior humeral circumflex
Subscapular
Deep brachial
Intercostals
Brachial
Superior ulnar collateral
Inferior ulnar collateral
Ulnar recurrent
Ulnar
Radial
Deep palmar arch
Superficial palmar arch
Digital arteries

Left common carotid
Brachiocephalic trunk
Subclavian
Aortic arch
Ascending aorta
Descending thoracic aorta
Heart
Internal thoracic
Descending abdominal aorta

FIGURE 22-3 Systemic arteries of the chest.

The Aortic Arch

Three vessels arise from the aortic arch, the **brachiocephalic trunk (innominate)**, the **left common carotid artery**, and the **left subclavian artery**. Locate and identify these vessels on the *torso model*. Use *Color Plates 69* and *75* for reference.

LOCATE

_____ Ascending aorta

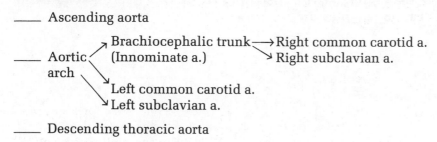

_____ Aortic arch

Brachiocephalic trunk (Innominate a.) → Right common carotid a. / Right subclavian a.

Left common carotid a. / Left subclavian a.

_____ Descending thoracic aorta

The Subclavian Arteries

The subclavian arteries supply blood to the skin and muscles of the thorax, upper extremities, and to the central nervous system. [∞ p583] The first branches of the subclavian artery are the **vertebral, internal thoracic,** and **thyrocervical trunk**. The right and left vertebral arteries pass through the right and left transverse foramina of the cervical vertebrae, merge into a single *basilar artery* in the cranial vault [see *Color Plates 82, 85,* and *66*], and eventually join with the internal carotids to supply the brain. The internal thoracic supplies the pericardium and anterior wall of the chest. The thyrocervical supplies blood to the skin, muscles, and other tissues of the back and shoulders.

After the above branches exit the subclavian arteries, these arteries become sequentially the **axillary** and then the **brachial**, which bifurcates into the **radial** and **ulnar**. Use *Color Plates 44, 46* to *48, 68,* and *70* for reference. Branches of the axillary artery supply the muscles of the pectoral region and axilla. The brachial artery supplies arterial branches to the skin and muscles of the arm. The radial artery supplies the radial side muscles of the forearm. The branches from the ulnar artery supply the ulnar side muscles of the forearm. (see *Fig 14-7, p208*) The radial and ulnar arteries anastomose into the **palmar arches** (superficial and deep) from which the **digital arteries** arise.

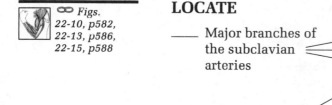

∞ *Figs.*
22-10, p582,
22-13, p586,
22-15, p588

LOCATE

_____ Major branches of the subclavian arteries → Thyrocervical trunk / Internal thoracic a. / Vertebral a.

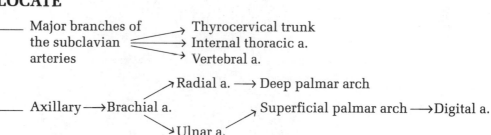

_____ Axillary → Brachial a. → Radial a. → Deep palmar arch / Ulnar a. → Superficial palmar arch → Digital a.

With the prosected cadaver in the supine position, expose the blood vessels of the axillary region and upper extremity. The pectoralis major (cut at insertion) must be reflected towards the midline to expose these blood vessels. All of the vessels previously described on the model may be identified on the prosected cadaver. Use *Color Plates 59, 83* and *84* for reference.

Carotid Arteries and Blood Supply to the Brain

Identify the **common carotid arteries** lateral to the trachea and trace one to the angle of the mandible. (*Figure 10-7, p118*) At this point, it bifurcates into an **external carotid artery** and an **internal carotid artery**. [∞ p583]

The external carotid arteries supply blood to the structures of the neck, pharynx, larynx, lower jaw, and face, while the internal carotid arteries supply blood to the brain. Use *Color Plates 36, 82, 85, 86, 66*, and *Figure 22-4* for reference. The expanded beginning portion of the internal carotid artery is the **carotid sinus.** It contains sensory receptors for detecting changes in blood pressure. Trace the internal carotid artery as it enters the cranial vault through the carotid canal and merges with the *basilar artery*. The basilar artery is formed by the fusion of the two vertebral arteries. These vessels lie on the basilar portion of the occipital bone, hence the name. Collectively, these vessels anastomose to form the **cerebral arterial circle** or **circle of Willis**, which serves as the origin of the arteries that supply blood to the brain at a constant pressure.

LOCATE

____ Branches of common carotid arteries → External carotid a.
→ Internal carotid a.

____ Carotid sinus

____ Branches of internal carotid arteries → Ophthalmic a.
→ Anterior cerebral a.
→ Middle cerebral a.

____ Vertebral arteries → Basilar a.

____ Branches of basilar artery → Posterior cerebral a.
→ Posterior communicating a.

____ Cerebral arterial circle (circle of Willis) formed by:
{ Anterior communicating a.
Anterior cerebral a.
Posterior communicating a.
Posterior cerebral a.

See: Color Plates 36, 82, 85, 86, 66, Fig. 22-4

∞ *Figs.*
22-13, p586
22-14, p587
22-15, p588

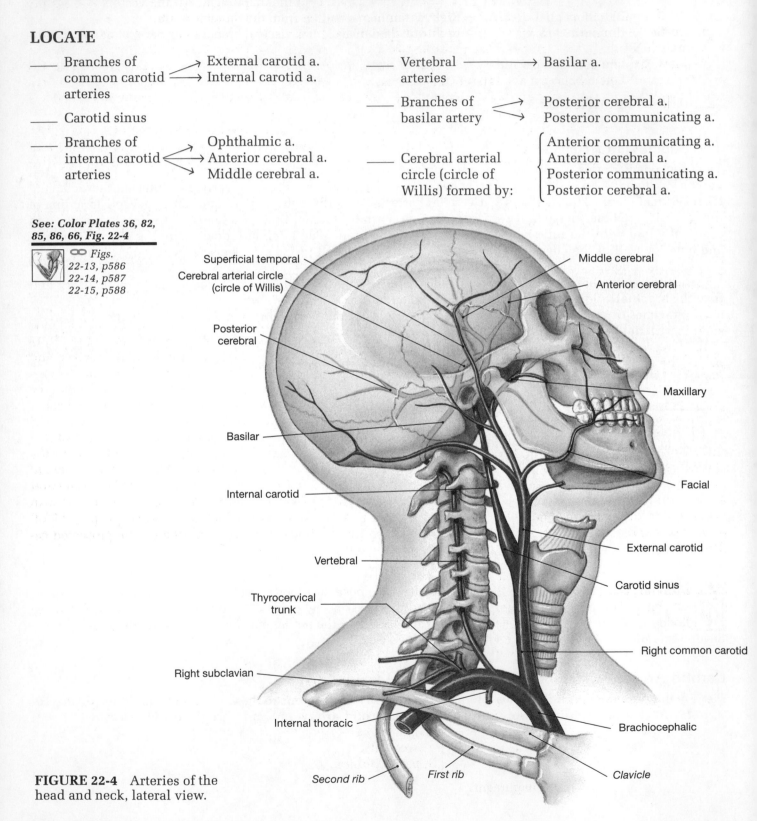

FIGURE 22-4 Arteries of the head and neck, lateral view.

Examine a demonstration specimen of the brain for the vertebral and basilar arteries, and for vessels of the cerebral arterial circle (circle of Willis). Since these vessels are typically not injected with colored latex, they will appear collapsed in most specimens. Refer to *Color Plates 66, 85, 86,* and *Figure 15-2b, p219* for reference.

The Descending Aorta

Trace the arch of the aorta to the point where it begins to descend toward the abdominal cavity dorsal to the heart and along the thoracic vertebrae. [∞ p589] Identify this segment of the aorta as the **descending thoracic aorta**. Use *Color Plates 69, 71, 87* and *Figure 22-5* for reference. Trace it to the diaphragm. All the vessels that supply the body wall of the thorax, trachea, bronchi, and esophagus branch from the thoracic aorta.

Branches of the thoracic aorta are grouped anatomically as either visceral branches or parietal branches. Visceral branches supply the organs of the chest: the **bronchial arteries** supply the nonrespiratory tissues of the lungs, **pericardial arteries** supply the pericardium, and **esophageal arteries** supply the esophagus. The parietal branches supply the chest wall: the **intercostal arteries** supply the chest muscles and the vertebral column area, and the **superior phrenic arteries** deliver blood to the superior surface of the muscular diaphragm that separates the thoracic and abdominopelvic cavities.

Continue to trace the thoracic aorta as it passes through the diaphragm into the abdominopelvic cavity. Use *Color Plates 71, 80,* and *121* for reference. Locate and identify the continuation of the aorta within the abdominal cavity as the **descending abdominal aorta**. Observe that it is just to the left of the vertebral column (the inferior vena cava will appear on the right) and note the numerous branches that arise to supply blood to all organs (e.g., digestive) and structures of the abdominopelvic cavity. At about L_4, the abdominal aorta terminates as it bifurcates into the left and the right *common iliac arteries* to supply blood to the pelvis and legs. An easy rule of thumb to remember as you trace the major branches of the aorta is that arteries to the visceral organs are unpaired and arise on the anterior surface of the abdominal aorta, whereas arteries to the abdominal wall, kidneys, gonads, and other structures are paired and originate on the lateral sides of the aorta. Keep in mind that arteries become progressively narrower as blood is distributed from the aorta to organs.

It is best to locate first the unpaired branches of the abdominal aorta and then the paired branches. Three unpaired branches can be distinguished easily in descending order on the anterior surface of the aorta. Locate and identify the first unpaired branch as the **celiac trunk artery**, which divides almost immediately into the **splenic artery** (to the spleen, stomach, and pancreas), the **left gastric artery** (to the stomach and inferior region of the esophagus), and the **common hepatic artery** (to the liver, stomach, gallbladder, and duodenum). Observe and identify the second unpaired branch as the **superior mesenteric artery** (to the pancreas, small intestine, and ascending and transverse colon). About two inches superior to the end of the aorta or bifurcation of the aorta, the third branch can be identified as the **inferior mesenteric artery** (to the transverse, descending, and sigmoid colon and rectum).

Now locate in descending order the four paired branches that arise on the lateral surfaces of the abdominal aorta. Return to the superior mesenteric artery and identify just inferior to it, the **suprarenal arteries** (to the adrenal glands). Next identify the large **renal arteries** (to the kidneys and adrenal glands), which can be observed as they pass dorsal to the peritoneal lining to reach the kidneys. Inferior to the renal arteries, identify the **gonadal arteries** (to the testes or ovaries). In females these arteries are called the *ovarian artery* (to the ovaries) and in males, the *testicular artery* (to testes). Along the entire length of the aorta arise **lumbar arteries** to supply blood to the body wall. All of the vessels previously described on the model may be identified on the prosected cadaver. Use *Color Plates 110, 111* and *123* for reference.

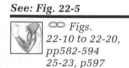

See: Fig. 22-5

∞ Figs.
22-10 to 22-20,
pp582-594
25-23, p597

LOCATE

_____ Aortic arch

_____ Divisions of descending → { Thoracic / Abdominal } aorta

_____ Branches of thoracic → { Bronchial a. / Pericardial a. / Esophageal a. / Intercostal a. / Superior phrenic a. } aorta

_____ Diaphragm

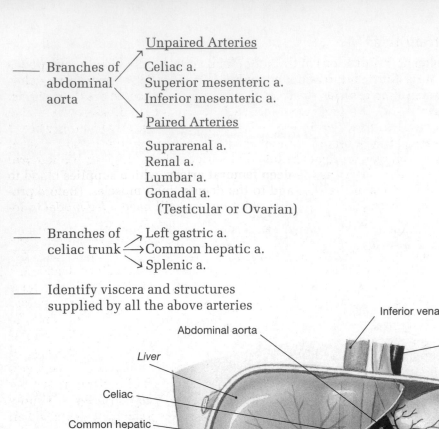

_____ Branches of
    abdominal
    aorta

<u>Unpaired Arteries</u>

Celiac a.
Superior mesenteric a.
Inferior mesenteric a.

<u>Paired Arteries</u>

Suprarenal a.
Renal a.
Lumbar a.
Gonadal a.
    (Testicular or Ovarian)

_____ Branches of
    celiac trunk

Left gastric a.
Common hepatic a.
Splenic a.

_____ Identify viscera and structures
    supplied by all the above arteries

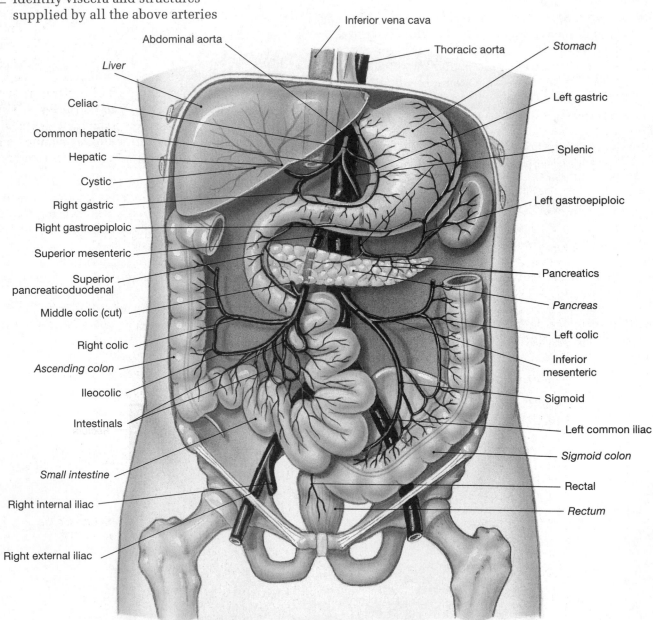

FIGURE 22-5   Arteries supplying the abdominal viscera.

Arteries of the Pelvis and Lower Extremities

The divisions (branches) of the terminal segment (bifurcation) of the abdominal aorta supply blood to the pelvis and lower extremities. Locate the terminal aorta (aortic bifurcation) at about L_4. Identify these terminal branches as the **right common iliac** and the **left common iliac arteries**. Trace each of these vessels to where they divide. Use *Color Plates 87, 88,* and *120,* and *Figures 22-6* to *22-8* for reference. Each common iliac divides into a large **external iliac artery**, to supply blood to the lower extremities, and a smaller **internal iliac artery** to supply blood to the pelvis and groin region. Locate the external iliac artery over the surface of the iliopsoas muscle and trace it out of the abdominopelvic cavity to where it emerges on the anteromedial surface of the thigh as the **femoral artery**. Observe the first major branch of the femoral artery as the **deep femoral artery**, which supplies blood to the ventral and lateral portions of the skin and to the deeper thigh muscles. (Before proceeding, check with your instructor.) From this point on you will need a *leg model* to locate the remaining branches of the femoral artery. Use *Color Plates 52* and *54*.

See: Figs. 22-6 to 22-8

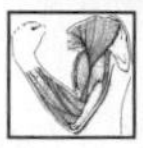

Figs.
22-17, p591
22-18, p592
22-19, p593
22-20, p594

Trace the femoral artery as it descends inferiorly and posterior to the femur. It becomes the **popliteal artery** as it reaches the popliteal fossa. As the popliteal artery descends, it branches into an **anterior tibial artery** and a **posterior tibial artery**. Locate the anterior tibial artery between the tibia and fibula and trace it inferiorly. Trace the posterior tibial artery as it passes inferiorly along the posterior surface of the tibia and observe its first branch as the **fibular artery**. The fibular artery continues to descend inferiorly along the medial surface of the fibula. The anterior tibial artery becomes the **dorsalis pedis artery**. The posterior tibial artery becomes the **medial** and **lateral plantar arteries** to supply blood to the foot. All of the vessels previously described on the models may be identified in a similar manner on the prosected cadaver. Use *Figures 22-6* to *22-8* for reference.

LOCATE

- ____ L_4 vertebra
- ____ Terminal branches of abdominal aorta → Right common iliac a.; → Left common iliac a.
- ____ Lumbosacral joint
- ____ Branches of common iliac arteries → External iliac a.; → Internal iliac a.

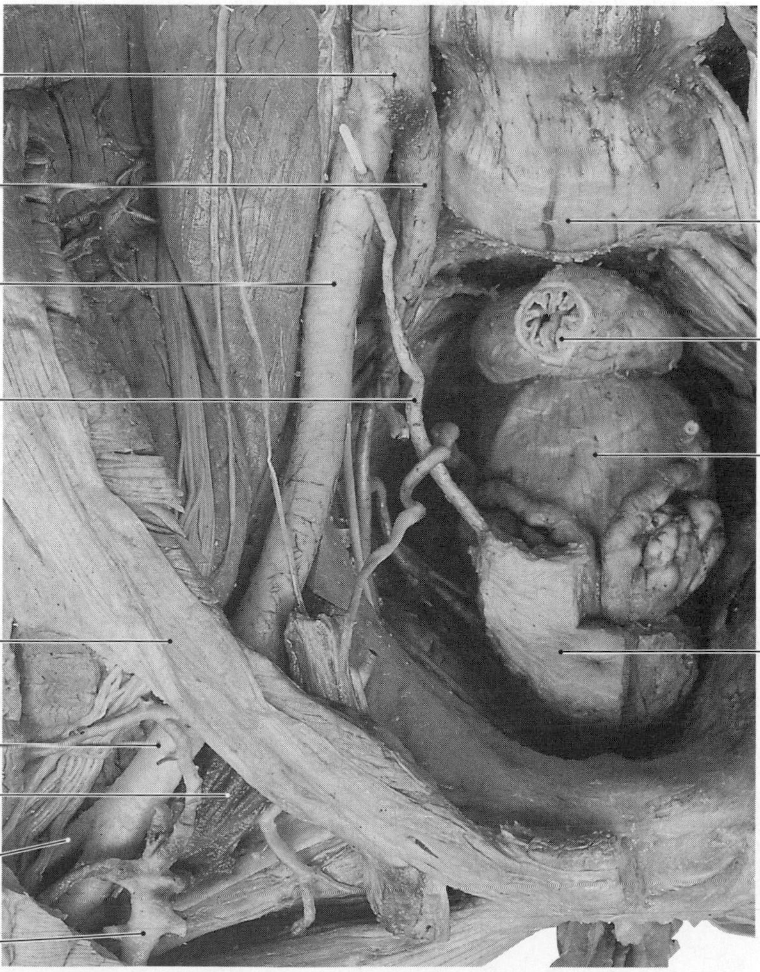
Right common iliac artery
Internal iliac artery
External iliac artery
Right ureter
Inguinal ligament
Femoral artery
Femoral vein
Deep femoral artery
Great saphenous vein
Fifth lumbar intervertebral disc
Sigmoid colon (cut)
Rectum
Urinary bladder (cut)

FIGURE 22-6 Major arteries of the pelvis.

Arteries of the Thigh and Leg

____ External iliac a.

____ Inguinal ligament

____ Femoral nerve

____ Branches of
femoral artery   → Deep femoral a.
  → Descending
genicular a.

____ Branches of
deep femoral
artery
  → Lateral femoral
circumflex a.
  → Medial femoral
circumflex a.

____ Popliteal fossa

____ Popliteal a.

____ Branches of
popliteal
artery
  → Posterior tibial a. → Fibular a.
  → Anterior tibial a.

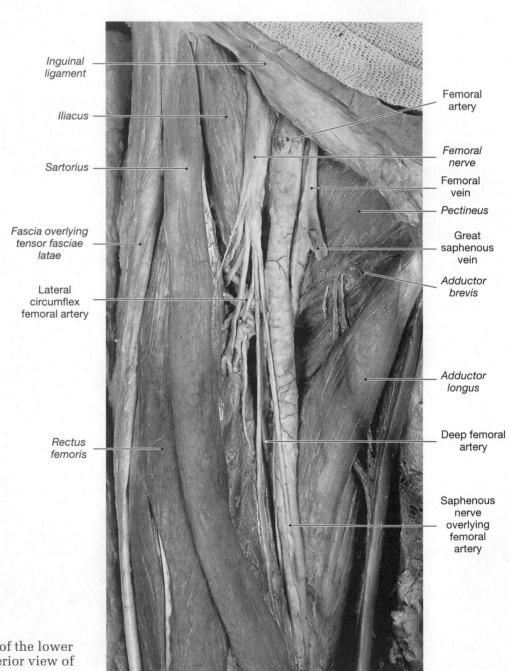

Inguinal ligament

Iliacus

Sartorius

Fascia overlying tensor fasciae latae

Lateral circumflex femoral artery

Rectus femoris

Femoral artery

Femoral nerve

Femoral vein

Pectineus

Great saphenous vein

Adductor brevis

Adductor longus

Deep femoral artery

Saphenous nerve overlying femoral artery

FIGURE 22-7 Major arteries of the lower extremity, anterior view of the right thigh.

Arteries of the Foot

_____ Anterior tibial a. becomes *dorsalis pedis artery* in the foot at the ankle

_____ Dorsalis pedis a. branches repeatedly in the ankle and dorsal portion of foot

_____ Branches of posterior tibial artery → Medial plantar a.
→ Lateral plantar a.

_____ Dorsal (arcuate) arch

_____ Plantar arch

Observation of the Systemic Circulation: Systemic Veins

Veins collect blood from the body's tissues and organs via a network that drains ultimately into the superior or inferior vena cavae before emptying into the right atrium of the heart. [∞ p594] The exception is the venous circulation of the heart, which drains into the coronary sinus. Keep in mind that the coronary veins are part of the systemic circulation. [∞ p560] Typically, veins in an area will run parallel to their artery counterpart and have the same name. Exceptions to this rule are the veins that drain the neck and extremities. Some of the veins in the neck and extremities run more superficially, with the corresponding arteries lying deep to the veins. This feature of superficial veins aids in regulation of body temperature, because excess body heat can be conducted out quickly at the skin surface. Because veins transport blood at such low pressure, the branching pattern of veins is more variable than that of arteries. This pattern of branching creates superficial veins whose names do not reflect their location. Keep in mind that veins get progressively larger as more blood drains into them as they return blood back to the venae cavae. The *superior vena cava (SVC)* receives blood from the tissues and organs of the head, neck, chest, and upper extremities. The *inferior vena cava (IVC)* collects almost all venous blood from structures inferior to the diaphragm.

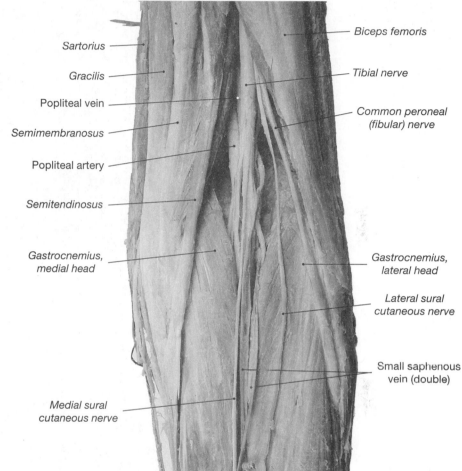

FIGURE 22-8 Vessels and nerves of the right popliteal fossa.

Procedure

Identify the relative locations of the **systemic veins** using first the *torso model,* then *arm* and *leg models* and then, a *cadaver specimen.*

Use *Color Plates 89, 90* and *Figure 22-9* to compare the distribution of the major veins to that of arteries. Identify and trace each **vein** as it collects blood from each organ or tissue area and **drains** into the **vena cavae.** All of the venous vessels described on the models may be identified in a similar manner on the prosected cadaver.

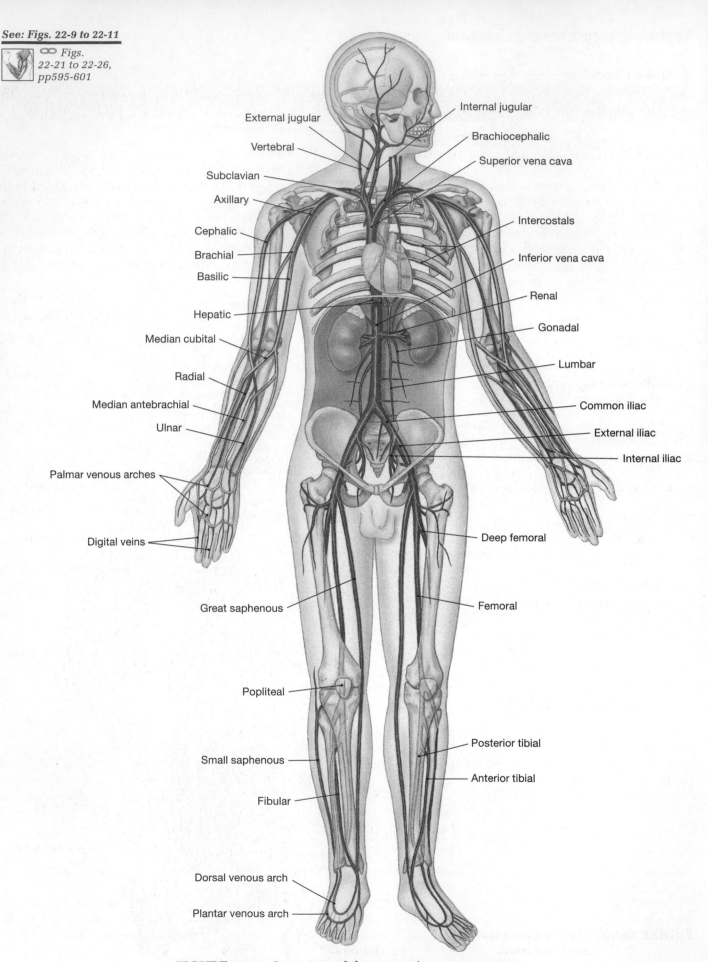

External jugular

Vertebral

Subclavian

Axillary

Cephalic

Brachial

Basilic

Hepatic

Median cubital

Radial

Median antebrachial

Ulnar

Palmar venous arches

Digital veins

Great saphenous

Popliteal

Small saphenous

Fibular

Dorsal venous arch

Plantar venous arch

Internal jugular

Brachiocephalic

Superior vena cava

Intercostals

Inferior vena cava

Renal

Gonadal

Lumbar

Common iliac

External iliac

Internal iliac

Deep femoral

Femoral

Posterior tibial

Anterior tibial

FIGURE 22-9 Overview of the systemic venous system.

Venous Return from the Cranium

✓ **Quick Check**

Before you begin to examine the sinuses and vessels responsible for the venous return of blood from the cranium, review the cranial meninges and the dural sinuses. (*Figure 15-2, pp218-219*)

The **superficial cerebral veins** drain into the superior sagittal sinus within the dura mater. [∞ p594] This vessel drains posteriorly and meets the straight sinus (which drains the structures deep within the brain) to join with the left and right transverse sinuses (lying with the transverse cerebral fissure) to drain collectively blood from the cranium by way of the sigmoid sinuses. The sigmoid sinuses emerge from the jugular foramina as the **internal jugular veins**. The internal jugulars collect blood from the cranium, face, and neck. Lateral to the internal jugulars identify the **external jugular veins.** These veins collect blood from the scalp, face, neck, and salivary glands. Between the internal and external jugulars identify the **vertebral veins**, which collect blood from the cranium, spinal cord, and vertebrae. These vessels drain into the **brachiocephalic veins.** Use *Color Plates 69, 89,* and *Figure 22-10* for reference.

See: *Color Plates 69, 89, Fig. 22-10*

∞ *Fig. 22-22, p596*

LOCATE

____ Superficial cerebral v.

____ Network of dural sinuses { Superior sagittal / Inferior sagittal / Left transverse / Right transverse

____ Dura mater

____ Internal cerebral v

____ Great cerebral v.

____ Straight sinus

____ Cavernous sinus

____ Sigmoid sinus

____ Internal jugular v.

____ Jugular foramen

____ Vertebral v.

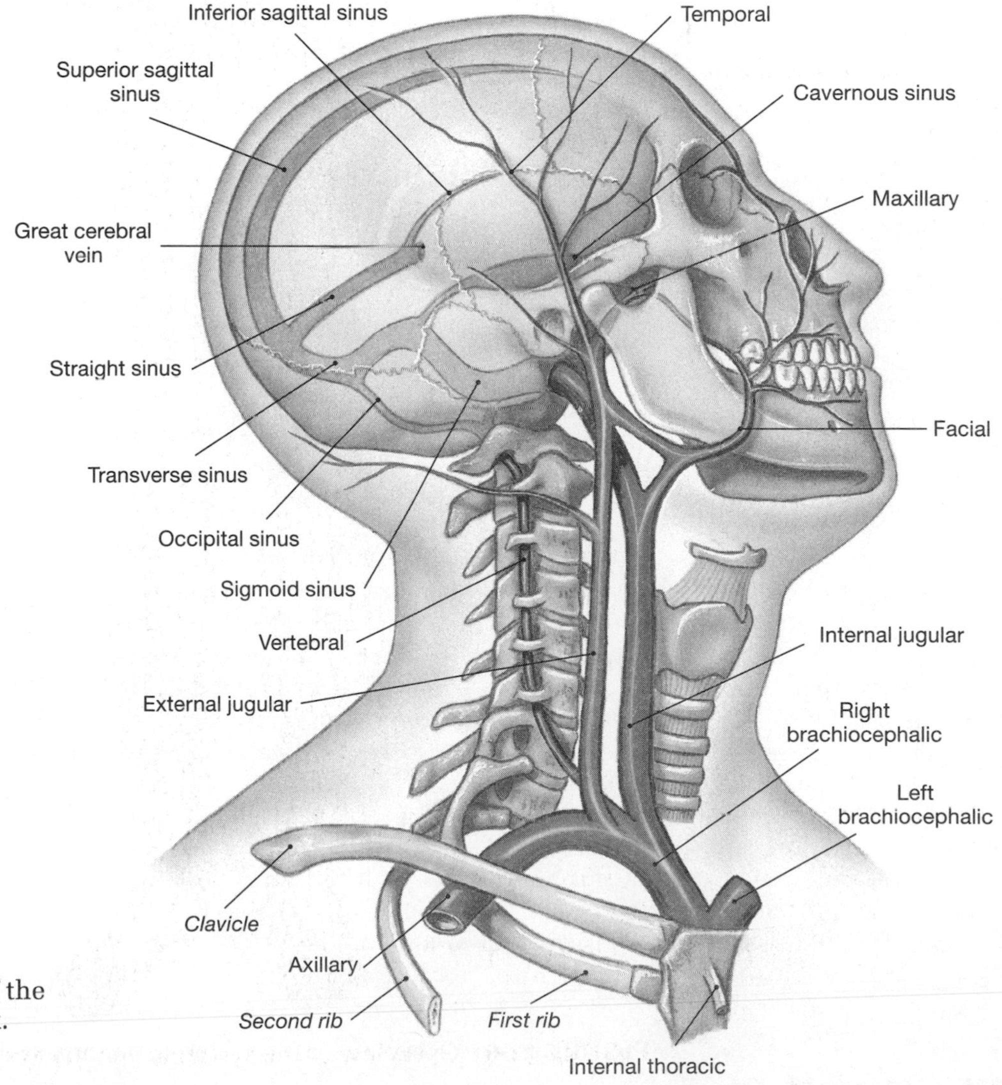

FIGURE 22-10 Major veins of the head and neck.

Superficial Veins of the Head and Neck

Most of the superficial veins of the head that drain into the external jugular vein merge and form two major vessels, the temporal and maxillary veins (*Figure 22-10*). The **temporal vein** can be identified superior to the ear on the surface of the auricularis muscle. The **maxillary vein** can be identified as it merges with the temporal vein at the ear opening. These veins merge into the external jugular vein. [∞ p596] Additionally, trace the **facial vein** across the body of the mandible and follow it deep to the sternocleidomastoid muscle where it drains into the internal jugular vein. (*Figure 10-7, p118*)

See: Fig. 22-10

∞ *Fig. 22-22, p596*

LOCATE

____ Temporal v.   ____ External jugular v.

____ Facial v.   ____ Subclavian v.

____ Maxillary v.

Venous Return from the Upper Extremities

To locate the veins of the upper extremity, view the arm in the anatomical position. Use *Color Plates 47, 48, 89* and *Figure 22-9* for reference. Identify the **digital veins** and observe how they drain blood into the **superficial** and **deep palmar veins** of the hand. These veins branch and interconnect within the palm to form the **palmar venous arches**. Trace the superficial arch to the wrist, and continue to ascend and trace along the radial side of the forearm to the **cephalic vein**. The cephalic vein continues to ascend along the radial side of the upper arm to the level of the clavicle. On the ulnar side of the forearm, identify the **median antebrachial vein** on the anterior surface of the ulna, and medial to it the **ulnar vein**. Trace the cephalic, the median antebrachial, and the ulnar veins to the cubital fossa and observe these vessels merging to form the **basilic vein**. The **median cubital vein** crosses the elbow medially and at an oblique angle to join with the basilic vein. In the clinical setting, blood samples are drawn typically from the median antebrachial or median cubital vein. The basilic vein ascends along the medial surface of the biceps brachii muscle. In the axilla it joins the **brachial vein** to form the **axillary vein**.

LOCATE

____ Digital v.   ____ Median cubital v.

____ Superficial palmar v.   ____ Basilic v.

____ Deep palmar v.   ____ Radial v.

____ Palmar venous arches   ____ Ulnar v.

____ Cephalic v.   ____ Brachial v.

____ Median antebrachial v.   ____ Axillary v.

Formation of the Superior Vena Cava

The cephalic vein joins the axillary vein on the outer surface of the first rib, forming the subclavian vein, which continues into the chest. This vessel passes superior to the surface of the first rib, deep to the clavicle and into the thoracic cavity. Shortly after entering the thoracic cavity, the subclavian vein merges with the internal jugular vein to create the **brachiocephalic vein**. The two brachiocephalic veins unite just to the right of the midline to form the **superior vena cava** (**SVC**). Close to the point of fusion the internal thoracic veins empty into the brachiocephalic veins. On the right side of the vertebral column, identify the **azygos vein** as it ascends to about the level of T$_2$. Ascending on the left side of the vertebral column is the smaller **hemiazygos vein**. The azygos receives blood from the hemiazygos vein. These vessels are the chief collecting vessels of the thorax and they receive some blood from two sources, (1) the **intercostal veins**, located between the ribs on the intercostal muscles and (2) the **esophageal, pericardial**, and **mediastinal veins**, which are located on the anterior surface of the esophagus. Collectively the azygos and hemiazygos veins and their tributaries drain blood from the thorax and abdomen into the superior vena cava.

Identify the following veins using *Color Plates 69, 71, 73, 89*, and *Figure 22-9* for reference.

See: Color Plates 69,
71, 73, 89, Fig. 22-10

∞ Fig.
22-23, p597

LOCATE

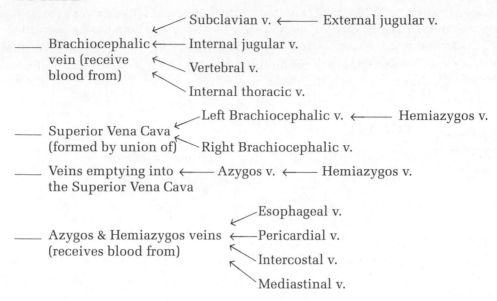

```
                          ┌─ Subclavian v. ←──── External jugular v.
___ Brachiocephalic ←──── Internal jugular v.
    vein (receive    ←─── Vertebral v.
    blood from)      ←─── Internal thoracic v.

                          ┌─ Left Brachiocephalic v. ←──── Hemiazygos v.
___ Superior Vena Cava ←──
    (formed by union of)  └─ Right Brachiocephalic v.

___ Veins emptying into ←──── Azygos v. ←──── Hemiazygos v.
    the Superior Vena Cava

                               ┌─ Esophageal v.
___ Azygos & Hemiazygos veins ←── Pericardial v.
    (receives blood from)     ←── Intercostal v.
                               └─ Mediastinal v.
```

The Inferior Vena Cava

The *inferior vena cava (IVC)* collects most of the venous blood from the lower extremities, pelvis, and abdominal organs. Identify and trace each **vein** as it collects blood and drains into the **inferior vena cava**. Generally, veins parallel the arterial distribution pattern.

Veins Draining the Lower Extremity

To locate the veins of the lower extremity, view the leg in the anatomical position. Use *Color Plates 71, 89, 90,* and *Figure 22-9* for reference. The venous drainage pattern of the legs is similar to the arterial distribution pattern. Identify the **great saphenous vein** as the vessel that ascends the medial leg and thigh from the ankle until it drains into the **femoral vein** at the femoral triangle. On the lateral side, ascending on the lateral and posterior surfaces of the gastrocnemius muscle, identify the **small saphenous vein**, which arises from the dorsal venous arch. (Note: Not all models show the saphenous veins). At mid-calf, deep to the gastrocnemius and soleus muscles, locate two vessels ascending parallel to each other. Identify the lateral vessel as the **fibular vein** and the medial vessel as the **posterior tibial vein**. Trace these vessels and note they merge into a single vessel about 3 inches inferior to the popliteal fossa. Just superior to this merging, in the popliteal fossa, identify the **anterior tibial vein**. The union of these three vessels forms the **popliteal vein**, which ascends along the posterior surface of the femur to become the femoral vein.

See: Color Plates 71,
89, 90, Fig. 22-9

∞ Fig.
22-25, p599

LOCATE

```
                           ┌─ Anterior Tibial v. ←──── Plantar v.
___ Popliteal v. ←── Posterior Tibial v.
                           └─ Fibular v. ←──────── Plantar v.


                        Great Saphenous v. ←──────── Dorsal venous arch
___ Femoral v. ←──── Popliteal v. ←──────── Small Saphenous v.
    Deep Femoral v.        Anterior Tibial v.
                           Posterior Tibial v.


___ External Iliac v. ←──── Femoral v.
```

Veins Draining the Pelvis

Locate and identify the veins draining the pelvis, viewing the torso model in the anatomical position with all abdominopelvic viscera removed. Use *Color Plates 89, 121,* and *Figure 22-9* for reference. The venous drainage pattern of the pelvis parallels the arterial distribution pattern. Identify the **internal iliac vein** and note that it joins the **external iliac vein** to form a single **common iliac vein**. Trace the left and right common iliac veins as they ascend medially at an oblique angle to merge at about the L_5 vertebra. The union of these two vessels forms the inferior vena cava.

See: Color Plates 89, 121, Fig. 22-9

∞ *Fig. 22-23, p597*

LOCATE

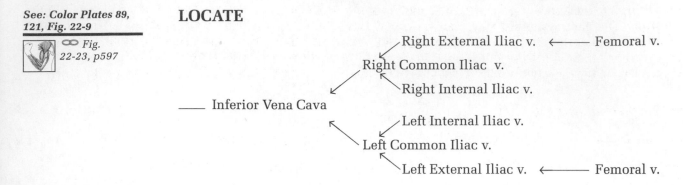

Veins Draining the Abdomen

Locate and identify the veins draining the abdomen, viewing the torso model with the liver, stomach, and intestines removed. Use *Color Plates 89, 121,* and *Figure 22-9* for reference. Locate the inferior vena cava at its origin and trace it along the vertebral column as it ascends parallel to the aorta. The inferior vena cava receives blood from six pairs of major veins. It is best to locate these vessels in ascending order. Locate the **lumbar veins**, which drain the lumbar portion of the abdominal wall. Observe the **gonadal veins** (ovarian or testicular), which drain blood from the ovaries or testes. (Note: The right gonadal vein ends in the IVC, while the left one drains into the left renal vein.) Identify the short **renal veins**, which collect blood from the kidneys and deliver it to the inferior vena cava. Capping each kidney is an adrenal gland. The right **suprarenal vein** collects and transports blood from the right adrenal gland into the IVC, while the left usually drains into the left renal vein. On the inferior surface of the diaphragm, locate the **phrenic veins**. The right phrenic vein drains blood from the diaphragm into the IVC, the left drains into the left renal vein. Locate and observe the **hepatic veins** as they exit from the superior surface of the liver to drain into the IVC at about T_{10}.

See: Color Plates 89, 121, Fig. 22-9

∞ *Fig. 22-23, p597*

LOCATE

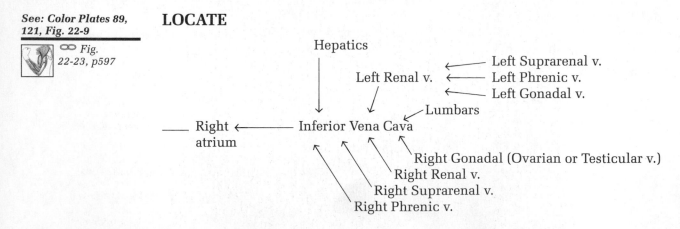

Observation of the Hepatic Portal System

The drainage pattern of the **hepatic portal system** is quite different from the pattern of veins just observed. All of the blood that is drained from the digestive organs does not enter the vena cava. Instead, blood flows into vessels of the hepatic portal system, which combine into one major vessel, the **hepatic portal vein**, leading into the liver. This design permits blood high in nutrients from capillaries of the digestive organs to pass directly to the capillaries of the liver and not into the inferior vena cava as one might predict. The liver is the only digestive organ drained by the inferior vena cava. Hence, this is a portal system because all of the blood flows between capillaries of the digestive organs and those of the liver via one vessel, the hepatic portal vein.[∞ p600]

Procedure

The hepatic portal system begins in the capillaries of the digestive organs. Identify the **digestive organs** an̲... the **veins** that form the **hepatic portal system** from the organs they drain to the liver, using a *torso model* and... *daver specimen*. The inferior portion of the large intestine is drained by the **inferior mesenteric vein**. The l̲... **colic** and **superior rectal veins** drain the *descending colon, sigmoid colon,* and *rectum* regions of the large intes- tine. Use *Color Plates 89, 102, 119,* and *Figure 22-11* for reference. The left colic drains blood into the inferior mesenteric vein. The inferior mesenteric vein drains into the **splenic vein** just inferior to the liver. The splenic vein is formed by the veins of the stomach, pancreas, and spleen. The **superior mesenteric vein** collects blood from veins draining the stomach, all the blood from the *small intestine* and about two-thirds of the blood from the large intestine. The **gastroepiploic veins** collect blood from the *lesser and greater curvatures of the stomach* (*Color Plate 108*) and drain into the splenic and superior mesenteric veins. Trace the superior mesenteric vein and the splenic veins to the point where they unite, and identify this vessel as the **hepatic portal vein**. Trace it to the liver and observe that the **gastric** and **cystic veins** drain into the hepatic portal vein prior to entering the liver.

See: Color Plates 89, 102, 108, 119, Fig. 22-11

∞ *Figs. 22-26, p601 25-11, p679 25-17, p687*

LOCATE

____ Inferior mesenteric v.

____ Left colic v.

____ Superior rectal v.

____ Splenic v.

____ Superior mesenteric v.

____ Gastroepiploic v.

____ Gastric v.

____ Hepatic portal v.

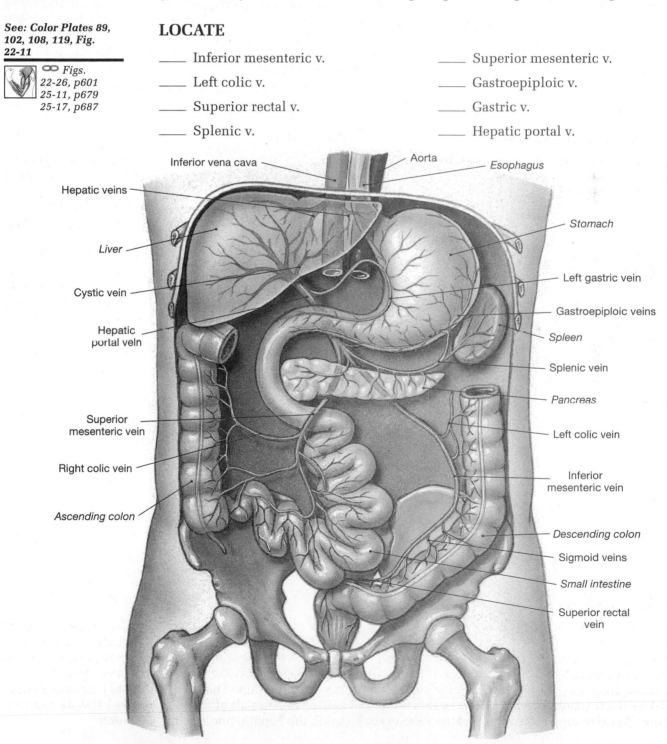

FIGURE 22-11 Vessels of the hepatic portal system.

rculatory systems are different. This difference reflects the source for: obtaining nutrients, ex-
and carbon dioxide, and excreting or removing waste. Before birth the lungs are collapsed and
g. Additionally the digestive system in the fetus has nothing to digest. Since only minimal
are required to maintain the lungs and digestive organs, these structures are for the most part
tal circuit. [∞ p601]

Procedure

Identify the structures and **vessels** that form the **fetal circuit** using a *model* depicting *fetal circulation*. Trace the
flow of blood from the placenta through the fetal circuit and returning to the placenta using *Figure 22-12* for reference. Identify the *placenta* and the *umbilical cord*. Note that the umbilical cord contains blood vessels to link
mother with fetus. Normally, maternal blood never comes in contact with fetal blood. All nutrients and oxygen
are supplied from maternal blood to the fetus via diffusion between circulatory systems across the placenta. Fetal waste products are returned across the placenta to the maternal circulation for removal. Identify the **umbilical vein** within the umbilical cord and trace it to the liver. Identify the **ductus venosus** as the fetal bypass that
collects blood from the liver and carries blood directly into the inferior vena cava. At birth, blood flow through
the umbilical vein ceases and shortly thereafter the ductus venosus degenerates into connective tissue.

Identify the two passageways that permit blood to bypass the fetal lungs. Locate the **foramen ovale** between
the two atria. This interatrial opening and its associated flap act as a valve to allow blood to flow only from right
to left atria, but not the reverse. At birth, the flap over the foramen ovale closes and becomes a shallow depression, the fossa ovalis. Observe the short **ductus arteriosus** between the pulmonary trunk and the aorta. This connection allows blood to bypass the gas exchange surfaces of the lungs by directing it from the pulmonary trunk
into the systemic circuit. At birth, this vessel degenerates and persists in the adult as a fibrous cord called the
ligamentum arteriosum.

Locate the internal iliac arteries and identify the two **umbilical arteries** that arise from them. Blood carrying
waste products and carbon dioxide is returned by these arteries to the placenta for waste removal via diffusion
into the maternal blood.

See: Fig. 22-12

∞ *Fig.
22-27, p602*

LOCATE

____ Placenta

____ Umbilical { Umbilical arteries
    cord   { Umbilical vein

____ Fetal liver

____ Ductus venosus

____ Inferior vena cava

____ Chambers of fetal heart

____ Foramen ovale

____ Ductus arteriosus

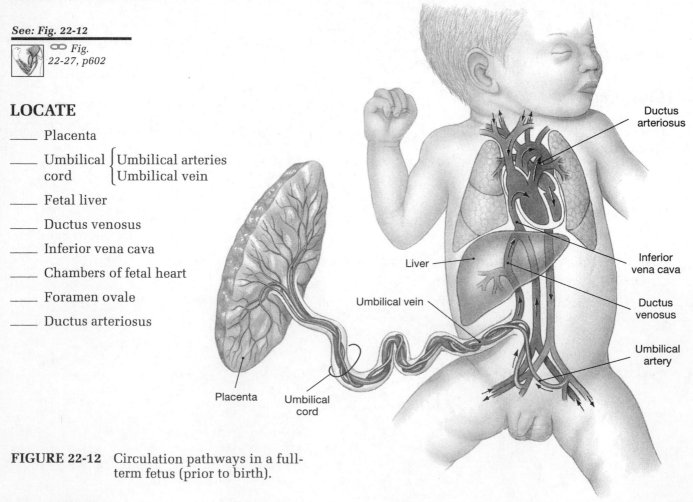

FIGURE 22-12 Circulation pathways in a full-
term fetus (prior to birth).

Anatomical Identification Review

Identify the arterial branches of the aortic arch. Fill in the blank boxes with the correct arterial branches.

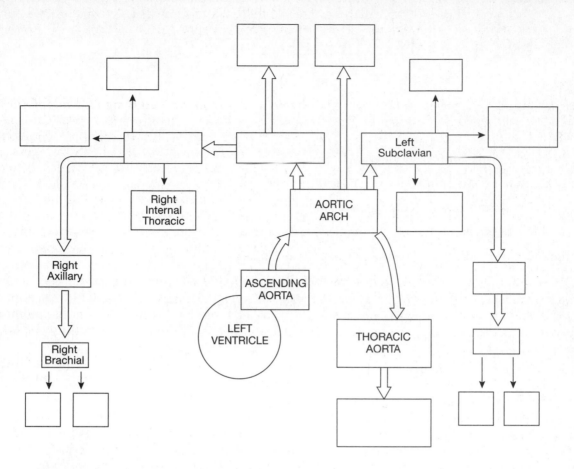

To Think About

1. By what structures is it ensured that the blood supply to the brain is maintained under as many conditions as possible?

2. What is the role of the azygos and its tributaries in returning blood to the vena cava?

3. What is the pathway of blood into, through, and out of the hepatic portal system?

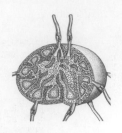

23

THE LYMPHATIC SYSTEM

Objectives

1. Describe the functions of the lymphatic system.

2. Identify the major components of the lymphatic system.

3. Identify the major clusters of lymph nodes found in the body.

4. Name and locate the major lymph collecting vessels, and describe their structure and function.

5. Describe the anatomical relationship between the lymphatic and circulatory systems.

6. Recognize under the microscope: lymphatic nodules, lymph nodes, and the thymus.

7. Describe the structure, function, and histological features of the spleen.

Of all the body systems, the **lymphatic system** is the least observable; yet it has a profound influence on the body's ability to protect and defend itself. The lymphatic system plays a major role in defending the body from invasions by bacteria, viruses, microorganisms, and the traumas of daily life. [∞ 613] The defensive role performed by the lymphatic system is the production, maintenance, and distribution of lymphocytes. Lymphocytes are the primary cells of the lymphatic system and are produced and stored in the bone marrow, thymus, and peripheral lymphatic tissues. They later differentiate or specialize within the bone marrow and thymus. Lymphocytes are vital to the immune response, which involves two different pathways: a direct attack of killing cells (mediated by NK and T-cells) or inactivation of an invading pathogen (an antibody attack mediated by the B-cells). The immune response is initiated by the presence of an antigen (such as bacteria or foreign particles like pollen).

The lymphatic system secondarily helps to maintain the normal blood volume of the body. The lymphatic system includes a network of vessels that collect interstitial fluid, transport it between lymphatic tissues, and ultimately return it to the blood. Recall that capillaries permit large amounts of fluid to escape from the cardiovascular system. With the aid of various pressure gradients, this interstitial fluid enters the lymphatic system and is transported back to the cardiovascular system through lymphatic vessels. In performing this role, consider the lymphatic system an accessory to the circulatory system.

Components of the Lymphatic System

The lymphatic system consists of specific tissues, organs, and vessels. *Lymphoid tissues* consist of a specialized form of reticular tissue that houses large numbers of lymphocytes. Small oval or bean-shaped lymphoid organs are known as **lymph nodes**. Lymph nodes are clustered in specific regions throughout the body. Nodes are concentrated outside the large joints and are closely associated with the digestive, circulatory, and respiratory systems. The *thymus* and *spleen* are the largest masses of lymphoid tissue. **Lymphatic fluid** or **lymph** is the interstitial fluid that has diffused into **lymphatic vessels**. Lymphatic vessels vary in size and are similar in structure to veins. The largest vessels contain valves. The smallest vessels are the lymphatic capillaries. These capillaries are designed to permit interstitial fluid to percolate through their walls and flow into lymphatic vessels. Lymphatic vessels carry lymph from peripheral tissues to the venous system.

Procedure

Identify the location of the following clusters of lymph nodes, lymphatic vessels, and structures of the **lymphatic system** using an *anatomical chart* and *torso model* of the *lymphatic system*. Lymph nodes tend to be located where there is a potential for the entry of pathogens into the body, such as through the mouth, nose, and digestive tract. They are also clustered near the large joints, but outside the joint capsule. Using *Color Plate 131* and *Figure 23-1* for reference, locate the **cervical lymph nodes** in the cervical region, the **axillary lymph nodes** in the axilla, the **abdominal lymph nodes** within the abdominopelvic cavity, and the **inguinal lymph nodes** within the femoral triangle and the pelvic basin. Occasionally, accumulated cellular debris and bacteria can cause lymph nodes within a specific region to enlarge. Any enlargement can be significant clinically and is important to the clinician in diagnosis.

Lymphatic vessels typically parallel the superficial veins. Trace the lymphatic vessels of the upper limb to the axillary lymph nodes. Now trace the lymphatic vessels from the lower limb to the inguinal lymph nodes. You have just traced the pattern of lymph flow from the extremities to their receiving lymph nodes. From the lower extremities, lymph collects into the inguinal lymph nodes and then flows into a sac-like chamber. Identify this chamber as the **cisterna chyli** near the L_2 vertebra and adjacent to the abdominal aorta (see *Figure 23-2*). Lymph from the lower extremities, pelvis, and lower abdomen collects here. The cisterna chyli represents the origin of the **thoracic duct**. The duct lies on the anterior surface of the lumbar and thoracic vertebrae and travels alongside the aorta. Trace the thoracic duct and aorta superiorly, observing their penetration through the diaphragm at the aortic hiatus, to enter the thoracic cavity, and ascend to the level of the left clavicle. Locate the point on the left internal jugular or subclavian veins where the thoracic duct enters the circulatory system. Just inferior to the right clavicle, identify the smaller **right lymphatic duct**, and note its relationship to the right subclavian vein. The thoracic duct and the right lymphatic duct are the only collecting ducts to drain lymph into the circulatory system. Locate the thymus (typically not depicted in torso models) and spleen. These organs are described under separate headings.

See: Color Plate 131,
Figs. 23-1, 23-2

 ∞ *Fig.*
23-1, p613
23-4, p616
23-5, p617

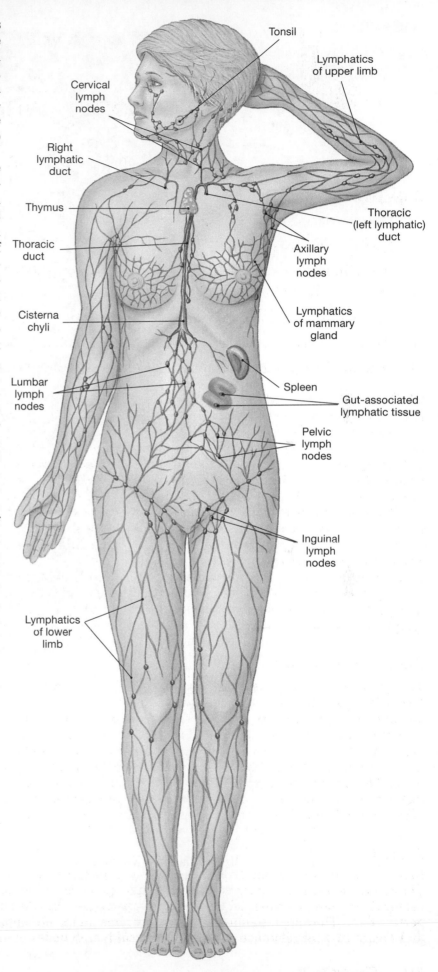

FIGURE 23-1 Components of the lymphatic system.

LOCATE

_____ Cervical lymph nodes

_____ Axillary lymph nodes

_____ Abdominal lymph nodes

_____ Inguinal lymph nodes

_____ Lymphatics of mammary gland

_____ Lymphatics of upper limb

_____ Lymphatics of lower limb

_____ Cisterna chyli

_____ Thoracic duct (left lymphatic)

_____ Right lymphatic duct

_____ Thymus

_____ Spleen

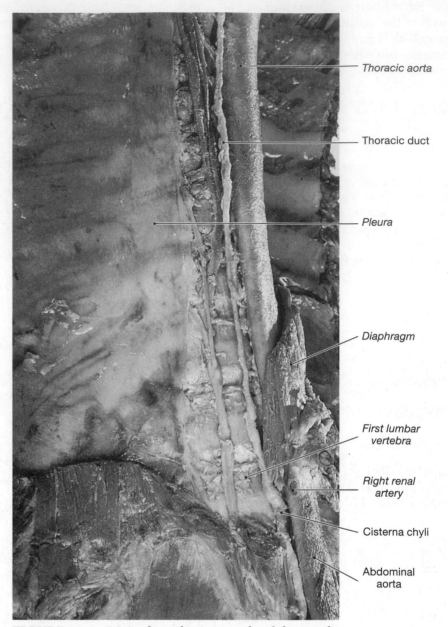

FIGURE 23-2   Major lymphatic vessels of the trunk (thoracic and abdominal organs removed).

In the prosected cadaver, lymphatic vessels are not as easily seen as blood vessels. Many lymphatics are often damaged inadvertently during the prosection. Keep in mind that lymphatics typically accompany veins. Trace and identify the nodes and the major collecting ducts as described above. Lymph nodes can be readily observed in cervical, axillary, and inguinal regions. In the thorax, the thoracic duct and right lymphatic duct can typically be observed. In life, these vessels lie posterior to the parietal pleura, but emerge at the level of the brachiocephalic veins to then join with their respective internal jugular or subclavian veins. The bodies of the lumbar vertebrae must be exposed to view the cisterna chyli and the origin of the thoracic duct. Expose them by reflecting the stomach and the small and large intestines to the right side. Identify the cisterna chyli and trace the thoracic duct as it ascends alongside the aorta to enter the thoracic cavity via the aortic hiatus.

Microscopic Identification of Lymphatic Vessels

Procedure

Identify the structure of a **lymphatic vessel** using the *microscope slide* provided, first by viewing it under a dissecting microscope and then under the low power magnification. Use *Figure 23-3* for reference. Lymphatic vessels are typically prepared as a "whole mount" type of slide. In this type of preparation, a small specimen of a lymph vessel has been stained and affixed to the slide. Lymphatic vessels have a structure similar to that of veins, but have thinner walls and very little, if any, smooth muscle. Observe the entire specimen with the dissecting microscope and locate a **lymphatic valve**. One or more valves can be seen within the lumen of the lymph vessel. Lymphatic valves prevent lymph from reversing its direction and returning toward the lymphatic capillaries. Valves ensure that lymph will flow toward the heart.

Now observe the vessel and valve with the lower power magnification of the compound microscope. The wall of the vessel is formed mostly by collagen and elastic fibers. The elastic fibers appear as short, dark, wavy pencil-thin lines. Notice at higher magnification the connective tissue leaflets that form the valve are actually arranged in pairs opposite each other within the vessel.

Slide # _____ Lymph Vessel, w.m.

See: Fig. 23-3

∞ *Fig.*
23-3, p615

LOCATE

_____ Lymphatic vessel

_____ Lymphatic valve

_____ Collagen and elastic fibers

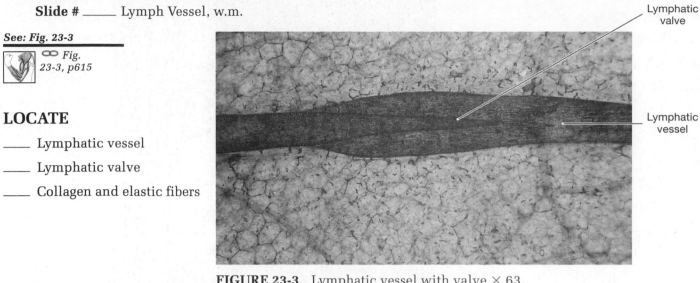

FIGURE 23-3 Lymphatic vessel with valve × 63.

Relationship of Lymphatic Ducts and Circulatory System

Procedure

✓ **Quick Check**

Before you begin to identify the relationship of the lymphatic ducts to the circulatory system, review the location and relationships of the following veins on the torso: subclavian, internal jugular, brachiocephalic, and superior vena cava. *Color Plates 69, 73, 89,* and *Figure 22-10, p327.*

Identify the **major lymphatic collecting ducts** and their relationship to the large veins of the circulatory system, using a *torso model* and *cadaver specimen*. This relationship will be observed in the thoracic cavity. For you to view this relationship, it will be necessary for you to remove the lungs from the torso. Use *Color Plates 71, 89,* and *Figure 23-4* for reference. Locate the veins identified in Quick Check. Ultimately all lymphatic vessels merge and lymph drains into either the thoracic duct or right lymphatic duct. Locate the large thoracic duct and identify the merging of the **left jugular trunk**, and the **left subclavian trunk** with it. The former drains the head and neck, and the latter the left upper chest and arm. Locate the right lymphatic duct and identify the **right jugular trunk** and the **right subclavian trunk** vessels. These and other small vessels form the right lymphatic duct. The right duct drains lymph only from the upper right quadrant of the body (right arm, upper chest, neck, and head), while the thoracic duct drains lymph from the remainder of the body. Identify the location at which both of these ducts enter near the union of the internal jugulars and subclavians on the appropriate sides. The thoracic duct and right lymphatic duct are the only lymphatic ducts that return the collected lymph to the blood. Along the lateral surfaces of the trachea and bronchi are a continuous chain of about 50 lymph nodes. Lymph flows superiorly through a series of nodes to form the **left** and **right bronchomediastinal lymph trunks**. Typically, these trunks drain lymph near the union of the internal jugular and subclavian veins on their respective sides.

See: Color Plates 71,
89, Fig. 23-4

∞ *Figs.*
23-4, p616
23-5, p617

LOCATE

_____ Thoracic duct

_____ Right lymphatic duct

_____ Left/right jugular trunks

_____ Left/right subclavian trunks

_____ Left/right internal jugular veins

_____ Left/right bronchomediastinal trunks

_____ Left/right subclavian veins

_____ Left/right brachiocephalic veins

_____ Superior vena cava

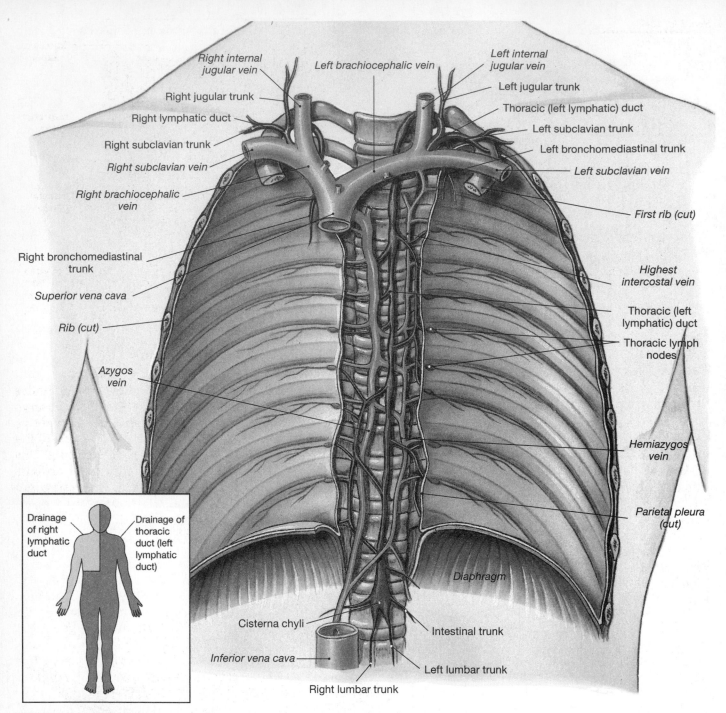

FIGURE 23-4 Relationship of lymphatic ducts and circulatory system.

Microscopic Identification of Lymphoid Nodules (Peyer's Patch)

Procedure

Observe the appearance and structure of a typical **lymphoid nodule** using the *microscope slide* provided, first by viewing under low power magnification and then under high power magnification. Use *Figure 23-5* for reference. Unencapsulated lymphoid nodules located beneath the epithelial lining of the small intestine are known as **Peyer's patches**. These are an example of **aggregate lymphoid nodules** that are found within selected internal organs. Observe the wall of the small intestine (ileum) under low power and locate a group of three to four flask-shaped, purple-staining structures. Each structure is a lymphoid nodule and can now be observed under high power. Identify the pale-staining central area of the lymphoid nodules as the **germinal center**. This area contains lymphocytes actively undergoing mitosis, thereby increasing the population of lymphocytes.

Slide # _____ Small Intestine (Ileum), sec.

See: Fig. 23-5

∞ *Fig.*
23-8, p620

LOCATE

____ Intestinal epithelium
containing Peyer's patches

____ Germinal center in each
lymphoid nodule containing
lymphocytes

The tonsils are also examples of large lymphoid nodules
and can be examined if time permits.

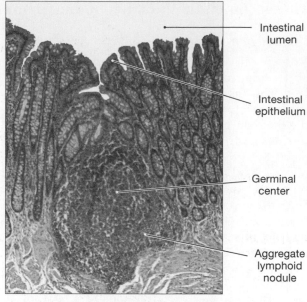

Intestinal
lumen

Intestinal
epithelium

Germinal
center

Aggregate
lymphoid
nodule

FIGURE 23-5 Lymphoid nodule,
Peyer's patch from
small intestine × 51.

Microscopic Identification of Lymph Node

Procedure

Identify the structure of a typical **lymph node** using the *microscope slide* provided, first by viewing under
low power magnification and then under high power magnification. Use *Figure 23-6* for reference. Observe
the kidney-bean shaped lymph node encapsulated by a dense connective tissue capsule. Some connective
tissue invades the interior of the node to form partitions called **trabeculae**. Locate the outer *cortex* portion of the
node. A lymph node is mostly lymphocytes supported by a loose meshwork of dark-staining reticular fibers. The
dark-staining lymphocytes are located chiefly in the cortex. The medulla lies central to the cortex and extends
into the hilus. Lymphocytes in the medulla lie in branching rows of cells called *medullary cords*, projecting into
an inner **medullary sinus**. Identify
pale-staining areas of lymphocytes
within the cortex as **germinal centers**. This area contains lymphocytes
actively undergoing mitosis, which
increases the lymphocyte population.

Slide # ____ Lymph node, sec.

See: Fig. 23-6

∞ *Fig.*
23-9, p621

LOCATE

____ Connective tissue capsule

____ Trabeculae

____ Cortex containing
lymphocytes

____ Medullary cords

____ Medullary sinus

____ Germinal centers

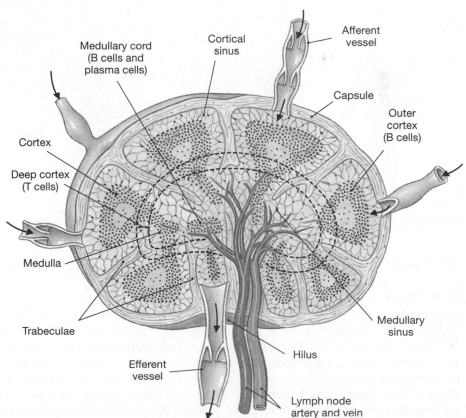

Medullary cord
(B cells and
plasma cells)

Cortical
sinus

Afferent
vessel

Capsule

Outer
cortex
(B cells)

Cortex

Deep cortex
(T cells)

Medulla

Trabeculae

Medullary
sinus

Efferent
vessel

Hilus

Lymph node
artery and vein

FIGURE 23-6 Structure of a lymph node.

Microscopic Identification of the Thymus

Procedure

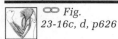Identify the **histological appearance** of the **thymus**, using the *microscope slide* provided first by viewing under low power magnification and then under high power magnification. Use *Figure 23-7* for reference.

The thymus is attached to the pericardium within the anterior portion of the mediastinum and lies just superior to the heart. It reaches maximum size by puberty, but gradually degenerates into a fibrous mass by early adulthood. The specimen on your slide is a young thymus. Under low power identify the connective tissue **septa** that divide the gland into distinct **lobules**. Each lobule is characterized by an outer dark **cortex** area and an inner lighter **medulla** area. Numerous blood vessels can be observed between the lobules. Observe the medulla of a lobule with high power and identify groups of epithelial cells arranged in concentric layers, forming structures known as **Hassall's corpuscles**. The thymus is involved in the differentiation of lymphocytes and the normal development of lymphoid tissues.

Slide # _____ Thymus

See: Fig. 23-7

∞ *Fig. 23-16c, d, p626*

LOCATE

____ Lobules

____ Septa

____ Cortex

____ Medulla

____ Blood vessels

____ Hassall's corpuscles

____ Lymphocytes

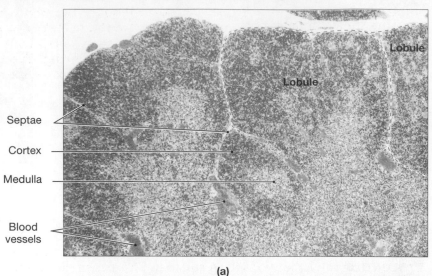

(a)

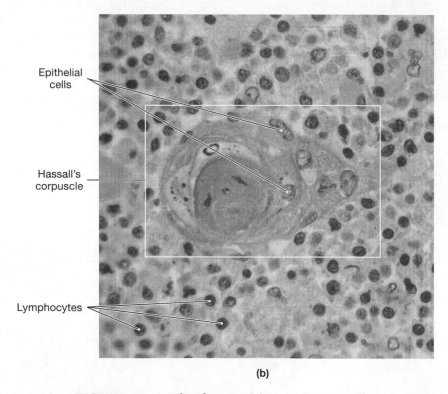

(b)

FIGURE 23-7 The thymus (a) × 43 (b) Hassall's corpuscle, x 700.

Observation of the Spleen

The spleen is located in the left hypochondriac region and is protected by the stomach, diaphragm, and the costal margin of the lower ribs. It is the largest lymphoid organ. In life, the spleen is easily damaged because of its highly vascular structure. Its vascular structure creates an organ that conforms to the space between the organs and the structures it contacts (stomach, left kidney, intestines, and diaphragm). In addition to its lymphoid function, macrophages in the spleen destroy old RBCs through phagocytosis and recycle their iron molecules and amino acids from their proteins.

Procedure

Identify the location of the **spleen** within the abdominopelvic cavity and examine its surface features, using a *torso model* or *cadaver specimen*. Use *Color Plates 102, 103, 109,* and *111*. Observe the location of the spleen within the abdominopelvic cavity of the intact torso model. The spleen is attached to the lateral border of the stomach by the **gastrosplenic ligament**. Remove the spleen and identify its **superior** and **inferior borders** and its surfaces. Identify the convex **diaphragmatic surface** as that area of the spleen that contacts the diaphragm and body wall. The concave **visceral surface** can be identified as the area that comes in contact with the stomach and left kidney. Within the visceral surface area, identify the impression for the stomach, termed the **gastric area**, and the impression for the left kidney, the **renal area**. Identify the groove, termed **hilus**, which separates these two areas and serves as the point of attachment for the **splenic artery** and **splenic vein**.

See: Color Plates 102, 103, 109, 111

Fig. 23-17, p627

LOCATE

_____ Gastrosplenic ligament

_____ Superior/inferior borders

_____ Diaphragmatic surface

_____ Visceral surface ⎰ Gastric area / Renal area

_____ Hilus {Splenic artery/vein

Microscopic Identification of the Spleen

Identify the histological appearance of the **spleen** using the *microscope slide* provided, first by viewing under low power magnification and then under high power magnification. Use *Figure 23-8* for reference. The spleen is very similar in structure to a large lymph node. Begin your observation at the outer surface of the specimen. This area can be identified by the presence of a connective tissue capsule. The cellular portion of the spleen is referred to as *pulp*. The bulk of the cells are contained in the **red pulp**. Identify the red pulp portion of the specimen as the area of lighter stain. Dark purple-staining areas, termed **white pulp**, are peppered throughout the lighter red pulp. Red pulp contains RBCs, lymphocytes, and macrophages. White pulp represents aggregations of lymphocytes, and macrophages. These cells can be observed under high power magnification. Associated with the white pulp are small **trabecular arteries**, which are branches of the splenic artery. Trabecular arteries continue to branch to capillaries that supply blood to red pulp. Most of the functions of the spleen take place in the red pulp.

Slide # _____ Spleen, sec.

See: Fig. 23-8

Fig. 23-17c, p627

LOCATE

_____ Connective tissue capsule

_____ White pulp containing lymphocytes

_____ Red pulp

_____ Small trabecular arteries

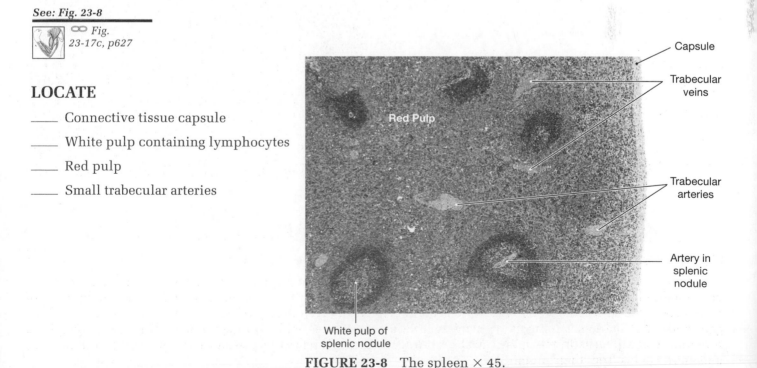

FIGURE 23-8 The spleen × 45.

Anatomical Identification Review

1. Name and identify on the accompanying illustration the major lymphatic vessels, lymphoid tissues, and lymphoid organs of the lymphatic system.

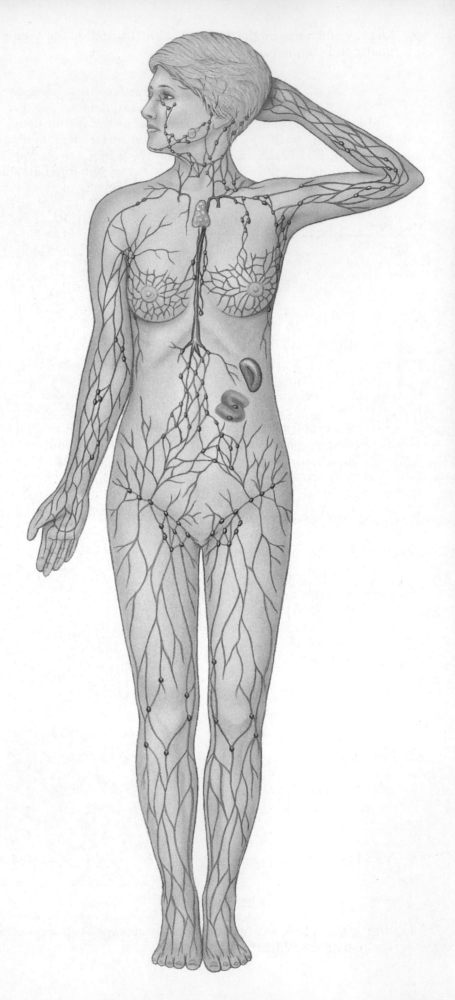

2. Identify and name on the accompanying illustration the anatomical structures associated with the lymphatic and circulatory systems.

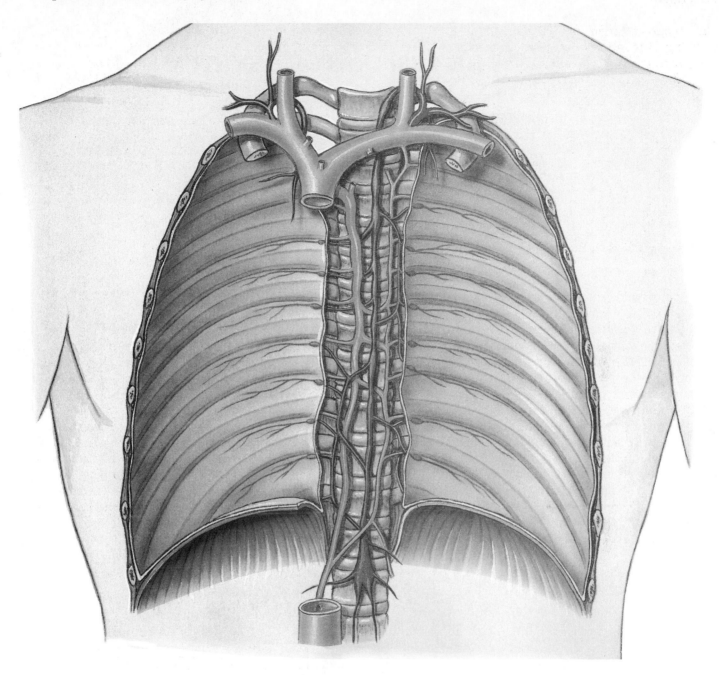

To Think About

1. Why are the lymphatic capillaries often involved in the spread of metastatic cancers?

2. Why are there a large number of lymphatic vessels associated with the large intestine?

3. Why is the spleen an organ particularly susceptible to rupture in an accident, and why is this such a serious potential problem?

24

THE RESPIRATORY SYSTEM

Objectives

1. Describe the major functions of the respiratory system.

2. Describe the structural organization of the respiratory system and identify its major organs.

3. Locate and identify the structure of each component of the upper and lower respiratory tracts.

4. Recognize the histology of the respiratory epithelium.

5. Locate and identify the structure of the pleural cavities and pleural membranes.

6. Identify and name the muscles of respiration and describe their role in breathing.

The **respiratory system** is responsible for the exchange of carbon dioxide and oxygen between the body and the air of our external environment. All but one of the following six functions of this system are vital to life processes: (1) provides an extensive surface area for gas exchange; (2) moves air to and from these surfaces; (3) protects these surfaces from desiccation and temperature changes; (4) defends the body tissues from pathogenic organisms; (5) assists other systems (such as the circulatory system) in the regulation of blood volume, blood pressure, and pH; and (6) assists in sound production. [∞ p636]

The respiratory system is divided into two divisions, an *upper respiratory tract* and a *lower respiratory tract*. The term tract is used to refer to those structures that move air to and from the surface areas within the lungs for gas exchange. The structures of the **upper respiratory tract** are the nose, nasal cavity, paranasal sinuses, and pharynx. These structures filter, moisten, and warm the incoming air. The **lower respiratory tract** consists of the larynx, trachea, bronchi, and the lungs, which contain the cellular areas for gas exchange. The lower respiratory tract terminates in the *alveoli* of the lungs, the point of exchange of oxygen and carbon dioxide between inhaled air and the body. The respiratory system consists of two types of structures, those that conduct air and those that provide a surface for gas exchange. [∞ p635]

Structures of the Respiratory System

Procedure

Identify on a *torso model* or *anatomical chart* the following structures of the **respiratory system** using *Color Plates 69, 70, 92,* and *Figure 24-1* for reference. The *nose* and *nasal cavity* are the primary entrance to the respiratory system. The *nasal conchae* are bony projections into the nasal cavity that promote air turbulence and serve to warm, moisten, and filter inhaled air. (*Figure 6-20, p53*) The *pharynx* is a common chamber into which the nasal and oral cavities open, and therefore is shared by both respiratory and digestive systems. The *larynx* functions in both sound production and separation of inhaled air into the *trachea* and of food/liquids into the *esophagus*. From the larynx, air passes into the trachea where it is conducted into the *left* and *right bronchi*. The rigid, tubular bronchi conduct air to the *lungs*. Within the lungs, the bronchi divide many times before air is delivered to the *respiratory membrane*, the site for exchange of gases. The major muscle used in respiration is the *diaphragm*.

See: Color Plates 69, 70, 92, Figs. 24-1

∞ *Fig. 24-1, p635*

LOCATE

____ Nose

____ Nasal conchae

____ Pharynx

____ Larynx

____ Trachea

____ Right/left bronchi

____ Lungs

____ Diaphragm

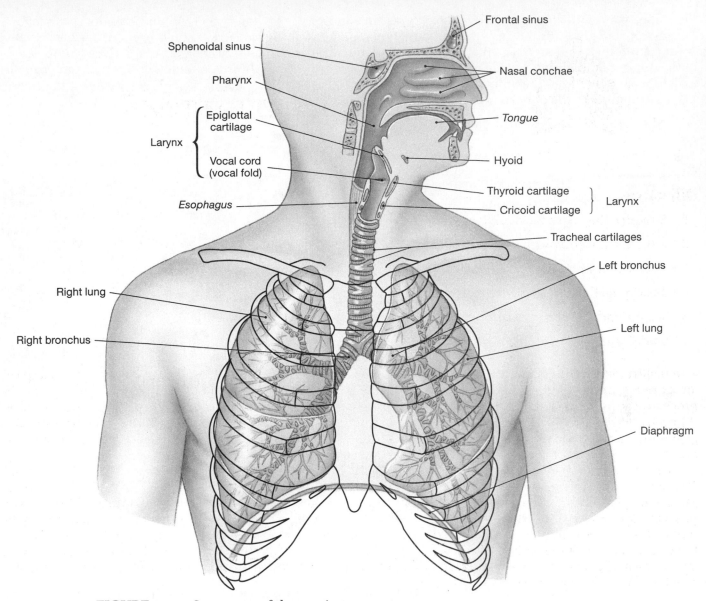

FIGURE 24-1 Structures of the respiratory system.

Labels (clockwise from upper right):
- Frontal sinus
- Nasal conchae
- Tongue
- Hyoid
- Thyroid cartilage — Larynx
- Cricoid cartilage — Larynx
- Tracheal cartilages
- Left bronchus
- Left lung
- Diaphragm
- Right bronchus
- Right lung
- Esophagus
- Larynx — Epiglottal cartilage / Vocal cord (vocal fold)
- Pharynx
- Sphenoidal sinus

Microscopic Identification of the Respiratory Epithelium

The upper respiratory system is lined by a **respiratory epithelium**, which is pseudostratified ciliated columnar epithelium with goblet cells. [∞ pp 57, 59] Inhaled air comes into contact with this epithelial surface of the conducting portion of the respiratory tract. Goblet cells produce mucus, which entraps particulate matter and prevents desiccation. Cilia continually move mucus-containing particulate matter out of the respiratory conducting passageways, just like a continuously moving escalator.

Procedure

✓ **Quick Check**

Before you begin to identify the respiratory epithelium, turn to *Chapter 3* and review the histological characteristics of pseudostratified ciliated columnar epithelium. (*Figure 3-9, p16*)

 Locate the **respiratory epithelium** using the *microscope slide* provided, first by viewing under low power magnification and then under high power magnification. Use *Color Plate 8* and *Figure 24-2* for reference. The specimen on the slide provided to you may be a section of either the trachea or nasal cavity. Typically, these specimens will also contain hyaline cartilage in addition to respiratory epithelium. Scan the specimen under low power to locate the respiratory epithelium. Under high power, identify the pseudostratified ciliated columnar epithelium that forms the respiratory epithelium. Identify the three characteristics (see Quick Check) of this type of epithelium: arrangement of nuclei in pseudostratified ciliated columnar epithelium, goblet cells, and cilia emerging from the apical surface of the cells. Note the attachment of the columnar cells at their basement membranes to the underlying lamina propria.

Slide # _____ Trachea, sec.

See: Color Plate 2,
Fig. 24-2

∞ *Fig.*
24-2, p636

LOCATE

_____ Columnar epithelial cells

_____ Cilia

_____ Goblet cells

_____ Basement membrane

_____ Lamina propria

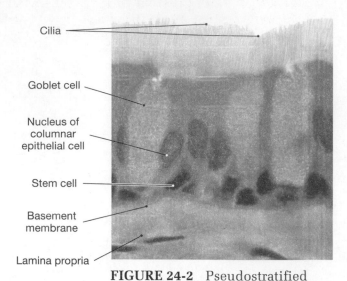

FIGURE 24-2 Pseudostratified ciliated columnar epithelium × 932.

Observation of the Nose, Nasal Cavity, and Pharynx

Procedure

✓ **Quick Check**

Before you begin to identify the nose and nasal cavity, review a sagittal section through a dry skull specimen and observe the bones and sinuses of the nasal complex. (*Color Plate 32* and *Figure 6-20, p53*)

Identify the structure of the nose, nasal cavity, and pharynx using a *torso head model* or *cadaver specimen*. View the model in sagittal section. Use *Color Plates 38, 91, 92,* and *Figures 24-3* and *24-4* for reference. Locate the bones and sinuses associated with the nose (see Quick Check). Identify the **external nares**, or nostrils, through which air enters into the **nasal cavity** to be warmed, moistened, and filtered by the respiratory epithelium. The external nares conduct air into the anterior portion of the nasal cavity, termed the **vestibule,** which is located in the fleshy, moveable portion of the nose. The posterior part of the nasal cavity becomes narrowed so that air can pass into the *pharynx.* Identify the **internal nares** as this reduced opening into the pharynx.

The nasal septum (vomer and perpendicular plate of ethmoid) divides the nasal cavity into two equal halves. Air turbulence is created by air movement over the nasal conchae (turbinate bones), which enhances the moistening and cleansing of incoming air and for olfaction. Additionally, the paranasal sinuses communicate with the nasal cavity and contribute to the moisturizing of air.

The nasal cavity is separated from the oral cavity by the *palate.* The palate forms the floor of the nasal cavity and the roof of the oral cavity. Identify the anterior part as the **hard palate**, which is formed by the fusion of the palatine processes of the maxillary bones and palatine bone. Identify the **soft palate** as the soft tissue portion that is posterior to the palatine bone.

Pharynx

The **pharynx** is a chamber shared by the respiratory and digestive systems. Identify the pharynx as the chamber area limited by the internal nares and the larynx. The pharynx is subdivided into three regions: nasopharynx, oropharynx, and laryngopharynx. [∞ p637]

1. Identify the **nasopharynx** as the area superior to the soft palate and posterior to the internal nares. On its posterior wall locate the **pharyngeal** (adenoid) **tonsil** and the openings to the pharyngotympanic (auditory or eustachian) tubes.

2. Locate the **oropharynx** as the area of the pharynx visible when yawning. This area extends from the nasopharynx to the base of the tongue at the level of the hyoid bone. Identify the dangling terminal end of the soft palate as the **uvula**. Just inferior to the uvula, on the lateral walls of the oropharynx, identify the **palatine tonsils**. Both oropharynx and oral cavity are lined with stratified squamous epithelium and not respiratory epithelium.

3. Identify the area between the hyoid bone and the openings of the esophagus and larynx as the **laryngopharynx**. This area represents the most inferior portion of the pharynx. Air must now pass through the *larynx*, the next structure on its way to the lungs.

See: Color Plates 38, 91, 92, Figs. 24-3, 24-4

∞ *Fig. 24-3, pp638-639*

LOCATE

____ Frontal sinus

____ External nares

____ Vestibule

____ Nasal cavity

____ Internal nares

____ Nasal septum

____ Nasal conchae { Superior / Middle / Inferior

____ Oral cavity { Hard palate / Soft palate

____ Pharynx { Nasopharynx / Oropharynx / Laryngopharynx

____ Tonsils { Pharyngeal / Palatine

____ Entrance to auditory tube

____ Trachea

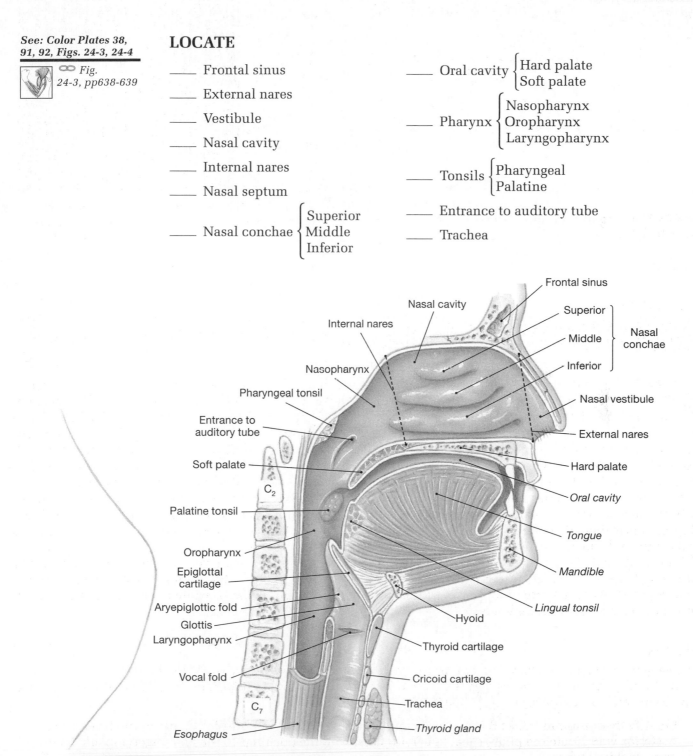

FIGURE 24-3 Sectional view of the head.

Observation of the Larynx

The hollow **larynx** is a visible, externally palpable bulge located inferior to the hyoid bone in the throat. It is very prominent during swallowing. The larynx is formed by nine pieces of cartilage, three pairs and three singles, held together by dense connective tissue ligaments. Collectively, these cartilages form the shape of the larynx. Respiratory epithelium lines the interior of the larynx (inferior to the vocal folds). At its superior border, identify the narrow slit-like opening, the **glottis**, through which air passes into the larynx. The larynx typically extends from the level of C_4 to C_7 vertebrae. [∞ p639]

The larynx provides a means for sound production. On the lateral walls of the larynx, identify the delicate **vocal folds**. These are highly elastic ligaments that are involved in sound production and are termed the **true vocal cords**. Air passing through the glottis vibrates the vocal folds and produces sound. Speech is produced by the mouth and tongue, with the nasal cavity and sinuses acting as resonating and amplifying chambers. [∞ p641]

Procedure

✓ **Quick Check**

Before you begin to identify the structure of the larynx, review the two major muscles that elevate the larynx, the digastric and stylohyoid. (*Table 6, p117* and *Figure 10-7, p118*)

Identify the following structures of the **larynx** using a *model* or *cadaver specimen* of the *human larynx*. Use *Color Plates 93* to *95* for reference. View the intact larynx model from the anterior to locate the single cartilages. Locate the hyoid bone and use it as a landmark. Inferior to it locate the largest and most prominent single cartilage, the **thyroid cartilage** (Adam's apple). Identify the blunt anterior edge in the midline of the thyroid cartilage as the **laryngeal prominence**. Identify the **thyrohyoid ligament**, which connects the thyroid cartilage to the hyoid bone. Now locate the **cricoid cartilage** immediately inferior to the thyroid cartilage. Observe how this cartilage is attached in the midline to the thyroid cartilage by the **cricothyroid ligament**. The trachea is attached to the cricoid cartilage of the larynx by the **cricotracheal ligament**. Identify the **epiglottal cartilage** of the **epiglottis** by its characteristic shoehorn-shape and the slit-like opening, the **glottis**. (See *Color Plates 91* and *92*). During swallowing, the epiglottis is pushed inferiorly and anteriorly, thereby covering the glottis and directing food and liquids into the opening of the esophagus.

The following paired cartilages are best viewed from the posterior. Identify the **arytenoid cartilages** just posterior to the thyroid cartilage and superior to the cricoid cartilage. Muscles attached to the arytenoid cartilages control the tension (either increasing or decreasing the tension) in the true vocal cords. Attached to the superior border of the arytenoid cartilages are the **corniculate cartilages**. The elongated, curving **cuneiform cartilages** are located within a fold of tissue between the arytenoid cartilage and epiglottis. (Note: These may or may not be depicted on your model and in the cadaver are difficult to display.)

LOCATE

See: Color Plates 91 to 93, 95

∞ *Fig. 24-4, p640 24-5, p641*

Viewed from Anterior and Sagittal

____ Hyoid bone

____ Larynx

____ Thyroid cartilage

____ Laryngeal prominence

____ Thyrohyoid ligament

____ Cricoid cartilage

____ Cricothyroid ligament

____ Trachea

____ Tracheal cartilages

____ Cricotracheal ligament

____ Epiglottal cartilage of the epiglottis

____ Glottis

Viewed from the Posterior

____ Cartilages $\begin{cases} \text{Arytenoid} \\ \text{Corniculate} \\ \text{Cuneiform} \end{cases}$

____ Epiglottis

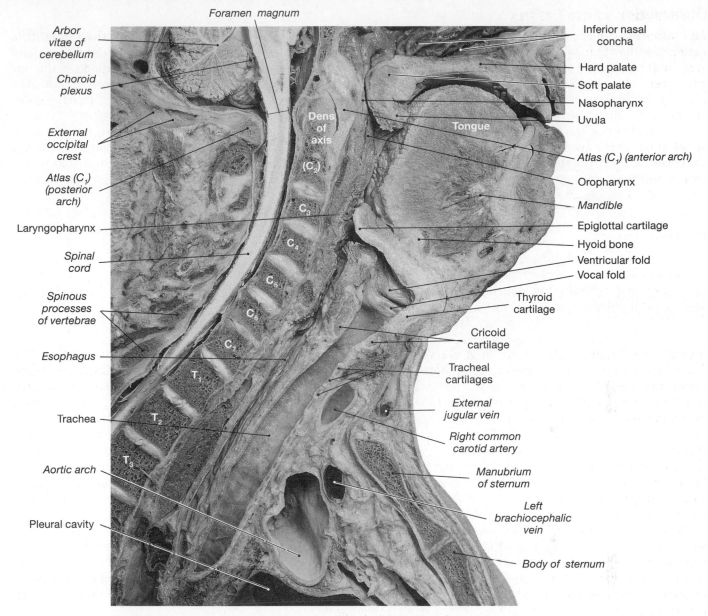

Arbor vitae of cerebellum

Choroid plexus

External occipital crest

Atlas (C₁) (posterior arch)

Laryngopharynx

Spinal cord

Spinous processes of vertebrae

Esophagus

Trachea

Aortic arch

Pleural cavity

Foramen magnum

Dens of axis

(C₂)

C₃

C₄

C₅

C₆

C₇

T₁

T₂

T₃

Tongue

Inferior nasal concha

Hard palate

Soft palate

Nasopharynx

Uvula

Atlas (C₁) (anterior arch)

Oropharynx

Mandible

Epiglottal cartilage

Hyoid bone

Ventricular fold

Vocal fold

Thyroid cartilage

Cricoid cartilage

Tracheal cartilages

External jugular vein

Right common carotid artery

Manubrium of sternum

Left brachiocephalic vein

Body of sternum

FIGURE 24-4 Sectional anatomy of the head and neck.

Observation of the Trachea and Primary Bronchi

The **trachea** is a tough, but flexible tube, that extends from the larynx to where it branches into the left and right *primary bronchi*. Air passes from the larynx into the trachea. The trachea is formed from 15 to 20 C-shaped hyaline **tracheal cartilage** rings. The trachea would collapse each time we inhaled if it was not held always open by these ring supports. The fact that the rings are C-shaped, or incomplete, permits the esophagus to bulge into the posterior surface of the trachea during the swallowing of food. At the base of the trachea, the respiratory tract branches into primary bronchi, which continue to divide into smaller branches as air is conducted into the lungs. The trachea and bronchi are lined with respiratory epithelium. [∞ p642]

Procedure

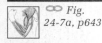 Identify the structures of the **trachea** and the divisions of the **primary bronchi** using a *torso*, *larynx/trachea/bronchi model*, or *cadaver specimen*. In the torso (chest plate must be removed), verify the hyoid bone, larynx, and attachment of the trachea to the larynx by the cricotracheal ligament. Use *Color Plates 70, 71, 73, 77*, and *Figure 24-5* for reference. Locate the trachea anterior to the vertebrae, beginning at the level of C$_7$ and extending to T$_5$. Note the relationship of the trachea to the esophagus and carotid arteries. On the larynx/trachea/bronchi model, identify the **annular ligaments** that connect each tracheal cartilage ring to its neighbor. Locate in the torso at about the level of T$_5$, the point at which the trachea bifurcates to form the **left** and **right primary bronchi**. Now verify these structures on the model. The point of separation between the two bronchi is marked by a ridge; identify this point as the **carina**. The right bronchus conveys air into the right lung and is straighter and larger in diameter than the left. Foreign bodies when inhaled often become lodged in this bronchus because it is straight and large. They may pass into the right lung.

Each primary bronchus divides as it enters the *hilus* portion of the lung. Locate the hilus of each lung and observe the bronchi branching into the lung tissue. Each lung is divided into lobes, the right lung has three and the left lung has two. Observe the right primary bronchus and identify three divisions: **superior lobar bronchus**, **middle lobar bronchus**, and **inferior lobar bronchus**. Observe the left primary bronchus and identify two divisions: **superior lobar bronchus** and **inferior lobar bronchus**. The lobar bronchi conduct air into their respective lung lobes and continue to divide within the lung tissue.

See: Color Plates 70, 71, 73, 77, Fig. 24-5

∞ *Fig. 24-7a, p643*

LOCATE

____ Hyoid

____ Larynx

____ Trachea

____ Annular ligaments

____ Carina

____ Right primary bronchus

____ Right lobar bronchi { Superior Middle Inferior

____ Left primary bronchus

____ Left lobar bronchi { Superior Inferior

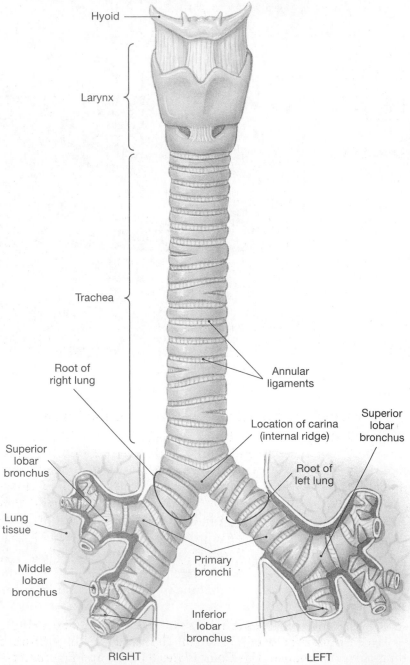

FIGURE 24-5 Anterior view of the trachea and primary bronchi.

Microscopic Identification of the Respiratory Epithelium of the Trachea

Procedure

Identify the **respiratory epithelium** of the **trachea** using the *microscope slide* provided, first by viewing with a dissecting microscope and then under low and high power magnifications. The specimen on the slide is a cross-section of the trachea. Your specimen may also contain the esophagus and carotid artery. Use *Figure 24-6* for reference as you examine the specimen under the dissecting microscope. Scan the specimen and locate the trachea and, if present, the esophagus and carotid arteries. Identify the **lumen of the trachea**, **tracheal cartilage**, and **respiratory epithelium**. Locate the ends of the tracheal cartilage and identify the **trachealis muscle**. This is a band of smooth muscle and elastic fibers, which under the influence of the ANS, controls the diameter of the trachea. Note the relationship of this muscle to the esophagus. Now observe your slide under the low power of the compound microscope and locate the respiratory epithelium. Under high power, identify the pseudostratified, ciliated columnar epithelium that forms the respiratory epithelium. Verify the three characteristics (see Quick Check) of this type of epithelium: arrangement of nuclei, goblet cells, and cilia emerging from the free surface of the cells. Numerous **mucous glands** can be identified in the submucosa layer. These glands secrete mucus onto the surface of the respiratory epithelium. Mucus entraps particulate matter as it is moved up and out.

Slide # _____ Trachea, complete section

See: Fig. 24-6

∞ *Fig. 24-7b,c, p643*

LOCATE

_____ Lumen of trachea

_____ Tracheal (hyaline) cartilage

_____ Respiratory ⎰ Goblet cells
epithelium ⎱ Lamina propria

_____ Relationship to esophagus

_____ Mucous glands

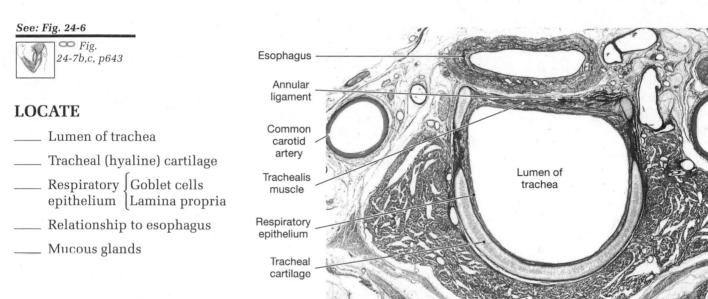

FIGURE 24-6 Cross-sectional views of the trachea × 60.

Observation of Superficial Anatomy of the Lungs

The left and right lungs lie in the thoracic cavity within the **pleural cavity**. Recall that the pleural membranes line the thoracic cavity and encase each lung. The **parietal pleura** lines the thoracic cavity, and the **visceral pleura** covers each lung. Collectively, these serous membranes secrete **pleural fluid** into the pleural cavity, which is the space between the parietal and visceral layers (it is only a potential space). [∞ p651]

The lungs are made up of millions of air sacs, termed **alveoli**. The capillaries of the pulmonary circuit and alveoli form a structure, termed the **respiratory membrane**, where gas exchange takes place. Diffusion of gases occurs at the respiratory membrane. Keep in mind that lung tissue is nourished by blood supplied from the systemic circuit and not the blood of the pulmonary circuit. The pulmonary circuit serves to exchange gases between the blood and alveoli of the lungs. [∞ p651]

Procedure

Identify the following **superficial structures** of the **lungs** using a *lung model* or *cadaver specimen*. In the torso (with chest plate removed), locate the right and left lungs within their respective pleural cavities. Notice that the lungs are cone-shaped. Use *Color Plates 68* to *73* and *Figures 24-7* to *24-11* for reference. Identify the pointed superior portion as the **apex** and the broad inferior portion that rests on the diaphragm as the **base**. Each lung presents three surfaces. Identify the **costal surface** as the convex lateral surface that conforms to the curve of the internal thoracic wall. The **mediastinal surface** can be identified as the irregular shaped surface containing the hilus region for the passage of all structures (pulmonary arteries and veins, nerves, lymphatics, and lobar bronchi) that must enter and exit the lungs. On the mediastinal surface of the right lung an impression or **groove for esophagus** can be identified. Identify on the mediastinal surface of the left lung both a pocket for the heart, termed the **cardiac notch**, and a **groove for aorta**. At the base portion of each lung, identify the inferior concave surface that rests upon the diaphragm as the **diaphragmatic surface**.

The lungs are subdivided into lobes, three in the right and two in the left. Each lobe is distinct and is separated by **fissures**. This arrangement permits the lungs to inflate to their maximum. Examine the right lung from the anterior and identify the **superior**, **middle**, and **inferior lobes**. Identify the **horizontal fissure** that divides superior and middle lobes. Observe the right lung from the lateral and identify the **oblique fissure** that separates superior and inferior lobes at an oblique angle from anterior to posterior. Examine the left lung from the anterior and identify the **superior** and **inferior lobe**, which are separated also by an oblique fissure.

Observe the right primary bronchus and its three divisions: superior, middle, and inferior lobar bronchi. Verify the left primary bronchus and its two divisions: superior and inferior lobar bronchi. Observe the branching pattern of the lobar bronchi within the lung tissue. The lobar bronchi of each lung divide into **segmental (tertiary) bronchi**.

See: Color Plates 68 to 73, Figs. 24-7, 24-8, 24-9, 24-10, 24-11

 ∞ *Fig. 24-8, p644*

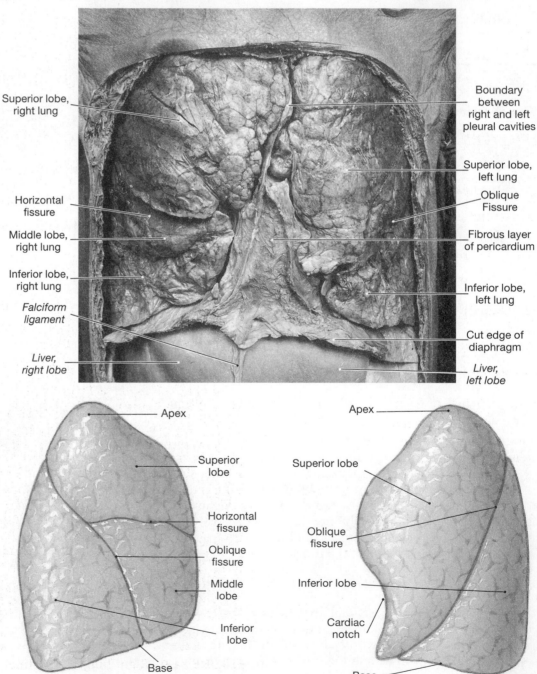

FIGURE 24-7 Thoracic viscera and views of right and left lungs, costal surface.

LOCATE

Viewed from Anterior and Lateral

____ Right/left lungs

____ Apex

____ Base

____ Surfaces $\begin{cases} \text{Costal} \\ \text{Mediastinal} \\ \text{Diaphragmatic} \end{cases}$

____ Cardiac notch

____ Right lung lobes $\begin{cases} \text{Superior} \\ \text{Middle} \\ \text{Inferior} \end{cases}$

____ Fissures $\begin{cases} \text{Horizontal (Rt. Lung only)} \\ \text{Oblique} \end{cases}$

____ Left lung lobes $\begin{cases} \text{Superior} \\ \text{Inferior} \end{cases}$

Viewed from Medial, Right Lung

____ Lobar bronchi $\begin{cases} \text{Superior} \\ \text{Middle} \\ \text{Inferior} \end{cases}$

____ Segmental (tertiary) bronchi

____ Hilus

____ Pulmonary $\begin{cases} \text{Arteries} \\ \text{Veins} \end{cases}$

____ Groove for esophagus

____ Fissures $\begin{cases} \text{Horizontal} \\ \text{Oblique} \end{cases}$

____ Cardiac impression

____ Diaphragmatic surface

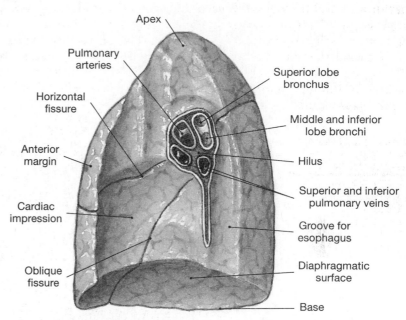

FIGURE 24-8 Right lung, mediastinal surface.

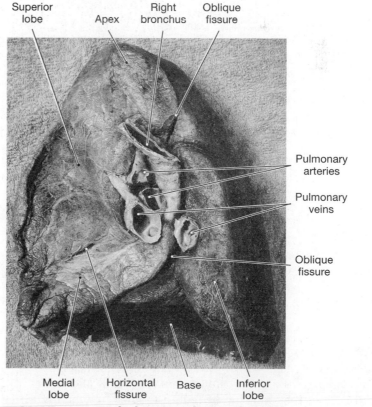

FIGURE 24-9 Right lung, mediastinal surface.

Viewed from Medial, Left Lung

_____ Apex

_____ Cardiac notch

_____ Hilus

_____ Pulmonary $\begin{cases} \text{Arteries} \\ \text{Veins} \end{cases}$

_____ Lobar bronchi $\begin{cases} \text{Superior} \\ \text{Inferior} \end{cases}$

_____ Segmental (tertiary) bronchi

_____ Groove for aorta

_____ Base

_____ Diaphragmatic surface

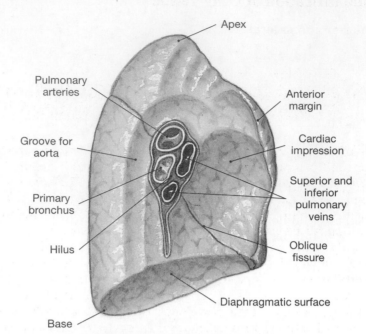

FIGURE 24-10   Left lung, mediastinal surface.

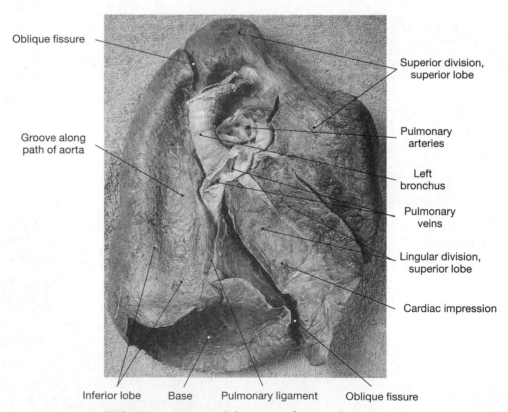

FIGURE 24-11   Left lung, mediastinal surface.

With the prosected cadaver in the supine position, remove the chest plate to observe the lungs in the thoracic cavity. You may be provided with a demonstration specimen of the human lung in lieu of examining the cadaver lungs. Examine the trachea, bronchi, and lungs in the same manner as described above. All of the structures described and identified on models may be identified on the prosected cadaver. Use _Color Plates 72, 73,_ and _97_ for reference.

Microscopic Identification of Lung Tissue

Procedure

✓ **Quick Check**

Before you begin to observe lung tissue, review the structure and appearance of small arteries and veins. (*Figure 22-1, pp315-316*)

Identify the following structures of **lung tissue** using the *microscope slide* provided, first by viewing under low power magnification and then under high power magnification. Scan the specimen with the 10x objective and locate an area containing both arteries and veins. Use *Figure 24-12* for reference. Identify the structure of a small artery and vein. Typically located in the same area will be a **segmental (tertiary) bronchus**. These bronchi have a structure similar to that of the trachea but have an abundance of smooth muscle fibers surrounding the cartilage. Identify **respiratory bronchioles** by the folded or wavy appearance of their cuboidal epithelium, which has only scattered cilia and no goblet cells, and by the smooth muscle fibers that encircle it. As bronchioles become smaller in diameter, epithelial cells change from columnar to cuboidal. Identify **alveolar ducts** as communicating spaces from respiratory bronchioles that end at **alveolar sacs**, which are common chambers connected to many individual **alveoli**. An alveolus is formed by a single layer of simple squamous epithelium that adheres to its basement membrane and lies back-to-back with the basement membrane of the endothelium of a capillary. Collectively these structures form the **respiratory membrane** through which oxygen and carbon dioxide diffuse. [∞ p651]

Slide # _____ Lung, Normal, sec.

See: Fig. 24-12

∞ *Fig. 24-11b,c p649*

LOCATE

_____ Arteriole/Venule

_____ Segmental (tertiary) bronchus

_____ Cartilage (hyaline) plate of segmental bronchus

_____ Epithelial cells (columnar or cuboidal) of respiratory bronchiole

_____ Alveolar duct

_____ Alveolar sac

_____ Alveolus (i)

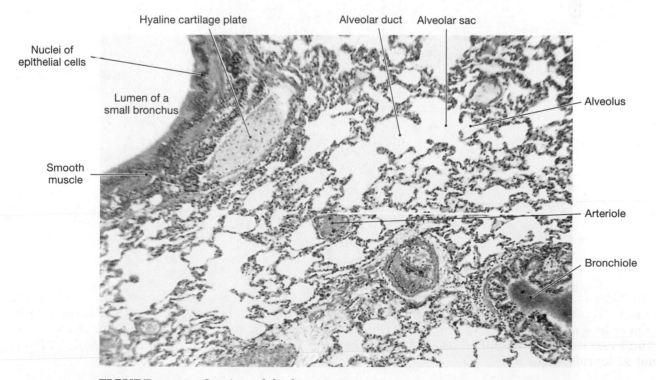

Nuclei of epithelial cells

Hyaline cartilage plate

Alveolar duct   Alveolar sac

Lumen of a small bronchus

Alveolus

Smooth muscle

Arteriole

Bronchiole

FIGURE 24-12   Section of the lung × 62.

Observation of the Bronchopulmonary Segments of the Lungs

Each lung lobe is divided into smaller functional units termed **bronchopulmonary segments**. Each segment is a structural and functional unit of lung tissue. The lungs contain a total of 20 bronchopulmonary segments. Each segment is named on the basis of the segmental bronchus that supplies it. [∞ p645]

Procedure

✓ **Quick Check** —————————————————————————————

Before you begin to identify the bronchopulmonary segments of the lungs, review the divisions of the primary bronchi. (*Color Plate 96*)

Identify the **bronchopulmonary segments** of the lungs using a *model*, *plastic cast*, *bronchogram* of the *bronchial tree*, or *dissected demonstration specimen*. Examination and identification of the bronchopulmonary segments of the lungs are performed typically on demonstration specimens. You may be provided with a demonstration specimen of the human lung(s). Check with your instructor before proceeding. Use the following identification pattern, *Color Plate 96* and *Figure 24-13* for reference to identify the divisions of the bronchi and bronchopulmonary segments of the lungs.

See: Color Plate 96, Fig. 24-13

∞ *Fig. 24-10, pp646-647*

LOCATE

___ Trachea

___ Primary bronchi

___ Lobar bronchi

Right Lung

___ Superior lobe
{ Apical
  Posterior
  Anterior

___ Middle lobe
{ Lateral
  Medial

___ Inferior lobe
{ Anterior basal
  Lateral basal
  Posterior basal
  Medial basal
  Superior

Left Lung

___ Superior lobe
{ Apical
  Posterior
  Anterior
  Superior lingular
  Inferior lingular

___ Inferior lobe
{ Superior
  Anterior basal
  Lateral basal
  Medial basal
  Posterior basal

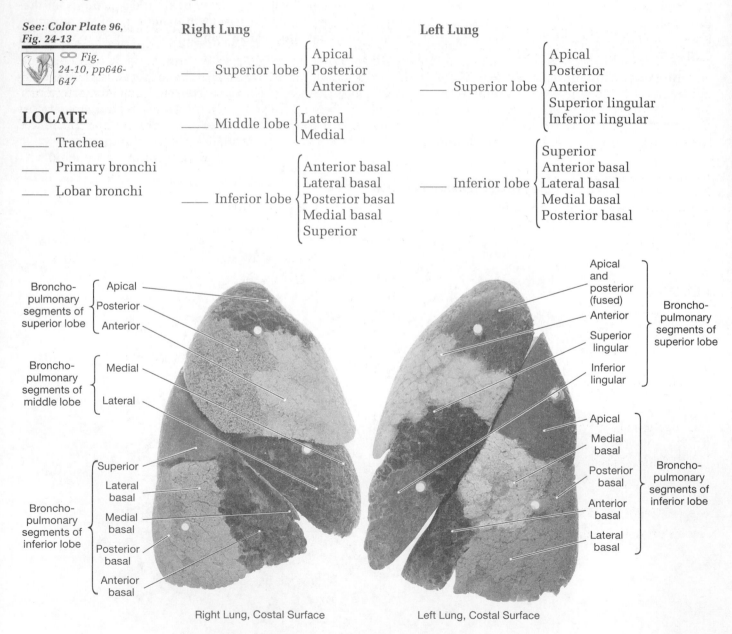

Bronchopulmonary segments of superior lobe
{ Apical
  Posterior
  Anterior

Bronchopulmonary segments of middle lobe
{ Medial
  Lateral

Bronchopulmonary segments of inferior lobe
{ Superior
  Lateral basal
  Medial basal
  Posterior basal
  Anterior basal

Apical and posterior (fused)
Anterior
Superior lingular
Inferior lingular
} Bronchopulmonary segments of superior lobe

Apical
Medial basal
Posterior basal
Anterior basal
Lateral basal
} Bronchopulmonary segments of inferior lobe

Right Lung, Costal Surface

Left Lung, Costal Surface

FIGURE 24-13 The bronchopulmonary segments.

Observation of the Respiratory Muscles

The process of **pulmonary ventilation**, or breathing, refers to the process of moving air into and out of the bronchial tree. [∞ p654] Moving air in is called *inhalation*, and moving air out is termed *exhalation*. Moving air is an active process that requires the participation of specific skeletal muscles. The muscles of respiration include the diaphragm, the external intercostals, the internal intercostals, and the scalenes. The major muscle involved in inhalation is the diaphragm. Its contraction flattens the thoracic cavity floor and increases the volume of the thoracic cavity. The external intercostal muscles elevate the ribs thereby increasing the volume of the thoracic cavity. To a lesser extent, the internal intercostals compress the ribs to decrease the volume of the thoracic cavity. During deep breathing, the four muscles of the abdominal wall participate. The act of normal quiet breathing is termed *eupnea* and the act of deep or forced breathing is called *hyperpnea*.

Breathing is maintained by the respiratory centers of the brain, which are located within the medulla oblongata and pons. [∞ p655] Three respiratory centers regulate breathing and control the degree of stretch of the lungs. Normal quiet rhythmic breathing is maintained by the *respiratory rhythmicity center* located in the medulla oblongata. The apneustic and pneumotaxic centers of the pons adjust the output of the rhythmicity center. The *pneumotaxic center* in the upper pons inhibits the apneustic center and promotes passive or active exhalation. The *apneustic center* in the lower pons provides continual stimulation to the inspiratory regions of the respiratory rhythmicity center. [∞ Figure 24-15, p656]

Procedure

Identify the **muscles of respiration** using a *torso model* or *cadaver specimen*. These muscles include the oblique and rectus groups and are identified in *Chapters 10* and *11*. Verify the muscles of the oblique group, which includes the scalenes, external intercostals, internal intercostals, external oblique, internal oblique, and transverse abdominis. Verify the muscles of the rectus group, which consists of the rectus abdominis and diaphragm. Even though the serratus anterior usually moves the shoulder girdle and the sternocleidomastoid usually flexes or turns the neck, during periods of forced breathing (such as running), these muscles help respiratory muscles to enlarge the thoracic cavity. The rectus abdominis and the sternocleidomastoid are not considered muscles of respiration. Refer to *Color Plates 34, 37, 41, Table 10-8, p122*, and *Figure 10-10, p124*.

See: Color Plates 34, 37, 41, Figs. 24-14, 24-15

∞ *Figs.*
10-13, p281
10-14, p282
24-14, p655

LOCATE

____ Scalenes { Anterior / Middle / Posterior

____ Intercostals { External / Internal

____ Obliques { External / Internal

____ Rectus abdominis

____ Diaphragm

____ Serratus anterior

____ Sternocleidomastoid

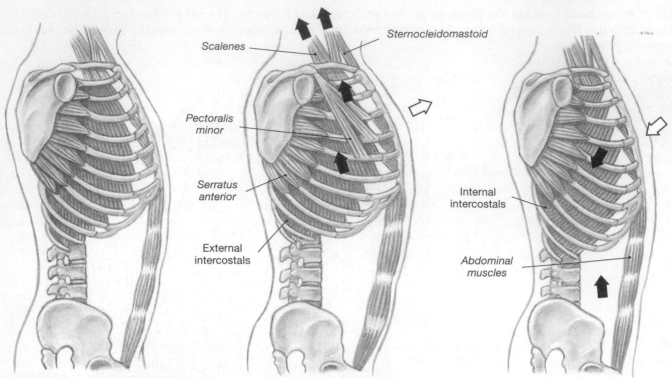

FIGURE 24-14 Respiratory muscles.

AT REST

INHALATION

EXHALATION

Scalenes

Sternocleidomastoid

Pectoralis minor

Serratus anterior

External intercostals

Internal intercostals

Abdominal muscles

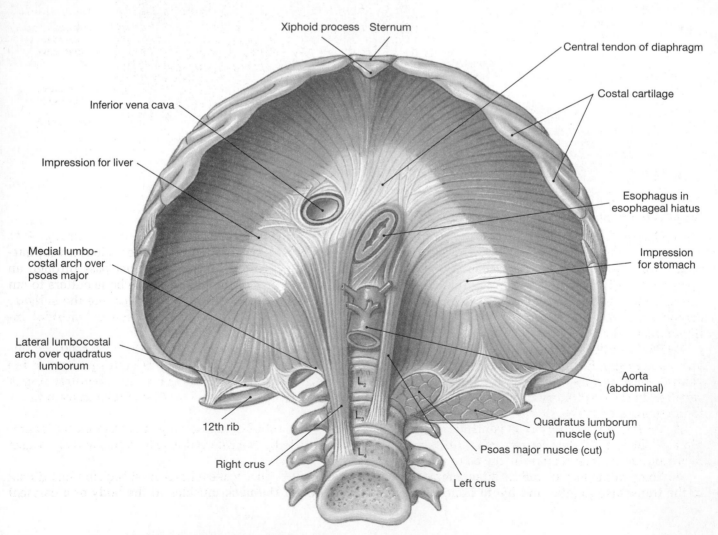

Xiphoid process   Sternum

Central tendon of diaphragm

Costal cartilage

Inferior vena cava

Esophagus in esophageal hiatus

Impression for liver

Impression for stomach

Medial lumbo-costal arch over psoas major

Lateral lumbocostal arch over quadratus lumborum

Aorta (abdominal)

Quadratus lumborum muscle (cut)

12th rib

Psoas major muscle (cut)

Right crus

Left crus

L₂

L₃

L₄

FIGURE 24-15 The diaphragm.

Dissection of the Respiratory System in the Cat

This exercise complements the study of the human respiratory system. The cat respiratory system is similar in structure to the human system. In your dissection of the cat respiratory system, trace the path of air through the respiratory tract.

See: Figs. 21-8, 24-16

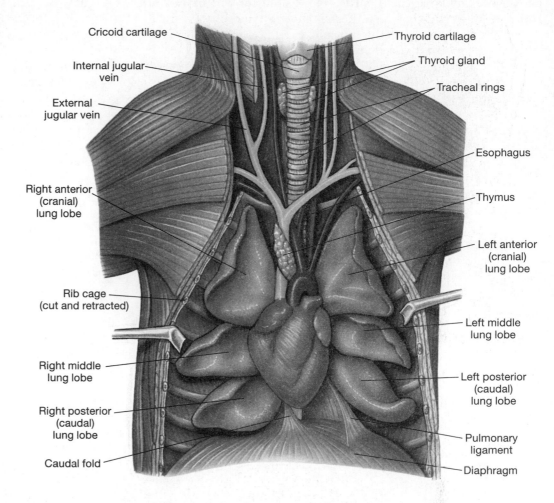

Cricoid cartilage
Internal jugular vein
External jugular vein
Right anterior (cranial) lung lobe
Rib cage (cut and retracted)
Right middle lung lobe
Right posterior (caudal) lung lobe
Caudal fold
Thyroid cartilage
Thyroid gland
Tracheal rings
Esophagus
Thymus
Left anterior (cranial) lung lobe
Left middle lung lobe
Left posterior (caudal) lung lobe
Pulmonary ligament
Diaphragm

FIGURE 24-16   Cat respiratory system.

Respiratory System

1. Observe the paired **external nares**. Air enters the **nasal cavity** via these openings and passes through it to the **internal nares (choanae)**. Air continues into the **pharynx,** which is partitioned into three regions: **nasopharynx, oropharynx,** and **laryngopharynx**, before it reaches the cartilaginous **larynx**. Air must pass through an opening, the **glottis,** prior to entering the larynx. To observe the regions of the pharynx, use bone cutters to cut through the angle of the mandible on one side only. Pull the jaw toward the other side, leaving the salivary glands intact on the uncut side. Carefully, use a scalpel to cut the soft tissues (connective tissue and muscle) until you reach the pharynx.

2. Four cartilages compose the larynx. The **epiglottis** is the flap-like cartilage covering the opening to the larynx, the **glottis,** during swallowing. The **thyroid cartilage** is the large prominent ventral cartilage deep to the ventral neck muscles. Caudal to the thyroid cartilage is the **cricoid cartilage**, which is the only complete ring of cartilage in the respiratory tract. The paired **arytenoid cartilages** occupy the dorsal surface of the larynx anterior to the thyroid cartilage.

3. Posteriorly, the larynx is continuous with the **trachea**. This tubular passageway is supported by incomplete, C-shaped cartilaginous rings and bordered on its dorsal side by the internal jugular veins and the vagus *nerve* and on its lateral sides by the common carotid arteries.

4. Remove the larynx and its attachments, the trachea and esophagus. Make a horizontal incision just dorsal to the transverse jugular and hyoid bone. Cut completely through the neck muscles to the body of a cervical

vertebra. Carefully cut through any remaining connective tissue that may be still securing the larynx. Care must be taken not to cut the common carotid arteries or vagus nerves. Pull the larynx with its attachments out of the neck. Identify the trachea and the collapsed tube, the *esophagus.* Probe the esophagus. Make a median incision on the dorsal surface of the larynx. Open the larynx to expose the vocal cords between the thyroid and arytenoid cartilages. The **false vocal cords** are the anterior pair of mucous membranes; the **true vocal cords** are the posterior pair. The space between these is the glottis. Probe the trachea through the glottis.

5. The trachea enters the **thoracic cavity** between the left and right **lungs**. At its posterior end, the trachea bifurcates to form, left and right **primary bronchi.**

6. The left lung is divided into three lobes; **cranial** (anterior), **middle**, and **caudal** (posterior).

7. The right lung is divided into four lobes. The large fourth lobe, termed **accessory** is located dorsal to the right caudal lobe.

8. In the thoracic cavity, two *pleural cavities* contain the lungs. Each lung is attached to other structures at its root. The **root** of the lung is formed by a bronchus, pulmonary artery, pulmonary veins, bronchial arteries and veins, nerves and lymphatic vessels. Collectively, they are covered by the serous membranes, termed **pleura**.

9. Each lung lobe is wrapped by the **visceral pleura**, and the thoracic cavity lined by the **parietal pleura**. These membranes can be observed as the glistening surface of the lungs and interior surface of the thoracic cavity. Between these layers is a potential space, the **pleural cavity**, which contains *pleural fluid* secreted by the pleura.

10. The floor of the thoracic cavity is formed by a skeletal muscle, the **diaphragm**. The diaphragm is a major muscle of respiration.

Anatomical Identification Review

1. Label the structures of the respiratory system on the accompanying diagram. As you label each structure, mentally define it.

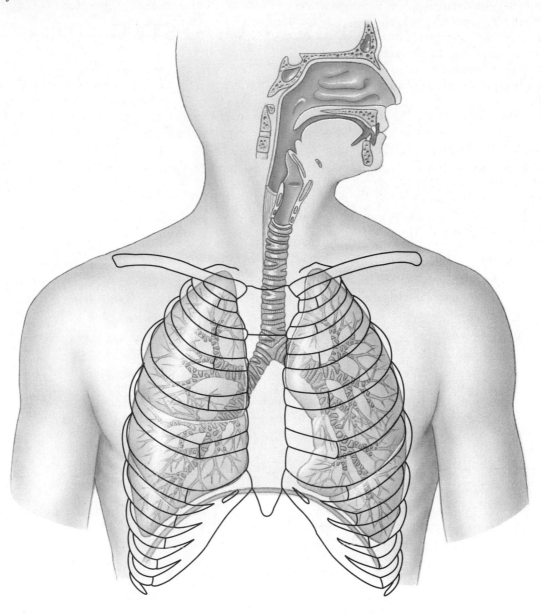

To Think About

1. If a child swallows a foreign object, and it lodges in the respiratory tract, where is it most likely to be found, and what symptoms are most likely to occur?

2. What is the significance of the hard and soft palates that separate the nasal cavity from the oral cavity?

3. What changes occur during vigorous exercise regarding breathing?

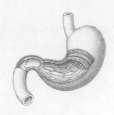

25

THE DIGESTIVE SYSTEM

Objectives

1. Summarize the functions of the digestive system.

2. Locate, identify, and list the functions of each structure that composes the digestive system.

3. Identify and describe the histological organization of the four layers in the digestive tube wall.

4. Observe and describe the serous membranes of the digestive tract.

5. Identify and describe both the gross and microscopic anatomy of the digestive system organs: tongue, teeth, pharynx, esophagus, stomach, small intestine, and large intestine.

6. Locate and identify the gross anatomy of these digestive system accessory organs, identify their microscopic anatomy, and describe their functional role in digestion: salivary glands, liver, gallbladder, and pancreas.

The **digestive system** is designed to mechanically and chemically process the foods that we eat. It absorbs the nutrients contained in the food and eliminates as waste all undigested materials. The structures of the digestive system perform collectively the following major functions: (1) the ingestion of foods and liquids, (2) the mechanical processing of solid foods (such as the grinding of food by the teeth and mixing with the secretions of the oral cavity by the tongue), (3) the chemical breakdown of foods via the process of enzymatic digestion (such as occurs in the stomach and small intestine), (4) the absorption and processing of the nutrients released from chemical breakdown of food, and (5) the compaction and elimination of undigested and non-digestible materials. [∞ p663]

Structures of the Digestive System

The digestive system consists of a hollow muscular tube that has pouches, a coil, and bends along its length. Additionally, it has *accessory organs* that are connected by ducts to this tube. Accessory organs include the *salivary glands, liver, gallbladder*, and *pancreas*. The accessory organs manufacture, store, or secrete fluids that contain water, enzymes, buffers, and other components that assist in preparing organic and inorganic nutrients for absorption. [∞ p663] The digestive system structures can be summarized as the following:

1. **Oral (buccal) cavity** contains the *teeth*, the *tongue*, and openings of ducts from salivary glands; it is the first portion of the digestive system. Grinding, moistening, and mixing of foods with secretions occur in the oral cavity. Taste, temperature, and even the texture of foods are analyzed in this region of the digestive tract.

2. **Salivary glands** secrete a lubricating and moistening fluid, termed *saliva*, that contains a digestive enzyme that begins the chemical breakdown of complex carbohydrates into disaccharide sugars.

3. **Pharynx** is a space that serves as a common passageway for solid food, liquids, and air. In the digestive system, it connects the oral cavity to the esophagus.

4. **Esophagus** transports food and liquids from the pharynx to the stomach. No digestion occurs in the esophagus.

5. **Stomach** performs several major functions: bulk storage of ingested food and liquids during the initial stages of digestion, the mechanical breakdown of ingested food, initiation of the chemical breakdown of food through the action of acid and enzymes, and the production of intrinsic factor.

6. **Small intestine** is the site of most of the enzymatic digestion and the absorption of water, nutrients, vitamins, and ions.

7. **Large intestine** functions primarily to remove water (dehydration), to absorb important vitamins produced by bacteria, and compact and store the indigestible and unabsorbable materials for elimination as *fecal matter*.

8. **Rectum** temporarily stores the fecal matter waste product prior to elimination.

9. **Pancreas** exocrine cells secrete buffers and digestive enzymes; endocrine cells secrete hormones [∞ Chapter 19, p519].

10. **Liver** secretes bile (required for emulsification and subsequent digestion of fats), stores nutrients and vitamins, and has many metabolic and regulatory functions.

11. **Gallbladder** stores and concentrates bile that it receives from the liver. Under hormonal and nervous system control, bile drains from the gallbladder into the small intestine.

Procedure

Identify the structures that compose the **digestive tract** and **accessory organs** on the *torso model*. Begin by removing the chest plate and the detachable head/neck section from the intact torso. Head and neck structures are now viewed in mid-sagittal section. Next remove the heart and lungs from the thoracic cavity. Use *Color Plates 99* and *103* for reference. Identify the **oral (buccal) cavity**, **salivary glands**, **esophagus**, **stomach**, **small intestine**, **large intestine**, and **rectum.** Now identify the accessory organs; **liver**, **gallbladder**, and **pancreas**.

Review the **abdominal regions** on the torso or cadaver using *Color Plate 103* and *Figure 25-1*. It is important to have a mental picture of the shape, size, and organization of the digestive organs within their respective abdominal regions, because descriptions of their location and relationships are based upon this regional terminology.

See: Color Plates 99, 103, Fig. 25-1

∞ *Figs.*
25-1, p664
25-12, p680

LOCATE

____ Oral (buccal) cavity

____ Salivary glands

____ Esophagus

____ Stomach

____ Small intestine

____ Large intestine

____ Rectum

____ Liver

____ Gallbladder

____ Pancreas

Abdominal Regions

____ Right/left hypochondriac regions

____ Epigastric region

____ Right/left lumbar regions

____ Umbilical

____ Right/left iliac regions

____ Hypogastric region

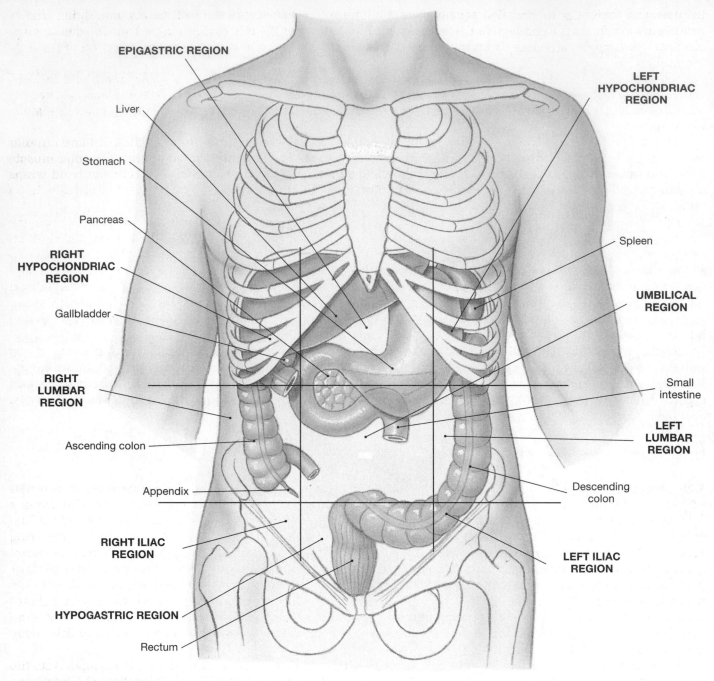

EPIGASTRIC REGION

LEFT
HYPOCHONDRIAC
REGION

Liver

Stomach

Pancreas

RIGHT
HYPOCHONDRIAC
REGION

Spleen

Gallbladder

UMBILICAL
REGION

RIGHT
LUMBAR
REGION

Small
intestine

Ascending colon

LEFT
LUMBAR
REGION

Appendix

Descending
colon

RIGHT ILIAC
REGION

LEFT ILIAC
REGION

HYPOGASTRIC REGION

Rectum

FIGURE 25-1 Abdominal regions.

Microscopic Identification of the Histological Organization of the Digestive Tube

The digestive tract is organized as a four-layered tube. [∞ p663] These layers are continuous along the entire length of the tube, but are modified within specified organs reflecting the special digestive activities that occur in these organs. These structural modifications determine the function of each of these digestive organs. Modifications create a division of labor among the digestive organs, with each organ playing a specific role in the overall digestive process. The four layers (tunics) of the digestive tube are best identified from inside to out.

Mucosa is the innermost layer that lines the lumen of the digestive tube. This layer is composed of an epithelium moistened by glandular secretions and sometimes organized into pleats and folds. Folds increase the surface area for digestion and absorption. They also permit expansion of the diameter of the digestive tube to accommodate foods for their passage through the tube lumen. The mucosa of the oral cavity and esophagus has a stratified squamous epithelium. It changes to columnar epithelium in the stomach, and small and large intestines, and then back to stratified squamous epithelium in the last part of the rectum. An underlying lamina propria always forms the connective tissue basement foundation for the mucosa. A narrow band of smooth muscle fibers, muscularis mucosae, is located in most areas of the digestive tract along the outer portion of the lamina propria.

Submucosa is the layer that contains dense connective tissue that serves to support the blood vessels, nerves, and lymphatics that service each structure of the digestive tube. In some regions, the submucosa has exocrine glands that secrete products into the lumen of the digestive tract to aid in digestion (e.g., mucous glands, esophagus).

Muscularis externa is the layer that contains smooth muscle fibers organized into longitudinal and circular bands for the movement of foodstuffs through the digestive tube. The longitudinal band forms the outer muscle layer and runs the length of the tube. When contracted, a portion of the tube shortens. The circular band wraps around the tube and forms the inner muscle layer. Contracting this band causes the diameter of the tube to be constricted. In the stomach only, an oblique band is the innermost smooth muscle layer.

Under the control of the ANS, the contractions of these smooth muscle bands promotes movement of chyme through the digestive tube and compartmentalizes it. It is the arrangement of smooth muscle layers, the contractile properties of smooth muscle cells, and control by the ANS that produces slow sustained contractions known as **peristalsis.** Peristalsis occurs because of the coordinated contractions of longitudinal and circular muscle layers, which results in the slow movement of contents through the lumen of the digestive tract. A more specialized movement, termed **segmentation**, occurs within the small and large intestines to mix foods with digestive secretions. Segmentation movements predominantly involve the circular muscle layer, serving not only to mix chyme but also to fragment and compact it.

Serosa is the layer that is the visceral peritoneum. The peritoneum is a serous membrane consisting of two portions: a visceral layer that covers each abdominal organ and a parietal layer that lines the abdominal cavity. The peritoneal fluid formed on the surfaces of these membranes prevents friction between the body walls and organ surfaces. A connective tissue layer of collagen fibers, termed the **adventitia**, attaches the oral cavity, esophagus, and rectum to surrounding structures.

Procedure

Identify the histological organization of the **digestive tube** using the *microscope slide* provided. View under a dissecting microscope, then under low power magnification of the compound microscope. Observing a cross section of the small intestine under the dissecting microscope will aid in the identification of the histological organization of the digestive tube. Locate from inside to out the four layers of the digestive tube using *Figure 25-2* for reference. Observe the lumen and general structure of the small intestine. Identify the dark-staining **mucous epithelium** that forms part of the **mucosa** layer. Note how the mucosa is arranged into folds (**plica**) or pleats. Identify the **submucosa** layer by the dense connective tissue that encircles the submucosa. Identify the **muscularis externa** layer by the light-staining bands of smooth muscle fibers that appear to encircle the submucosa. The **serosa** layer can be identified by the appearance of loose connective tissue that is attached to the muscularis externa. In the preparation of this microscope slide, the serosa layer often appears either broken away from the muscularis or in fragments.

View each of these layers in the same manner with the 10x objective of the compound microscope. Note the simple columnar epithelium and goblet cells that compose the mucous epithelium. (*Color Plate 6*) Underlying the epithelium, identify both the lamina propria as the loose connective tissue layer deep to the epithelium, and the **muscularis mucosae** as the two thin layers of smooth muscle fibers deep to the lamina propria. (*Color Plate 20*) These muscle fibers aid in the constricting of the lumen. Note the numerous small arteries and veins that characterize the submucosa layer. Two layers of smooth muscle fibers form the muscularis externa layer. Identify the inner circular layer and the outer longitudinal muscle layers. The loose connective tissue and mesothelium of the serosa can be distinguished clearly from the muscularis.

Slide # _____ Small Intestine, sec.

See: Color Plates 6, 20, Fig. 25-2

∞ *Fig. 25-2, p665*

LOCATE

_____ Mucosa
{ Mucous epithelium
Lamina propria
Muscularis mucosae

_____ Plica (mucosal fold)

_____ Submucosa

_____ Muscularis externa
{ Circular muscular layer (inner)
Longitudinal muscle layer (outer)

_____ Serosa

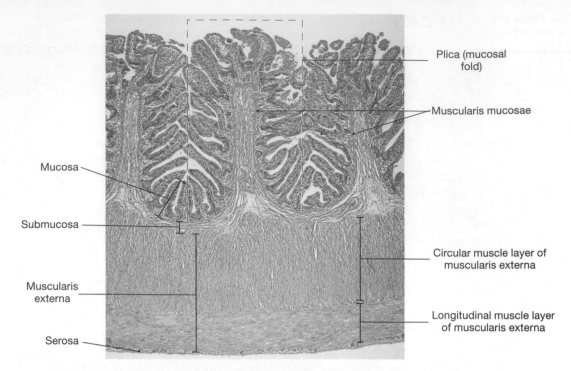

Mucosa

Submucosa

Muscularis
externa

Serosa

Plica (mucosal
fold)

Muscularis mucosae

Circular muscle layer of
muscularis externa

Longitudinal muscle layer
of muscularis externa

FIGURE 25-2 General histological organization of the ileum × 160.

Observations of the Serous (Peritoneal) Membranes

Segments of the digestive tract are suspended within the peritoneal cavity by the visceral peritoneum (serosa) covering these segments. Laminating the visceral and parietal peritoneal membranes together creates the **serous membranes**. [∞ p667] Serous membranes are named according to the location and organs suspended. For example, the serous membrane connected with the large intestine is termed *mesocolon*. The serous membrane of the small intestine, termed **mesentery**, serves to attach it to the posterior body wall and to connect its coils together thereby preventing strangulation by the twisting of blood vessels, lymphatics, and nerves that supply the small intestine.

Procedure

Identify the **serous membranes** that support the digestive tract and attach the visceral organs to the posterior peritoneal wall using a *torso model* or *cadaver specimen*. Use *Color Plates 104, 105, 110* and *Figure 25-3* for reference. Note, a torso model will not depict all of the serous membranes; only a cadaver specimen contains all serous membranes.

Identify the serosa lining of the abdominal wall as the *parietal peritoneum*. Observe the fold of serous membrane between the liver and the stomach as the **lesser omentum**. (Typically not present on models.) Locate the greater curvature of the stomach and identify the **greater omentum** as the curtain of serous membrane with fat that drapes over the intestines. (Only a portion is depicted on models.) The greater omentum is a large fold of the dorsal mesentery of the stomach that drapes and hangs anterior to the intestines. "Pot bellies" are the result of the accumulation of large amounts of fat within the greater omentum. Recall that the abdominal wall is formed only by skin and four layers of muscle. It does not contain any bony areas for protection (except for L_{1-5} vertebrae and pelvis). Both omenta serve to protect the abdominal organs and provide a storage area for fat until it is required for metabolism.

Locate the diaphragm and liver. Identify the **falciform ligament** as the serous membrane that suspends the liver from the anterior abdominal wall and inferior surface of the diaphragm. It appears as a white line on the anterior surface of the liver, dividing it into right and left lobes. Trace the ligament superiorly and note that it splits. Identify the right division as the **coronary ligament** of the liver (appears as a white line on all models). (*Color Plates 116 and 118*) In life, the coronary ligament encases the right lobe of the liver. Locate the small intestine and identify the **mesentery proper**, which suspends the small intestine from the posterior body wall via the **root of the mesentery proper**. Locate the large intestine and identify that portion that passes in a transverse plane as the *transverse mesocolon*. The **mesocolon** suspends this portion of the large intestine at its dorsal surface to the posterior body wall. The remaining portions are also suspended by mesocolon and are described with the large intestine.

LOCATE

___ Parietal peritoneum

___ Lesser omentum

___ Greater omentum

___ Falciform ligament

___ Coronary ligament of liver

___ Mesocolon

___ Mesentery (proper) of the sm
intestine

___ Root of the mesentery proper

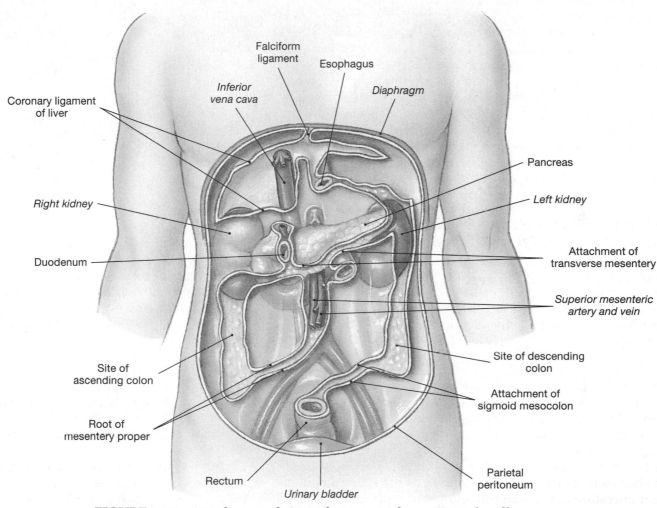

FIGURE 25-3 Attachment of visceral organs to the peritoneal wall.

It is best to observe the serous membranes in the prosected cadaver specimen during your study of each digestive organ. After you have identified the structure of each digestive organ, return to this text section and follow the above identification directions. Use *Color Plates 104, 105, 109* and *110* for reference. Check with your instructor before proceeding.

Observation of Oral (Buccal) Cavity Structures and Pharynx

As described earlier, the oral (buccal) cavity contains the teeth, openings of ducts to the salivary glands, and tongue. It is the first part of the digestive system and is lined by the **oral mucosa**, which is a stratified squamous epithelium. [∞ p667] The oral cavity is continuous with the *pharynx*, which is the common passageway for foods, liquids, and air. [∞ p637]

ify the following structures of the **oral (buccal) cavity** using your *lab partner's body*, *torso model*, and *daver specimen*. View through the open mouth the structures of the oral cavity on your partner or use a *mirror* to observe them on yourself. Use *Color Plate 92*, *Figures 25-4* and *25-5* for reference. Identify the h, floor, and roof of the oral cavity. The floor contains the tongue and the openings of the ducts of most of the alivary glands. Differentiate between the **hard** and **soft palates** by running the tip of your tongue over the roof of the oral cavity. The palatine process of the maxillary bone and the palatine bone form the hard palate. The soft palate begins immediately posterior to the palatine bone and can be felt as the "soft" portion of the roof. Identify a pendulous extension in the midline, the **uvula,** at the posterior of the soft palate. Identify the **palatopharyngeal arches** as muscular arches of the soft palate, which frame the oral cavity and form a boundary between it and the oropharynx. Posterior to the arches a pair of *palatine tonsils* can be identified. These tonsils are typically visible. If not, they may have been surgically removed.

The majority of the space in the oral cavity is occupied by the **tongue**. Recall that the muscular tongue is covered by stratified squamous epithelium. Observe the superior surface of the tongue, the **dorsum,** and note the numerous papillae (*Figure 18-5, p257*) that appear only on this surface. Identify the portion of the tongue anterior to the palatoglossal arch as the **body**, and the portion located posterior to the arch as the **root**. In the midline, on the inferior surface of the tongue, identify a fold of mucous membrane, the **lingual frenulum**, which secures the tongue to the floor of the oral cavity. Identify on the floor lateral to the frenulum the *openings of submandibular ducts*. These ducts transport saliva from the *submandibular glands* to the floor of the oral cavity.

Remove the detachable head/neck section from the intact torso. Head and neck structures are viewed now in mid-sagittal section. Observe all of the above described structures from a sectional view (submandibular ducts not visible). The relationship between oral and nasal cavities is observed easily from this view. Review the divisions of the pharynx: nasopharynx, oropharynx, and laryngopharynx. Identify the lateral walls of the oral cavity as the **cheeks**, and at the anterior wall, the **lips**, which create or close the opening to the oral cavity (mouth). Between the cheeks, lips, and teeth is a space, the **vestibule** (your toothbrush enters this space). The oral mucosa that covers the alveoli and surrounds part of each tooth as gums can be identified as the **gingivae**. In the vestibule, at the level of the last (2nd or 3rd) upper molars, the **openings of the parotid ducts** are easily identified. These ducts carry saliva from the *parotid glands* to the vestibule of the oral cavity.

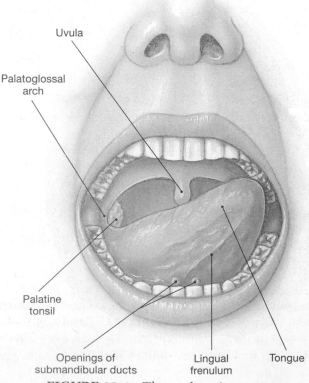

FIGURE 25-4 The oral cavity as seen through the open mouth.

See: Color Plate 92,
Figs. 25-4, 25-5

 ∞ *Fig.*
25-5, p669

LOCATE

Viewed through the Open Mouth

_____ Teeth

_____ Hard/soft palates

_____ Uvula

_____ Palatoglossal arches

_____ Palatine tonsils

_____ Tongue { Dorsum / Body / Root

_____ Papillae

_____ Lingual frenulum

_____ Openings of submandibular ducts

Viewed in Sagittal Section

____ All of the above structures may be seen in section, except submandibular ducts

____ Nasal/oral cavities

____ Pharynx { Nasopharynx
Oropharynx
Laryngopharynx

____ Cheek (oral mucosa)

____ Lips

____ Vestibule

____ Gingiva (gums)

____ Opening to parotid duct

____ Entrance to auditory tube

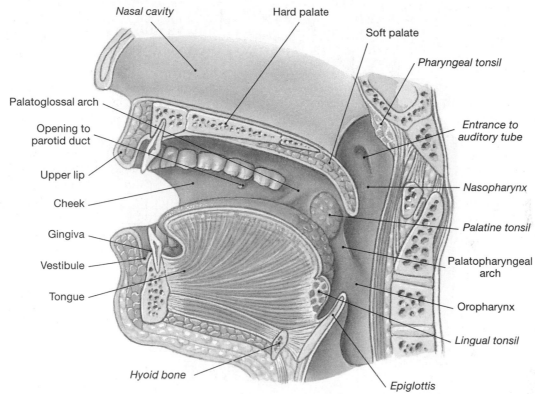

FIGURE 25-5 The oral cavity, sagittal section.

Observation of the Salivary Glands

Three salivary glands produce and secrete *saliva* into the oral cavity. Saliva initiates the digestive function of the chemical breakdown of starch, lubricates and moistens food so that tasting and swallowing can occur, and dilutes acidic foods (such as eating salad marinated in vinegar and oil dressing). [∞ p672]

Procedure

Identify the **salivary glands** and their **associated ducts** using a *torso head model* or *cadaver specimen*. Use *Color Plates 35, 37, 39* and *Figure 10-7, p118* for reference. A pair of **parotid glands** can be identified just inferior to the zygomatic, anterior to the ear openings, and overlying the posterior region of the masseter muscles. A **parotid duct** may be identified on the surface of the masseter muscle and traced into the oral cavity where it opens near the 2nd or 3rd upper molar teeth. The duct transports saliva from the gland to the oral cavity. A pair of **submandibular glands** can be identified inferior to the mucous membrane floor of the oral cavity at the mandibular angle and along its inner surface. Locate the **openings of the submandibular ducts** on either side of the frenulum. Identify a collection of **sublingual glands** located inferior to the mucous membrane floor of the oral cavity, positioned in the midline, inferior to the body of the tongue. Numerous **sublingual ducts** can be identified opening lateral (on either side) to the frenulum.

See: Color Plates 35,
37, 99, Fig. 10-7

∞ Fig.
25-6, p672
10-4, p270
10-11, p277

LOCATE

____ Parotid gland

____ Parotid duct

____ Submandibular glands

____ Submandibular ducts

____ Sublingual gland

____ Sublingual ducts

In the prosected cadaver, all of the salivary glands may be observed. Use *Figure 10-7, p118* for reference. If not damaged during dissection, the parotid duct can be identified on the surface of the masseter muscle. The parotid duct often bifurcates into several accessory ducts. The sublingual glands and their ducts are difficult to observe in most specimens.

Microscopic Identification of the Salivary Glands

Procedure

Identify the histological organization of the **salivary glands** using the *microscope slide* provided, first by viewing under low power magnification and then under high power magnification. Use *Figure 25-6* for reference. Scan your slide under low power magnification to observe both the mixture of cells that form the gland and the numerous blood vessels and ducts associated with it. Under high power, identify the abundant darker-staining **serous cells**. These cells appear in groups and produce a thin watery fluid containing the enzyme salivary amylase that begins the digestion of starches. Identify the faint-staining cells that appear to contain lipid droplets as **mucous cells.** These cells secrete mucins that serve to bind food together for swallowing. A network of **ducts** leads these products out of the gland. Ducts are identified easily both by their lumens and by cuboidal cells that form the epithelial lining of the duct wall. Distinguish arteries and veins from ducts.

Slide # ____ Submandibular (or Parotid), sec.

See: Fig. 25-6

∞ Fig.
25-6b, p672

LOCATE

____ Serous cells

____ Mucous cells

____ Ducts

____ Blood vessels

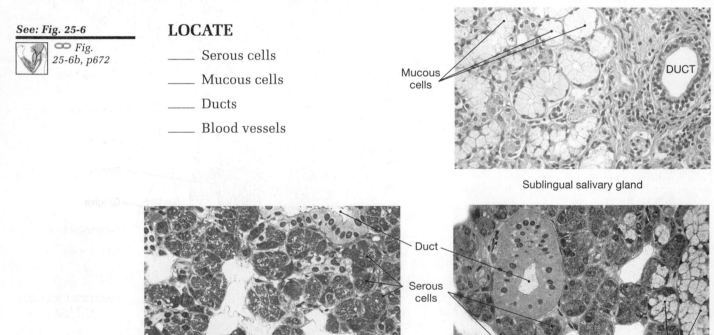

FIGURE 25-6 Sublingual, parotid (× 316), and submandibular (× 303) salivary glands.

Observation of Teeth (Gross Structure)

All of the teeth are located in the oral cavity. The act of chewing, termed *mastication*, is the major function of teeth. Teeth occur in an assortment of sizes and shapes to perform this function. The muscles of mastication (masseter, pterygoids, and temporalis muscles) close the jaws and cause the rocking and grinding motions associated with chewing. The major portion of a tooth is a non-cellular mineral matrix, termed **dentin**. Dentin surrounds the cellular or living portion of a tooth, called the **pulp cavity**. [∞ p673]

Procedure

Identify the structure of a **typical tooth** and the **types** of **teeth** using *mandible/jaw* with *teeth*, and *tooth models*. On a mandible model with teeth, observe the appearance and orientation of the teeth. Use *Color Plates 100, 101*, and *Figure 25-7* for reference. Each tooth is contained in an alveolar socket. Begin at the anterior of the jaw and proceed posteriorly to identify by shape the four types of teeth: **incisors** as blade-shaped, **cuspids** (canines) as sharp and conical shaped, **bicuspids** (premolars) by their large flat grinding surface, and **molars** as a very large tooth with a flat grinding surface.

To observe both the external and internal structures of a typical tooth, use a model that shows a sectional view of the tooth. Use *Color Plate 101* and *Figure 25-8* for reference. Identify the visible portion of a tooth as the **crown**, the portion that is contained in the alveolar socket as the **root**, and the region between the two as the **neck**. Identify the gum indentation surrounding the neck of the tooth as the **gingival sulcus**. It aids in sealing the tooth within the alveolar socket and preventing the entry of food and bacteria. Locate the **periodontal ligament** (membrane) which connects the root of the tooth to the alveolar socket.

See: Color Plates 100, 101, Figs. 25-7, 25-8

 ∞ *Fig. 25-7, p674*

LOCATE

____ Incisors (medial/lateral)

____ Cuspids (canines)

____ Bicuspids (premolars)

____ Molars

____ Crown

____ Root

____ Neck

____ Enamel

____ Dentin

____ Pulp cavity

____ Gingiva

____ Gingival sulcus

____ Periodontal ligament (membrane)

____ Bone of alveolus

____ Root canal

____ Apical foramen

____ Branches of alveolar vessels and nerve

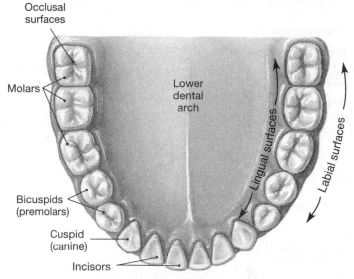

FIGURE 25-7 The adult lower jaw teeth.

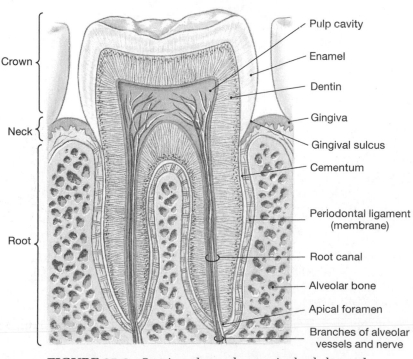

FIGURE 25-8 Section through a typical adult tooth.

Observation of the Esophagus

The esophagus is a hollow, muscular tube that transports the swallowed food bolus between the pharynx and stomach. It lies posterior to the heart and passes through the diaphragm at the esophageal hiatus to join with the superior portion of the stomach. [∞ p676]

Procedure

Identify the **esophagus** within the thoracic and abdominal cavities, using a *torso model* and *cadaver specimen*. In the torso (chest plate must be removed, heart and lungs removed), verify the laryngopharynx, trachea, aortic arch, and descending thoracic aorta. Use *Color Plates 77* and *99* for reference. Identify the **esophagus** posterior to the heart, great vessels, and trachea. Trace the esophagus from its origin at the level of the cricoid cartilage as it descends along the bodies of the thoracic vertebrae to the diaphragm. Observe the point, termed **esophageal hiatus**, where the esophagus perforates the diaphragm to enter the abdominal cavity. The esophagus joins to and opens into the superior portion of the stomach. In addition to the esophagus, right and left vagus nerves also pass through the hiatus. Esophageal arteries supply blood to the esophagus and esophageal veins drain blood from the esophagus into the azygos vein. The azygos vein collects blood from the thorax and drains into the superior vena cava.

LOCATE

See: Color Plates 77, 99, 108

∞ *Figs.*
22-16, p589
22-23, p597
25-10, p678
25-11, p679
24-3d, p639

____ Laryngopharynx

____ Trachea

____ Esophagus

____ Diaphragm

____ Esophageal hiatus

____ Stomach

____ Vagus nerves

____ Blood vessels { Esophageal a./v. Azygos v.

In the prosected cadaver, the esophagus appears as a collapsed muscular tube. Use the above directions to observe the esophagus. Use *Color Plate 108* for reference. To examine the esophagus, retract the heart and left lung. Verify the left and right vagus nerves lateral to the esophagus in the thoracic cavity. From the heart, the left vagus crosses the left side of the aortic arch and continues to descend posterior to the left bronchus, pulmonary artery, and pulmonary vein. At this point, the vagal fibers spread out over the surface of the esophagus as the **esophageal plexus**. Just superior to the diaphragm, these fibers (anterior and posterior) merge to form the **vagal trunks**. The **left phrenic nerve** crosses the left side of the aortic arch but descends anterior to the left bronchus. The right vagus can be located on the dorsal surface of the esophagus. To observe the esophagus and its relation to the stomach, reflect the abdominal muscles and liver to the right. Pull the stomach inferiorly and inspect the esophagus as it perforates the esophageal hiatus. Again, verify the positions of vagus nerves adjacent to the esophagus. Small esophageal arteries and veins and the large azygos vein may be observed.

Microscopic Identification of the Esophagus

Procedure

Identify the histological organization of the **esophagus** using the *microscope slide* provided. View first under a dissecting microscope, then under low power magnification of the compound microscope. Observe a cross section of the esophagus under the dissecting microscope. Identify, from lumen to outer wall, the four layers of the digestive tube using *Figure 25-9* for reference. The esophagus conforms to the histological plan of the digestive tube, but instead of containing a serosa layer, it has a fibrous adventitia layer. Notice how both mucosa and submucosa layers are thrown into folds. Stratified squamous epithelium (see *Color Plate 2*) forms the lining of the mucosa. Within the submucosa identify the **esophageal glands**, which produce a mucous secretion to aid in the lubrication and movement of the food bolus during swallowing. The muscularis externa is most obvious in the esophagus. It is organized such that the upper one-third of the esophagus is skeletal muscle, the lower one-third is smooth muscle, and the middle one-third is mixed with both muscle types.

Slide # _____ Esophagus, sec.

See: Fig. 25-9

∞ Figs.
25-9, p677
25-2, p665

LOCATE

_____ Mucosa { Stratified squamous epithelium
Lamina propria
Muscularis mucosae

_____ Submucosa { Esophageal glands
Blood vessels

_____ Muscularis externa { <u>Arrangement Only in Lower Third</u>
Circular muscle (inner) layer
Longitudinal muscle (outer) layer

_____ Adventitia

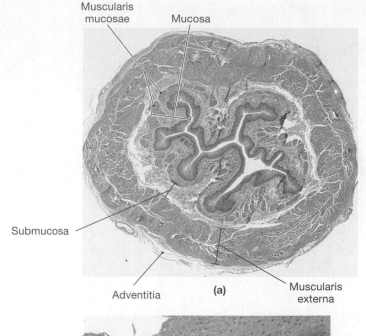

(a)

(b)

FIGURE 25-9 The esophagus (a) low-power section × 16 (b) esophageal × 77.

Gross Anatomy of the Stomach

A food bolus and liquids move by peristalsis down the length of the esophagus and enter the stomach. The stomach is located in the upper left quadrant of the abdomen and is shaped like an expanded letter **J**. The stomach acts both as a storage hopper and a mixing bowl. This design enables us to eat several meals a day, rather than requiring continuous eating. A bolus of food is processed in small batches within the stomach. Processing is in the form of mechanically churning the food bolus, mixing it with the acidic digestive secretions of the stomach, and starting chemical digestion. This mixture is called *chyme*. [∞ p677]

Procedure

✓ **Quick Check**

Before you begin to identify the structure of the stomach, review the blood vessels that supply blood to the stomach and drain blood from it; namely, the branches of the celiac artery (*Figure 22-5, p322*) and the corresponding veins that drain blood. (*Figure 22-11, p331, Color Plates 87* and *89*)

Identify the structure of the **stomach** using a *torso, stomach model,* and *cadaver specimen*. Using *Color Plates 104* to *111* and *Figure 25-10* for reference. Verify on the torso the position of the stomach and its relationship to the diaphragm, esophagus, liver, and intestines. Locate anterior, posterior, medial, and lateral surfaces of the stomach. Identify the curve along the inferior-lateral surface as the **greater curvature** and the curve on the superior-medial surface as the **lesser curvature**. The greater and lesser omenta attach to these surfaces, respectively. The lesser omentum is formed by the **hepatoduodenal** and **hepatogastric ligaments**. It attaches the stomach to the liver and provides an access route for blood vessels entering and exiting the stomach. (Note—ligaments are typically not visible on all models.) The outer surface on the model represents the serosa.

The stomach is divided into four anatomical regions, which are easily distinguished on the stomach exterior. Use *Color Plate 106* for reference. Verify the esophagus, and identify the most superior portion of the stomach (the region in contact with the inferior surface of the diaphragm) as the **fundus**. The fundus is the expanded region superior to the point at which the esophagus enters the stomach. Identify the **cardia** as the area where the esophagus opens into the stomach. The **body** is the largest region of the stomach and is the area between the fundus and the curve of the **J**. Storage and churning of food occurs in this region. Identify the remaining region of the stomach as the **pylorus**. The mixing of chyme with the digestive secretions of the stomach occurs chiefly in this region. The pylorus connects to and is continuous with the first region of the small intestine, the *duodenum*. The three branches of the celiac artery (left gastric, common hepatic, and splenic arteries) supply blood to the stomach. The splenic, superior mesenteric, and gastric veins drain blood into the hepatic portal vein. The blood supply to the stomach is described in Chapter 22.

See: Color Plates 104 to 111, Figs. 25-10, 25-11

∞ *Figs. 25-10, p678 25-11, p679*

LOCATE
External Anatomy

____ Relation of stomach to { Diaphragm, Esophagus, Liver }

____ Greater omentum

____ Surfaces of stomach { Anterior, posterior, medial, and lateral, Greater curvature, Lesser curvature }

____ Stomach regions { Fundus, Cardia, Body, Pylorus }

____ Lesser omentum { Hepatoduodenal ligament, Hepatogastric ligament }

____ Celiac Artery → Left gastric a. → Common hepatic a. → Splenic a.

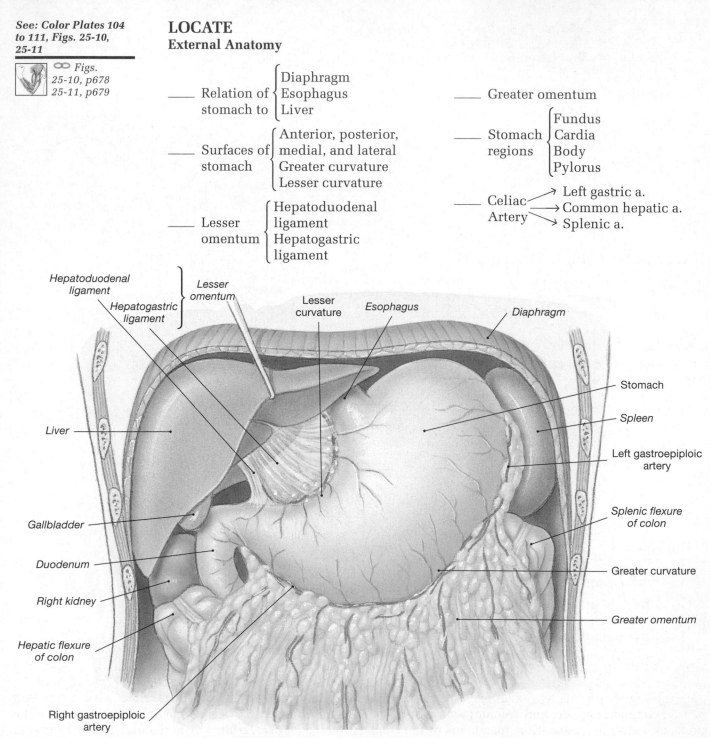

FIGURE 25-10 Surface anatomy of the stomach.

Disassemble the stomach model to observe the internal anatomy. Use *Color Plate 107* and *Figure 25-11* for reference. From the interior, observe all four regions. At the cardia, identify the opening of the esophagus into the stomach as the **cardiac orifice**. At this junction, the smooth muscle of the muscularis externa is arranged in concentric rings to form a **cardiac (esophagus) sphincter**. This sphincter prevents the stomach contents, including gases that result from digestion, from entering the esophagus. Otherwise the acidic digestive secretions of the stomach would destroy the esophageal lining quickly. Identify the **pyloric sphincter** as the terminal portion of the pylorus. Smooth muscle fibers of the muscularis externa are arranged in a ring pattern to form the sphincter, which controls the emptying of the stomach contents into the duodenum. Observe the wrinkled or pleated appearance of the surface of the mucosa. These pleats are termed **rugae** and run the length of the organ. Three layers of muscle of the muscularis externa form the bulk of the wall of the stomach. Identify from outside to inside of the muscular externa, the outer **longitudinal** layer, the thick middle **circular** layer, and the inner **oblique** layer.

Internal Anatomy

____ Stomach regions

____ Cardiac orifice and esophageal lumen

____ Pyloric sphincter

____ Pyloric orifice

____ Rugae

____ Muscular layers { Longitudinal (outer) / Circular (middle) / Oblique (inner) }

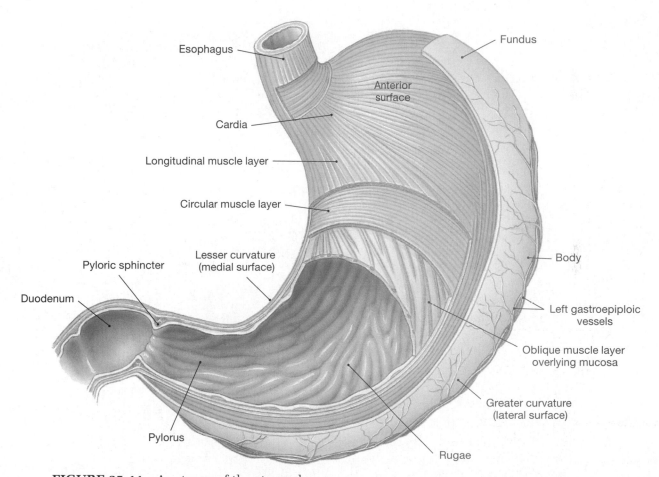

FIGURE 25-11 Anatomy of the stomach.

In the prosected cadaver, reflect the abdominal muscles to observe the stomach and its relationship to the other digestive organs. Identify the anatomy of the stomach using the above directions and *Color Plates 104, 105, 108* to *111* for reference. Reflect the liver superiorly and to the right to identify the surfaces, curvatures, and omenta. Simultaneously, pull the diaphragm superiorly and the stomach inferiorly to inspect the esophagus, esophageal hiatus, cardia, and fundus. Verify the vagus nerves adjacent to the esophagus. If possible, trace the vagus nerve (N X) along the lesser curvature of the stomach. Verify the branches of the celiac artery (left gastric, common hepatic, and splenic) that supply blood to the stomach. To observe the internal anatomy of the stomach, either demonstration specimens of the stomach sealed in preservative or horizontal sections through the abdomen may be available to you for study. Check with your instructor.

Microscopic Identification of the Stomach

Procedure

Identify the histological organization of the **stomach** using the *microscope slide* provided. View under a dissecting microscope, then under low and high power magnifications with the compound microscope. Observe a section of the stomach under the dissecting microscope. Identify from the lumen to the outer wall and the four layers of the digestive tube using *Figure 25-12* for reference. The stomach follows the histological plan of the rest of the digestive tube. Note the rugae or wrinkled appearance of the inner stomach wall. Now observe the mucosa layer with the compound microscope and identify the simple columnar epithelium. Digestive secretions empty into shallow depressions, **gastric pits**, which are identified as clear, unstained pockets lined by the simple columnar epithelium. Each gastric pit communicates with several gastric glands that extend under into the mucosa. Glands are dominated by two types of cells, which release their secretions into a gastric pit. Observe a gastric gland under high power and identify the abundant dark-staining cells as **chief cells**, and the lighter-staining cells as **parietal cells**. Chief cells secrete the digestive enzymes of the stomach and parietal cells secrete both hydrochloric acid and intrinsic factor. [∞ p680] The circular and longitudinal layers of the muscularis externa are easily identified. Depending upon the specimen, the serosa may or may not be present.

Slide # _____ Stomach, sec.

See: Fig. 25-12

∞ Figs.
25-2, p665
25-13, p681

LOCATE

_____ Mucosa
- Simple columnar epithelium
- Mucous epithelial cells
- Muscularis mucosae
- Gastric pits
- Gastric glands
- Parietal cells
- Chief cells

_____ Submucosa

_____ Muscularis externa
- Circular layer
- Longitudinal layer

_____ Serosa

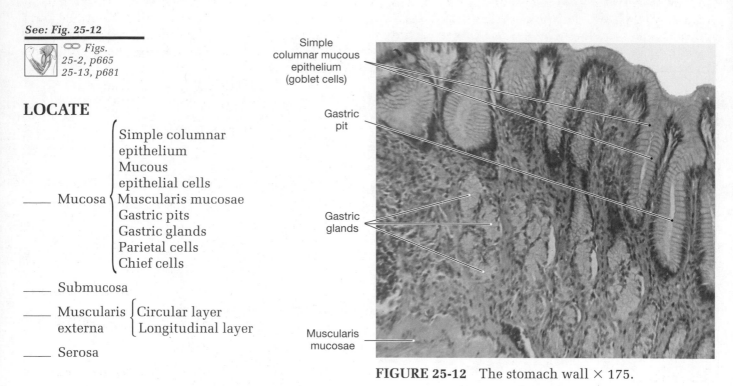

Simple columnar mucous epithelium (goblet cells)

Gastric pit

Gastric glands

Muscularis mucosae

FIGURE 25-12   The stomach wall × 175.

Gross Anatomy of the Small Intestine

The small intestine is the longest portion (in the cadaver 15–25 ft) of the digestive tube and it fills most of the abdominal cavity as a coiled tube. Most chemical digestion of chyme and absorption of nutrients occurs here. The small intestine is divided into three regions: (1) *duodenum*, the first region (about 10 inches), which receives both chyme from the stomach and secretions from the pancreas, liver, and gall bladder; (2) *jejunum*, the middle portion (about 8 ft), where the majority of chemical digestion and nutrient absorption occurs; and (3) *ileum*, the terminal and longest portion (about 12 ft), where digestion and absorption continues. This ends at a sphincter that opens into the large intestine and controls the flow of chyme into the cecum of the large intestine. These dimensions are shorter in life (as seen in barium X-rays). Death causes relaxation of the muscularis layer of the digestive tube and a resulting increase in length of the tube. The movement of chyme through the small intestine is due to peristaltic activity resulting from parasympathetic innervation by the vagus (N X) nerve. Digestive secretions are regulated by both the CNS and hormones. [∞ p686]

Procedure

✓ Quick Check

Before you begin to identify the structure of the small intestine, review the blood supply via the superior mesenteric artery (*Figure 22-5, p322, Color Plates 87* and *110*) and the corresponding drainage by the superior mesenteric vein. (*Figure 22-11, p331* and *Color Plate 89*)

Identify the regions, mesenteries, and structures of the **small intestine** using a *torso model* or *cadaver specimen*. Locate on the torso the position of the small intestine in the abdominopelvic cavity and its relationship to the stomach, liver, and large intestine using *Color Plates 102*, and *103* for reference. Verify the pylorus and pyloric sphincter. Identify the regions of the small intestine from the pyloric sphincter inferiorly: the **duodenum** as the **C**-shaped region; the **jejunum,** forming a series of superior loops and is located inferior to the *transverse colon*; and the **ileum**, forming a series of inferior loops that ends at the **ileocecal valve**, which enters the large intestine at the *cecum*. This valve controls the entry of chyme into the large intestine. In the duodenum, the ducts of the pancreas and gall bladder enter its medial wall.

Identify the **mesentery proper**, which both supports the network of blood vessels and nerves to the small intestine and prevents its loops from tangling during digestive movements (See *Figure 25-3*). Part of this mesentery (visceral layer) is the serosa layer and is represented on the model as the outer visible surface. The origin of attachment of the mesentery proper at the dorsal body wall is the root of the mesentery. Locate the abdominal aorta and verify the superior mesenteric artery and celiac trunk, which supply blood to the small intestine. Verify the superior mesenteric vein (contained in the mesentery proper), which drains blood into the hepatic portal vein. The blood supply to the small intestine is described in *Chapter 22*.

See: Color Plates 102, 103, Fig. 25-13

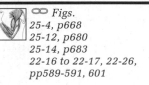

∞ *Figs.*
25-4, p668
25-12, p680
25-14, p683
22-16 to 22-17, 22-26, pp589-591, 601

LOCATE

____ Position of small intestine in abdominopelvic cavity and relation to other abdominal organs

____ Pyloric sphincter of stomach

____ Duodenum

____ Jejunum

____ Ileum

____ Ileocecal valve

Mesentery

____ Mesentery proper

____ Root of mesentery

Blood Vessels

____ Abdominal aorta

____ Superior mesenteric artery/vein contained in mesentery proper

____ Hepatic portal vein

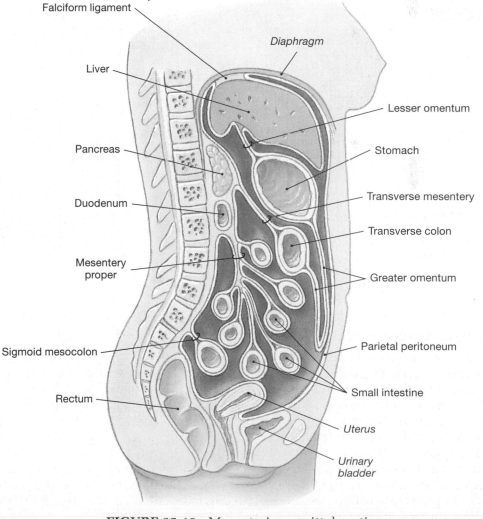

FIGURE 25-13 Mesenteries, sagittal section.

In the prosected cadaver, reflect the abdominal muscles to observe the small intestine and its relationship to the other digestive organs. Identify the anatomy of the small intestine using the above directions and *Color Plates 104, 105,* and *108* to *113* for reference.

Reflect superiorly the greater omentum to view the small intestine. Shift the large intestine as needed to view each portion of the small intestine. Trace the ileum to the ileocecal valve and inspect it internally as shown in *Color Plate 114.* Identify the mesentery proper and trace it posteriorly to its origin of attachment, the root of the mesentery at the dorsal body wall. Locate the abdominal aorta and observe both the superior mesenteric artery within the mesentery proper, tracing it to the small intestine, and the superior mesenteric vein, tracing it to its drainage into the hepatic portal vein. Demonstration specimens sealed in preservative of horizontal sections through the abdomen may be available for you to observe the internal anatomy of the duodenum and ileocecal valve. Check with your instructor.

Microscopic Identification of the Small Intestine Wall

Procedure

✓ **Quick Check**

Previously you observed a microscope slide of the small intestine (ileum) as an example of the histological organization of the digestive tube. Before proceeding, review the mucosa, submucosa, muscularis, and serosa layers. (See *Figure 25-2*)

 Review the histological organization of the **small intestine** using the *microscope slide* provided. Observe a section of the small intestine under the dissecting microscope. Observe the four layers of the digestive tube, from the lumen to the outer wall, using *Figure 25-14* for reference. Identify the folds of the mucosa and submucosa as the **plicae**.

Observe the mucosa under high power of the compound microscope. Identify each "finger-like" projection of the mucosa as a **villus**. Identify the simple columnar epithelium and goblet cells (see *Color Plate 6*) that form col-

Slide # _____ Small Intestine (Showing all three regions), sec.

See: Color Plate 6, Fig. 25-14

∞ *Fig. 25-15, p684*

LOCATE

____ Mucosa

____ Plicae

____ Villus { Simple columnar epithelium
Microvilli
Goblet cells
Lacteals
Intestinal glands (crypts of Lieberkuhn) }

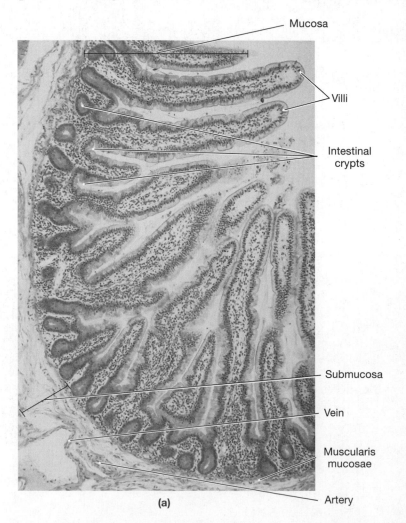

(a)

FIGURE 25-14 The intestinal wall (a) showing layers × 40

lectively the surface of villi. **Microvilli** appear as the "fuzzy," dark-staining, apical surface of columnar cells. The center of each villus contains a lymphatic capillary, termed **lacteal**, surrounded by a network of blood capillaries. The lacteal and capillary structures are not distinct in all villi. To identify these structures, select a villus that looks similar to *Figure 25-14*.

Locate the clear area between the bases of adjacent villi as the entrance to **intestinal glands** termed **intestinal crypts**. These glands extend deep into the mucosa. At their base, stem cell divisions constantly renew the simple columnar epithelium of the small intestine.

Characteristic of the submucosa are arrays of dense connective tissue, lymphatics, and numerous small blood vessels. Under low power, mucus secreting glands can be observed in the submucosa of the duodenum, termed **Brunner's glands**, and groups of small lymph nodes can be observed in the submucosa of the ileum, termed **Peyer's patches**. Identify these glands by their simple cuboidal epithelium and the nodes by their dark stain and appearance. They may appear as single nodes or in groups. The smooth muscle of the inner circular layer and outer longitudinal layer of the muscularis externa can be distinguished easily. The duodenum, jejunum, and ileum regions can be distinguished under the microscope. Villi in the duodenum are low and wide; tall and slender in the jejunum; and short and few in numbers in the ileum. Brunner's glands are found only in the duodenum. Goblet cells are most numerous in the ileum and the only location of Peyer's patches.

_____ Submucosa { Submucosal (Brunner's glands- only in duodenum region) glands
Blood vessels
Peyer's patches (ileum only)

_____ Muscularis externa { Circular layer (inner)
Longitudinal layer (outer)

_____ Serosa

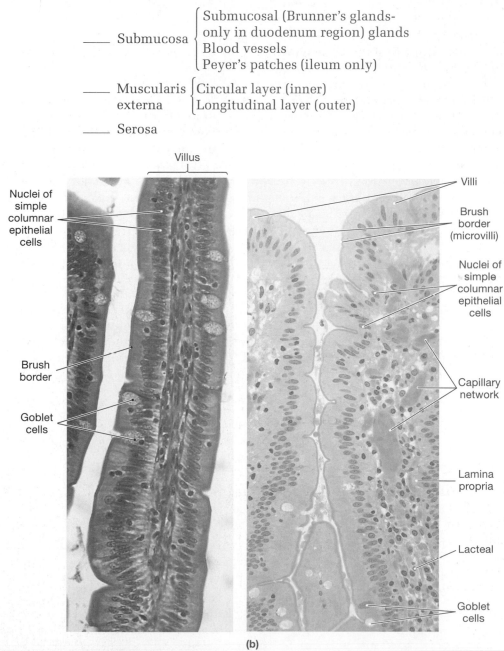

FIGURE 25-14 cont. (b) villi from the jejunum × 360, × 620.

Observation of the Large Intestine

The large intestine or bowel has a distinctive horseshoe shape that frames and helps position the small intestine. Since the majority of nutrients and substances have been absorbed by the small intestine, the large intestine acts as the final point for absorbing additional water, electrolytes, and some nutrients. The large intestine is responsible for compacting the unabsorbed and indigestible food residues into the fecal mass to be stored until defecation. Compared to the small intestine, the large intestine is much shorter in length (5 ft), much larger in diameter (as large as 3 inches), and less active. The large intestine is divided into the *cecum*, *colon*, and *rectum*. The cecum receives chyme from the small intestine, the colon performs the absorptive and compaction functions, and the rectum the storage function. [∞ p686]

Procedure

✓ Quick Check

Before you begin to identify the structure of the large intestine, review the blood supply to this organ through the superior and inferior mesenteric arteries (*Color Plates 87, 119,* and *Chapter 22, p321*) and drainage through the corresponding mesenteric veins. (*Color Plate 89* and *Chapter 22, p331*)

Identify the regions and structure of the **large intestine** using a *torso model* or *cadaver specimen* of the *large intestine*. Use *Color Plates 103* to *105, 109* to *111,* and *119* for reference. Locate on the torso the position of the large intestine in the abdominopelvic cavity and note its relationship to the surrounding organs. Notice the characteristic bubble or bag-like appearance of the large intestine. Identify these bags as **haustrae**, which serve to increase the lumen and epithelial surface area of the large intestine. On the outer surface, also observe three distinct ribbons of smooth muscle, the **taenia coli**. Their contractions aid in compaction activities. Locate the ileum and observe the ileocecal sphincter within the first region of the large intestine, the pouch-like **cecum**. Identify the **vermiform appendix** that hangs from the medial surface of the inferior part of the cecum. With your finger trace the path of the large intestine. Identify tear drops of fat, the **epiploic appendages** on the outer surface of the colon (serosa). These are not found on the small intestine. Identify the **ascending colon**, which makes a right-angled bend medially, the **right colic (hepatic) flexure**, just inferior to the liver. It continues horizontally as the **transverse colon**. Notice that the greater omentum contacts the anterior wall of the transverse colon. Posterior to the stomach and inferior to the spleen, the transverse colon makes a right-angled bend inferiorly, the **left colic (splenic) flexure,** and continues inferiorly as the **descending colon**. Trace it inferiorly to the iliac fossa and notice that it now bends posteriorly and then inferiorly as the **sigmoid colon**. The sigmoid colon opens into the expansible **rectum**. Within the true pelvis, the terminal portion (about the last 6 inches) of the large intestine is the rectum. Internally, the last few centimeters (about 1 inch) is termed the **anal canal**. If possible, examine the internal structures of the rectum. Identify the longitudinal folds as the **anal columns**, the circular band of smooth muscle as the **internal anal sphincter**, and a ring of skeletal muscle fibers as the **external anal sphincter** (note internal structures may not be shown on all models). Identify the terminal portion of the rectum as the **anus** and the opening as the **anal orifice**.

See: *Color Plates 103 to 105, 109 to 111, 119*

∞ *Figs.*
25-12, p680
25-17, p687
25-18, p688
22-16, p589
22-17b, p591
22-26, p601

LOCATE

____ Large intestine

____ Epiploic appendages

____ Haustra

____ Taenia coli

____ Ileocecal valve

____ Cecum

____ Appendix

____ Ascending colon

____ Right colic (hepatic) flexure

____ Transverse colon

____ Left colic (splenic) flexure

____ Descending colon

____ Sigmoid colon

____ Rectum

____ Anal canal

____ Anal columns

____ Internal/external anal sphincters

____ Anus

____ Anal orifice

Blood Supply

____ Superior mesenteric artery/vein

____ Inferior mesenteric artery/vein

In the prosected cadaver, reflect the abdominal muscles to observe the large intestine and its relationship to the other digestive organs. Identify the regions and structure of the large intestine using the above directions and *Color Plates 104, 105*, and *109* to *111* for reference. Pull and reflect superiorly the greater omentum. Note its attachment to the anterior wall of the transverse colon. Shift the small intestine and other organs as needed to view each portion of the large intestine. Observe both the ileocecal valve and its relation to the cecum. Trace the path of the colon and identify each region and structure as noted above. Identify the pendulous tear drops of fat, the epiploic appendages. On the posterior and superior surfaces of the transverse colon, identify the **transverse mesocolon**. Trace it posteriorly to its origin of attachment, the dorsal body wall. Locate the abdominal aorta and observe the superior mesenteric artery within the mesentery proper. Observe the superior mesenteric vein as it enters the hepatic portal vein. Demonstration specimens of horizontal sections through the abdomen may be available also for you to observe the internal anatomy of the small and large intestines. Check with your instructor.

Microscopic Identification of the Large Intestine

Procedure

Identify the histological organization of the **large intestine** using the *microscope slide* provided. Observe a section of the large intestine first under the dissecting microscope. Locate the four layers of the digestive tube from the lumen to the outer wall. The large intestine can be distinguished from the small intestine by lack of villi, abundance of intestinal glands, and excess of goblet cells.

Observe the mucosa under high power of the compound microscope. Use *Figure 25-15* for reference. Identify the characteristic mucus-secreting **intestinal glands** (crypts of Lieberkuhn) and a well developed muscularis mucosae. Excessive amounts of mucus are produced to lubricate the unabsorbed and undigested material and protect the mucosa lining. The inner circular layer and outer longitudinal layer (taenia coli) of the muscularis externa can be distinguished easily.

Slide # _____ Large Intestine, sec.

See: Fig. 25-15

∞ *Fig. 25-19, p689*

LOCATE

_____ Mucosa { Columnar epithelium
Intestinal glands (crypts of Lieberkuhn)
Goblet cells
Muscularis mucosae

_____ Submucosa

_____ Muscularis externa { Circular layer (inner)
Longitudinal layer (outer) as taenia coli

_____ Serosa

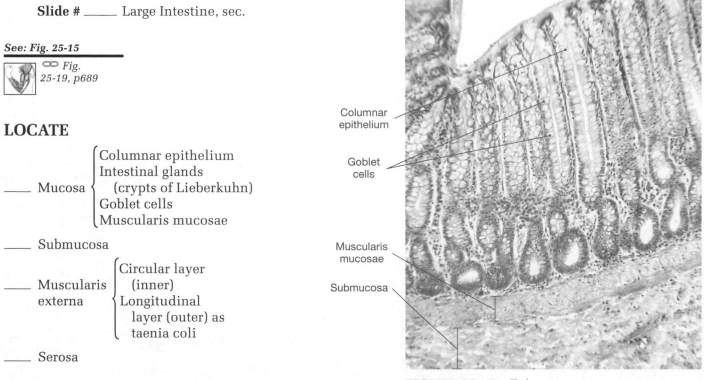

Columnar epithelium

Goblet cells

Muscularis mucosae

Submucosa

FIGURE 25-15 Colon × 104.

Observation of the Liver and Gallbladder

The largest organ (after the skin) in the body is the liver. It weighs over three pounds and occupies the upper right quadrant (right hypochondriac and epigastric regions). The liver is divided into four *lobes*. Each lobe contains many small *lobules*. Lobules are the basic structural and functional units of the liver. Liver cells, termed **hepatocytes**, perform most of the metabolic and secretory functions characteristic of the liver. The liver manufactures and secretes bile. Bile collects into *bile ducts* to be transported out of the liver either to the duodenum for fat digestion or to the *gallbladder* for storage. The gallbladder lies in a pocket on the (visceral) surface of the right liver lobe. The functions of the gallbladder are bile storage and concentration of bile. Bile is released from the gallbladder in response to the hormone cholecystokinin from the duodenum. ∞ [p693]

Procedure

✓ **Quick Check**

> Before you begin to identify the structure of the liver review the blood supply to this organ: hepatic artery (*Figure 22-5, p322*) and hepatic portal vein deliver blood to the liver and the hepatic veins drain blood into the inferior vena cava. (*Color Plate 89* and *Figure 22-11, p331*)

Identify the position, surfaces, and structure of the **liver** using a *torso model*, *liver model*, and *cadaver specimen*. On the torso locate the position of the liver in the abdominopelvic cavity and its relationship to the stomach and intestines using *Color Plates 102, 116, 118* and *Figures 25-16 to 25-18* for reference. To observe the anatomy of the liver and gallbladder, remove the liver from the torso or use the liver model provided. Observe the **falciform** and **coronary ligaments** that secure and position the liver. The falciform ligament separates the liver into **left** and **right lobes**. On the right side, identify the small **caudate lobe** (adjacent to the inferior vena cava) and the **quadrate lobe** between the left lobe and the gallbladder. Observe the smooth **anterior** (parietal) **surface** that contacts the abdominal wall and the **posterior** (visceral) **surface** that contacts the abdominal organs, and bears impressions of the stomach, right kidney, and intestines. Identify the **gallbladder** within a pocket or fossa on the visceral surface of the right liver lobe. Notice how the inferior vena cava is embedded in a fossa of the right liver lobe. Ducts and blood vessels enter and exit at the hilus, called the **porta hepatis** region of the liver.

Locate the gallbladder and identify the **fundus**, **body**, and **neck** regions. Observe the rounded fundus within the fossa of the right liver lobe as the region superior to the **cystic duct**. The bulk of the organ is the body. The neck is the constricted area between the body and the cystic duct. The duct is not a hollow tube, but rather an internal, spiral-like connection of tissue, termed the **spiral valve** (of Heister), which assists in controlling the flow of bile. Bile passes in both directions through the duct and valve. In the cadaver, this valve is not readily visible. Identify on the visceral surface of the left and right lobes, on either side of the falciform ligament, the **left** and the **right hepatic ducts**. Bile is drained from the liver by these ducts, which unite to form a single **common hepatic duct**. At the porta hepatis the common hepatic duct and cystic duct unite to form a single **common bile duct**. This duct accompanies the hepatic artery and portal vein as they pass through the lesser omentum. The common bile duct passes through a portion of the pancreas and with the *main pancreatic duct* penetrates the wall of the duodenum as a raised projection, the **major duodenal papilla**. Within the papilla the common bile duct terminates as the **hepatopancreatic sphincter** (of Oddi). This sphincter is typically not visible in models.

See: Color Plates 102, 116, 118, Figs. 25-16, 25-17, 25-18

 ∞ *Fig. 25-20, p691 25-22, p693*

LOCATE

Viewed from the Anterior (Parietal) Surface

____ Right/left liver lobes

____ Falciform ligament

____ Coronary ligament

____ Round ligament (ligamentum teres)

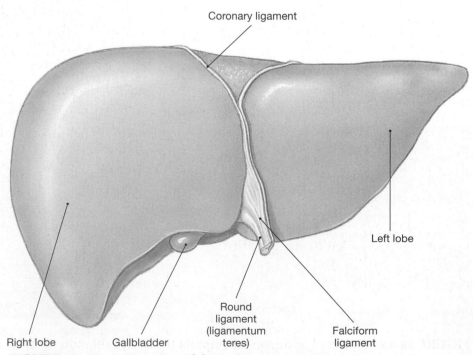

FIGURE 25-16 Liver, viewed from the anterior (parietal) surface.

Viewed from the Posterior (Visceral) Surface

____ Falciform ligament

____ Right liver lobe { Caudate lobe / Quadrate lobe

____ Left liver lobe

____ Porta hepatis

____ Inferior vena cava

____ Hepatic artery proper

____ Branches of hepatic portal vein

Gallbladder

____ Regions of gallbladder { Fundus / Body / Neck

____ Cystic duct

____ Spiral valve (of Heister—not readily visible)

____ Left/right hepatic ducts

____ Common hepatic duct

____ Common bile duct

____ Duodenal papilla within duodenum { Common bile duct / Pancreatic duct / Duodenal (major) ampulla / Hepatopancreatic sphincter

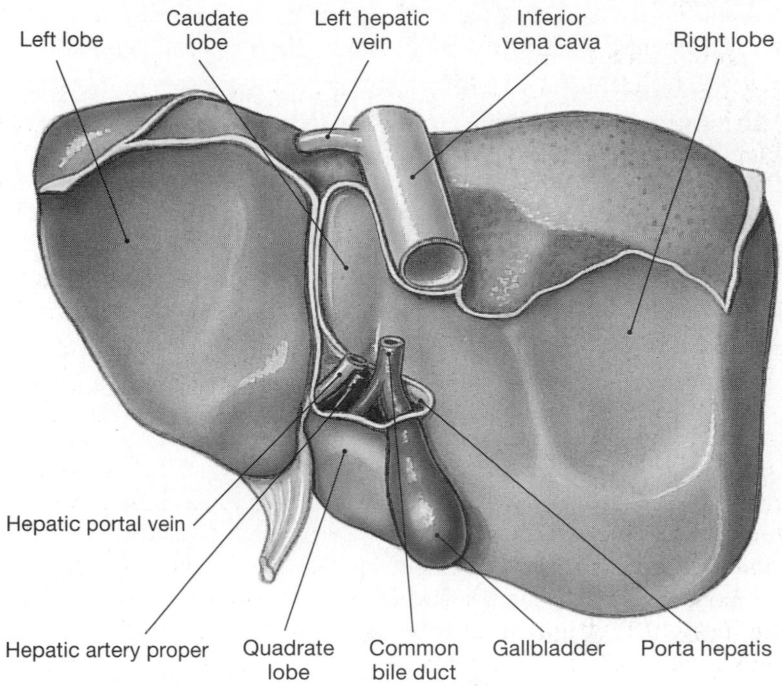

FIGURE 25-17 Liver, viewed from the posterior (visceral) surface.

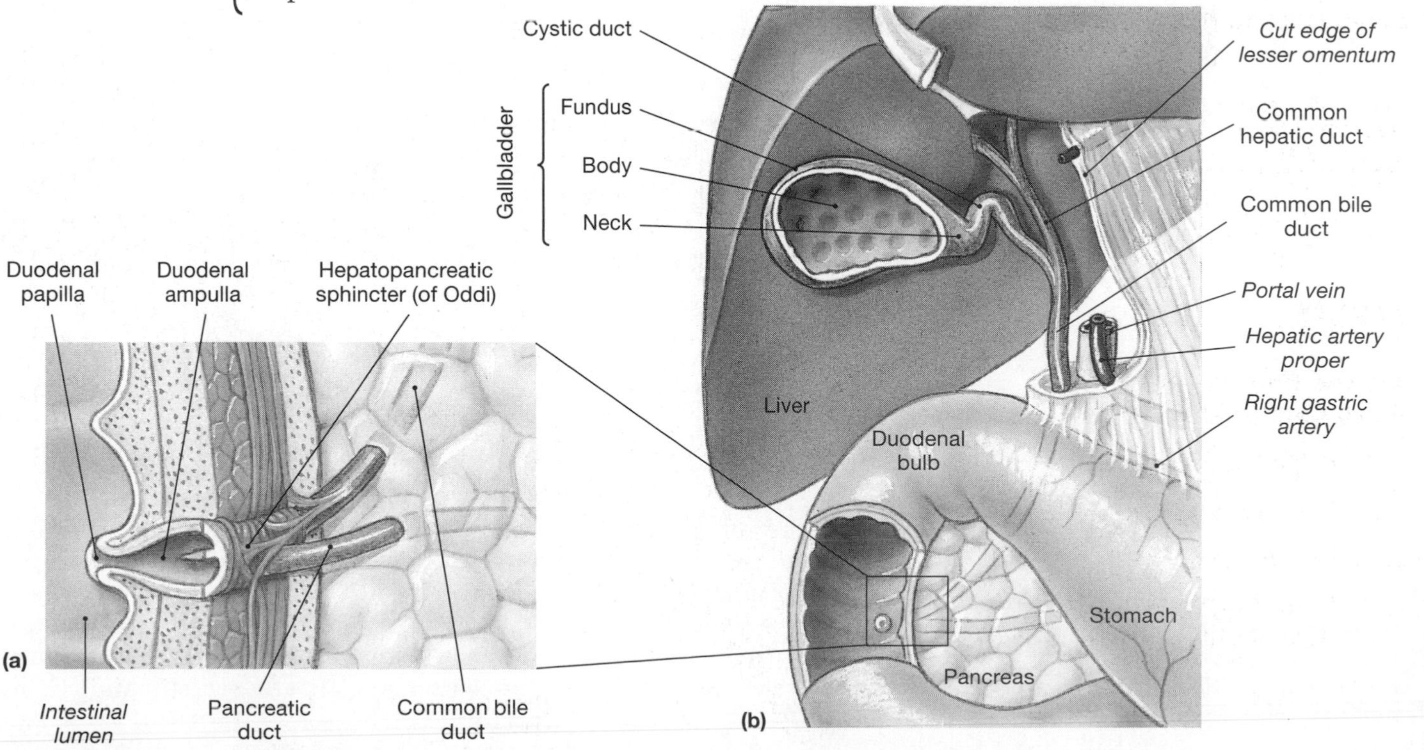

FIGURE 25-18 (a) Detail of duodenal ampulla (b) liver, gallbladder, and bile ducts.

In the prosected cadaver, observe the liver and its relationship to the other digestive organs. Identify the liver, gallbladder, and associated structures using the above directions. Use *Color Plates 104, 105, 108* to *111, 115* and *117* for reference. Pull and reflect superiorly the diaphragm. Note the attachment of the liver to the diaphragm and dorsal body wall by the falciform ligament. Shift the intestines and other organs as needed to view the gallbladder, pancreas, and associated ducts. Locate the gallbladder and pancreas and trace the path of the associated ducts. Demonstration specimens of horizontal sections through the abdomen may be available also for you to observe the internal anatomy of the liver and associated digestive structures. Check with your instructor.

Microscopic Identification of the Organization of the Liver

Procedure

Identify the histological organization and structure of a typical liver lobule using the *microscope slide* provided, first by viewing under the dissecting microscope, then with the compound microscope. Observe a section of the liver under the dissecting microscope. Use *Figure 25-19* for reference. Identify **liver lobules** by their hexagon shape and **central vein**. Each lobule is partitioned from neighbor lobules by a connective tissue **interlobular septum**. Observe how **hepatocytes** radiate in rows from the central vein with **sinusoids** separating each row. Located at the periphery of each lobule at the corners of the hexagon within the connective tissue septum are the **portal areas** (hepatic triads). Contained in each portal area is a branch of the hepatic artery, hepatic portal vein, and one or more bile ductules. Observe a portal area under high power of the compound microscope and identify these vessels. Then locate a central vein and observe hepatocytes and sinusoids under high power.

Slide # _____ Liver, sec.

See: Fig. 25-19

∞ *Fig. 25-21, p692*

LOCATE

____ Lobule

____ Interlobular septum (connective tissue)

____ Central vein

____ Hepatocytes

____ Sinusoid

____ Portal area (hepatic triad)
{ Branch of:
Hepatic artery
Hepatic portal vein
Bile ductule

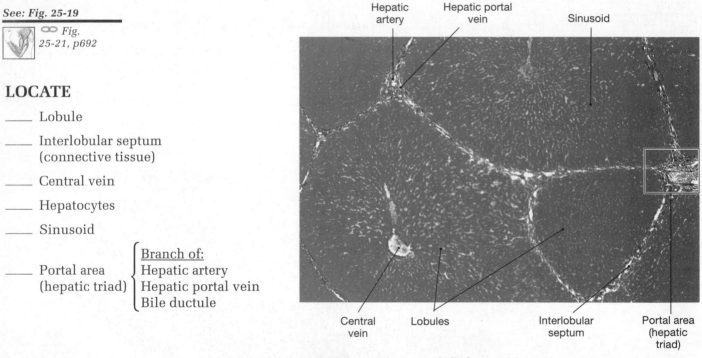

Hepatic artery   Hepatic portal vein   Sinusoid

Central vein   Lobules   Interlobular septum   Portal area (hepatic triad)

FIGURE 25-19   Liver lobules and hepatic triad × 47.

Observation of the Pancreas

At the posterior of the stomach and tucked into the curvature of the duodenum is the pancreas. The pancreas is a tongue-shaped organ that has both endocrine and exocrine (digestive) functions. The endocrine portion of the pancreas consists of *pancreatic islets* or *islets of Langerhans* and was described in *Chapter 19*. [∞ p519] Pancreatic juice (digestive enzymes and buffers) secreted by the pancreas are released into the duodenum via the pancreatic duct. These enzymes digest lipids, carbohydrates, and proteins down to their smallest (monomer) units. Pancreatic juice is secreted in response to hormones from the duodenum, specifically secretin and cholecystokinin. [∞ p693]

Procedure

Identify the following structures of the **pancreas** using a *torso model*, *pancreas model*, and *cadaver specimen*. Remove the liver, large intestine, and spleen to observe the pancreas in the torso. Use *Color Plates 103, 106,* and *Figure 25-20* for reference. Observe the relationship of the pancreas to both the stomach and the duodenum. The pancreas extends transversely from its position posterior to the stomach. If a model of the pancreas/ duodenum is available, use it to observe the structure of the pancreas and relationships of the pancreatic ducts to the duodenum. The pancreas is divided into three regions and is shaped like your tongue. Identify the **body** and **tail** regions of the pancreas along the posterior border of the inferior surface of the stomach, with the tail forming the tip of the organ. Identify the **head** within the **C**-curvature of the duodenum.

Within pancreatic tissue, locate the main **pancreatic duct** (duct of Wirsung) and trace it from tail to head regions. In the head region, a small **accessory duct** (duct of Santorini) often may be observed branching from the pancreatic duct. The main pancreatic duct perforates the wall of the duodenum along with the common bile duct as the raised duodenal papilla. When present, the accessory duct perforates the duodenal wall just superior to the papilla. Within the papilla, the main pancreatic and common bile ducts open into the **duodenal ampulla**. (Note: typically not visible in models).

See: Color Plates 103, 106, Fig. 25-20

 Figs. 25-23, p694

LOCATE

_____ Regions of pancreas { Head / Body / Tail

_____ Pancreatic duct (duct of Wirsung)

_____ Accessory duct (duct of Santorini)

_____ Hepatopancreatic sphincter (of Oddi)

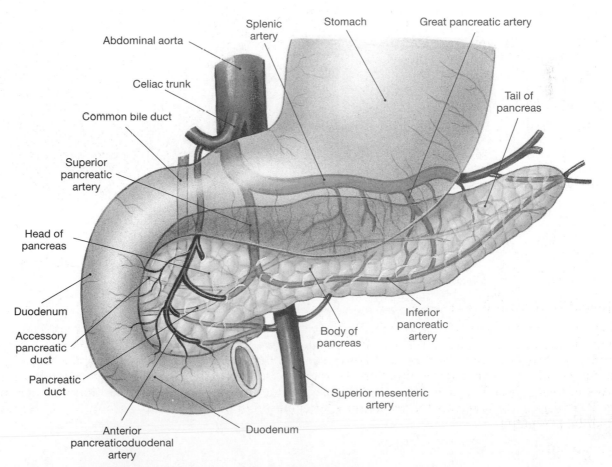

FIGURE 25-20 Gross anatomy of the pancreas.

In the prosected cadaver, observe the pancreas by shifting the small intestine and other abdominal organs as needed. Verify the location of the pancreas in reference to the duodenum and stomach. Use *Color Plates 109* and *111* for reference. Trace the pancreas and identify each region and duct as noted above. Reflect the wall of the duodenum and you may be able to locate the duodenal papilla. Often this area is colored green, as a result of bile leaking out of the common bile duct after death. Demonstration specimens showing the ducts entering the wall of the duodenum may also be available for you to observe. Check with your instructor.

Microscopic Identification of the Pancreas

Procedure

✓ **Quick Check**

The endocrine portion of the pancreas is described in *Chapter 19* under the heading "Observation of the Endocrine Pancreas" and should be reviewed before proceeding. (*pp286* and *287*)

Identify the structures of the **exocrine pancreas** using the *microscope slide* provided, first by viewing under low power magnification and then under high power magnification. Under low magnification, observe the general appearance of the pancreas. Use *Figure 25-21* for reference. The bulk of the pancreas contains the exocrine glandular tissue necessary to secrete digestive enzymes. Identify the exocrine portion of the gland as the darker-stained **pancreatic acini (exocrine) cells**. The endocrine portion can be seen scattered throughout the pancreas and is made up of *pancreatic islets* (islets of Langerhans). Identify the pancreatic islets as the circular-to-oval masses of pale-staining cells surrounded by the darker-staining exocrine tissue. Small ducts and blood vessels will also be observed. Now observe both endocrine and exocrine cells under high power.

Slide # _____ Pancreas, sec.

See: Fig. 25-21

∞ *Fig. 25-23b,c, p694*

LOCATE

_____ Pancreatic acinar cells (exocrine)

_____ Pancreatic islets (endocrine)

_____ Pancreatic duct

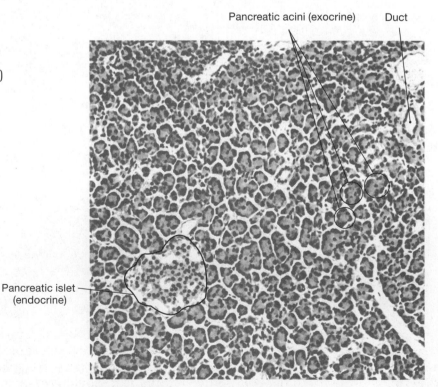

FIGURE 25-21 The pancreas acini × 120.

Dissection of the Digestive System in the Cat

This exercise complements the study of the human digestive system. The digestive system of the cat is very similar to the human. During our examination of cat digestive structures, some differences are described.

If the thoracic and abdominal cavities have not been opened, expose the thoracic and abdominal organs by making a longitudinal midline incision with scissors through the muscles of the neck, the bones and muscles of the midline of the thorax, and the muscles of the abdominal wall. Avoid cutting the diaphragm in order to identify the structures that pass between thoracic and abdominal cavities. Make lateral wall incisions in the thoracic cavity, just cranial to the diaphragm and in the abdominal cavity caudal to the diaphragm. In your continuing dissection of the digestive system, trace the path of ingested materials through the body.

See: Fig. 25-22

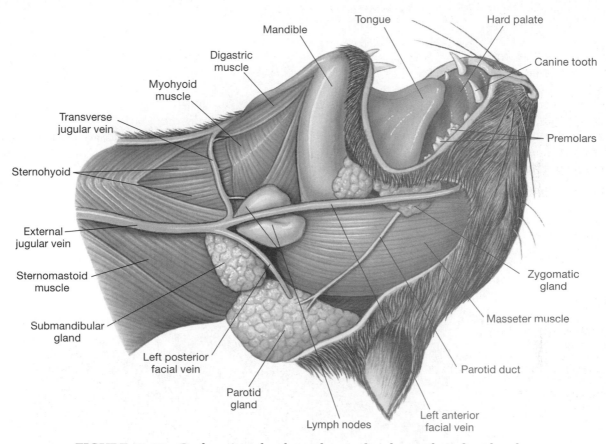

FIGURE 25-22   Oral cavity, glands, and superficial muscles of cat head and neck.

Oral Cavity, Glands, and Superficial Muscles of Head and Neck

1. The **oral cavity** consists of the **vestibule** and the oral cavity proper. The vestibule is surrounded by the lips and cheeks. The **tongue** and **teeth** are located within the oral cavity proper. With the aid of a magnifying lens, locate the papillae on the tongue. **Filiform papillae** are anterior, numerous, and pointed. **Fungiform papillae** are small, rounded, and posterior to the filiform papillae. A few **circumvallate papillae**, which are large and rounded, are located near the back of the tongue. Also, observe that the tongue is attached to the floor of the mouth by a mucous membrane, the **lingual frenulum**.

2. The roof of the mouth consists of a bony **hard palate** and its posterior extension, the **soft palate**. At the posterior region of the oral cavity, locate the **pharynx**. It is the cavity that occupies the space from dorsal to the soft palate to the larynx. There are three regions of the pharynx: the **nasopharynx** is dorsal to the soft palate; the **oropharynx** is posterior to the oral cavity; and the **laryngopharynx** is the region of the epiglottis and the opening to the esophagus. The **larynx** forms the passageway between the pharynx and trachea.

3. The salivary glands secrete saliva directly into the mouth. Do not confuse the small dark kidney-bean shaped lymph nodes with the oatmeal-colored and textured salivary glands. Locate the large **parotid gland** inferior to the ear on the surface of the masseter muscle. The **parotid duct** passes over the surface of this muscle and enters the oral cavity. The **submandibular gland** lies inferior to the parotid and deep to the point of merger between the facial and external jugular veins. The **sublingual gland** is anterior to the submandibular gland. Ducts of both glands open into the floor of the mouth, but typically only the submandibular duct can be traced.

4. Return to the pharynx and identify the opening into the **esophagus** posterior to the epiglottis of the larynx. The esophagus is the flattened tube dorsal to the trachea. It connects the laryngopharynx to the stomach. Reflect the organs of the thoracic cavity, and trace the esophagus through the **diaphragm** into the abdominal cavity where it joins the stomach.

See: Fig. 25-23

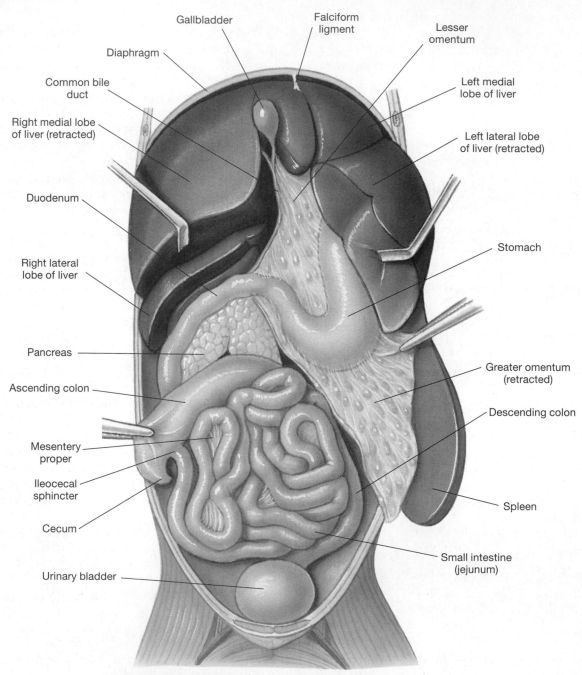

FIGURE 25-23 Ventral view of the cat abdominal viscera.

Abdominal Viscera, Ventral View

5. The inside of the abdominal cavity and the outer surface of the viscera are lined by a serous membrane, the **peritoneum**. The **visceral peritoneum** covers the viscera and the **parietal peritoneum** lines the inside of the abdominal body wall. A potential space, the **peritoneal cavity**, is found between these membranes. A serous fluid, called **peritoneal fluid** is produced by these membranes. It permits the organs to slide past each other. The peritoneal sheets extending between the body wall and the viscera are termed *mesenteries, omenta,* and *ligaments.* These membranous sheets provide support for vessels, nerves, and lymphatics that supply the viscera.

6. Locate the **stomach** on the left side of the abdominal cavity. Identify the four regions of the stomach: the **cardiac region**, at the entrance of the esophagus; the **fundus,** the dome-shaped pouch that rises dorsal to the esophagus; the **body**, the main portion of the stomach; and the **pyloric region**, the posterior region of the stomach, which ends at the **pyloric sphincter** where the digestive tube continues as the duodenum. Make an incision through the wall of the stomach along the greater curvature, continuing this incision two inches past the pyloric valve. Reflect the stomach wall to observe the pyloric valve and its inner surface. Large folds of the stomach mucosa, termed **rugae,** may be observed in the empty stomach.

7. The outer (left) margin of the stomach is convex in shape and called the **greater curvature**. The medial (right) margin of the stomach is concave in shape and called the **lesser curvature**. Attached to the greater curvature is a curtain of fat, the **greater omentum,** which covers the ventral surface of the abdominal organs. The **lesser omentum** connects the lesser curvature of the stomach to the liver.

8. Observe the large, brown **liver** posterior to the diaphragm. It is the largest organ in the abdominal cavity, and is divided into five lobes: *right/left medial*, *right/left lateral*, and a *caudate (posterior) lobe.* Identify a dark green elongated pear-shaped sac, the **gallbladder**, within a fossa in the right medial liver lobe. The **falciform ligament** is a delicate membrane that attaches the liver anteriorly to the diaphragm and dorsally to the abdominal wall. This ligament can be located between right and left medial liver lobes.

9. Identify the **bile duct** and trace its course anteriorly to where it splits into the **cystic duct**, to the gallbladder, and the **common hepatic duct**, to the liver. Locate these ducts by carefully teasing away the connective tissue.

10. Posterior to the stomach observe a large, dark brown organ, the **spleen**.

11. The three regions of the **small intestine** can be traced from its proximal end to its distal end. The first region, termed **duodenum,** is the C-shaped portion immediately distal to the pyloric valve. It is about six inches in length and receives the pancreatic and the common bile ducts. The **jejunum** composes the bulk of the remaining length of the small intestine. The **ileum** is the last region of the small intestine and merges with the large intestine. The ileum terminates at the **ileocecal sphincter (valve)**, which controls the flow of chyme into the large intestine. This valve can be felt as a firm ring at the end of the ileum where it projects into the large intestine.

12. The **large intestine** is divided into *cecum, colon,* and *rectum* portions. To view the large intestine, it will be necessary to pull the loops of the small intestine to the left and let them drape out of the body cavity.

13. The initial expanded region of the large intestine is the **cecum**. The ileocecal valve projects into the cecum. The small intestine enters the large intestine at about one-half inch from the beginning of the large intestine. This passageway from the small intestine to the large intestine is termed the **ileocecal sphincter**. Make an incision along the lateral border of the cecum in order to view the sphincter.

14. Trace the short **ascending, transverse**, and **descending colon** portions. The ascending colon lies on the right side of the abdominal cavity and begins just anterior to the ileocecal valve; the transverse colon extends transversely across the abdominal cavity; and the descending colon descends on the left side of the posterior abdominal wall. It is continuous with the terminal portion of the large intestine, the **rectum**, whose opening to the exterior is the **anus**. The membranous structure that supports the colon and attaches it to the posterior body wall is the **mesocolon**.

15. Posterior to the stomach and within the curvature of the duodenum locate the **pancreas**, the major glandular organ of the digestive system. In the cat, the pancreas has two regions, *head* and *tail*. Identify the region within the curvature of the duodenum as the **head** of the pancreas and the portion passing along the posterior surface of the stomach as the **tail** of the pancreas.

16. Locate the **pancreatic duct** (Duct of Wirsung) as it perforates the duodenum at an enlarged area, the **hepatopancreatic ampulla**. Identify the duct and ampulla by using a teasing needle probe to scrape away the pancreatic tissue of the head portion. Trace the duct to the ampulla. The pancreatic and bile ducts are adjacent to each other.

17. Return to the cut segment of the duodenum, reflect it and secure it open with dissecting pins. Now use a hand lens to observe the *villi, ampulla,* and the opening of the bile duct.

Anatomical Identification Review

Identify the structures of the digestive system, together with their primary functions.

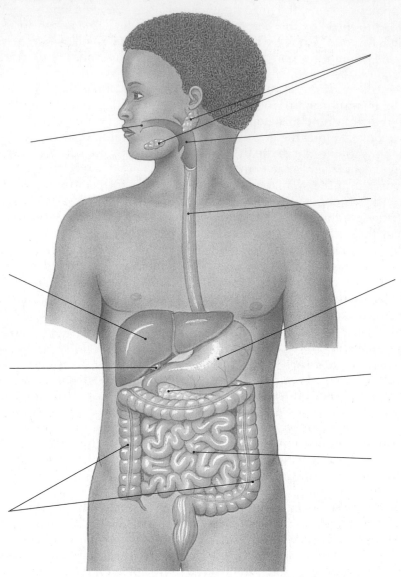

To Think About

1. Peter is suffering from an obstruction in his colon. He notices that when he urinates, the color of his urine is much darker than normal, and he wonders if there is any relationship between the color of his urine and his intestinal obstruction. Based upon your knowledge of anatomy, what would you tell him?

2. Diane followed her doctors advice and had her gallbladder removed after suffering a painful bout of acute cholecystitis. Several months after her surgery she began to experience upper abdominal pain similar to that from her gallbladder. These attacks were accompanied by jaundice and fever with pale-colored stools and very dark urine. Diane was hospitalized and diagnosed as having acute pancreatitis. Based upon your knowledge of the anatomy of the digestive system, would you say that Diane's case of pancreatitis was co-incidence, or did the gallstones or even the surgery cause this condition? Include anatomical structures in your explanation.

26

THE URINARY SYSTEM

Objectives

1. Summarize the functions of the urinary system and describe its relationship to the other excretory organs.

2. Describe the organization of the urinary system, identify its structures, and list their functions.

3. Name and identify the blood vessels that comprise the pattern of renal circulation.

4. Locate and identify the structures of the nephron.

5. Identify and compare both the gross and microscopic anatomy of the following urinary system structures: kidneys, ureters, urinary bladder, and urethra.

The **urinary system** has many functions, including the formation of urine and the subsequent elimination of the organic waste products formed by body cells. [∞ p703] Other functions include: regulating blood volume and blood pressure; helping to maintain normal blood pH; stabilizing the plasma concentration of sodium, potassium, chloride, and calcium ions; and conserving valuable nutrients. The urinary system is the major regulator of ions and many of the organic molecules (i.e., proteins) transported by the blood. Retaining or eliminating these substances is part of the work of the kidneys. The functions of the kidneys are indispensable to life. Other organs with excretory activities with organs (skin, lungs, large intestine) cannot take over its functions. The urinary system consists of a pair of *kidneys*, a pair of *ureters*, a *urinary bladder,* and a *urethra*. The urinary system is similar in men and women. The functions of the urinary system are exactly the same in both sexes, only the length and function of the structure (urethra) for the drainage of urine differ.

Organization of the Urinary System

The **kidneys** are the organs of the urinary system, which remove waste products and stabilize both the volume and composition of the plasma. The modified, final waste product produced by the kidneys is termed **urine**. The kidneys are just as vital to life as the heart. Humans can survive either damage to one kidney or its removal, but loss of both kidneys is life-threatening. The bean-shaped kidneys lie on the posterior of the abdominopelvic cavity, between the dorsal body wall muscles and the parietal peritoneum. Since the kidneys are positioned behind the peritoneum, the term *retroperitoneal* is used to describe their location. Urine produced by the kidneys is transported through the **ureters**, a pair of muscular tubes that extend inferiorly from the kidneys, to a storage organ, the **urinary bladder**. It is held here until it can be eliminated from the body. The urinary bladder is located anterior to the rectum in the pelvic cavity. Urine is drained from the urinary bladder by a single tube, the **urethra**. In females the urethra is short and serves only to convey urine out of the body. In males the urethra is much longer, passing through the length of the penis. The urethra in males shares both excretory and reproductive functions. [∞ p703]

Procedure

✓ **Quick Check**

Before you begin to locate the structures of the urinary system, review the lumbar vertebrae and the floating 12th rib, (*Figure 6-41, p 66*) diaphragm, (*Figure 10-11, p 125*) and psoas major muscle. (*Figure 11-14, p 164*)

Identify the structures of the **urinary system** within the abdominopelvic cavity, using first a *torso model* and then a *cadaver specimen*. To observe the urinary system in the torso, remove all the abdominopelvic organs. Use *Color Plates 120, 121, 123*, and *Figure 26-1* for reference. Identify left and right **kidneys** against the dorsal body wall at the level of the floating 12th rib. Notice that the left kidney is positioned higher (at the 11th rib) than the right. Observe the relative relationships of the left kidney to the spleen and the right kidney to the liver. In life, the kidneys are embedded within a connective tissue fat pad, the *perirenal fat*, and lie retroperitoneally (behind the parietal peritoneum). In torso models, only a portion of this fat is shown and the peritoneum is not shown. Identify all kidney surfaces (ventral, dorsal, superior, inferior, lateral, and medial). Verify the renal arteries and trace them from their origin at the abdominal aorta to the indented medial surface of each kidney. Now verify the renal veins, tracing them from the medial surface of each kidney to the inferior vena cava. Identify left and right **ureters** as they emerge from the medial surface of each kidney. This indented area where the renal vessels pass and the ureter exits is termed the **hilus**. Notice that the ureters are posterior to the renal arteries. Trace the ureters as they descend inferiorly and medially along the dorsal body wall over the psoas major muscles to their penetration into the posterior wall of the **urinary bladder**. Identify the urinary bladder on the floor of the pelvic cavity. Its location in males is anterior to the rectum and posterior to the symphysis pubis. In females, the bladder lies anterior to the uterus, anterosuperior to the vagina, and posterior to the pubic symphysis. We will examine each urinary structure in separate observations.

See: Color Plates 120, 121, 123, Fig. 26-1

Figs. 26-1, p704 26-2, p705

LOCATE

____ Right/left kidneys

____ Right/left ureters

____ Urinary bladder

____ Urethra (male and female)

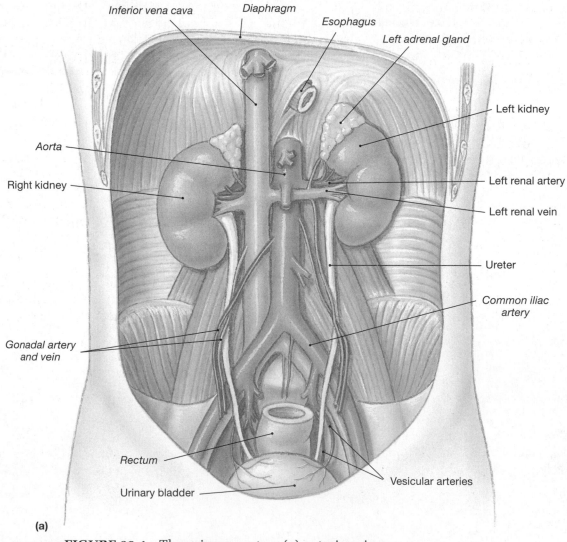

(a)

FIGURE 26-1 The urinary system (a) anterior view.

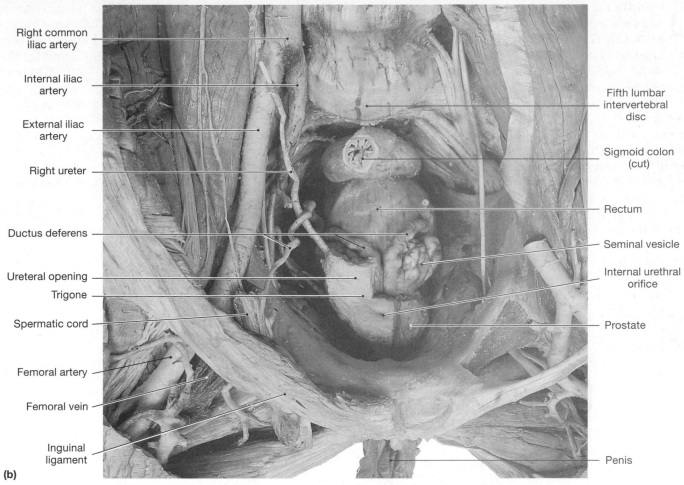

Right common iliac artery

Internal iliac artery

External iliac artery

Right ureter

Ductus deferens

Ureteral opening

Trigone

Spermatic cord

Femoral artery

Femoral vein

Inguinal ligament

Fifth lumbar intervertebral disc

Sigmoid colon (cut)

Rectum

Seminal vesicle

Internal urethral orifice

Prostate

Penis

(b)

FIGURE 26-1 cont.   The urinary system (b) in gross dissection.

Identify the components of the urinary system in the cadaver using the above directions and *Color Plate 123* for reference. The abdominopelvic organs first must be reflected and maintained in this position. Observe the dorsal and superior surfaces of the kidneys as they contact the diaphragm and psoas major muscle. Be careful as you examine the urinary system structures not to sever or separate the kidneys from their vessels. Depending upon the dissection, you may be able to observe the retroperitoneal position (typically only on one side) of the kidneys, ureters, and urinary bladder. In life, the crescent-shaped *suprarenal (adrenal) glands* cap the superior and medial portions of each kidney. Because these glands undergo rapid deterioration at death, they are typically degenerated in most cadaver specimens. At the hilus of the kidney, identify the renal artery, renal vein, and ureter. Observe the ureters as they pass anterior to the external iliac arteries and penetrate the posterior wall of the bladder. To observe the bladder and urethra, you will need to manipulate the viscera.

Demonstration models or specimens of horizontal sections through the abdomen (at about L_1) may be available also for you to observe the position of the kidneys in relation to the dorsal body wall, digestive structures, and associated blood vessels. Check with your instructor. Use *Figure 26-2* and *Color Plate 98* for reference.

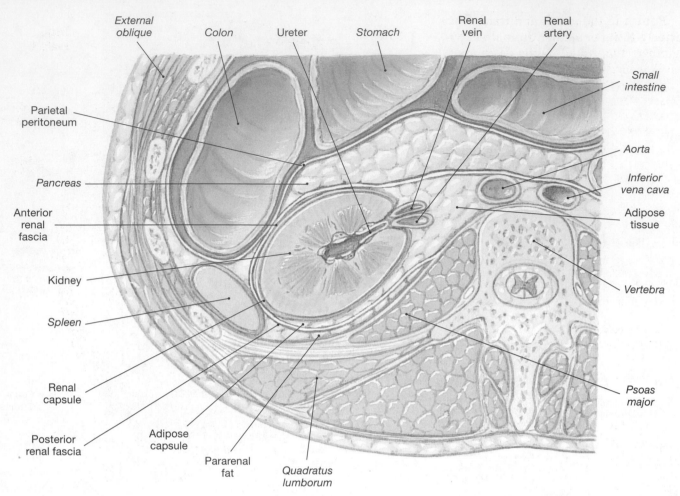

FIGURE 26-2   Sectional view of the trunk at about L_1 showing the position of the kidneys.

Labels on figure:
External oblique · Colon · Ureter · Stomach · Renal vein · Renal artery · Parietal peritoneum · Small intestine · Pancreas · Aorta · Anterior renal fascia · Inferior vena cava · Adipose tissue · Kidney · Spleen · Vertebra · Renal capsule · Psoas major · Posterior renal fascia · Adipose capsule · Pararenal fat · Quadratus lumborum

Observation of Kidney Structure

Each reddish-brown kidney is about the size of your fist. The characteristic color is due to the internal structure of tubules and extensive network of blood vessels. The kidneys are maintained in a retroperitoneal position against the dorsal body wall by means of three concentric layers of connective supporting tissue. First, a fibrous connective tissue layer, the *renal capsule*, wraps the surface of the kidney; second, a middle, *adipose capsule,* covers this layer; and third, an outer layer of dense collagen fibers forms a tough *renal fascia,* that attaches the kidney to the muscle of the dorsal body wall and to the peritoneum. The kidneys have a distinct outer *cortex* and inner *medulla* structure similar to those of some organs described in other sections of this text. [∞ p703] If the kidney structure is compared to that of a peach, then the peach skin = renal capsule, fleshy fruit = cortex, and the inner pit = medulla. The functional units of the kidney occupy cortical and medullary regions.

Procedure

Identify the following **kidney structures** using a *model* or *specimen* of the *kidney*. It is best to observe kidney structure first on a model, then on a specimen. Use *Color Plate 122* and *Figure 26-3* for reference. Externally, identify the **renal capsule**, hilus, and kidney surfaces. Verify the renal artery, renal vein, and ureter. View the kidney in section and observe the internal structures. Identify **cortex** and **medulla** regions. The renal cortex is the outer layer of the kidney in contact with the capsule. The medulla consists of numerous dark pie-shaped or triangular structures, **renal pyramids**. The base of each pyramid faces the cortex. Each renal pyramid consists of a collection of ducts that transport and process filtrate, which is then collected as urine. Columns of cortical tissue, called **renal columns**, can be identified between each renal pyramid. The tip region of each pyramid points inward to a hollow space within the kidney, the **renal sinus**. The renal sinus is filled both with branches of the renal blood vessels and by the expanded collecting regions of the ureter. Identify the tips of the renal pyramids as **renal papillae**.

Return to the ureter and trace it superiorly to its expanded funnel-shaped chamber, the **renal pelvis**. The renal pelvis is anchored by connective tissue within the renal sinus of the kidney. Observe two branches from the renal pelvis. Each branch is termed a **major calyx**. Four or five short cup-like tubes, each termed a **minor calyx**, merge to form a major calyx. From 4 to 13 minor calyces may be identified in each kidney. Each renal pyramid drains into a single minor calyx. It is important to note that calyces and the renal pelvis are modifications of the ureter and are not kidney tissue.

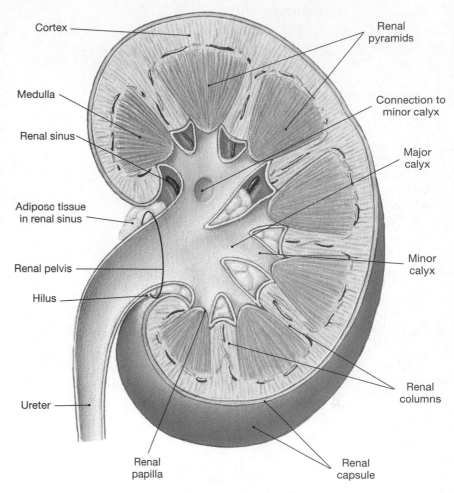

FIGURE 26-3 Frontal section showing the structure of the kidney.

See: Color Plate 122, Figs. 26-3, 26-4, 26-5

 ∞ *Fig. 26-3, p706*

LOCATE

____ Renal capsule

____ Hilus

____ Renal artery/vein

____ Ureter

____ Cortex

____ Medulla

____ Renal pyramids

____ Renal columns

____ Renal sinus

____ Renal papillae

____ Renal pelvis

____ Major calyx

____ Minor calyx

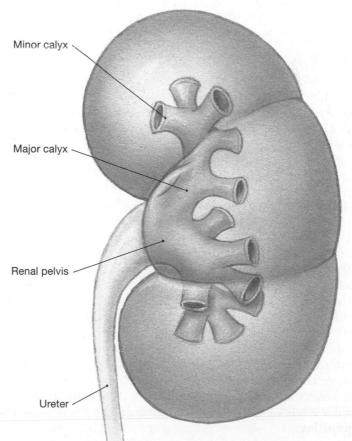

FIGURE 26-4 Kidney showing calyces and renal pelvis.

Kidney Specimen Dissection

Dissect and identify kidney structures using the above descriptions and the following directions as aids. Use *Figure 26-5* for reference. First, distinguish the surfaces of the kidney and the renal capsule. Each kidney has six surfaces: dorsal, ventral, superior, inferior, lateral, and medial. The renal capsule appears as a wax-paper like covering over the kidney. Identify first the hilus and then probe the renal sinus area. Observe the renal artery, renal vein, and ureter. To view the internal structures, you must prepare a frontal section of the kidney. Place your kidney specimen on a dissecting tray and slice the specimen from superior to inferior, such that the kidney is split into anterior and posterior halves for viewing. Identify the cortex by its light, grainy sand-colored appearance. Deep to the cortex, identify the medulla by its pie-shaped renal pyramids that display multiple striations. Probe to identify the tips of the renal pyramids and the renal papillae.

Trace the ureter to the renal pelvis, which is located within the renal sinus region of the kidney. Notice how the renal pelvis is firmly attached within the renal sinus by connective tissue and padded with fat. Identify two major calyces branching from the renal pelvis. Identify the minor calyces and several renal papillae. This identification will require probing and cleaning of the surrounding connective tissue.

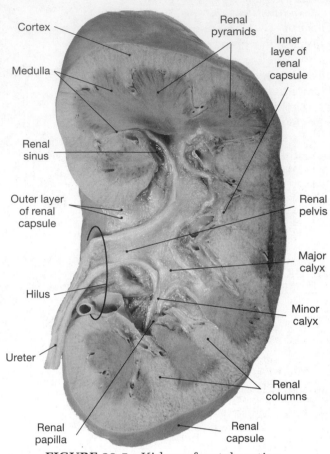

FIGURE 26-5 Kidney, frontal section showing internal structures.

Observation of the Circulation Pattern in the Kidney

The pattern of arterial blood supply and venous return in the kidney is designed to create and maintain a large volume of blood delivered at a high pressure to the functional units of the kidney, the nephrons. This is necessary for the kidneys to perform their functions. They receive approximately 25% of the total volume of blood ejected from the left ventricle. Blood flows from the renal artery through the hilus into vessels that carry blood to the boundary between the cortex and medulla. Further branching of arteries supplies blood to the functional units of the kidney. The venous return follows the arteries. Venous blood drains from the kidneys via the renal veins to the inferior vena cava. [∞ p705]

Procedure

Identify the vessels that supply **blood** to the **kidneys** using a *torso, kidney model,* or specimen. Remove all the abdominopelvic organs from the torso and verify the abdominal aorta, renal arteries, kidneys, renal veins, and inferior vena cava. Use *Color Plates 121, 122, Figures 26-6,* and *26-7* for reference. Observe the same structures in the cadaver. Accessory renal arteries occur in about 25% of the population.

Observe the pattern of renal circulation using a kidney model or demonstration specimen. Locate on the model the renal artery as it enters the hilus. Identify the branches of the renal artery as the **segmental arteries**. Trace these arteries toward the renal pyramids. The segmental arteries branch into **lobar** and then **interlobar arteries**, which radiate outward through the renal columns between the pyramids to arch over the base of the renal pyramids as the **arcuate arteries**. These arteries lie at the border between the cortex and medulla. Each arcuate artery gives off a number of very small **interlobular arteries** into the cortex, which branch into numerous **afferent arterioles**. Afferent arterioles supply blood to each *glomerulus* (a capillary knot—appearing like a tangled ball of yarn), which is the site where filtration occurs, the first step in the production of urine. From the glomerulus, an **efferent arteriole** emerges and branches into two other capillary networks. A series of peritubular capillaries and vasa recta capillaries form around different regions of the nephron tubules. These capillaries represent the sites for the reabsorption of substances back into the blood. From the peritubular capillaries and vasa recta, blood collects into **interlobular veins**, then **arcuate veins**, **interlobar veins**, and finally **segmental veins**. Segmental veins drain into the renal vein, which drains into the inferior vena cava.

See: Color Plates 121, 122, Figs. 26-6, 26-7

 ∞ *Figs.*
26-2, p705
26-4, p707
26-5, p707

LOCATE

____ Abdominal aorta

____ Renal artery

____ Segmental arteries

____ Interlobar arteries

____ Arcuate arteries

____ Interlobular arteries

____ Afferent arterioles

____ Glomerulus

____ Efferent arterioles

____ Peritubular capillaries

____ Vasa recta capillaries

____ Interlobular veins

____ Arcuate veins

____ Interlobar veins

____ Segmental veins

____ Renal vein

____ Inferior vena cava

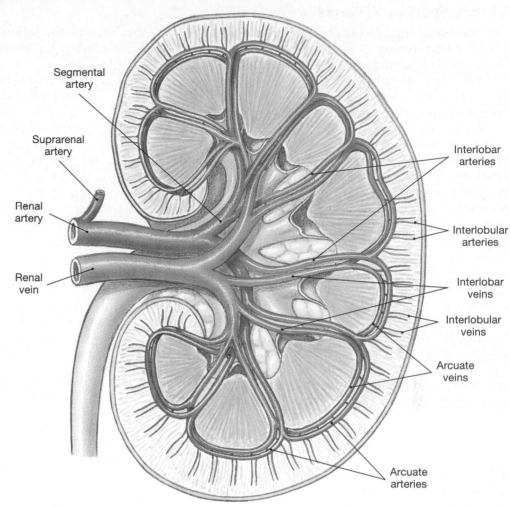

FIGURE 26-6 Kidney, sectional view showing major arteries and veins.

If available, view a corrosion cast demonstration specimen of the kidney. This type of preparation is a plastic cast of the actual blood supply, renal pelvis, and ureter. Check with your instructor. Use *Figure 26-7* for reference.

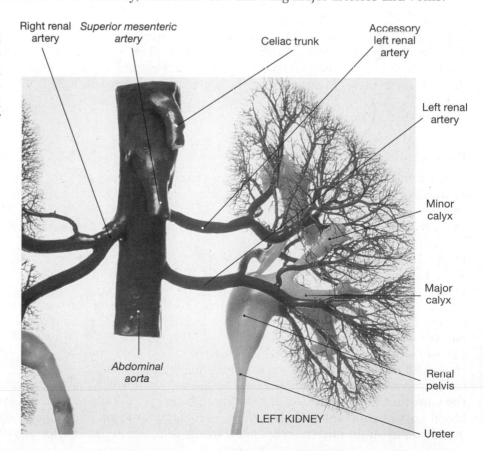

FIGURE 26-7 Corrosion cast of the circulation and conducting passageways of the kidney.

Observation of the Nephron of the Kidney

The functional unit of the kidney is the **nephron**. Here blood is filtered and urine is formed. A nephron is composed of a *renal corpuscle, proximal convoluted tubule, loop of Henle,* and *distal convoluted tubule.* [∞ p708] The **renal corpuscle** contains a knot of capillaries, the **glomerulus**, through which the blood plasma is filtered, and a collecting chamber called the **Bowman's capsule**. The blood that flows through this capillary knot enters through an afferent arteriole and exits the glomerulus through an efferent arteriole. This arrangement allows for precise control of the blood pressure within the glomerulus. A protein-free solution, termed *filtrate*, is collected within the capsular space of the renal corpuscle. Filtrate enters the nephron tubule passageway, where it is modified and altered. To modify the filtrate into urine, the nephron tubules are subdivided into three structural and functional regions. These regions are: (1) **proximal convoluted tubule**, the site of active and passive reabsorption of materials from the filtrate back to the blood plasma; (2) the **loop of Henle**, the site used to establish a salt gradient in the interstitial fluid of the medulla of the kidney, to aid in water and salt retention during urine formation; and (3) **distal convoluted tubule**, the site of active secretion of ions and acids and the selective reabsorption of sodium ions from the filtrate. The loop of Henle is subdivided into **descending** and **ascending limb** segments. The descending limb is permeable to water while the ascending limb contains active transport mechanisms to transport sodium and chloride ions from the filtrate. The water and ions re-enter the circulatory system. The distal convoluted tubule functions in water retention in response to the hormone aldosterone. The filtrate passes from the distal convoluted tubule into a **collecting tubule**, then a **collecting duct**, eventually to drain into the calyces, renal pelvis, and on to the bladder for storage. Under the influence of ADH, the permeability of collecting ducts to water is controlled. This plays a very important role in concentrating urine.

Procedure

Identify the following structures of the **nephron** and **collecting system** using the nephron *portion of a kidney model*. Use *Color Plates 124* and *125* for reference. Identify the structures in this order: renal corpuscle, glomerulus, glomerular (Bowman's) capsule, proximal convoluted tubule, loop of Henle, distal convoluted tubule, collecting tubule, and collecting duct. Return to the loop of Henle and locate the descending and ascending limb regions. Now observe the details of the renal corpuscle. Use *Figure 26-8* for reference. Identify the capsular epithelium of the renal corpuscle, which surrounds the capsular space. Observe how blood arrives at the glomerulus via the afferent arteriole and exits via efferent arterioles.

See: Color Plates 124, 125, Fig. 26-8

∞ Figs.
26-6, p709
26-8, p711

LOCATE

____ Nephron

____ Renal corpuscle { Glomerulus
Glomerular (Bowman's) capsule

____ Proximal convoluted tubule

____ Loop of Henle { Descending limb
Ascending limb

____ Distal convoluted tubule

____ Collecting tubule

____ Collecting duct

The Renal Corpuscle

____ Glomerulus

____ Capsular space

____ Capsular epithelium

____ Arterioles { Afferent
Efferent

____ Podocyte cells

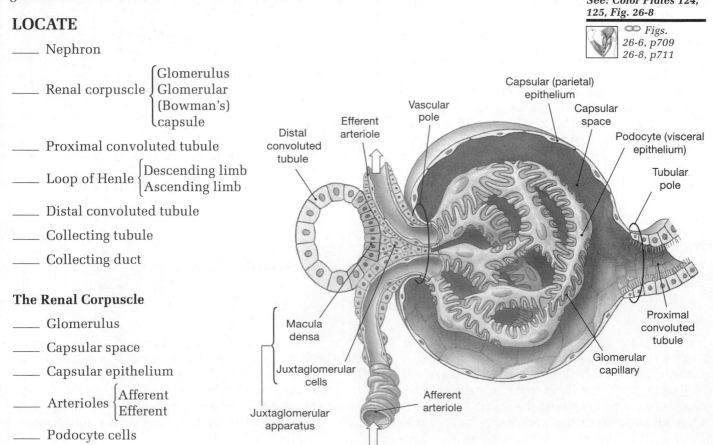

FIGURE 26-8   The renal corpuscle.

Microscopic Identification of Cortical and Juxtamedullary Nephrons

Procedure

Identify the histology of the **kidney** using the *microscope slide* provided by first viewing under a dissecting microscope, then with the compound microscope. Use *Figure 26-9* for reference. Observe a section of the kidney under the dissecting microscope and identify the cortex, medulla, and renal pelvis regions. Light-staining cells (sections of tubules) surrounding darker-staining cellular masses (sections of renal corpuscles) characterize the cortex. Elongated light-staining tubules characterize the medulla. Using the compound microscope, again locate the cortical and medullary regions. Under low power, locate dark-staining rounded cellular masses within the cortex. Identify these structures as glomeruli within the renal corpuscles. The renal corpuscles located in the outer cortex are called *cortical*, while those located near the border between cortex and medulla are termed *juxtamedullary*. Locate and observe a glomerulus under high dry power and then under oil. Identify the capsular space surrounding the glomerulus and the thin capsular epithelium of the Bowman's capsule.

Identify convoluted tubules adjacent to a renal corpuscle by the cuboidal cells that line the characteristic oval lumen. Proximal convoluted tubules are larger in diameter than distal tubules and can be distinguished by large light-staining cells containing microvilli that line a star-shaped lumen. Return to the medullary region and locate under low power the tubular rows of collecting ducts. Distinguish the ducts under high power by their very large lumens. The walls of the collecting ducts vary from cuboidal to low columnar epithelium. Nuclei stain dark, cytoplasm stains light, and cell walls are easily distinguished.

Slide # _____ Kidney, l.s.

See: Fig. 26-9

∞ Fig. 26-7, p710

LOCATE

Cortical Region

_____ Nephrons (cortical and juxtamedullary)

_____ Renal corpuscle

_____ Capsular space

_____ Glomerulus

_____ Capsular epithelium

_____ Distal convoluted tubules

_____ Proximal convoluted tubules

Medullary Region

_____ Collecting duct

_____ Capillaries

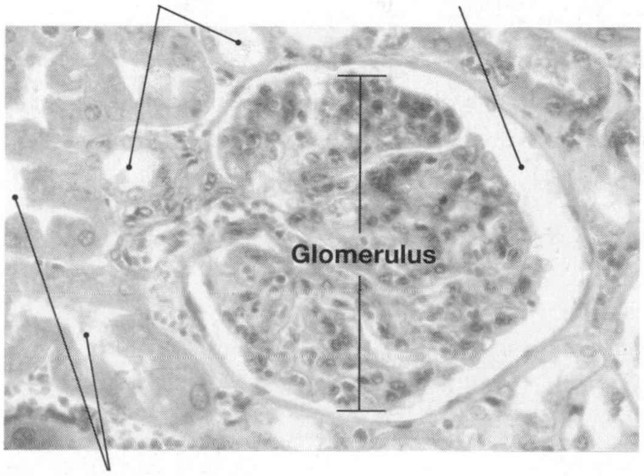

FIGURE 26-9 Proximal and distal convoluted tubules to a glomerulus × 980.

Organs Responsible for the Conduction and Storage of Urine

The ureters, urinary bladder, and urethra are responsible for the transport, storage, and removal of urine from the body. [∞ p712]

Ureters

The ureters are thick-walled muscular tubes, about 25–30 cm (12 inches) long, designed to transport urine from the kidneys to the urinary bladder. The ureters descend retroperitoneally over the psoas muscles, penetrating the posterior wall of the urinary bladder. In some individuals, the ureters may actually form a kink as they pass over the iliac crests. Urine flows from the calyces and collects in the renal pelvis draining into the ureters. Recall that the renal pelvis is the funnel-shaped initial region of the ureter. Urine flows down the ureters primarily due to the weak peristaltic action of this hollow muscular tube, but also due in part to gravity. Any interference with the flow of urine through the ureter shortens the life of the kidney. Over time, if not corrected medically, the pressure of urine retention will destroy the nephrons of the affected kidney. [∞ p714]

Urinary Bladder

The urinary bladder is an expansible, muscular sac positioned on the floor of the pelvic cavity. Its superior surface is covered by parietal peritoneum, thus it is retroperitoneal. (In males it is located anterior to the rectum and posterior to the symphysis pubis. But in females the bladder lies anterior to the uterus and anterosuperior to the vagina.) When filled with urine, the bladder rises from the pelvic cavity into the abdominal cavity. When this hollow muscular organ contracts, urine is forced into the urethra for expulsion from the body. The desire to urinate is initiated by the *micturition reflex*. Located in the muscular wall of the bladder are stretch receptors that sense distention and initiate the conscious desire to *urinate*, which in healthy individuals is controlled voluntarily. Orders from the ANS and cerebral cortex coordinate *urination* or the *voiding* of urine. Compression of the bladder by the abdominal muscles assists in the rapid expulsion of urine. [∞ p717]

Urethra

A single urethra drains the bladder and permits urine to flow from the body. As the urethra passes through the anterior triangle region of the urogenital diaphragm, a circular band of skeletal muscle forms the sphincter urethrae or external urethral sphincter. The structure and the function of the urethra differs significantly between females and males. In females, the urethra serves only to transport urine, but in males both urine and semen are conveyed out of the body via this single tube. [∞ p714]

Procedure

✓ **Quick Check** ——————————————————————————————

Before you begin to identify the position of the ureters, urinary bladder, and urethra, review the vertebrae, (*Figures 6-36, 6-37, p63*) the pelvis, (*Figure 7-11, p78*), and the perineum (muscles of the pelvic floor and associated structures). (*Figures 10-12 and 10-13, p128*)

Identify the ureters, urinary bladder, urethra, and associated structures in both males and females using *torso models* or cadaver *specimens*. Remove all of the digestive organs to observe these structures in the torso. Use *Color Plates 121, 128* and *Figure 26-1* for reference. Locate the hilus of the kidneys and identify the renal pelvis. Trace the ureters to the urinary bladder, observing how they descend medially and at an oblique angle over the psoas muscles. Examine the ureters as they penetrate the posterior and inferior wall of the urinary bladder. Open the bladder model and examine its interior. Use *Figure 26-10* for reference. Distinguish the pleats of the mucosa as **rugae**. The wall consists of three layers of smooth muscles that collectively form the **detrusor** muscle.

Within the wall of the bladder, identify the terminal portions of the ureters as the **ureteral openings**. Also, identify the opening to the urethra. The **sphincter vesicae** (*internal urethral sphincter*) controls the entry of urine into the urethra. Identify the smooth triangular area bounded by these openings as the **trigone**. When the detrusor muscle contracts, the trigone area is formed into the shape of a funnel, which aids in the rapid emptying of urine into the urethra.

The relationships of the urinary bladder and other organs are observed easily in sagittal sections of male and female pelvic models. Refer to *Color Plates 126* and *129*. Two ligaments stabilize and position the urinary bladder. Identify the **median umbilical ligament (urachus)**, which extends from the anterior and superior surface of the bladder to stabilize the anterior and superior borders. Identify the **lateral umbilical ligaments**, which pass along the lateral surface from anterior to posterior to stabilize the lateral borders. These ligaments are typically not shown on models, but are identifiable in the prosected cadaver.

The urethra in both sexes passes through the urogenital diaphragm and is observed best in sagittal section. Identify the short **female urethra** and its opening to the outside as the **external urethral meatus**. In the male, identify the **prostatic region** and the **penile urethra region**. A circular band of skeletal muscle surrounds the urethra to form the **sphincter urethrae** (*external urethral sphincter*), which is under voluntary control via the pudendal nerve.

See: *Color Plates 121,
123, 126 to 130,
Fig. 26-10*

 ∞ *Figs.
26-2, p705
26-9c, p713
26-10, p715*

LOCATE

____ Peritoneum

____ Ureter

____ Urinary bladder

____ Rugae

____ Ureteral openings

____ Sphincter vesicae (internal urethral sphincter)

____ Trigone

____ Median umbilical ligament (urachus)

____ Lateral umbilical ligaments

____ Female urethra

____ External urethral meatus

____ Male urethra $\begin{cases} \text{Prostatic region} \\ \text{Penile region} \end{cases}$

____ Sphincter urethrae (external sphincter) in urogenital diaphragm

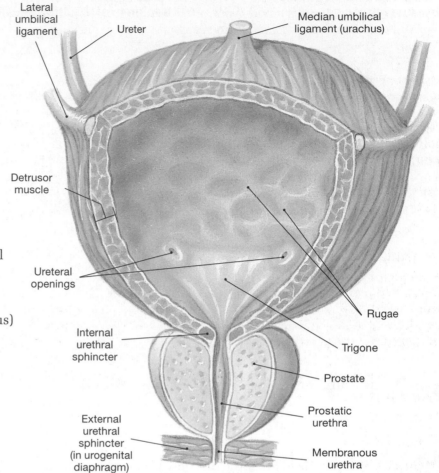

FIGURE 26-10   The urinary bladder.

Identify the anatomy of the ureters, urinary bladder, and urethra in the prosected cadaver using the above directions. Before you begin to identify these structures, verify the following landmarks: diaphragm, lumbar vertebrae, (L_1 to L_4), the floating 12th ribs, and psoas major muscle. Use *Color Plates 109* and *123*. The abdominopelvic organs must be reflected first and maintained in that position. Examination of the urinary structures will require repositioning of the digestive organs.

Demonstration specimens of sagittal sections of male and female pelvis may be available also for you to observe the position of the urinary structures in relation to the dorsal body wall, digestive structures, and pelvic floor. Check with your instructor. Use *Color Plates 127* and *130* for reference.

Microscopic Identification of the Ureter

Procedure

✓ **Quick Check**

Before you begin to identify the histology of the ureter, urinary bladder, and urethra, recall the histology of the digestive tube: mucosa, submucosa, muscularis, and serosa layers. (Chapter 25, p364) The ureter has a similar organizational plan.

Identify the histology of the **ureter** using the *microscope slide* provided, first by viewing under a dissecting microscope, then under low power magnification of the compound microscope. Observe a cross section of the ureter under the dissecting microscope. Identify from the lumen to the outer wall the three layers of this muscular tube using *Figure 26-11* for reference. The ureter conforms to the histological plan of the digestive tube, with two exceptions. It lacks a submucosa, and the bands of smooth muscle that form the muscularis layer are in reverse order to that observed in the digestive system. Inner longitudinal and outer circular smooth muscle layers are found in the ureter. Notice how the mucosa is thrown into folds, presenting a very irregular lumen (often star-shaped). Identify the transitional epithelium of the mucosa, (see *Color Plate 5*) and identify the muscularis layer by the obvious smooth muscle fibers, peripheral to the mucosa. The outer connective tissue layer (adventitia) is continuous with the fibrous capsule and the peritoneum.

Slide # _____ Ureter, sec.

See: Color Plate 5, Fig. 26-11

∞ *Fig. 26-11, p716*

LOCATE

_____ Lumen of ureter

_____ Mucosa of transitional epithelium

_____ Muscularis $\Big\{$ Smooth muscle $\Big\{$ Longitudinal (inner) layer / Circular (outer) layer

_____ Adventitia

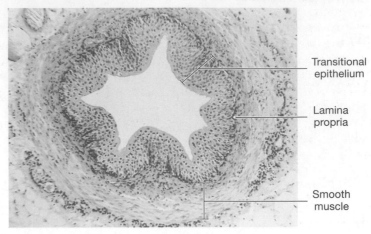

FIGURE 26-11   Ureter, transverse section × 65.

Labels: Transitional epithelium; Lamina propria; Smooth muscle

Microscopic Identification of the Urinary Bladder

Procedure

 Identify the histology of the **urinary bladder** using the *microscope slide* provided first by viewing under a dissecting microscope, then under low power magnification of the compound microscope. Observe a section of the urinary bladder under the dissecting microscope. Identify from the lumen to the outer wall the four layers that compose the wall of the urinary bladder, using *Figure 26-12* for reference. Identify the transitional epithelium of the mucosa. (*Color Plate 5*) Notice how the mucosa and submucosa are thrown into folds or rugae. The lamina propria of the mucosa blends with the submucosa to form an unusually thick layer of fibers. Identify the thick muscularis layer by the obvious smooth muscle fibers. Three layers of smooth muscle fibers compose the muscularis; inner longitudinal, middle circular and outer longitudinal. Collectively, these layers form the **detrusor muscle**. An outer serosa layer can be identified only on the superior surface of the bladder, since it is the only portion that is continuous with the parietal peritoneum.

Slide # _____ Urinary Bladder, sec.

See: Fig. 26-12

∞ *Fig. 26-11, p716*

LOCATE

_____ Lumen of urinary bladder

_____ Mucosa $\Big\{$ Transitional epithelium / Lamina propria

_____ Submucosa

_____ Muscularis $\Big\{$ Smooth muscle $\Big\{$ Detrusor Muscle / Longitudinal (inner) and (outer) layers / Circular (middle) layer

_____ Serosa (Visceral peritoneum) (visible only on superior surface)

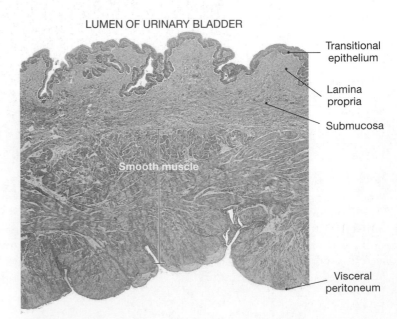

LUMEN OF URINARY BLADDER

FIGURE 26-12   The wall of the urinary bladder × 65.

Labels: Transitional epithelium; Lamina propria; Submucosa; Smooth muscle; Visceral peritoneum

Microscopic Identification of the Urethra

Procedure

 Identify the histology of the **urethra** using the *microscope slide* provided by first viewing under a dissecting microscope, then under low power magnification of the compound microscope. Observe a cross section of the female urethra under the dissecting microscope. Identify the layers from the lumen to the outer wall using *Figure 26-13* for reference. A collapsed, crescent-shaped lumen is characteristic of the female urethra. The mucosa near the bladder consists of transitional epithelium, but changes progressively to pseudostratified and finally to stratified squamous epithelium. You can determine the location of the urethral specimen on your slide by the type of epithelium observed. Identify mucous epithelial glands within the lamina propria of the mucosa. The muscularis layer consists of obvious smooth muscle fibers arranged in a circular pattern around the mucosa. Identify the two layers of smooth muscle fibers composing the muscularis layer: a longitudinal (inner) layer and a circular (outer) layer. A serosa layer is not present. The microscopic identification of the male urethra is described in Chapter 27.

Slide # _____ Urethra, sec.

See: Fig. 26-13

∞ *Fig.*
26-11, p716

LOCATE

_____ Lumen of urethra

_____ Mucosa
$\begin{cases} \text{Various epithelial types} \\ \text{Lamina propria containing} \\ \text{mucous epithelial glands} \end{cases}$

_____ Muscularis $\begin{cases} \text{Smooth} \\ \text{muscle} \end{cases} \begin{cases} \text{Longitudinal (inner) layer} \\ \text{Circular (outer) layer} \end{cases}$

_____ Serosa

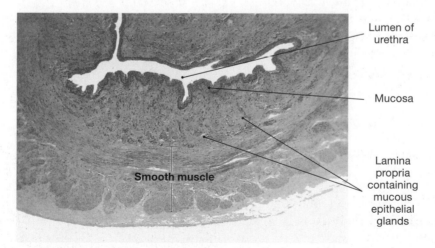

FIGURE 26-13   Female urethra, transverse section × 61.

Dissection of the Urinary System in the Cat

See: Fig. 26-14

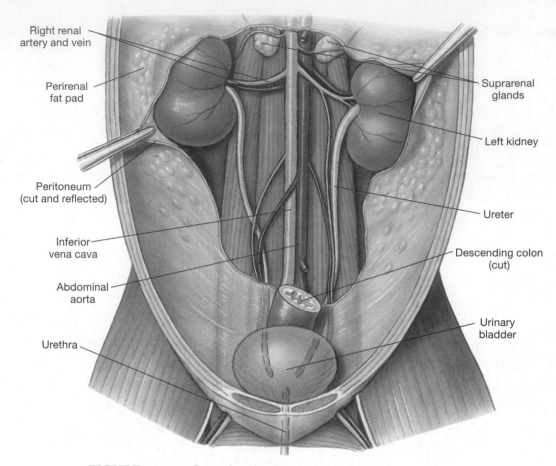

FIGURE 26-14 Cat urinary system.

Urinary System

This exercise complements the study of the human urinary system. The urinary system of the cat is similar to humans. In your dissection of the cat urinary system, trace the pathway of urine from its site of formation in the kidney, through its passage to the urinary bladder, through the urethra to the exterior of the body.

1. Reflect the abdominal viscera to the opposite side of the abdominal cavity to locate the large, bean-shaped **kidneys**. They are retroperitoneal and do not appear symmetrical on the dorsal body wall. Both kidneys are surrounded by pads of perirenal fat and are enclosed in a tough fibrous **renal capsule**.

2. Locate the **adrenal glands**, superior to the kidneys and close to the aorta. They are supplied by the suprarenal arteries.

3. Free each kidney by carefully removing the **peritoneum** and the perirenal fat pads.

4. Identify the following structures that pass through the **hilum**, the concave medial surface of the kidney: renal artery, renal vein, and the **ureter**.

5. The ureters are cream-colored tubes that descend posteriorly along the dorsal body wall to drain urine into the **urinary bladder**. The bladder is a muscular sac connected to both lateral and ventral walls by suspensory ligaments.

6. The large expanded region of the bladder is the **fundus**. Posteriorly, the bladder narrows into the **neck**, and continues as the **urethra**, through which urine passes to the exterior of the body. The bladder must be pulled anteriorly to observe the neck and urethra portions.

7. With your scalpel blade, make a frontal (coronal) section through the kidney, such that the section passes through the middle of the hilum. Locate the outer, lighter **cortex** and the inner, darker **medulla**.

8. Observe the funnel-shaped expansion of the ureter, the **renal pelvis**, which occupies the hollow interior of the kidney, the **renal sinus**, internal to the opening of the hilum. Locate a single **renal papilla**, the rounded projection of the **renal pyramid** of the medulla in the renal pelvis.

Anatomical Identification Review

Label the major structures of the kidney on the accompanying drawing. Mentally describe the function of each structure.

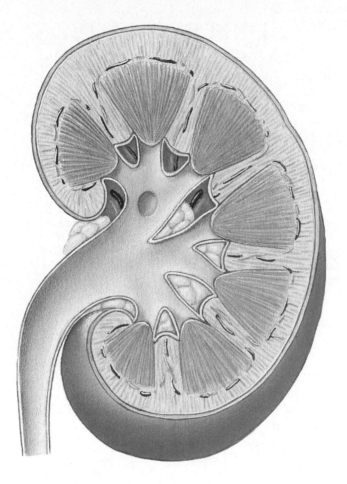

To Think About

1. Why does a pregnant woman have a greater need to urinate more frequently than she does when she is not pregnant?

2. Trace the path of a drop of urine from the point it leaves the renal pyramid to exit the body. List in order the structures urine must pass through.

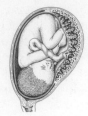

27

THE REPRODUCTIVE SYSTEM

Objectives

1. Describe the organization and function of the male and female reproductive systems.

2. Identify the structures that compose both the male and female reproductive systems and describe the functions of each structure.

3. Observe both the gross and microscopic anatomy of the following male reproductive system structures: testis, spermatic cord, epididymis, ductus deferens, seminal vesicles, prostate gland, bulbourethral gland, and penis.

4. Observe both the gross and microscopic anatomy of the following female reproductive system structures: ovaries, uterine tubes, uterus, vagina, external genitalia, and mammary glands.

The **reproductive system** functions to perpetuate the species. It does this by producing and supporting reproductive cells called **gametes**. [∞ pp725, 737] Male and female reproductive systems differ structurally and functionally because of their roles in the perpetuation of the species. Male and female reproductive organs, termed **gonads**, produce both gametes and hormones. The gonads of the male are the testes and those of the female are the ovaries. [∞ pp725, 737]

The reproductive system of the male is responsible for the production of hormones and the formation, nutrition, storage, transport, and delivery of the male gamete, called **sperm**. After production and storage, mature sperm cells pass through a long and tortuous system of ducts where they are mixed with the secretions of accessory glands, termed *seminal fluid*, producing a final fluid mixture known as **semen**. The process of expelling semen from the male reproductive tract is termed **ejaculation**. The structures of the male reproductive system include: *testis, epididymis, ductus deferens, spermatic cord, seminal vesicles, prostate gland, bulbourethral glands, urethra*, and *penis*.

The reproductive system of the female is responsible for the production of hormones and the formation, support, transport, and delivery of the female gamete, termed **ovum**. Additionally, the female reproductive system protects and supports the developing embryo and nourishes the growing fetus. The female system structures include: the *ovaries, uterine tubes, uterus* and *vagina*.

Observation of the Male Reproductive System

The **external genitalia** of the male include the *scrotum* and the *penis*. The male reproductive tract and urinary tract share a common passageway, the urethra. During fetal development, the testes form inside the body cavity adjacent to the kidneys. In the seventh month of pregnancy, the fetus's testes begin to descend through the inferior body wall into the scrotum. As each testis descends into the scrotum, it is accompanied by the spermatic cord, which contains blood vessels, nerves, and lymphatics that service the testes, plus the reproductive duct (ductus deferens) that carries sperm to the urethra. [∞ p725]

Procedure

✓ **Quick Check** ──

Before you begin to identify the structures of the male reproductive system, review on a dry skeleton the pelvis (*Figures 7-11* and *7-12, pp78-79*) and muscles of the male pelvic floor. (*Table 10-9, p126* and *Figure 10-12, p128*)

Locate the components of the **male reproductive system** using a *male pelvis model* or *cadaver*. To observe the shared reproductive and urinary tract structures, the male reproductive system must be examined from both sagittal and anterior views. Use *Color Plates 126, 127,* and *Figure 27-1* for sagittal views of the male pelvis. For anterior views, use *Color Plate 121* and *Figure 27-2* for reference. Remove the chest plate and all abdominal viscera from the intact torso. Locate the male external genitalia, the scrotum and the penis. The male gonads or **testes** are paired organs that produce both gametes and the male hormones. Each walnut-sized testis is contained within a sac of skin, the **scrotum**. Through the penis, the male urethra both conveys urine out of the body and introduces semen into the female vagina during sexual intercourse.

Each testis is supplied with nerves from the lumbar plexus, lymphatic ducts, a spermatic artery and vein, and a duct for the transport of semen to the urethra, termed the *ductus deferens*. These nerves and vessels must communicate between the abdominopelvic cavity and the testes within the scrotum. They are bundled into a tubular cable, the **spermatic cord**. Identify left and right spermatic cords at the deep inguinal ring. Here each cord penetrates the abdominal wall. The cord exits at the superficial ring to descend into the scrotal cavity to a testis.

The glandular structures of the male reproductive system that are involved with the maturation, nutrition, transport, and storage of spermatozoa are the *seminal vesicles, prostate gland,* and *bulbourethral glands.* Locate these glands in the neck area of the urinary bladder and on its posterior surface. Observe the glands from both anterior and sagittal views. Each gland is described in greater detail under its own heading.

See: Color Plates 121, 126, 127, Figs. 27-1, 27-2

∞ *Figs. 26-10, p715 27-1, p726 27-3, p728 27-4, p729 27-8a, p734*

LOCATE

_____ Testis

_____ Scrotum

_____ Penis

_____ Ductus deferens (vas deferens)

_____ Spermatic cord

_____ Seminal vesicle

_____ Prostate gland

_____ Bulbourethral gland

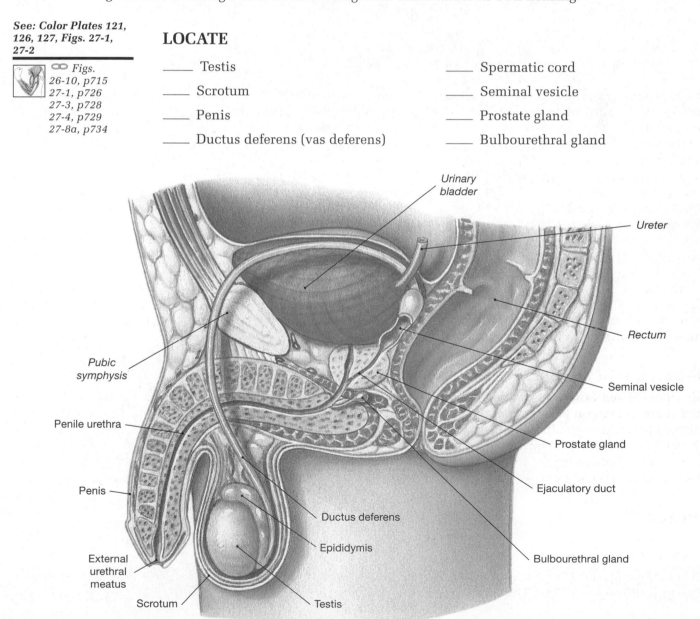

FIGURE 27-1 Male reproductive system as seen in sagittal view.

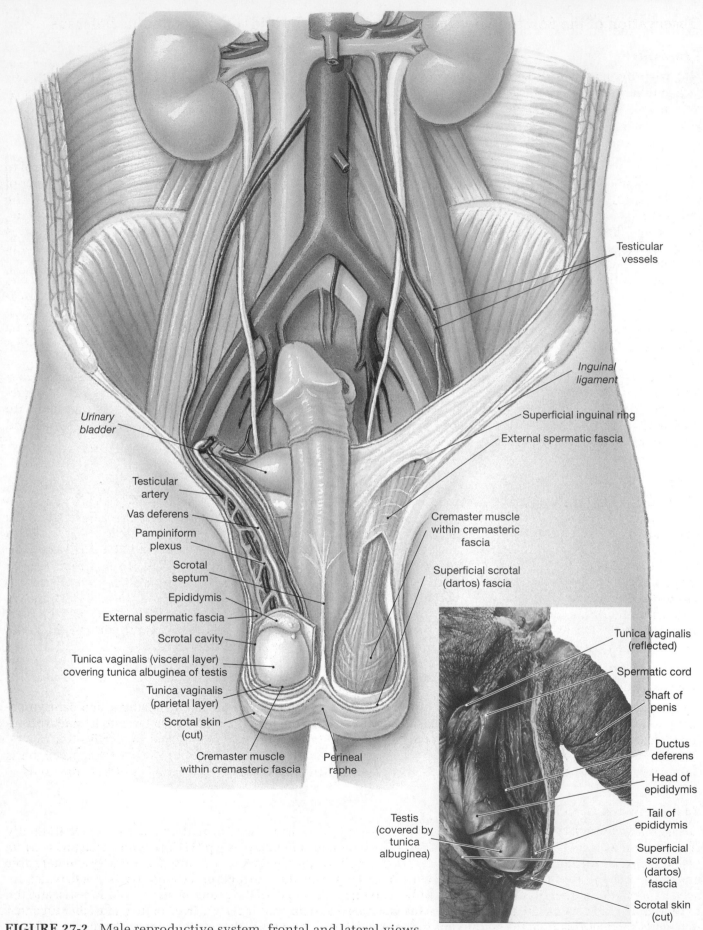

Testicular
vessels

*Inguinal
ligament*

Superficial inguinal ring

External spermatic fascia

Cremaster muscle
within cremasteric
fascia

Superficial scrotal
(dartos) fascia

*Urinary
bladder*

Testicular
artery

Vas deferens

Pampiniform
plexus

Scrotal
septum

Epididymis

External spermatic fascia

Scrotal cavity

Tunica vaginalis (visceral layer)
covering tunica albuginea of testis

Tunica vaginalis
(parietal layer)

Scrotal skin
(cut)

Cremaster muscle
within cremasteric fascia

Perineal
raphe

Tunica vaginalis
(reflected)

Spermatic cord

Shaft of
penis

Ductus
deferens

Head of
epididymis

Tail of
epididymis

Superficial
scrotal
(dartos)
fascia

Scrotal skin
(cut)

Testis
(covered by
tunica
albuginea)

FIGURE 27-2 Male reproductive system, frontal and lateral views.

Observation of the Scrotum, Spermatic Cord, Testes, Epididymis, and Ductus Deferens

Procedure

 Identify the **scrotum**, **spermatic cord**, structural organization of the **testes**, and their **association** with the **epididymis** and **ductus deferens** using a *male pelvis model* or *cadaver*. Use *Color Plates 126, 127,* and *Figures 27-1 to 27-4* for reference.

Scrotum

The dermis of the scrotum contains the **dartos muscle**, a thin smooth muscle that contracts under ANS control to give the characteristic wrinkled appearance to the scrotal sac. [∞ p728] Interior to the dartos lies a thin layer of skeletal muscle, the **cremaster muscle**, which surrounds each testis and the structures of the spermatic cord. The cremaster is continuous with the fibers of the internal oblique muscle. For the testes to produce viable sperm, the temperature in the testes must be maintained two degrees lower than body temperature. To maintain this temperature difference, the cremaster muscle contracts and relaxes under involuntary control in response to changes either in environment or body temperature, as well as during sexual excitement.

Examine the interior of the scrotum (from a coronal or horizontal view) and identify two separate chambers, the **scrotal cavities**, which are separated by an internal septum. This septum is marked by a raised midline thickening on the external scrotal surface, termed the **scrotal (perineal) raphe**. Observe that each scrotal cavity contains a testis and is lined by a delicate serous membrane, the **tunica vaginalis**. This tunic is derived from the parietal peritoneum of the abdomen during descent of the testes from the abdominal cavity in the later part of fetal development. It aids in reducing friction between testes and scrotum during body movements.

Blood is supplied to the scrotum and portions of the testes via three vessels: (1) the internal pudendal arteries (from the internal iliac arteries), (2) the external pudendal arteries (from the femoral arteries), and (3) the cremasteric branch of the inferior epigastric arteries (from the external iliac arteries). The scrotum is supplied with both sensory and motor nerves from the hypogastric plexus and branches of the ilioinguinal, genitofemoral, and pudendal nerves. (*Table 14-3, p209*)

Testes

The **tunica albuginea** is observed upon removal of the tunica vaginalis of the testis. [∞ p728] This tunic contains numerous collagen fibers, thereby making it tougher than the tunica vaginalis. Examine a cut section of the testis and observe that connective tissue *septa* partition the testis into 200 to 300 lobules. Each lobule contains about one to three tightly coiled tubes, termed **seminiferous tubules**. A testis may contain 400 to 600 seminiferous tubules. Sperm cells are called **spermatozoa**, a single sperm cell is termed a spermatozoon. The production of sperm cells occurs in the seminiferous tubules by spermatogenesis. Note how the septa converge at the **mediastinum**, which forms an internal partition to provide support for the vessels and nerves of the testis. Within the mediastinum, each seminiferous tubule connects to a **straight tubule**. The straight tubule forms a maze of tubes, the **rete testis**. Large **efferent ducts** connect the rete testis to the epididymis. Blood is supplied to the testes by the **testicular artery** and is drained by a collection of veins known as the **pampiniform plexus**.

Spermatic Cord

A testicular artery, the pampiniform plexus of veins, nerves from the lumbar plexus, sympathetic and parasympathetic nerves, lymphatic ducts, and the *ductus deferens* are surrounded and bundled together into a tubular structure, termed the spermatic cord, (see *Figure 26-1, p393*) by both the cremaster muscle and connective tissue. [∞ p725] Verify the deep inguinal ring and observe how each cord penetrates the abdominal wall, passing through the inguinal canal, exiting at the superficial ring, descending and entering left and right scrotal cavities to the testes.

Epididymis and Ductus Deferens

Locate the comma-shaped **epididymis** on the posterior surface of the testis. The epididymis is a 7 m (23 ft) tightly coiled, comma-shaped tube that begins superiorly and continues inferiorly. [∞ p732] The tunica albuginea covers the epididymis and attaches it to the testis. Within the epididymis, spermatozoa mature functionally and defective ones are either absorbed or persist in ejaculated semen. Locate the three regions of the epididymis: **head**, which receives spermatozoa from the efferent ducts; **body**, extending the length of the posterior surface of the testis; and the twisted **tail**, which opens into the tubular **ductus deferens** (vas deferens). This is a 16 to 18 inch muscular tube that transports spermatozoa from the epididymis to the ejaculatory ducts within the prostate gland. The ductus deferens lies retroperitoneal throughout its course. Locate the ductus deferens as they loop superior to the ureters and descend inferiorly along the posterior surface of the urinary bladder. (*Figures 26-1, p393* and *27-8, p413*) Each ductus enlarges to form an **ampulla** region. Both regions then uniting with the duct of the seminal vesicles, within the prostate gland, to form the two **ejaculatory ducts**. These ducts open into the prostatic urethra.

See: Color Plates 126, 127, Figs. 27-3, 27-4

∞ *Figs. 27-4a, p729 27-7a, p733*

LOCATE

_____ Scrotum

_____ Scrotal cavity

_____ Tunica vaginalis

_____ Scrotal (Perineal) raphe

_____ Testis

_____ Tunica albuginea

_____ Septa

_____ Lobules

_____ Seminiferous tubules

_____ Mediastinum

_____ Straight tubules

_____ Rete testis

_____ Efferent ducts

_____ Inguinal rings { Superficial / Deep

_____ Inguinal canal

_____ Spermatic cord {
External spermatic fascia
Cremaster muscle/ fascia
Superficial scrotal (dartos) fascia
Vas deferens
Arteries/Veins
  Testicular
  Pampiniform plexus
  Internal/external pudendal
  Inferior epigastric
}

_____ Epididymis { Head / Body / Tail

_____ Ductus deferens (vas deferens) { Ampulla / Ejaculatory duct

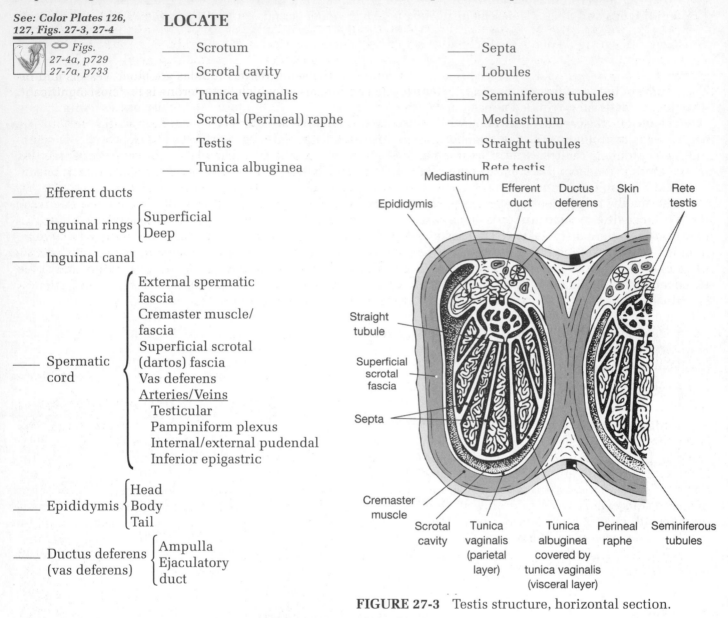

FIGURE 27-3 Testis structure, horizontal section.

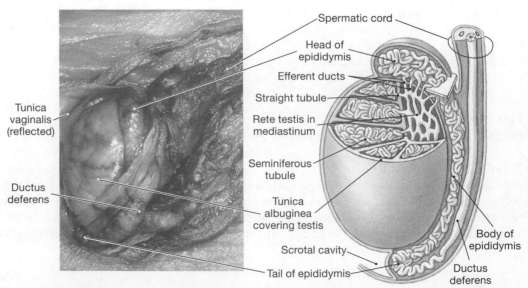

FIGURE 27-4 Testis and epididymis as seen in gross dissection.

Microscopic Identification of the Testes

Procedure

Identify the histology of the **testis** using the *microscope slide* provided, first by viewing under a dissecting microscope, then with the compound microscope. Use *Figure 27-5* for reference. Histologically, the testes have two major types of tissue, one for producing sperm cells, the other for producing the male hormone **testosterone**. Each lobule of the testis contains a tightly coiled mass of seminiferous tubules for the production of spermatozoa. Sperm cells are produced by the process of **spermatogenesis**. These male gametes contain one-half the normal number (haploid) of chromosomes. Surrounding the seminiferous tubules are blood vessels and **interstitial cells**. These cells produce sex hormones called androgens, of which testosterone is the most significant. [∞ p729]

Observe a section of the testis under the dissecting microscope and identify the connective tissue septa that extend centrally from the tunica albuginea, dividing the testis into lobules. Seminiferous tubules occupy the lobules. The cut tubules may appear in a variety of forms: circles, ovals, question marks, or the shapes of the letters "C" or "U." Use the low power of the compound microscope to locate several tubules, then switch to high power. Under high power, locate one seminiferous tubule section in which cells can be distinguished clearly. To observe cells in different stages of spermatogenesis, begin your observation at the outside of the tubule and work towards the lumen. Locate at the periphery actively dividing stem cells, the large **spermatogonia**. Working towards the lumen, identify the large **primary spermatocytes** and the smaller **secondary spermatocytes** by their prominent dividing nuclei. Tall, irregularly-shaped **sustentacular cells** can be observed supporting these dividing cells. Within the lumen, identify torpedo-shaped cells with dark-staining nuclei as **spermatids** and those with a tail as **spermatozoa**. Located in the spaces between the seminiferous tubules are small clusters of **interstitial cells** (*cells of Leydig*). These are large oval-shaped cells with a large prominent nucleus.

Slide # _____ Testis, sec.

See: Fig. 27-5

∞ *Figs.*
27-4b, p729
27-5, p730

LOCATE

_____ Septa

_____ Seminiferous tubules ⎰ Spermatogonia
⎱ Dividing spermatocytes
⎱ Spermatids
⎱ Spermatozoa

_____ Interstitial cells (cells of Leydig)

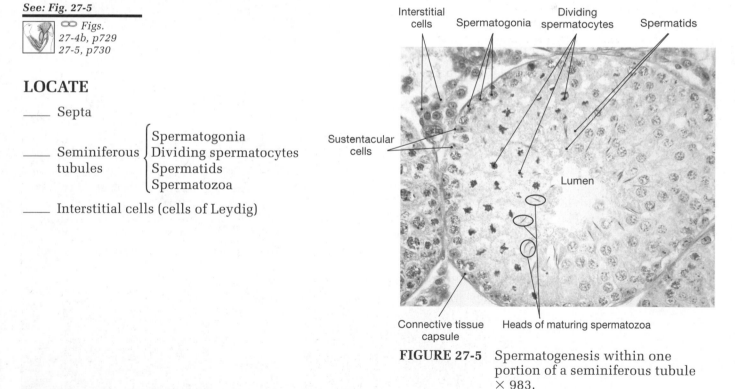

Interstitial cells | Spermatogonia | Dividing spermatocytes | Spermatids

Sustentacular cells

Lumen

Connective tissue capsule | Heads of maturing spermatozoa

FIGURE 27-5 Spermatogenesis within one portion of a seminiferous tubule × 983.

Microscopic Identification of the Epididymis

Procedure

Identify the histology of the **epididymis** using the *microscope slide* provided, first by viewing under a dissecting microscope, then with the compound microscope. Use *Figure 27-6* for reference. Observe a section of the epididymis under the dissecting microscope and identify the tubules of the coiled epididymis. The cut tubules of the epididymis may appear in the same shapes as the seminiferous tubules of the testis. Use the low power of the compound microscope to locate several cut tubules, then switch to high power. Under high power, observe extremely large numbers of spermatozoa with flagella that fill the lumen of each cut tubule. Observe pseudostratified columnar epithelium (see *Color Plate 8*) with stereocilia lining the lumen of the epididymal tube.

Slide # _____ Epididymis, sec.

See: Fig. 27-6
∞ *Fig.
27-7b, c, p733*

LOCATE

_____ Coiled epididymis containing spermatozoa

_____ Epithelium of epididymis with stereocilia

Microscopic Identification of Spermatozoa

Procedure

 Identify the basic structure of **spermatozoa** using the *microscope slide* provided, first by viewing under low power magnification and then under oil immersion. Use *Figure 27-7* for reference. Align the center of the cover slip of the slide directly under the low power objective. Focus under low power, then switch to high dry magnification. Use the oil immersion objective to observe the detail of a spermatozoon. (If your slide has been stained with iron hematoxylin, spermatozoa will appear blackish/gray.) A spermatozoon consists of three regions: *head, middle piece*, and *tail*. The head appears round to torpedo-shaped and contains an acrosome and chromosomes within a nucleus. The constricted and slightly elongated middle piece contains mitochondria and microtubules for propelling the tail. The tail is a **flagellum.** (It will not be possible for you to identify mitochondria or microtubules because greater magnification is required.) Microtubules extend into the flagellum and move it by a corkscrew pattern of motion.

Slide # _____ Human Sperm Smear

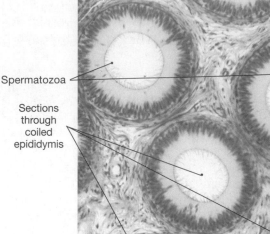
See: Fig. 27-7
∞ *Fig.
27-6b, p731*

LOCATE

_____ Head (containing nucleus and acrosome)

_____ Middle piece

_____ Tail (flagellum)

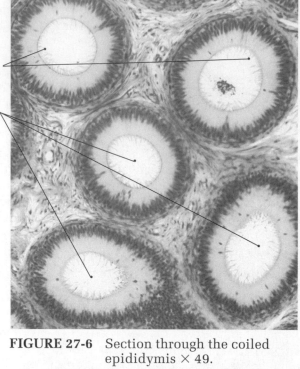

FIGURE 27-6 Section through the coiled epididymis × 49.

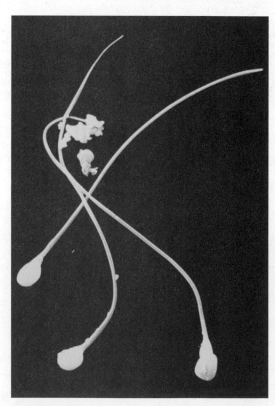

FIGURE 27-7 Human spermatozoa × 1688.

Observation of the Accessory Glands: Seminal Vesicles, Prostate, and Bulbourethral

Seminal fluid is a mixture of the secretions of three accessory glands. [∞ p733] The **seminal vesicles** are a pair of small glands that lie on the posterior surface of the urinary bladder and secrete directly into the ejaculatory duct. These secretions furnish nutrients to the spermatozoa, make up about 60% of the seminal fluid, and help to neutralize the acid environment of the urethra and vagina. The **prostate gland** surrounds the urethra at the neck of the urinary bladder. The secretion of this gland accounts for more than 20-30% of seminal fluid. Secretions are directed into the prostatic portion of the urethra. This secretion provides nutrients to the slightly acidic spermatozoa. The **bulbourethral glands** (*Cowper's glands*) are located adjacent to the urethra within the urogenital diaphragm. These glands contribute a viscid, alkaline fluid that lubricates the penis and neutralizes the acid environment of the male urethra.

Procedure

Identify the **accessory glands** of the male reproductive system using a *male pelvis model* or *cadaver specimen*. The glands must be examined from anterior, posterior, and sagittal views to observe their relationship with urinary tract structures. Use *Color Plates 126, 127*, and *Figure 27-1* for sagittal views of the male. For anterior and posterior views, use *Color Plate 121* and *Figures 27-2, 27-8*, and *27-9* for reference. The accessory glands may be viewed also from the floor of the perineum. Remove all abdominal viscera from the intact torso. Verify first urinary bladder, ureters, urethra, and rectum.

Seminal Vesicles

Locate the seminal vesicles embedded in connective tissue between the urinary bladder and rectum. (*Figure 26-1, p393*) The 2½ inches long seminal vesicles are cone-shaped sacs with irregular pouches that empty into the ejaculatory duct at the posterior surface of the prostate.

Prostate

Locate the prostate gland between the neck of the urinary bladder and the pelvic floor. (*Figure 26-1b, p393*) The slightly pyramid-shaped prostate completely encircles the prostatic urethra. The secretions of the prostate are conveyed by 12 to 20 prostatic ducts into the prostatic urethra.

Bulbourethral

The bulbourethral glands are very difficult to locate in the cadaver. Locate these glands adjacent to the membranous urethra within the urogenital diaphragm. Their secretions are conveyed directly into the penile urethra.

See: Color Plates 121, 126, 127, Figs. 27-8, 27-9

∞ Fig.
26-10a, c, p715
27-8a, p734

LOCATE

____ Urinary bladder

____ Urethra segments { Prostatic Membranous Penile

____ Ductus deferens

____ Ejaculatory ducts

____ Seminal vesicles

____ Prostate gland

____ Bulbourethral glands

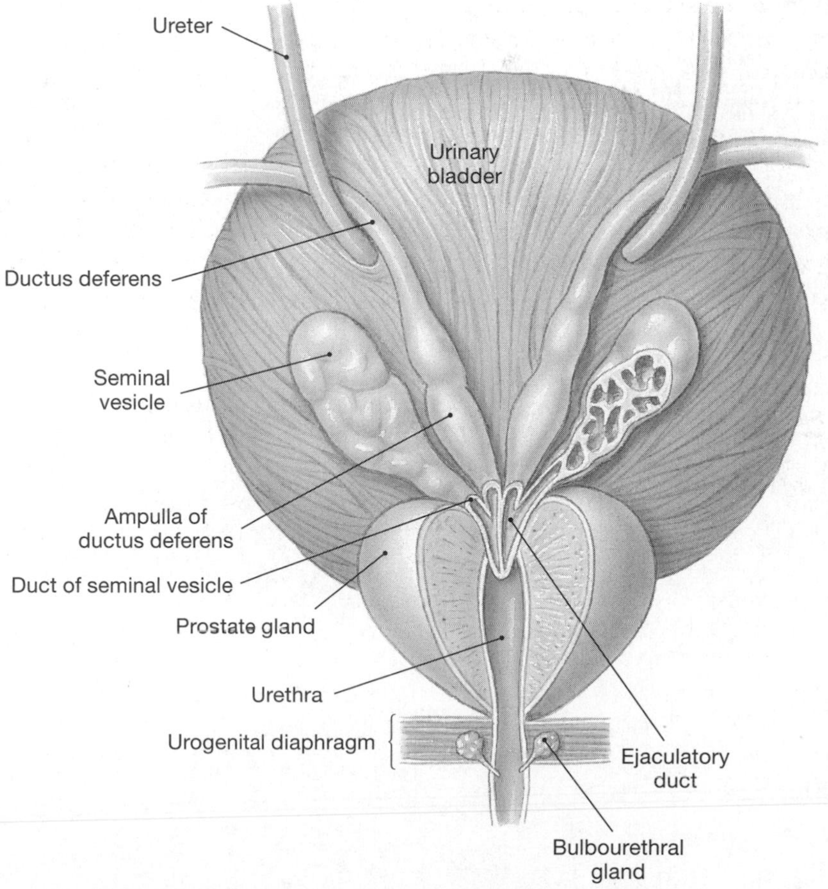

FIGURE 27-8 The accessory glands, posterior view.

Microscopic Identification of the Accessory Glands: Seminal Vesicle, Prostate, and Bulbourethral

Procedure

Identify the histology of each **accessory gland** using the *microscope slide* provided, first by viewing with low power magnification and then under high power magnification.

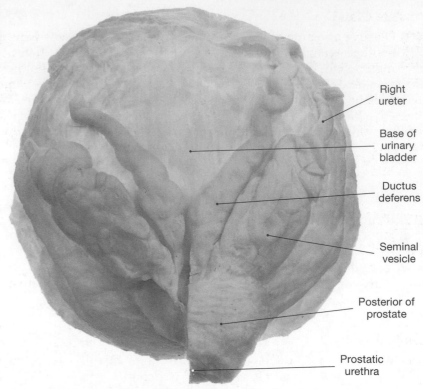

FIGURE 27-9 Male urinary bladder and accessory glands, as seen from the posterior.

Seminal Vesicle

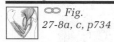 Using the compound microscope, observe a section of the seminal vesicle under low power and note the many folds of the mucosa that extend into the lumen. Switch to high power. Use *Figure 27-10* for reference. Under high power, observe the extensive glandular epithelium, showing areas of *secretory pockets*. Most of the seminal fluid is produced by these glands. Pseudostratified columnar epithelium is characteristic. (*Color Plate 8*) A muscularis layer of smooth muscle can be observed externally, and an adventitia layer surrounds the muscularis layer. Peristaltic contractions of the smooth muscle discharge seminal fluid into the ductus deferens at emission.

Slide # _____ Seminal Vesicle, sec.

See: Fig. 27-10

∞ *Fig. 27-8a, c, p734*

LOCATE

_____ General appearance of the gland

_____ Pseudostratified columnar epithelium

_____ Secretory pockets

_____ Muscularis layer (smooth muscle fibers)

_____ Adventitia

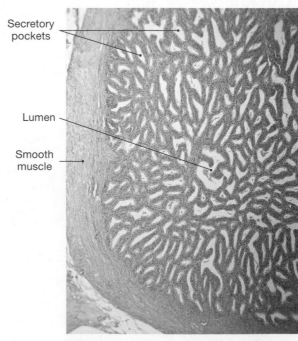

FIGURE 27-10 Seminal vesicles × 57.

Prostate Gland

Observe a section of the prostate gland under low power using the compound microscope and note the general honeycombed appearance of the gland. Switch to high power. Use *Figure 27-11* for reference. Under high power, observe numerous clusters of compound tubuloalveolar glands. The epithelium characteristic of these glands is simple columnar to psuedostratified. (*Color Plates 6* and *8*) Groups of these glands are surrounded and separated by connective tissue and smooth muscle fibers. Peristaltic contractions of these muscle fibers discharge the *prostatic fluid* into the urethra at emission. Muscle and connective tissue constitute more than one-third of the prostate. After middle-age the glandular structures atrophy, while the connective tissue undergoes hypertrophy, causing prostatic enlargement.

Slide # _____ Prostate Gland, sec.

See: Fig. 27-11

∞ Fig. 27-8 a, e, p734

LOCATE

_____ General appearance of the gland

_____ Glands of the prostate (compound tubuloalveolar glands)

_____ Smooth muscle/connective tissue

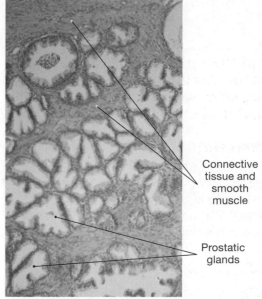

Connective tissue and smooth muscle

Prostatic glands

FIGURE 27-11   Prostate gland × 51.

Bulbourethral (Cowper's) Gland

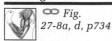

Observe a section of the bulbourethral gland under low power using the compound microscope and note the general appearance of the gland. Switch to high power and use *Figure 27-12* for reference. Under high power, observe the numerous tubuloalveolar mucous glands. (*Color Plates 3* and *7*) Columnar to cuboidal epithelium is characteristic of this gland. The bulbourethral gland produces a thick mucus secretion that is released into the penile urethra at emission for lubrication of the penis tip. Both an external muscularis layer of skeletal and smooth muscle and a connective tissue capsule can be observed surrounding the gland. Peristaltic contractions of these muscle fibers discharge the mucus secretion.

Slide # _____ Bulbourethral Gland, sec.

See: Fig. 27-12

∞ Fig. 27-8a, d, p734

LOCATE

_____ General appearance of gland

_____ Mucous glands

_____ Muscularis layer (smooth muscle fibers)

_____ Connective tissue capsule

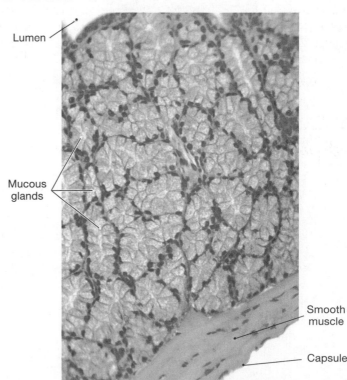

Lumen

Mucous glands

Smooth muscle

Capsule

FIGURE 27-12   Bulbourethral gland × 233.

The Penis

The penis contains the penile urethra and is the male organ for depositing sperm into the female vagina. Most of the penis consists of **erectile tissue**. Under parasympathetic stimulation, the erectile tissues of the penis become engorged with blood, stiffening the penis and producing an *erection*. The rhythmic wave of powerful contractions of the bulbocavernosus and ischiocavernosus muscles forcibly propel semen toward the external urethral orifice, causing *ejaculation* of semen. The male urethra carries urine, and at ejaculation, semen to the exterior of the body. [∞ p735]

Procedure

Identify the structure of the **penis** using a *male pelvis model* or *cadaver specimen*. The skin of the penis is hairless and very thin. It does not contain fat in the hypodermis. To view the structure of the penis, the skin and underlying fascia must be removed. Use *Color Plate 127, Figures 27-13* and *27-14* for reference. The penis is divided into three regions: the **root** (*crura and bulb of the penis*), fixed to the inferior rami of the ischium; the **body** (*shaft*), the tubular, pendulous region composed of erectile tissue; and the **glans** (*head*), the expanded terminal end surrounding the external urethral meatus. The body consists of three cylindrical columns of erectile tissue: on the dorsal surface are two **corpora cavernosa**, and on the ventral, a single column mass, the **corpus spongiosum**, which surrounds the urethra. The base of each corpora cavernosa is the root (crura) of the penis. It contains a tendon that attaches the penis to the inferior rami of the ischium. If possible, observe prostatic, membranous, and penile portions of the urethra. Externally, the **prepuce** (*foreskin*) covers the glans and may have been removed by circumcision for hygienic or religious reasons. The glans is an enlargement of the corpus spongiosum.

Branches of the internal pudendal artery (**dorsal arteries**) supply blood to the erectile tissue of the penis. The **dorsal veins** drain it. Vessels and nerves pierce the urogenital diaphragm to enter at the root. The dorsal blood vessels can be located dorsal to the corpora cavernosa. The relationship of the blood vessels, cavernous bodies, and urethra should also be observed in cross section. Observe bulbocavernosus (insertion—dorsum and sides of penis) and ischiocavernosus (insertion—crura of penis) muscles. (*Figure 10-12, p128*)

See: Figs. 27-1, 27-13, 27-14

∞ Figs.
27-1, p726
27-8a, p734
27-9, p736

LOCATE

_____ Penis { Root (Crura and bulb)
Body (Shaft)
Glans

_____ Corpora cavernosa

_____ Corpus spongiosum

_____ Urethra { Prostatic
Membranous
Penile

_____ External urethral meatus

_____ Prepuce

_____ Dorsal blood vessels

_____ Muscles { Ischiocavernosus
Bulbocavernosus

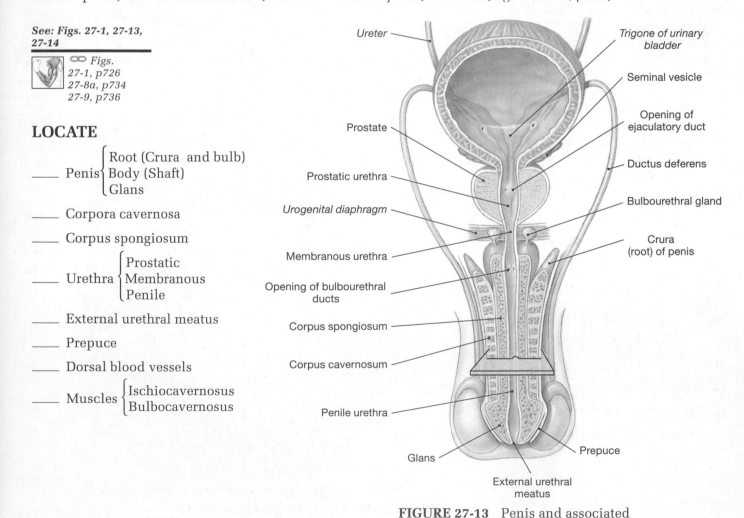

FIGURE 27-13 Penis and associated structures, frontal view.

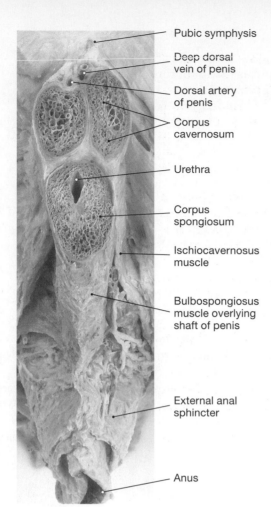

Pubic symphysis

Deep dorsal vein of penis

Dorsal artery of penis

Corpus cavernosum

Urethra

Corpus spongiosum

Ischiocavernosus muscle

Bulbospongiosus muscle overlying shaft of penis

External anal sphincter

Anus

FIGURE 27-14 Cross section of penis and associated structures.

See: Fig. 27-15

∞ *Fig. 27-9b, p736*

LOCATE

____ Corpora cavernosa

____ Collagenous sheath

____ Dorsal blood vessels

____ Corpus spongiosum

____ Penile urethra

Microscopic Identification of the Penis

Procedure

Identify the histology of the **penis** using the *microscope slide* provided, first by viewing under the dissecting microscope, then using the low scanning power magnification of the compound microscope. Use *Figure 27-15* for reference. The relationship of three cylindrical, cavernous bodies are observed easily in cross section. The two large round, dorsal (anterior) masses of tissue can be identified as the corpora cavernosa. A connective tissue sheath of collagen (the tunica albuginea) surrounds and separates each cavernous body. Several dorsal blood vessels can be located in the subcutaneous tissue dorsal to the corpora cavernosa. Just ventral to the corpora cavernosa lies the single corpus spongiosum, which surrounds the penile urethra.

Slide # _____ Human Penis

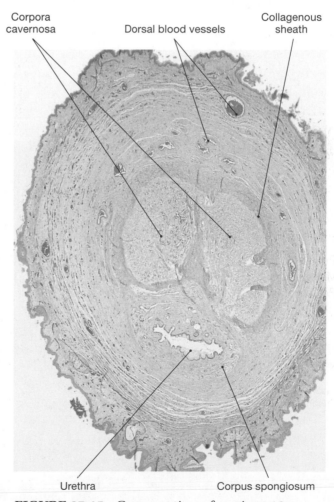

Corpora cavernosa   Dorsal blood vessels   Collagenous sheath

Urethra   Corpus spongiosum

FIGURE 27-15 Cross section of penis × 12.

Observation of the Female Reproductive System

The anatomy of the female reproductive system differs from the male in two major respects. First, female reproductive tract and urinary tract structures do not share common passageways. Second, female reproductive tract structures provide nourishment, protection, and support to a developing embryo/fetus.

In males, spermatogenesis begins at puberty and can continue almost for the rest of the life of the male. In females, the process of oogenesis begins before birth, but then stops. At puberty, oogenesis commences again periodically and continues until menopause. [∞ p737]

Procedure

✓ **Quick Check**

Before you begin to identify the structures of the female reproductive system, review on a skeleton the bone markings of the pelvis (*Figures 7-11* and *7-12, pp78-79*) and muscles of the female pelvic floor. (*Table 10-9, p126* and *Figure 10-13, p128*)

Locate the structures of the **female reproductive system** using a *female pelvis model* or *female cadaver specimen*. Examine the female reproductive system from both anterior and sagittal views. Use *Color Plates 128* for the anterior view, *129* and *130* for the sagittal view, and *Figures 27-17* and *27-18* for reference. Begin by removing all abdominal viscera from the intact torso model. Verify the urinary bladder, ureters, and rectum. Locate the ovaries, uterine tubes, uterus, and vagina in the female pelvis. As the peritoneum drapes over the posterior surface of the urinary bladder and the anterior surface of the uterus, a pocket or space is formed. Identify this pocket as the **vesicouterine pouch**. A second pocket is formed between the posterior wall of the uterus and the anterior wall of the rectum. Identify it as the **rectouterine pouch**.

See: Color Plates 129, 130, Fig. 27-16

∞ *Fig. 27-10, p738*

LOCATE

____ Ovary	____ Vagina
____ Uterine tube	____ Vesicouterine pouch
____ Uterus	____ Rectouterine pouch

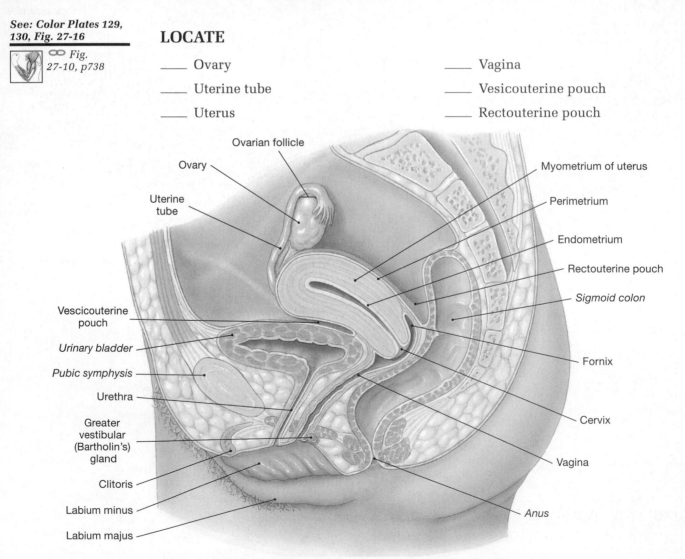

FIGURE 27-16 Female reproductive system as seen in sagittal view.

Observation of the Ovaries and Uterine Tubes

The production of female gametes (ova) occurs within the **ovaries** by the process of *oogenesis*. [∞ p740] The ovaries lie near the lateral walls of the pelvic cavity and are supported and stabilized by several ligaments. At puberty, the maturation of female gametes occurs on a regular basis in response to changes in hormone levels. A secondary oocyte is discharged into the peritoneal cavity from an ovary at ovulation. The uterine tubes are hollow muscular tubes that function to capture the secondary oocyte after ovulation, and to provide for and coordinate the transport and meeting of a secondary oocyte and spermatozoa for fertilization. Fertilization normally takes place within the uterine tubes. [∞ p742]

Procedure

 Identify the structure of the **ovaries**, **uterine tubes**, and associated structures using a *female pelvis model* or *cadaver specimen*. Use *Color Plates 128* to *130* and *Figures 27-17* and *27-18* for reference.

The Ovaries

Locate the walnut-sized ovaries within the pelvic cavity. The surface of each ovary is covered by the visceral peritoneum, which overlies a layer of dense connective tissue called the **tunica albuginea**. Observe how each ovary is firmly suspended from the posterior by a mesenteric sheet, the **broad ligament**, and a fold of this mesentery, the **mesovarium**. Locate on the lateral surface of the ovary the **suspensory ligament** and trace it to the pelvic brim. Blood is supplied to the ovaries chiefly via the ovarian arteries and drained by the ovarian veins. These vessels and nerves are contained in this ligament and can be traced to enter the ovary at the **ovarian hilum**. Identify on the medial surface of the ovary the **ovarian ligament**, which extends from the ovary to the uterus. Suspension of the ovaries permits great mobility within the pelvic cavity. Observe that the anterior and lateral surfaces of the ovaries are in contact with the *uterine tubes*.

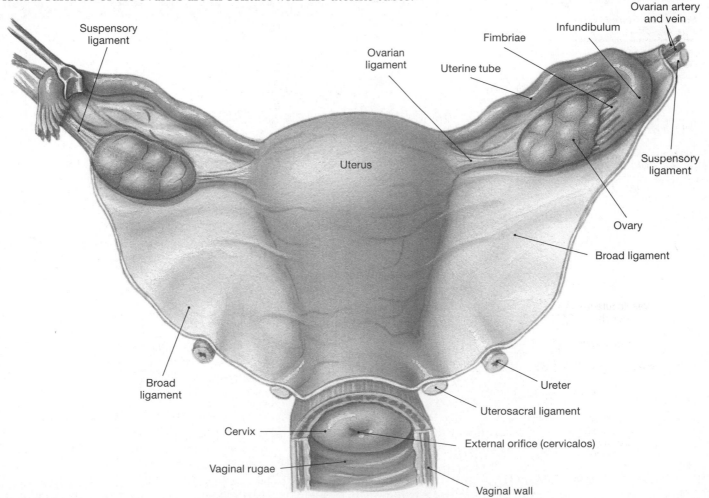

FIGURE 27-17 Posterior view of the ovaries, uterine tubes, and uterus.

The Uterine Tubes

Observe the uterine tubes. Trace each tube from ovary to uterus. Each uterine tube is about 10 cm (4 inches) long. The uterine tubes are divided into three regions. Beginning at the ovary, identify the **infundibulum** as the trumpet-shaped portion of the tube containing fringe or finger-like projections, the **fimbriae**. Observe the point where the infundibulum narrows. Identify this narrowed portion as the **ampulla**, which makes up about two-thirds of the uterine tube. Trace the ampulla to the point at which it narrows and identify this region as the **isthmus**, which represents the remaining one-third of the tube. The point at which the uterine lumen passes through the uterine wall is the **intramural portion**. Observe the mesovarium and suspensory ligaments. Note how the broad ligament connects the uterine tubes to the ovaries. Blood is supplied to the uterine tubes by branches of the uterine arteries and drained by uterine veins.

See: Color Plates 128 to 130, Figs. 27-17, 27-18

∞ Figs. 27-11, p739
27-14, p743

LOCATE

____ Ovary

____ Tunica albuginea

____ Broad ligament

____ Mesovarium

____ Suspensory ligament

____ Ovarian arteries/veins

____ Ovarian hilum

____ Ovarian ligament

____ Uterine tube { Infundibulum {Fimbriae
Ampulla
Isthmus
Intramural

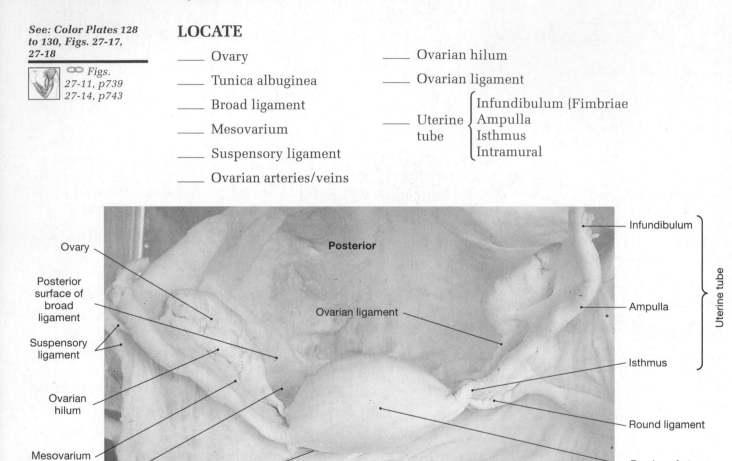

FIGURE 27-18 Superior and frontal view of the ovaries, uterine tubes, and uterus.

Microscopic Identification of the Ovary

Procedure

Trace **follicular development** using the *microscope slide* provided, first by viewing the slide under low power and then under high power magnification to see the different types of **follicles**. The surface of the ovary is covered by the visceral peritoneum. Deep to this lies the dense connective tissue of the tunica albuginea. The tissue of the ovaries consist of an outer *cortex* and inner *medulla*. The process of oogenesis and follicular development occurs in the germinal epithelium of the cortex. The vessels that supply the ovary enter at the hilum. Look within the cortex for follicular development. Use *Figure 27-19* [∞ Figure 27-12, p740-741] as an aid to the identification of follicular development. Identify as many of the following stages as possible on your slide.

Slide # ____ Ovary, sec.

See: Fig. 27-19

⚭ Fig.
27-12, p740-741

LOCATE

____ Ovary {Cortex / Medulla}

____ Egg nest contains {Primordial follicles (which are at oocyte stage)}

____ Primary follicle {Follicle cells / Oocyte}

____ Secondary follicle {Follicle cells / Nucleus of oocyte / Zona pellucida}

____ Tertiary follicle {Oocyte / Corona radiata / Follicle cells / Antrum containing follicular fluid}

____ Corpus luteum

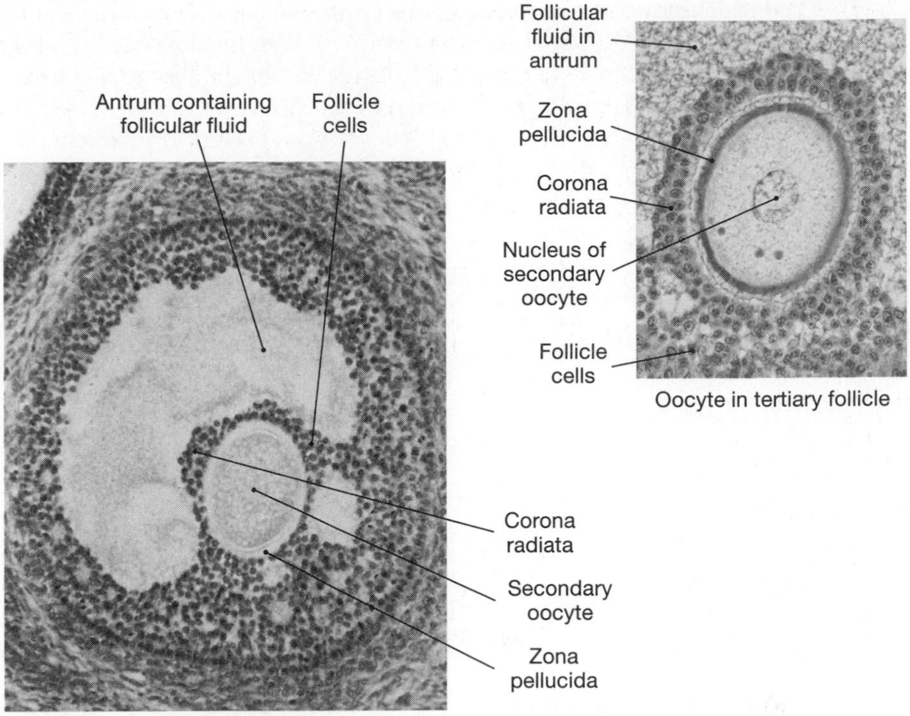

FIGURE 27-19 Tertiary follicle × 136 and secondary oocyte × 1052.

Microscopic Identification of the Uterine Tube

Procedure

Identify the histology of the **uterine tube** using the *microscope slide* provided, first by viewing under a dissecting and then a compound microscope. Observe a cross section of the uterine tube under the dissecting microscope. Note the general appearance of the uterine wall. The wall becomes thickest as the tube approaches the uterus. Locate from inside to out the following three layers of this tube using *Figure 27-20* for reference. Under low power magnification of the compound microscope, locate the irregular lumen and identify the dark-staining simple columnar epithelium with scattered mucus-secreting cells that form the mucosa layer. Observe the muscularis layer with its light-staining bands of smooth muscle fibers that appear to encircle the mucosa. Some slides may show a serosa layer containing flattened mesothelial cells.

Slide # _____ Uterine Tube, sec.

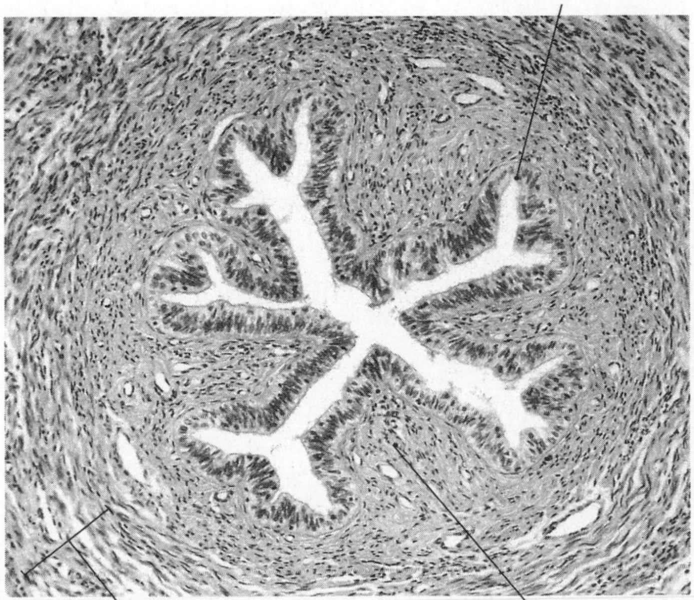

See: Fig. 27-20

⚭ Fig.
27-14b, p743

LOCATE

____ General appearance of uterine wall

____ Mucosa layer {Simple columnar epithelial cells / Lamina propria}

____ Muscularis layer {Smooth muscle fibers}

____ Serosa layer

FIGURE 27-20 Uterine tube of isthmus, sectional view × 122.

Observation of the Uterus and Vagina

The hollow pear-shaped **uterus** acts as the organ for gestation of the embryo. This organ serves as the site for nutritional support, protection, and waste removal for the developing embryo/fetus. [∞ p743] The term *embryo* designates development from weeks 1 through 8, while the term *fetus* designates development from weeks 9 through delivery. The non-pregnant uterus is approximately 8 cm (3 inches) long and is located superior and posterior to the urinary bladder and anterior to the rectum. It lies bent in an anterior flexed (**antiflexed**) position over the superior and posterior surfaces of the urinary bladder. Failure to maintain this antiflexed position is a cause of infertility in women. Ligaments secure and support the uterus in this position. During pregnancy, the uterus becomes quite large and requires support by these ligaments. Extending from the uterus to the exterior of the body is a highly muscular and distensible 10 cm (4 inch) tube, the **vagina**. The vagina serves as the organ for copulation (receives the erect male penis during intercourse), outlet for menstrual fluids, and as the birth canal during childbirth. [∞ p747]

Procedure

Identify from anterior, superior, and sagittal views the **structure** of the **uterus** and **vagina**, and the **ligaments** that maintain the position of these organs within the pelvic cavity, using a *female pelvis model* or *cadaver*. Use *Color Plates 128* to *130*, and *Figures 27-21* and *27-22* for reference. In the female cadaver, the internal structure of the vagina and uterus can be viewed also from the perineum. Perform this observation after examining the "Female External Genitalia." Keep in mind that these hollow organs will be collapsed in the female cadaver.

Uterus

Verify the uterus in its position between the urinary bladder and rectum. Note its antiflexed position. The uterus is divided into two regions. Identify the large, upper rounded region as the **body** and the narrow, inferior neck region as the **cervix**. Identify the domed portion of the body, just superior to the attachment of the uterine tubes as the **fundus**. Observe the distal end of the cervix, tapering and projecting into the vagina. Note the triangular shape of the **uterine cavity**, which receives the uterine tubes. Trace the cavity of the uterus inferiorly through the **internal os** and into the **cervical canal**. It opens through the **external os (external orifice)** into the vagina. View the uterus from a superior position and locate the ligaments that stabilize it. Viewing from superior to inferior, identify the **round ligaments of the uterus**, the **broad ligament**, the **uterosacral ligaments**, and the **lateral (cardinal) ligaments**.

The uterine wall is made up of a thin inner glandular portion, the *endometrium*, and thick outer muscular portion, the *myometrium*. The histology of the uterine wall will be identified later. Blood is supplied to the uterus via the paired uterine arteries and drained by the uterine veins.

Vagina

Verify the vagina. The cervical os of the cervix opens into the cavity of the vagina, the **vaginal canal**. Identify the **fornix** as the shallow recess that surrounds the cervix within the vaginal canal. Internally, observe the folded appearance of the vaginal wall. Identify these transverse folds as **vaginal rugae.** The vagina opens exteriorly into the **vestibule**. The vestibule is concealed by the inner skin folds, the *labia minora*. By spreading the labia minora, you can locate the opening of the vagina, the **vaginal orifice**. Blood is supplied to the vagina via the ovarian, vaginal, and uterine arteries and drained by the uterine and vaginal veins. Most of the vagina receives its innervation from the pudendal nerves.

See: Color Plates 128 to 130, Figs. 27-21, 27-22

∞ *Figs. 27-11, p739 27-15, p744 27-20, p749*

LOCATE

____ Urinary bladder

____ Uterus
- Fundus
- Body
- Isthmus
- Cervix
- Uterine cavity
- Cervical canal
- External os

____ Round ligaments of uterus

____ Broad ligament

____ Uterosacral ligaments

____ Lateral (cardinal) ligaments

____ Uterine arteries/veins

Internal Anatomy

____ Uterine cavity

____ Endometrium

____ Myometrium

____ Visceral peritoneum or serosal
layer (Perimetrium)

____ Cervix

____ External os

____ Cervical canal

Vagina

____ Vaginal canal

____ Fornix

____ Vaginal rugae

____ Vestibule

____ Vaginal orifice

____ Arteries/veins { Ovarian / Vaginal / Uterine }

____ Pudendal nerve

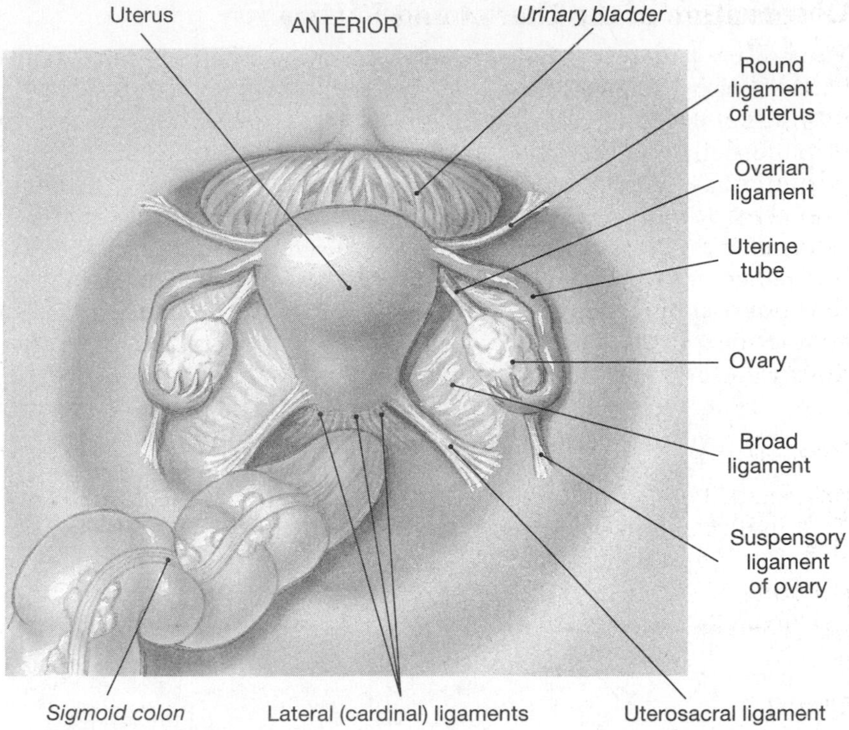

FIGURE 27-21 Uterus and ligaments as seen from superior.

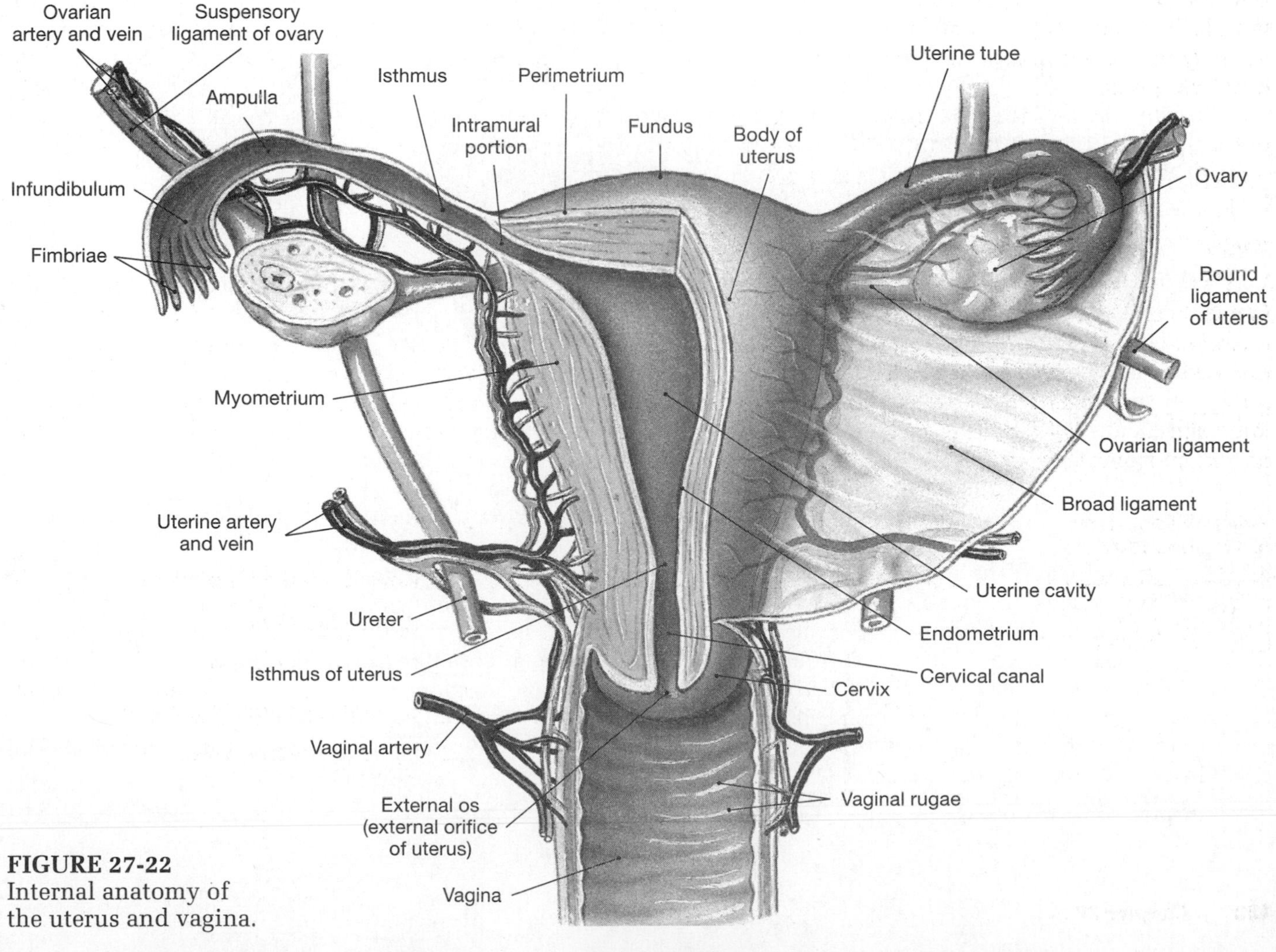

FIGURE 27-22
Internal anatomy of
the uterus and vagina.

Microscopic Identification of the Uterus

Procedure

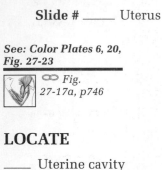 Identify the histology of the **uterine wall** by viewing the *microscope slide* provided under a dissecting microscope. Using *Figure 27-23* for reference, locate from lumen to the outer wall, the three layers of the uterus. Identify the dark-staining, simple columnar epithelium (see *Color Plate 6*) of the inner **endometrium** that forms the thick glandular layer of the mucosa. Identify the **myometrium** layer by the light-staining bands of smooth muscle fibers (see *Color Plate 20*) and numerous blood vessels. The myometrium can be distinguished clearly from the endometrium by its structure. Locate the outer **perimetrium** (or peritoneum) by the appearance of loose connective tissue attached to the myometrium. Return to the endometrial layer. The endometrium is divided into two glandular zones (layers): termed functionalis and basalis. Identify the layer closest to the lumen of the uterine cavity as the **functional zone** and the layer contacting the myometrium as the **basilar zone**. Observe the numerous **endometrial glands**, which support the physiological demands of the developing embryo. The endometrium undergoes monthly, cyclic changes in thickness and appearance in response to changing estrogen levels, which cause the uterine (menstrual) cycle.

Slide # _____ Uterus

See: Color Plates 6, 20,
Fig. 27-23

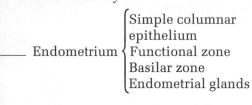 ∞ *Fig.*
27-17a, p746

LOCATE

_____ Uterine cavity

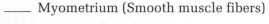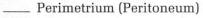

_____ Endometrium {
Simple columnar epithelium
Functional zone
Basilar zone
Endometrial glands
}

_____ Myometrium (Smooth muscle fibers)

_____ Perimetrium (Peritoneum)

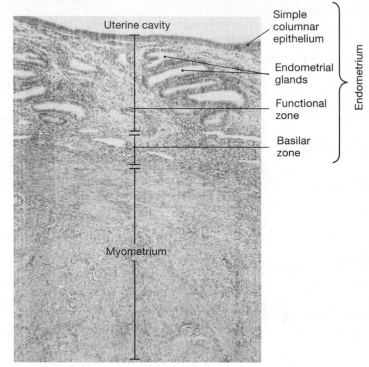

FIGURE 27-23   Uterine wall × 32.

Microscopic Identification of the Vagina

Procedure

Identify the histology of the **vaginal wall** by viewing the *microscope slide* provided under low power of the compound microscope. Use *Figure 27-24* for reference. Locate from lumen to outer wall, the three layers of the vaginal wall. The organization of the wall of the vagina resembles that of the esophagus. Identify the dark-staining, noncornified, stratified squamous epithelium (see *Color Plate 2*) that forms the mucosa layer. Note how the mucosa is arranged into irregular folds (vaginal rugae). Identify the underlying lamina propria by the connective tissue that encircles the epithelium. Peripheral to the mucosa, the muscularis layer may be observed by the light-staining bands of smooth muscle fibers that appear to encircle the mucosa. The adventitia layer can be identified peripheral to the muscularis by the appearance of loose connective tissue attached to the muscularis. In the preparation of this microscope slide, the adventitia layer often appears to be in fragments or broken away from the muscularis.

Slide # _____ Vagina, l.s.

See: Color Plate 2, Fig. 27-24

∞ *Fig. 27-19, p748*

LOCATE

_____ Mucosa layer { Stratified squamous epithelium (noncornified)

_____ Muscularis (Smooth muscle fibers)

_____ Adventitia

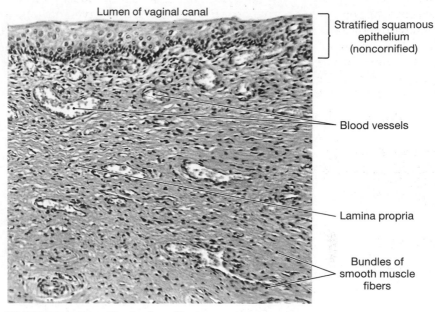

Lumen of vaginal canal

Stratified squamous epithelium (noncornified)

Blood vessels

Lamina propria

Bundles of smooth muscle fibers

FIGURE 27-24   Vaginal wall × 36.

The Female External Genitalia

The region enclosing the external genitalia of the female is called the **vulva** (pudendum). The vulva consists of: *mons pubis, labia minora, labia majora, vestibule,* and *clitoris*. The area bounded by the inferior margin of the pelvis is called the perineum. The perineum is divided into **anal** and **urogenital triangles** *(∞ p283).* The urogenital triangle contains the structures of the vulva. [∞ p749]

Procedure

Identify the structures that compose the **vulva** using a *female pelvis model* or *cadaver specimen.* Observe the vulva in the model and cadaver both from the external perineum and in sectional view. Use *Figure 27-25* for the external reference. For sectional views, see *Color Plates 129, 130,* and *Figure 27-16* for reference. Locate the structures of the vulva first on the model. Observe the vulva from the perineum in the female cadaver. With the cadaver in the supine position and the thighs widely spread, the perineal region is now exposed. *(Figure 10-13, p128)* Maintain the cadaver in this position by restraining the lower limbs in this position.

Verify the anal orifice and external urethral orifice. Locate the **mons pubis** as the bulge of skin created by fat pads that lie over the pubic symphysis and pubic bones. Inferior to the mons pubis identify the broad flaps of fleshy skin that encircle the vestibule and run anterior to posterior as the **labia majora**. These folds of skin help conceal the underlying structures of the vulva. Just medial to the labia majora locate smaller and more delicate flaps of skin, the **labia minora**. The **vestibule** is a central space concealed between the folds of the labia minora. The *vagina* opens into the vestibule. By spreading the labia minora you can locate the opening of the vagina, the **vaginal orifice.** Identify the **clitoris** within the vestibule, just anterior to the external urethral orifice. The clitoris is analogous to the male penis, but only contains two columns of erectile tissue instead of three. The labia minora divide into two pairs of folds, uniting anteriorly to form the *glans of the clitoris* and *the prepuce of the clitoris.* The **greater vestibular glands** (Bartholin's) are homologous to the bulbourethral glands in the male. Typically it is not possible to identify these glands. During sexual intercourse these pea-sized glands secrete mucus into the vestibule to lubricate the clitoris and vagina.

See: Color Plates 129,
130, Fig. 27-25

∞ Figs.
27-20, p749
10-15, p284

LOCATE

____ Female Perineum

____ Urogenital region

____ Mons pubis

____ Labia $\begin{cases} \text{Majora} \\ \text{Minora} \end{cases}$

____ Vestibule

____ Vaginal orifice

____ Clitoris $\begin{cases} \text{Glans} \\ \text{Prepuce} \end{cases}$

____ Greater vestibular
(Bartholin's) glands

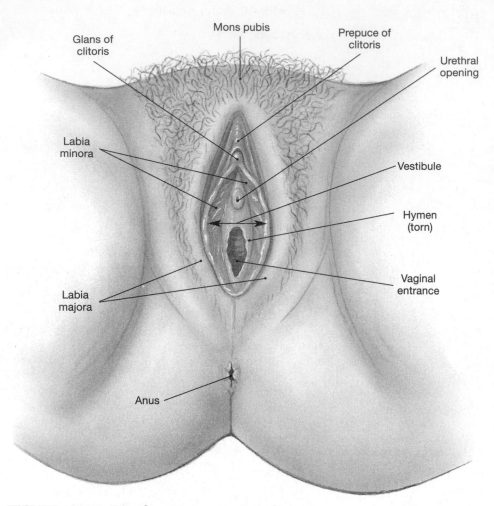

FIGURE 27-25 Female perineum, external view.

The Mammary Glands

The **mammary glands** are accessory glands of the female reproductive system. They lie in the subcutaneous tissue of the **pectoral fat pad** interior to the skin of the breast. The mammary glands are normally active only during pregnancy and during nursing. Mammary gland development and milk production (*lactation*) are stimulated by the hormone prolactin. During pregnancy, the mammary glands are stimulated to develop and become active in milk production influenced by a collection of hormones. This activity enlarges the female breasts. [∞ p749]

Procedure

Identify the structures that compose the **female breast** using a *female breast model* or *female cadaver specimen*. Observe the relationship of the breast to the pectoralis major muscle. (*Color Plate 40*) Use *Figure 27-26* and *Color Plates 131* and *132* for reference. Locate the pectoral fat pad and note how the **mammary glands** of the breast are supported by it. The mammary glands are also supported and connected to the pectoralis major muscle by **suspensory ligaments**. Observe the ligaments as they emerge from the dermis, branching over the glandular tissue and connecting it to the surface of the pectoralis muscle. Glandular tissue is divided into lobules, with each lobule being drained by a single **lactiferous duct**. Identify lactiferous ducts and observe the ducts merging and opening into a common expanded chamber, the **lactiferous sinus**. Each lactiferous sinus is a collecting area for milk prior to ejection through the pigmented projection, the **nipple**. About 20 sinuses open onto the surface of the nipple. Identify the circular pigmented area surrounding the nipple as the **areola**. The ducts of sebaceous glands can be observed as minute projections that open onto the surface of the areola. Oil is secreted by these glands to protect the area against chafing during nursing.

See: Color Plates 131, 132, Fig. 27-26

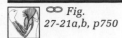
∞ *Fig. 27-21a,b, p750*

LOCATE

____ Pectoralis major muscle

____ Pectoral fat pad

____ Suspensory ligaments

____ Mammary glands

____ Lactiferous ducts

____ Lactiferous sinus

____ Nipple

____ Areola

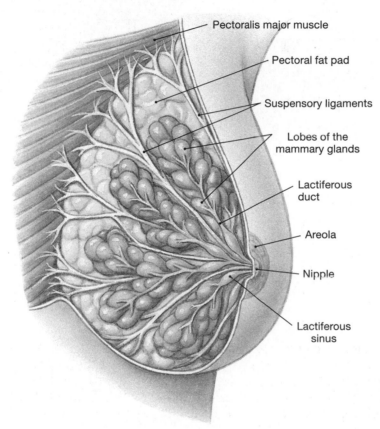

Pectoralis major muscle

Pectoral fat pad

Suspensory ligaments

Lobes of the mammary glands

Lactiferous duct

Areola

Nipple

Lactiferous sinus

FIGURE 27-26 Gross anatomy of the breast.

Dissection of the Reproductive System in the Cat

This exercise complements the study of the human reproductive system. The female cat reproductive system is different from that found in humans. Some differences will be noted during the description of the system.

See: Fig. 27-27

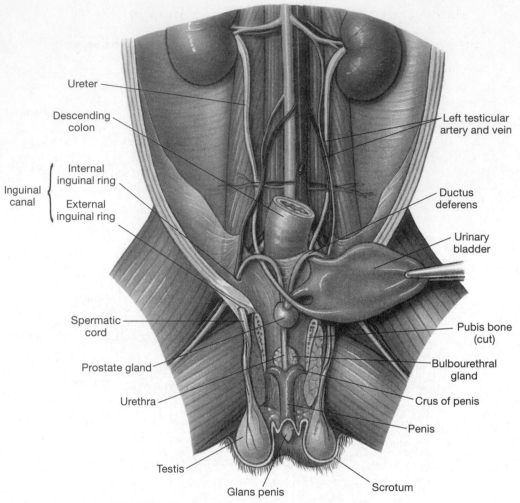

Ureter

Descending colon

Left testicular artery and vein

Internal inguinal ring

Inguinal canal

External inguinal ring

Ductus deferens

Urinary bladder

Spermatic cord

Pubis bone (cut)

Prostate gland

Bulbourethral gland

Urethra

Crus of penis

Penis

Testis

Scrotum

Glans penis

FIGURE 27-27 Reproductive system of the male cat.

Male Cat

1. Identify the skin covering the testes, a double-chambered sac located ventral to the anus, termed the **scrotum**. Identify the **penis**, ventral to the scrotum. Carefully make an incision through the skin and expose the **testes**. The testes produce the male gamete, **sperm**, and are covered by a peritoneal capsule, the **tunica vaginalis**.

2. On the lateral surface of each testis, extending from the cranial to the caudal end of each testis, lies the **epididymis**, a comma-shaped convoluted tubule that stores sperm.

3. Locate and identify the connective tissue-covered **spermatic cord**, which consists of the **spermatic artery, spermatic vein, spermatic nerve**, and the **ductus deferens (vas deferens)**. The ductus deferens carries sperm from the epididymis to the urethra, for transport out of the body.

4. Trace the spermatic cord through the **inguinal canal** and **ring** into the abdominal cavity.

5. If the spermatic cord components are loosened from their connective tissue wrapping, the ductus deferens is observed to loop over the ureter and pass posterior to the bladder.

NOTE: To observe the remaining structures, you must make a midline cut through the pelvic muscles and the pubic symphysis of the pelvic bone with bone cutters. Cut carefully, since the urethra is immediately dorsal to the bone. After cutting, split the pubic bone apart by spreading the thighs apart. This action exposes the structures within the pelvic cavity. Carefully tease and remove any excess connective tissue.

6. Locate the **prostate gland**, a large, hard mass of tissue surrounding the urethra. Relocate the ductus deferens and trace it to the prostate. Compared to human males, cats lack seminal vesicles.

7. Trace the urethra to the proximal end of the penis. The urethra consists of three parts: the **prostatic urethra** (passing through the prostate), the **membranous urethra** (between the prostate gland and the penis), and the **spongy** or **penile urethra** (passing through the penis).

8. The **bulbourethral glands** are located on either side of the membranous urethra.

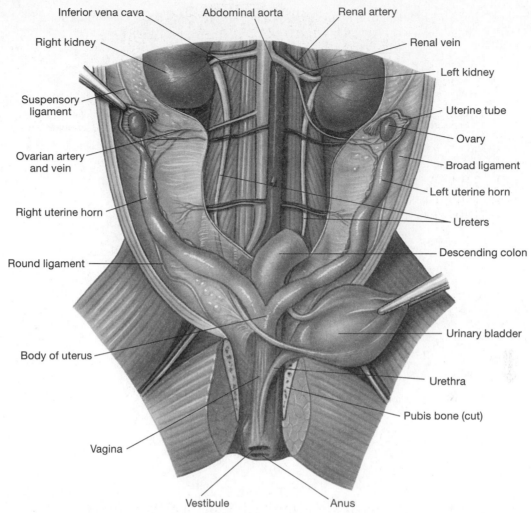

FIGURE 27-28 Reproductive system of the female cat.

Female Cat

1. Reflect to one side the abdominal viscera and locate the paired, oval **ovaries**, which lie on the dorsal body wall lateral to the kidneys.

2. Locate on the surface of the ovaries the small coiled **uterine tubes**. Their funnel-like opening, the **infundibulum**, curves around the ovary to partially cover it. This assures the capture of ova released from the ovary.

3. Each uterine tube leads into a single large tube, the **uterine horn**. The horns of the uterus are the location for fertilized egg development. Unlike the pear-shaped uterus of the human, the uterus of the cat is Y-shaped (bicornate) and consists of two uterine horns that merge into a single **uterine body**.

4. Identify the **broad ligament** that aids in anchoring the uterine horns to the body wall.

NOTE: To observe the remaining structures, you must make a midline cut through the pelvic muscles and the pubic symphysis of the pelvic bone with bone cutters. Cut carefully, since the urethra and vagina are immediately dorsal to the bone. After cutting, split the pubic bone apart by spreading the thighs apart. This action exposes the structures within the pelvic cavity. Tease and remove any excess connective tissue.

5. Return to the uterine body, and follow it caudally into the pelvic cavity where it is continuous with the **vagina**.

6. Locate the urethra, which has emerged from the bladder. The vagina is dorsal to the urethra. At its posterior end, the vagina and urethra unite at the **urethral orifice** to form the **urogenital sinus (vestibule)**, the common passage for the urinary and reproductive systems.

7. The urogenital sinus opens to the outside at the urogenital aperture. It is bordered by skin folds, **labia majora**. Together, the urogenital aperture and labia constitute the **vulva**, external genitalia.

Anatomical Identification Review

1. Identify by coloring and labeling the structures of the male reproductive system as seen in sectional view.

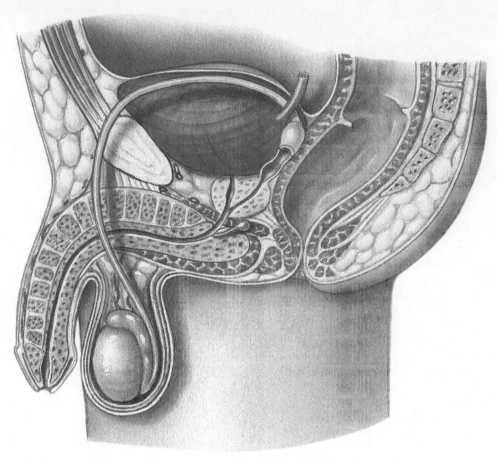

2. Identify by coloring and labeling the structures of the female reproductive system as seen in sectional view.

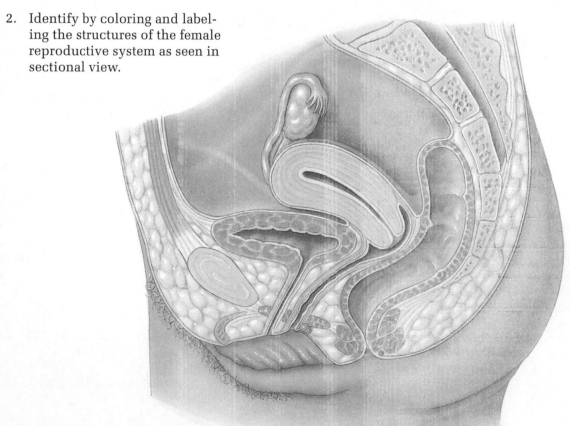

To Think About

1. What is the function of the scrotum?

2. Describe the anatomical structures involved and the mechanism that permits the penis to become erect.

3. Describe the difference between the non-pregnant uterus and one shortly prior to the birth of a baby.

1   An Introduction to Anatomy

Laboratory Review Questions

Multiple Choice

_____ 1. A body cavity is:
   (a) an internal chamber, or space
   (b) a body space in which internal organs are suspended
   (c) present in several areas of the body
   (d) a space that allows internal organs to change volume or position
   (e) all of the above

_____ 2. Structures contained within the mediastinum include the:
   (a) thymus
   (b) trachea
   (c) esophagus
   (d) large arteries and veins attached to the heart
   (e) all of the above

_____ 3. The lining of the abdominopelvic cavity is the:
   (a) pleura
   (b) pericardium
   (c) mesentery
   (d) mucous membrane
   (e) peritoneum

_____ 4. In the anatomical position:
   (a) the palms face posteriorly
   (b) the legs are spread comfortably, with feet at least 18" apart
   (c) the subject stands, hands at the sides, palms facing anteriorly
   (d) the head is turned to the left, and the eyes are closed
   (e) all of the above

_____ 5. In anatomical position, a hand that is pronated:
   (a) has the fingers spread open
   (b) has the palms facing posteriorly
   (c) is bent laterally at the wrist
   (d) has the fingers fully flexed
   (e) has the characteristics of both b and c

_____ 6. The heart occupies a space called the:
   (a) mediastinum
   (b) pleural cavity
   (c) pericardial cavity
   (d) dorsal cavity
   (e) none of the above

_____ 7. Structures that occupy the dorsal cavity include the:
   (a) heart
   (b) liver
   (c) intestines
   (d) brain
   (e) two of the above

_____ 8. The cavities that contain the lungs are the:
   (a) pericardial
   (b) abdominopelvic
   (c) dorsal
   (d) visceral
   (e) pleural

_____ 9. Which of the following is a view of the human body at right angles to the long axis?
   (a) transverse (horizontal)
   (b) frontal
   (c) sagittal
   (d) coronal
   (e) oblique

_____ 10. A transverse (horizontal) section would pass through which body cavity if it were made at the level of the umbilicus?
   (a) pleural cavity
   (b) thoracic cavity
   (c) abdominal cavity
   (d) pelvic cavity
   (e) none of the above

Matching

Match each of the anatomical directional terms in the left column with its opposite from the column on the right.

_____ 11. medial   a. dorsal

_____ 12. proximal   b. inferior

_____ 13. ventral   c. lateral

_____ 14. superior   d. posterior

_____ 15. anterior   e. distal

Fill in the Blank

16. The elbow is _____ to the wrist of the brachium.

17. The point of the hip (the anterior, superior, iliac spine) is _____ and _____ to the umbilicus.

18. Organs such as the liver, stomach, and pancreas are located within the _____ cavity.

19. The portion of the ventral body cavity that lies inferior to a line between the lowest spinal vertebra posteriorly and inferior to the superior margin of the pelvic girdle anteriorly is the _____ cavity.

20. The cranial and spinal cavities are part of the dorsal body cavity, while the thoracic, abdominal, and pelvic cavities are part of the _____ .

2 The Cell

Laboratory Review Questions

Multiple Choice

_____ 1. Study of the structure and function of cells is:
   (a) biochemistry
   (b) gross anatomy
   (c) cytology
   (d) electron microscopy
   (e) phrenology

_____ 2. A structural barrier that separates the inside of the cell from the surrounding extracellular fluid is:
   (a) the plasma (cell) membrane
   (b) a glycolipid
   (c) structural proteins
   (d) the glycocalyx
   (e) none of the above

_____ 3. The cell membrane controls:
   (a) entry and exit of water
   (b) entry of ions and nutrients
   (c) the position of the cell within the tissue of which it is a part
   (d) the size of the cellular nucleus
   (e) all of the above

_____ 4. The general functions of the cell membrane include:
   (a) physical isolation of cell contents from the extracellular fluid
   (b) regulation of material exchange between the cell and extracellular fluid
   (c) sensitivity to changes in the extracellular fluid
   (d) structural support of the cell
   (e) all of the above

_____ 5. All of the following are true of organelles *except*:
   (a) they are structures that perform generalized functions within the cell
   (b) they occur in all body cells
   (c) some may be nonmembranous
   (d) some may be surrounded by limiting membranes that separate their contents from the cytosol
   (e) no exceptions; all of the above are true

_____ 6. The functions of microfilaments include:
   (a) holding open gated channels in the cell membrane
   (b) anchoring the cytoskeleton to integral proteins of the cell membrane
   (c) acting with other proteins to produce active movement of a portion of the cell or a change in cell shape
   (d) forming small, finger-shaped projections from the cell membrane
   (e) two of the above are true

_____ 7. Microvilli:
   (a) are usually composed of actin
   (b) are composed of relatively massive filament strands of myosin protein subunits
   (c) can be disassembled to provide a mechanism for changing the shape of the cell
   (d) occur in cells that are actively engaged in absorbing materials from the extracellular fluid
   (e) form the primary components of the cellular cytoskeleton

_____ 8. If a cell lacks ribosomes, it could not:
   (a) manufacture proteins
   (b) produce ATP
   (c) package proteins manufactured by the fixed ribosomes
   (d) move through the extracellular fluid
   (e) reproduce itself

_____ 9. During cell division, if a cell lacked centrioles, it would be unable to:
  (a) direct the movement of DNA strands during cell division
  (b) move through the surrounding fluid
  (c) replicate its own DNA
  (d) manufacture proteins
  (e) move fluids or solutes across the cell membranes

_____ 10. The nucleus:
  (a) communicates by chemical means with the cytosol
  (b) is surrounded by an envelope that separates the nucleoplasm from the cytosol
  (c) houses the DNA content of the cell
  (d) all of the above are true of the nucleus
  (e) a and c are true

Matching

Match each of the statements in the left column with the appropriate structure or organelle from the right column.

_____ 11. Structures that control movement of the chromosomes during cell division

_____ 12. Structures responsible for extracellular removal of damaged organelles
  or pathogens

_____ 13. Organelles that produce 95% of the ATP required by the cell

_____ 14. Organelles responsible for the synthesis of secretory products;
  provides for intracellular storage and transport

_____ 15. Organelles that provide for storage, alteration, and packaging of
  secretory products and lysosomes

(a) mitochondria
(b) centrioles
(c) Golgi apparatus
(d) endoplasmic reticulum
(e) lysosomes

Fill in the Blank

16. Microfilaments are slender protein strands, usually composed of the protein _____ .
17. The cytoskeleton of a cell is an internal protein framework that gives the cytoplasm _____ and _____ .
18. The nucleus directs the processes that take place in the _____ .
19. The structures composed of DNA and protein that are housed in the nucleus are the _____ .
20. Structures that form the primary components of the cytoskeleton and that give the cell strength and rigidity and anchor the position of the major organelles are _____ .

3 The Tissue Level of Organization

Laboratory Review Questions

Multiple Choice

_____ 1. How many primary tissue types are there?
   (a) trillions
   (b) 200
   (c) 4
   (d) 17
   (e) 6

_____ 2. Primary tissue types exclude which of the following?
   (a) epithelial tissue
   (b) germinative tissue
   (c) connective tissue
   (d) muscle tissue
   (e) neural tissue

_____ 3. Which of the following is *not* characteristic of epithelial cells?
   (a) they may consist of a single or multiple cell layer
   (b) they always have a free surface exposed to the external environment or some inner chamber or passageway
   (c) they are avascular
   (d) they consist of a few cells but have a large amount of extracellular material

_____ 4. The type of epithelium that is present where mechanical stresses are severe (such as the palm of your hand) is:
   (a) stratified squamous epithelium
   (b) columnar epithelium
   (c) cuboidal epithelium
   (d) endothelium
   (e) simple epithelium

_____ 5. Connective tissues form the internal framework of the body, and does which of the following:
   (a) provide strength and stability
   (b) maintain the relative position of internal organs
   (c) provide a route for the distribution of blood vessels, lymphatics, and nerves
   (d) all of the above
   (e) b and c from above

_____ 6. Muscle tissue types include all of the following *except*:
   (a) smooth muscle
   (b) crenulated muscle
   (c) skeletal muscle
   (d) cardiac muscle

_____ 7. Properties of smooth muscle include all of the following *except*:
   (a) having small cells with tapering ends
   (b) having many, irregularly shaped nuclei
   (c) the ability to divide and regenerate after an injury
   (d) a lack of striations
   (e) the ability to contract on their own, or to be stimulated by nervous activity

_____ 8. Unique features of cardiac muscle include:
   (a) a location only in the heart
   (b) a single nucleus for each cell
   (c) cylindrical appearance
   (d) intercalated disks
   (e) a and d only are unique features

_____ 9. The skin is an example of a(an):
   (a) simple squamous epithelium
   (b) stratified epithelium
   (c) columnar epithelium
   (d) mesothelium
   (e) endothelium

_____ 10. A type of epithelium that allows stretching and lines the renal pelvis, ureters, and urinary bladder is:
   (a) cuboidal epithelium
   (b) columnar epithelium
   (c) transitional epithelium
   (d) squamous epithelium
   (e) glandular epithelium

_____ 11. Cells responsible for storing lipids are called:
   (a) adipocytes
   (b) melanocytes
   (c) macrophages
   (d) mast cells
   (e) lymphocytes

_____ 12. In correct order, select the list of connective tissues that go from least to most flexible in providing structural support for body tissues:
   (a) adipose tissue, tendon, blood, lymph, bone
   (b) bone, cartilage
   (c) tendon, blood, lymph, fixed macrophages
   (d) mesenchymal cells, tendon, bone, cartilage
   (e) none of the above

_____ 13. What type of supporting tissue would you expect to find within the external lobe of the ear and the nose?
   (a) bone
   (b) fibrocartilage
   (c) elastic cartilage
   (d) hyaline cartilage
   (e) all of the above

_____ 14. Adults who lose a large amount of weight rapidly will often feel cold temperature more than previously because of the:
   (a) loss of padding of body contours
   (b) loss of shock absorption capacity, which results in feeling the cold more
   (c) increased metabolism
   (d) loss of the volume of insulating adipose tissue
   (e) loss of weight, which decreases the body's ability to shiver

_____ 15. Uncontrolled muscle contractions or skeletal movements are most likely to break which type of structure:
   (a) ligaments
   (b) tendons
   (c) the bone
   (d) the muscle
   (e) a and b from above are the most likely to be broken

Matching

Match each of the items in the left column with the best response from the right column.

_____ 16. A tissue type with thin, flat, and somewhat irregularly shaped cells that covers the surface

_____ 17. A single cell layered tissue that lines the ventral body cavities

_____ 18. A tissue type found where mechanical stresses are severe and made of multiple layers of cells

_____ 19. A tissue type with cells that resemble little hexagonal boxes, which appear square (equal in height and width)

_____ 20. A tissue type with cells that are taller and more slender in cross section than other types, with nuclei crowded into a narrow band close to the basement membrane

(a) columnar epithelium
(b) stratified squamous epithelium
(c) squamous epithelium
(d) cuboidal epithelium
(e) mesothelium

4 The Integumentary System

Laboratory Review Questions

Multiple Choice

_____ 1. The integument is separated from the deep fascia around organs by the:
(a) epidermis
(b) dermis
(c) hypodermis
(d) cutaneous membrane
(e) none of the above

_____ 2. Accessory structures of the integument:
(a) are located in the dermis
(b) include hair
(c) include nails
(d) include multicellular exocrine glands
(e) include all of the above

_____ 3. A section of thick skin includes five layers. In correct order from the basement membrane to the surface, these are:
1. stratum granulosum   (a) 1,2,3,4,5
2. stratum lucidum   (b) 4,5,1,2,3
3. stratum corneum   (c) 5,4,3,2,1
4. stratum germinativum   (d) 3,2,5,4,1
5. stratum spinosum   (e) 1,3,5,4,2

_____ 4. The structural difference between thick and thin skin is in the:
(a) thickness of the papillary layer
(b) thickness of the dermis
(c) thickness of the epidermis
(d) thickness of the stratum germinativum
(e) none of the above

_____ 5. The epidermis is composed of a:
(a) stratified squamous epithelium
(b) cuboidal epithelium
(c) simple squamous epithelium
(d) transitional epithelium
(e) none of the above

_____ 6. The normal length of time that an individual cell spends as a part of the epidermis is:
(a) five to eight hours
(b) two weeks
(c) four weeks
(d) at least one year
(e) for the entire life of the individual

_____ 7. Sensory nerves in the skin are responsible for monitoring:
(a) touch
(b) temperature
(c) pain
(d) pressure
(e) all of the above, and more sensations as well

_____ 8. The typical color of the epidermis is caused by:
(a) the dermal blood supply
(b) pigment composition
(c) pigment concentration
(d) all of the above
(e) b and c from above

_____ 9. Cells that produce pigment are called:
   (a) melanocytes
   (b) melanin
   (c) carotene
   (d) lymphocytes
   (e) none of the above

_____ 10. The blood supply to the skin is called the:
   (a) brachial plexus
   (b) cutaneous plexus
   (c) hypodermis
   (d) a and c from above
   (e) none of the above

_____ 11. Specialized skin sensory receptors include all of the following *except*:
   (a) Meissner's corpuscles
   (b) Ruffini corpuscles
   (c) Pacinian corpuscles
   (d) melanocytes
   (e) no exceptions; all of the above are included

_____ 12. Choose the correct definition for the root of a hair:
   (a) it extends from the hair bulb to the point where the internal organization of the hair is complete
   (b) it extends distally from the beginning of the shaft of the hair
   (c) it encompasses all of the hair deep to the surface of the skin
   (d) it includes all of the structures of the follicle
   (e) none of the above are correct

_____ 13. Two types of exocrine glands in the skin are:
   (a) sebaceous and sweat
   (b) apocrine and endocrine
   (c) holocrine and endocrine
   (d) mammary and ceruminous
   (e) none of the above

_____ 14. Sebaceous follicles include each of the following *except*:
   (a) large sebaceous glands
   (b) they have openings that communicate directly with the epidermis
   (c) they never produce hairs
   (d) they are located on the face, back, chest, nipples, and male sex organs
   (e) no exceptions; all of the above are true

_____ 15. The most widely distributed sweat gland on the body is a(an):
   (a) apocrine sweat gland
   (b) merocrine sweat gland
   (c) myoepithelial gland
   (d) none of the above
   (e) a and b from above

_____ 16. The structure of a hair from the starting point of growth within the dermis toward the outside of the skin in correct order is:
   (a) cortex, medulla, shaft
   (b) bulb, root, shaft
   (c) follicle, bulb, cuticle, and cortex
   (d) hard keratin, bulb, root
   (e) none of the above

_____ 17. In older individuals, wrinkling of the skin occurs because of:
   (a) thinning of the dermis
   (b) loss of the subcutaneous fat in the distribution pattern caused by the secondary sexual characteristics
   (c) exposure to ultraviolet radiation
   (d) all of the above are true
   (e) a and c from above are true

_____ 18. Which of the following are attributes of the papillary layer?
   (a) it consists of loose connective tissue
   (b) it contains capillaries and sensory neurons
   (c) it has dermal papillae that project between the epidermal ridges
   (d) all of the above are attributes of the papillary layer
   (e) a and b from above are attributes of the papillary layer

_____ 19. The function of the nerve fibers in the skin include all of the following *except* the:
   (a) control of the blood flow
   (b) adjustment of glandular secretion rates
   (c) monitoring sensory neuron receptors in the dermis and deeper layers of the epidermis
   (d) changing the temperature of the skin
   (e) a and b from above

_____ 20. Why is the skin capable of repair, even after serious damage?
   (a) individual skin cells have a high metabolic rate
   (b) stem cells persist in both the epithelial and connective tissue components
   (c) construction of scar tissue is easier than making normal skin
   (d) all layers of the skin have an excellent blood supply
   (e) none of the above are true

5   The Skeletal System: Osseous Tissue and Skeletal Structure

Laboratory Review Questions

Multiple Choice

_____ 1. Which is a function of the skeletal system?
 (a) support of the body
 (b) formation of blood cells
 (c) storage of minerals
 (d) provision of levers for muscle action
 (e) all of the above are functions of the skeletal system

_____ 2. Protective functions of the skeleton include:
 (a) surrounding delicate tissues and organs with a bony framework
 (b) absorbing the shock of unexpected rapid body movements
 (c) facilitating transmission of nerve impulses
 (d) surrounding air cells, such as those in the mastoid region
 (e) all of the above

_____ 3. Bone tissue can be classified as:
 (a) dense, irregular connective tissue
 (b) fluid connective tissue
 (c) supporting connective tissue
 (d) inert tissue
 (e) none of the above

_____ 4. Materials included in the matrix of bone are:
 (a) calcium phosphate
 (b) collagen fibers
 (c) calcium carbonate
 (d) all of the above
 (e) a and b from above

_____ 5. The basic functional unit of mature compact bone is the:
 (a) osteon
 (b) canaliculus
 (c) lamella
 (d) osteoclast
 (e) none of the above

_____ 6. The two types of osseous tissue are:
 (a) Haversian and lamellar bone
 (b) spongy (cancellous) and compact (dense) bone
 (c) trabecular and osteoclastic
 (d) spicular and trabecular
 (e) none of the above

_____ 7. The function of an osteoblast in osseous tissue is:
 (a) to produce new bone matrix
 (b) to break down old bone matrix and release calcium salts
 (c) to produce new perichondrium for endochondral bone to displace
 (d) there is no known function of an osteoblast in osseous tissue
 (e) none of the above are correct

_____ 8. Bones are classified according to their:
 (a) length to breadth ratios
 (b) individual shapes
 (c) proportions of compact to spongy bone
 (d) developmental history
 (e) none of the above

_____ 9. One example of a long bone is:
   (a) humerus
   (b) carpus
   (c) occipital
   (d) patella
   (e) none of the above

_____ 10. Important characteristics of a long bone include:
   (a) a marrow cavity
   (b) a flat, short, ridged surface
   (c) irregular shapes
   (d) an outer layer of compact bone, but the inner region entirely of spongy bone
   (e) none of the above

_____ 11. All of the following are true of a sesamoid bone _except_:
   (a) they always form within tendons
   (b) the patella is a sesamoid bone
   (c) sesamoid bones form wherever there is a tendon passing over a joint
   (d) they are small and irregularly shaped
   (e) no exceptions; all of the above are correct

_____ 12. Extra bones that develop between the usual bones of the skull are:
   (a) sesamoid bones
   (b) wormian bones
   (c) flat bones
   (d) endochondral bones
   (e) none of the above

_____ 13. A thoracic vertebra is an example of what type of bone?
   (a) long
   (b) flat
   (c) irregular
   (d) wormian
   (e) sutural

_____ 14. Processes that are formed where ligaments attach to a bone include:
   (a) condyles and trochleas
   (b) fossae, sulci, and foramina
   (c) rami and heads
   (d) trochanters, tuberosities, and tubercles
   (e) all of the above

_____ 15. The component of bone structure that occupies the lacunae is the:
   (a) nerve that innervates the spongy bone
   (b) osteon
   (c) osteoblast
   (d) osteocyte
   (e) none of the above

Fill in the Blank

16. Blood vessels that supply nutrients to osteocytes in an osteon are located within the _____ .

17. Any projection or bump on a bone may be called a _____ .

18. A large passageway through a bone that permits blood vessels and/or nerves to pass is a _____ .

19. A chamber within a bone, normally filled with air, is a _____ .

20. The expanded articular end of an epiphysis, separated from the shaft by a narrower neck, is typically the

_____ .

6   The Skeletal System: Axial Division

Laboratory Review Questions

Multiple Choice

_____ 1. The divisions of the skeletal system include the:
   (a) dorsal and ventral
   (b) axial and appendicular
   (c) proximal and distal
   (d) cranial, caudal, and anterior
   (e) none of the above

_____ 2. The axial skeleton consists of the bones of the:
   (a) skull, thorax, and vertebral column
   (b) pectoral and pelvic girdles and limb bones
   (c) pectoral and pelvic girdles and skull
   (d) skull, thorax, and girdles
   (e) none of the above

_____ 3. Which of the following is a function of the axial skeleton?
   (a) supporting the organs of the dorsal and ventral body cavities
   (b) protecting the organs of the dorsal and ventral body cavities
   (c) providing surface area for attachment of muscles
   (d) acting with axial muscles to provide respiratory movements
   (e) no exceptions; all of the above are true

_____ 4. Which of the following are true of the ethmoid bone?
   (a) it contains the crista galli
   (b) it contains the cribriform plate
   (c) it serves as the anterior attachment of the falx cerebri
   (d) all of the above are true

_____ 5. The occipital bone is or does all of the following *except*:
   (a) covers the occipital lobes of the brain
   (b) contains the foramen magnum
   (c) bears the condyles
   (d) is paired in the adult

_____ 6. Which of the following does *not* apply to the sella turcica?
   (a) it supports and protects the pituitary gland
   (b) it is surrounded by the cavernous sinus
   (c) it is located within the body of the sphenoid bone
   (d) similar to mastoid process and air cells, does not develop until after birth
   (e) it permits passage of the optic nerves

_____ 7. Which of the following features does *not* belong to the frontal bone?
   (a) supraorbital margins
   (b) supraciliary arches
   (c) metopic suture
   (d) lacrimal fossa
   (e) no exceptions; all of the features belong to the frontal bone

_____ 8. The bony orbit is formed by which of the following?
   (a) maxillary, zygomatic, lacrimal, and palatine
   (b) ethmoid, sphenoid, frontal, lacrimal, maxillary, zygomatic, and palatine
   (c) zygomatic, nasal, frontal, and sphenoid
   (d) zygomatic, nasal, frontal, sphenoid, palatine, lacrimal, and temporal
   (e) none of the above are correct

_____ 9. Bones of the facial region include:
   (a) frontals, nasals, parietals, and occipital
   (b) maxillae, palatines, mandible, zygomatics, lacrimals, inferior conchae, and the vomer
   (c) sphenoid, ethmoid, maxillae, and mandible
   (d) inferior conchae, vomer, ethmoid, nasal, lacrimals, and sphenoid
   (e) none of the above combinations is correct

_____ 10. Paranasal sinuses occur in which of the following bones?
   1. frontal   (a) 1,2,3,4,5
   2. sphenoid   (b) 1,3,5
   3. ethmoid   (c) 1,2,3
   4. maxillary   (d) 3,4,5
   5. nasal   (e) 1,2,3,4

_____ 11. The smallest bones in the skull are the:
   (a) ethmoids   (d) nasals
   (b) lacrimals   (e) vomer
   (c) zygomatics

_____ 12. The "soft spot" in an infant's head:
   (a) occurs where the parietal and frontal bones have not yet grown together
   (b) is one of several fontanels in an infant skull
   (c) is covered by fibrous connective tissue
   (d) all of the above are true
   (e) all of the above are true and, additionally, these spots may reappear in the skulls of aged individuals

_____ 13. Which of the following is among the curvatures of the vertebral column?
   (a) sacral   (d) lumbar
   (b) thoracic   (e) no exceptions; all of the above are vertebral curvatures
   (c) cervical

_____ 14. Functions of the curves of the vertebral column include all of the following _except_:
   (a) accommodation of the thoracic and abdominopelvic viscera
   (b) aligning the weight of the trunk over the legs
   (c) transmitting the weight of the body to the hips and lower extremities
   (d) increasing the ability to extend the limbs

_____ 15. The part of the vertebra that transfers weight along the long axis of the vertebral column is the:
   (a) spinous process   (d) vertebral arch
   (b) transverse process   (e) zygapophysis
   (c) vertebral body

_____ 16. Which of the following vertebra has two transverse foramen?
   (a) coccygeal   (d) lumbar
   (b) sacral   (e) thoracic
   (c) cervical

_____ 17. The dens is a part of the:
   (a) atlas   (d) axis
   (b) vertebra prominens   (e) none of the above
   (c) anticlinal vertebra

_____ 18. The true ribs:
   (a) consist of twelve pairs
   (b) are the only ribs that are unpaired
   (c) attach to the sternum by separate cartilaginous extensions
   (d) attach only to the vertebral column
   (e) b and d from above are true

_____ 19. Bones articulating at the coronal suture are:
   1. frontal   (a) 1,2,3
   2. occipital   (b) 1,3
   3. nasal   (c) 1,5
   4. vomer   (d) 3,4,5
   5. parietal   (e) 3,5

_____ 20. The temporomandibular joint is:
   (a) the location of the insertion of the temporalis muscle
   (b) the suture between the temporal and maxillary bones
   (c) the site of the articulation between the cranium and mandible
   (d) the first suture to be completely ossified in the adult skull
   (e) none of the above

7   The Skeletal System: Appendicular Division

Laboratory Review Questions

Multiple Choice

_____ 1. The pelvic girdle consists of:
   (a) the os coxae and innominate bones
   (b) ilium, ischium, and pubis
   (c) ala, ilium, and femur
   (d) acetabulum, femoral head, and pubis
   (e) all of the bones in a and c as well as the sacrum

_____ 2. Which of the pairs of the following terms is correctly associated?
   (a) pubic symphysis — pectoral girdle
   (b) femur — linea aspera
   (c) anterior superior iliac spine — sacrum
   (d) patella — endochondral bone
   (e) a and c from above are true

_____ 3. Which of the following is true of the distal end of the fibula?
   (a) it articulates with the distal end of the ulna
   (b) it has the linea aspera
   (c) it has an epicondylar foramen
   (d) it bears the lateral malleolus
   (e) a and c from above are true

_____ 4. The expanded end of the spine of the scapula is the:
   (a) olecranon
   (b) coracoid
   (c) xiphoid
   (d) superior angle
   (e) acromion

_____ 5. The pectoral girdle consists of:
   (a) scapula and clavicle
   (b) scapula and humerus
   (c) humerus, radius, ulna, carpal, and hand bones
   (d) sternum and ribs
   (e) none of the above

_____ 6. The point at which the pectoral girdle is attached to the axial skeleton is:
   (a) clavicle-acromion process
   (b) glenohumeral joint
   (c) sternoclavicular joint
   (d) clavicle and first rib
   (e) a and c from above

_____ 7. Which of the following are *not* features of the scapula?
   (a) acromion process
   (b) spine
   (c) coracoid process
   (d) infraspinous fossa
   (e) all of the above are features of the scapula

_____ 8. The pelvic girdle differs from the pectoral girdle in that:
   (a) the pelvic girdle is strongly fused to the axial skeleton, while the pectoral girdle is not fused to the axial skeleton
   (b) the pelvic girdle is more robust, for weight bearing
   (c) there is no difference in the number of bones in each; both the pectoral and pelvic girdles consist of three bones
   (d) the pelvic girdle is attached to the axial skeleton by a hinge joint, while the attachment of the pectoral girdle is a ball and socket joint
   (e) a and b from above are true

_____ 9. The superior margin of the hip bone is the:
   (a) symphysis pubis
   (b) iliac spine
   (c) acetabulum
   (d) iliac crest
   (e) superior sciatic notch

_____ 10. The socket that receives the head of the femur is the:
   (a) glenoid cavity
   (b) popliteal fossa
   (c) cubital fossa
   (d) acetabulum
   (e) sciatic notch

_____ 11. The distal protuberance of the tibia that can be palpated at the inside of the ankle is the:
   (a) medial malleolus
   (b) fibular notch
   (c) tibial tuberosity
   (d) popliteal fossa
   (e) none of the above

_____ 12. As compared to a male pelvis, the female pelvis is or has:
   (a) a larger birth canal
   (b) a relatively wider false pelvis
   (c) a wider pubic arch
   (d) all of the above are true
   (e) b and c from above are true

_____ 13. Bone marking characteristics that suggest a skeletal element as female include:
   (a) a bone approximately 10% heavier than many others of the same size
   (b) a bone with smaller prominences and a smoother surface
   (c) larger cranial sinuses
   (d) a long narrow triangular shaped sacrum with prominent sacral curvature
   (e) all of the above are female characteristics

_____ 14. The bone responsible for permitting pronation and supination of the forearm is the:
   (a) humerus
   (b) ulna
   (c) radius
   (d) scapula
   (e) the proximal row of carpal bones

Fill in the Blank

15. The bone that articulates with the distal end of the femur is the _____ .

16. The only direct bony connection between the pectoral girdle and the axial skeleton is between the clavicle and the _____ .

17. The expansions of the medial and lateral sides of the distal aspect of the humerus are the _____ .

18. The disc-shaped radial head articulates with the _____ of the humerus.

19. Of the two bones of the forearm, only the distal aspect of the _____ articulates with the carpal bones of the wrist.

20. The smooth curved surface of the acetabular fossa which articulates with the head of the femur is the _____ .

8 The Skeletal System: Articulations

Laboratory Review Questions

Multiple Choice

_____ 1. A bursa is:
  (a) a fluid-filled sac that reduces the friction of structures that would otherwise rub against one another
  (b) often located within the area of a synovial joint
  (c) a source of joint pain called bursitis
  (d) all of the above are true
  (e) a and b from above are true

_____ 2. The joint in the body with the greatest range of movement is the:
  (a) humeroulnar joint
  (b) glenohumeral joint
  (c) sternoclavicular joint
  (d) carpometacarpal joint
  (e) radioulnar joint

_____ 3. If a joint has great stability, it will also have:
  (a) great mobility
  (b) great strength
  (c) little strength
  (d) little mobility
  (e) a and b from above

_____ 4. Structures and functions that the elbow and knee joints have in common include:
  (a) a large sesamoid bone on the extensor surface
  (b) abduction and adduction
  (c) acts almost entirely as a hinge
  (d) for both joints a single proximal element articulates with two distal elements
  (e) none of the above

_____ 5. A joint that does not permit movement is a(an):
  (a) synarthrosis
  (b) hinge joint
  (c) synovial joint
  (d) diarthrosis
  (e) amphiarthrosis

_____ 6. A joint that permits a slight amount of movement is a:
  (a) gomphosis
  (b) suture
  (c) syndesmosis
  (d) synchondrosis
  (e) synostosis

_____ 7. A small, synovial fluid-filled sac in connective tissue and or muscle, which may be connected to a joint cavity is a(an):
  (a) fat pad
  (b) bursa
  (c) acromion process
  (d) gomphosis
  (e) ball and socket joint

_____ 8. Of the following which are not true of sutures:
  (a) they occur only between the bones of the skull
  (b) the edges of the bones forming this joint interlock
  (c) they are synarthroses
  (d) they may become completely fused later in life
  (e) no exceptions; all of the above are true

_____ 9. All of the following are true of the movement capabilities of joints *except*:
  (a) great stability decreases mobility
  (b) they may be directed or restricted to certain directions by the shapes of articulating surfaces
  (c) they may be modified by the presence of accessory ligaments and collagen fibers of the joint capsule
  (d) the strength of the joint is determined by the strength of the muscles that attach to it and its joint capsule
  (e) no exceptions; all of the above are true

_____ 10. Examples of angular motion include all of the following *except*:
  (a) flexion
  (b) adduction
  (c) extension
  (d) rotation
  (e) no exceptions; all of the above are included

_____ 11. A movement away from the longitudinal axis of the body in the frontal plane is:
  (a) abduction
  (b) flexion
  (c) extension
  (d) rotation
  (e) gliding

_____ 12. Rotation of the distal end of the radius across the anterior surface of the ulna is:
  (a) adduction
  (b) pronation
  (c) gliding
  (d) supination
  (e) extension

_____ 13. All of the following factors contribute to the integrity and normal functioning of the glenohumeral joint *except*:
  (a) the glenoid labrum
  (b) tendons
  (c) ligaments
  (d) sternoclavicular joint
  (e) no exceptions;
      all of the above contribute

_____ 14. Which of the following is correctly paired with regard to the humeroulnar joint?
  (a) humeral trochlear notch — radial tuberosity
  (b) humeral capitulum — radial head
  (c) radial collateral ligament — medial epicondyle
  (d) olecranon — radial notch
  (e) none of the above

_____ 15. What is the role in the stability of the radiocarpal joint of the carpal bone surfaces that do *not* participate in articulation?
  (a) they mainly contain grooves for passage of nerves and blood vessels
  (b) they are roughened by the attachments of ligaments
  (c) they are marked by the passage of tendons
  (d) they have no role in carpal stabilization
  (e) b and c from above are true

_____ 16. All of the following are ligaments that stabilize the hip joint *except* the:
  (a) iliofemoral ligament
  (b) pubofemoral ligament
  (c) ischiofemoral ligament
  (d) transverse acetabular ligament
  (e) popliteal ligament

_____ 17. Which ligament(s) of the knee is (are) responsible for limiting the anterior-posterior movement of the femur and maintaining the alignment of the femoral and tibial condyles?
  (a) anterior and posterior cruciate ligaments
  (b) patellar ligament
  (c) popliteal ligaments
  (d) tibial and fibular collateral ligaments
  (e) none of the above

_____ 18. Which knee ligament(s) function(s) to reinforce the medial and lateral surfaces of the joint, tighten only at full extension of the joint, and in this position, act to stabilize the joint?
  (a) patellar ligament
  (b) tibial and fibular collateral ligaments
  (c) popliteal ligaments
  (d) anterior and posterior cruciate ligaments
  (e) all of the above

_____ 19. The metatarsophalangeal joints of the foot resemble which joints of the hand?
  (a) carpometacarpal joints
  (b) metacarpophalangeal joints
  (c) interphalangeal joints
  (d) radiocarpal joints
  (e) a and c from above

_____ 20. What is most likely to be the anatomical problem of a person experiencing a "slipped disc"?
  (a) the menisci between the articular surfaces of the knee are no longer correctly aligned
  (b) the fibrocartilaginous pad between the pubic bones has moved out of place
  (c) the nucleus pulposus of an intervertebral disc has extruded through the annulus fibrosus
  (d) this is not a problem; it occurs normally as people age
  (e) none of the above

9 The Muscular System: Skeletal Muscle Tissue

Laboratory Review Questions

Multiple Choice

_____ 1. Skeletal muscles perform all of the following functions *except*:
   (a) production of skeletal movement
   (b) maintain posture and balance
   (c) support soft tissues
   (d) guard entrances and exits of the digestive and urinary tracts
   (e) no exceptions; all of the above are correct

_____ 2. The layers of connective tissue surrounding a skeletal muscle from the outside to the inside in correct order are:
   (a) endomysium, perimysium, and epimysium
   (b) epimysium, perimysium, and endomysium
   (c) sarcomere, fascicle, and endothelium
   (d) tendon, aponeurosis, and muscle fibers
   (e) none of the above are correct

_____ 3. At each end of a muscle, the collagen fibers of the epimysium often converge to form:
   (a) the endomysium
   (b) the muscle fibers
   (c) a tendon
   (d) the perimysium
   (e) none of the above

_____ 4. Why are skeletal muscles called voluntary?
   (a) the individual can make a conscious decision to contract these muscles
   (b) these muscles maintain a constant rhythm of contraction
   (c) these muscles are stimulated to contract by the motor nerves of the somatic nervous system
   (d) a and c from above are correct
   (e) none of the above

_____ 5. The function of the neuromuscular junction is to:
   (a) generate new muscle fibers if the muscle is damaged
   (b) facilitate chemical communication between the nerve fiber and the muscle fiber with which it is associated
   (c) unite nerves from different muscle fibers to one another
   (d) unknown
   (e) give feedback about muscle activity to nerves

_____ 6. The *most* important function of a skeletal muscle is to:
   (a) contract to push a bone from one position to another
   (b) burn energy to maintain the body metabolic rate
   (c) yield a supply of calcium to strengthen the bones
   (d) maintain the blood vessels at the proper diameter to ensure the correct amount of blood flow to all tissues at all times
   (e) do none of the above

_____ 7. The "Zone of Overlap" in a skeletal muscle fiber is:
   (a) where the tendon passes over onto the surface of the muscle fiber
   (b) where the innervation of one muscle fiber intermeshes with that of another
   (c) where the thin filaments pass between the thick filaments
   (d) where the A band crosses the thick filaments
   (e) none of the above

_____ 8. The repeating functional units of thick and thin filaments within a myofibril are called:
   (a) muscle fibers
   (b) sarcomeres
   (c) tropomyosin and troponin
   (d) terminal cisternae
   (e) a or d from above

_____ 9. When a sarcomere contracts, which of the following band(s) get(s) smaller?
   (a) A band
   (b) I band
   (c) Z band
   (d) A and Z bands
   (e) M band

_____ 10. Events that occur in a contracting muscle include all of the following *except*:
   (a) exerting a tension, or pull
   (b) thickening in width of the belly
   (c) shortening in length
   (d) alternating tension and relaxation of the tendon depending on the state of contraction
   (e) no exceptions; all of the above happen to a contracting muscle

Matching
Match each of the statements in the left column with the appropriate structure or term from the right column.

_____ 11. This expanded portion of the axonal branch faces a region of the sarcolemma.

_____ 12. This region of the sarcolemma faces an expanded portion of an axonal branch.

_____ 13. The space that separates the motor end plate from the synaptic knob.

_____ 14. The cytoplasm of the synaptic knob contains mitochondria and these small membrane structures.

_____ 15. Both the synaptic cleft and the motor end plate contain this enzyme.

(a) synaptic cleft
(b) synaptic vesicles
(c) synaptic knob
(d) acetylcholinesterase
(e) motor end plate

Fill in the Blank
16. A dense fibrous connective tissue layer that surrounds the entire skeletal muscle is the _____ .
17. Tendons that are formed into flattened sheets are called _____ .
18 Thick (myosin) and thin (actin) filaments together form _____ .
19. The smallest functional units of the muscle fibers are the _____ .
20. Skeletal muscle fiber contractions are caused by the interaction between the _____ and _____ of the sarcomeres.

10 The Muscular System:
Muscle Organization and the Axial Musculature

Laboratory Review Questions

Multiple Choice

_____ 1. Two factors that interact to determine the effect of the contraction of a skeletal muscle are:

    1. the arrangement of the muscle fibers   (a) 1 and 2
    2. the structural arrangement by which muscles attach to the skeleton   (b) 2 and 3
    3. the size of a muscle   (c) 3 and 4
    4. the amount of connective tissue present within a muscle   (d) 4 and 5
    5. the number of tendons each muscle has   (e) 1 and 3

_____ 2. The fleshy mass at the middle of a muscle, which has a cord-like tendon or an aponeurosis at each end, is termed a:
    (a) fascicle
    (b) muscle belly
    (c) muscle sheath
    (d) convergent muscle
    (e) none of the above

_____ 3. A visible slender band of collagen fibers that often serves as the centrally located site of attachment for the two sides of a bilaterally symmetrical muscle is called a(n):
    (a) raphe
    (b) aponeurosis
    (c) tendon
    (d) deep fascia
    (e) sarcoplasm

_____ 4. A muscle with all of its fibers located on a single side of the tendon is called a(an):
    (a) circular muscle
    (b) sphincter muscle
    (c) convergent muscle
    (d) multipennate muscle
    (e) unipennate muscle

_____ 5. A muscle that has fibers on both sides of a single tendon is called a(an):
    (a) convergent muscle
    (b) bipennate muscle
    (c) circular muscle
    (d) unipennate muscle
    (e) multipennate muscle

_____ 6. Regarding a muscle, which of the following is or are true?
    (a) the origin of the muscle remains stationary
    (b) the tendon is longer than the muscle belly
    (c) muscles of the class I lever type are the most common in the body
    (d) the insertion is proximal to the origin
    (e) b and d are true

_____ 7. Regarding a muscle insertion, which of the following is true?
    (a) the insertion moves more than does the origin
    (b) the insertion is proximal to the origin
    (c) as the speed of the muscle movement increases, the insertion moves further from the origin
    (d) as the force of muscle contraction increases, the insertion moves toward the origin
    (e) all of the above are true

_____ 8. Contraction of a muscle produces a specific:
    (a) site of origin
    (b) site of insertion
    (c) action
    (d) lever system
    (e) a and b from above

_____ 9. A muscle always:
  (a) has an origin, but may not have an insertion
  (b) belongs to a specific type of lever system
  (c) causes flexion when it contracts
  (d) brings the origin closer to the insertion when it contracts
  (e) none of the above are correct

_____ 10. A muscle whose contraction is chiefly responsible for producing a particular movement is a(n):
  (a) flexor
  (b) prime mover
  (c) adductor
  (d) antagonist
  (e) example of a first class lever

_____ 11. Muscles have been named for all of the following reasons *except*:
  (a) size of muscle
  (b) shape of muscle
  (c) parallel versus pennate fibered muscles
  (d) orientation of muscle fibers
  (e) location of the muscle

_____ 12. If part of the name of a muscle is maximus, it means that the muscle:
  (a) has at least three parts
  (b) is the largest in the region of a group of similar muscles
  (c) is superficial
  (d) is always an agonist
  (e) is none of the above

_____ 13. What does the term "rectus" tell about the muscle with that term as a part of the name?
  (a) the muscle location in the body
  (b) the muscle size
  (c) the manner in which the fibers of the muscle are oriented
  (d) the direction in which the muscle fibers are oriented with respect to the long axis of the body
  (e) c and d from above

_____ 14. What information does the name of the "pronator teres" provide about that muscle?
  (a) it is a short, round arm muscle
  (b) it is a superficial muscle
  (c) it has a long, flat aponeurosis and is multipennate
  (d) it is a muscle that moves the hand from a palm up to a palm down orientation
  (e) a and b from above are true

_____ 15. What can be determined about the vastus lateralis muscle from its name?
  (a) it is a flat muscle
  (b) it is a thigh muscle
  (c) it is located at the lateral aspect of the body
  (d) it is a large muscle
  (e) c and d are true

_____ 16. Which of the following are true of the axial musculature?
  (a) it arises and inserts on the axial skeleton
  (b) it arises, but does not insert on the axial skeleton
  (c) it arises elsewhere, but inserts on the axial skeleton
  (d) it stabilizes the pectoral and pelvic girdles
  (e) it makes up 40% of the body musculature

_____ 17. All of the following are actions of the axial musculature *except*:
  (a) positioning the head
  (b) moving the rib cage
  (c) positioning the spinal column
  (d) stabilizing the pectoral girdle to allow more precise forelimb movements
  (e) no exceptions; all of the above are true

_____ 18. The axial muscles of the spine control the position of the:
  (a) head, neck, and pectoral girdle
  (b) head, neck, and spinal column
  (c) spinal column only
  (d) spinal column and pectoral and pelvic girdles
  (e) b and d are correct

_____ 19. The oblique and rectus axial muscles are or do all of the following *except*:
  (a) are partially responsible for positioning the limb girdles
  (b) form the walls of the thoracic cavity
  (c) occur in the neck
  (d) provide walls for the abdominal cavity
  (e) support the walls of the pelvic cavity

_____ 20. When a muscle is injured near its insertion, this injury is most likely to be:
  (a) distal
  (b) in the tendon part of the muscle that is more mobile
  (c) at 90 degrees to the structure onto which the muscle inserts
  (d) combined with injuries to bones
  (e) only a and b are true

11 The Muscular System: The Appendicular Musculature

Laboratory Review Questions

Multiple Choice

_____ 1. The functions of the appendicular musculature do *not* include:
  (a) stabilizing the pectoral and pelvic girdles
  (b) assisting in positioning the head
  (c) moving the upper and lower limbs
  (d) composing approximately 40% of the skeletal muscles in the body
  (e) any of the activities listed above

_____ 2. Skeletal elements controlled by the appendicular musculature include:
  (a) pectoral and pelvic girdles and limb bones
  (b) the vertebral column
  (c) the cranial bones
  (d) the ribs and sternum
  (e) none of the above

_____ 3. The major groups of appendicular muscles include those of the:
  (a) vertebral column
  (b) posterior cervical region
  (c) shoulder and upper limb region
  (d) pelvic girdle and lower extremities
  (e) c and d from above

_____ 4. Which of the following do *not* apply to the trapezius muscle?
  (a) it covers the back of the neck and base of the skull
  (b) it is innervated by more than one nerve
  (c) it is diamond-shaped
  (d) it has specific regions that can contract separately
  (e) no exceptions; all of the above apply

_____ 5. The muscles that form the rotator cuff include:
  (a) supraspinatus, infraspinatus, deltoid, and teres major
  (b) biceps brachii and deltoid
  (c) teres minor, infraspinatus, supraspinatus, and subscapularis
  (d) pectoralis major and pectoralis minor and subscapularis
  (e) subclavius, deltoid, and pectoralis minor

_____ 6. The arm muscles that are the primary flexors of the elbow include the:
  (a) biceps brachii, brachialis, and brachioradialis
  (b) triceps brachii and anconeus
  (c) coracobrachialis and brachioradialis
  (d) muscles that are innervated by the musculocutaneous nerve
  (e) none of the above

_____ 7. Muscles that cause pronation of the forearm include:
  (a) pronator quadratus and pronator teres
  (b) supinator and teres major
  (c) biceps brachii and supinator
  (d) brachialis and coracobrachialis
  (e) none of the above

_____ 8. What is the main function of the extensor retinaculum?
  (a) it separates the forearm flexors from the extensors
  (b) it separates the elongate tendon sheaths from the nerves and blood vessels entering the hand
  (c) it transmits the tendons of the hand and finger flexors
  (d) a and b are true
  (e) none of the above are true

_____ 9. Muscles that exert the fine control of the hand and fingers do not originate on the:
(a) phalanges
(b) carpals
(c) metacarpals
(d) retinacula
(e) radius

_____ 10. All of the following muscles belong to the groups of the lower extremity except:
(a) muscles that move the hip
(b) muscles that move the thigh
(c) muscles that move the leg
(d) muscles that move the foot and the toes
(e) no exceptions; all of the above are real groups

_____ 11. The function of the iliotibial tract is to:
(a) stabilize the posterior aspect of the gluteus maximus
(b) assist in extension of the thigh
(c) assist in rotation of the thigh
(d) form a lateral brace for the knee, especially when standing on one foot
(e) none of the above

_____ 12. The flexors of the leg include all of the following _except_ the:
(a) biceps femoris
(b) rectus femoris
(c) semimembranosus
(d) semitendinosus
(e) sartorius

_____ 13. The extensor muscles of the leg include all of the following _except_:
(a) rectus femoris
(b) vastus intermedius
(c) vastus medialis
(d) vastus lateralis
(e) no exceptions; all of the above are extensors

_____ 14. The muscles of the quadriceps femoris group insert on the:
(a) femoral condyles
(b) head of the fibula
(c) patellar tendon
(d) anterior inferior iliac spine
(e) none of the above

_____ 15. A pairs figure skater who was accidentally dropped by her partner suffers a broken clavicle in the fall. Which of the following muscles are the most likely to be damaged in this injury?
(a) subclavius
(b) deltoid
(c) subscapularis
(d) infraspinatus
(e) teres minor

Fill in the Blank

16. The paired muscle that covers the upper back, extends to the base of the skull, originates on the middle of the neck and back and inserts on the clavicles and scapular spines forming a diamond shape is the _____ .

17. The most important muscle that abducts the arm is the _____ .

18. The major extensor of the arm that takes its origin on the scapula as does the long head of the biceps brachii is the

_____.

19. The common tendon shared by the gastrocnemius and the soleus muscles is the _____ .

20. Foot muscles that originate from the sides of the metatarsals and abduct the toes are the _____ .

12 Surface Anatomy

Laboratory Review Questions

Multiple Choice

_____ 1. Surface anatomy is used to study:
   (a) the epidermis
   (b) anatomical landmarks on the exterior of the human body
   (c) the texture of the surface of the body
   (d) the movement patterns of structures at or near the surface of the body
   (e) all of the above

_____ 2. Surface anatomy primarily reveals information about which of the following systems?
   (a) the digestive system
   (b) the reproductive system
   (c) the muscular system
   (d) the skeletal system
   (e) only c and d are correct

_____ 3. The study of surface anatomy attempts to relate, among other things:
   (a) the structure and function of the skeletal and muscular system
   (b) size and strength of the muscles near or at the surface of the body
   (c) structure to the age of the individual
   (d) physical and physiological conditions in an individual
   (e) none of the above

_____ 4. The best approach to use in studying surface anatomy is:
   (a) system by system
   (b) region by region
   (c) studying the most superficial structures first, then the intermediate and, finally, the deepest structures
   (d) use of articulated skeletons in the laboratory
   (e) a and c from above are correct

_____ 5. The muscle that can be palpated as a prominent ridge extending superiorly and inferiorly posterior to the ear, especially when the head is turned, is the:
   (a) levator scapulae
   (b) scalenus
   (c) subclavius
   (d) sternocleidomastoid
   (e) temporalis

_____ 6. The muscle that shapes the form of the lower posterior neck, where it merges into the shoulder is the:
   (a) trapezius
   (b) sternocleidomastoid
   (c) levator scapulae
   (d) anterior scalenus
   (e) pectoralis major

_____ 7. The structure that forms the high point of the shoulder is the:
   (a) head of the clavicle
   (b) manubrium of the sternum
   (c) acromion
   (d) coracoid process
   (e) body of the sternum

_____ 8. The muscle in the anterior thoracic and abdominal body wall that has prominent tendinous inscriptions is the:
   (a) transversus abdominis
   (b) serratus anterior
   (c) external oblique
   (d) internal oblique
   (e) rectus abdominis

_____ 9. A muscle or muscles of the chest that are visible in the anterior view is/are the:
   (a) pectoralis major
   (b) rectus abdominis
   (c) external oblique
   (d) serratus anterior
   (e) all of the above

_____ 10. A favorite site in infants and children for administration of an intramuscular injection is superior to the bulge of the gluteus maximus muscle. The muscle that is used for this injection is the:
   (a) gluteus medius
   (b) gluteus minimus
   (c) iliotibial tract
   (d) tensor fasciae latae
   (e) rectus femoris

Matching

Match each choice from the column on the left with the best response from the column on the right.

_____ 11. Muscles that form a bulge on the posterior aspect of the thigh are the

_____ 12. The structure that forms the raised ridge on the medial aspect of the foot when the great toe is dorsiflexed is the

_____ 13. The projection on the lateral aspect of the ankle is the

_____ 14. The structure that rides in the groove posterior to the lateral malleolus of the fibula is the

_____ 15. The four ridges that appear on the dorsal surface of the foot when the toes are dorsiflexed are caused by the

   (a) tendon of the fibularis longus
   (b) lateral malleolus of the fibula
   (c) hamstrings
   (d) tendons of the extensor digitorum longus
   (e) tendon of the extensor hallucis longus

Fill in the Blank

16. In addition to using visual observation, surface anatomy can be understood by _____ of the structures under study.

17. In medical practice, a knowledge of _____ is essential for both invasive and noninvasive procedures.

18. A muscle that forms the widest part of the back and appears on the lateral surface of the lower thoracic and upper abdominal region is the _____ .

19. The arch-shaped depressions that occur across the lower thoracic region when a person inhales and contracts the abdominal wall musculature reveal the _____ .

20. The prominent soft tissue bulge on the posterior aspect of the arm when seen in lateral view is caused by the _____ .

13 The Nervous System: Neural Tissue

Laboratory Review Questions

Multiple Choice

_____ 1. Functions of the nervous system include all of the following *except*:
   (a) providing information about the internal and external environments
   (b) integrating sensory function
   (c) coordinating voluntary and involuntary motor activities
   (d) controlling and regulating other tissues and systems
   (e) all of the above are functions

_____ 2. The anatomical subdivisions of the nervous system are the:
   (a) dorsal and ventral nervous systems
   (b) central and peripheral nervous systems
   (c) appendicular and axial nervous systems
   (d) controlling and regulating nervous systems
   (e) none of the above

_____ 3. Neuroglia in the nervous system function to:
   (a) carry nerve impulses
   (b) process information in the nervous system
   (c) support the neurons
   (d) transfer nerve impulses from the brain to the spinal cord
   (e) all of the above are correct

_____ 4. Components of a neuron include all of the following *except*:
   (a) soma
   (b) one or more dendrites
   (c) an elongated axon
   (d) one or more synaptic terminals
   (e) no exceptions; all of the above are included

_____ 5. Neurons that have several dendrites and a single axon are called:
   (a) anaxonic
   (b) unipolar
   (c) bipolar
   (d) tripolar
   (e) multipolar

_____ 6. The site of intercellular communication between neurons is the:
   (a) telodendria
   (b) synaptic knob
   (c) collaterals
   (d) axon hillock
   (e) synapse

_____ 7. The cytoplasm that surrounds the nucleus of a neuron is termed the:
   (a) soma
   (b) glial
   (c) perikaryon
   (d) Nissl bodies
   (e) neurofibrils

_____ 8. The most important function of the soma of a neuron is to:
   (a) allow communication with another neuron
   (b) support the neuroglial cells
   (c) generate an electrical charge
   (d) house organelles that produce energy and synthesize organic molecules
   (e) all of the above are correct

_____ 9. Most neurons lack centrosomes. This fact explains:
    (a) why neurons grow such long axons
    (b) why neurons cannot undergo mitosis
    (c) the conducting ability of neurons
    (d) the ability of neurons to communicate with each other
    (e) the ability of neurons to produce an action potential

_____ 10. Bundles of axons in the spinal cord are called:
    (a) nerves
    (b) tracts
    (c) centers
    (d) nuclei
    (e) ganglia

_____ 11. Which of the following nervous system activities do *not* apply to the neuroglia?
    (a) they isolate neurons
    (b) they provide a supportive framework for the nervous tissues
    (c) they transmit electrical signals to the axons
    (d) they act as phagocytes
    (e) a and b from above

_____ 12. What structures constitute the gray matter of nervous tissue?
    (a) neurofilaments and neurotubules
    (b) no structures; only the light transmitted through the cells during study makes them appear gray
    (c) have their cell body in the central nervous system
    (d) the centrosome complex causes the gray color
    (e) a and d from above cause the gray color

_____ 13. Sensory neurons:
    (a) form the efferent neurons
    (b) deliver sensory information to the central nervous system
    (c) are a portion of the central nervous system
    (d) b and c from above are true
    (e) none of the above are true

_____ 14. The cell bodies of unipolar sensory neurons are located in:
    (a) the brain
    (b) the spinal cord
    (c) sensory receptors
    (d) ganglia outside of the spinal cord
    (e) none of the above

_____ 15. The white matter of the CNS is largely made up of:
    (a) nerve cell bodies
    (b) axons
    (c) nuclei
    (d) sensory neurons only
    (e) none of the above

Fill in the Blank

16. All of the nervous tissue outside of the central nervous system composes the _____ nervous system.

17. _____ carry instructions from the CNS to peripheral effectors.

18. _____ are nerves that communicate directly with the spinal cord.

19. _____ are nerves that connect directly to the brain.

20. Areas of the nervous system that are dominated by nerve cell bodies are called the _____ .

14 The Nervous System: The Spinal Cord and Spinal Nerves

Laboratory Review Questions

Multiple Choice

_____ 1. Components of the central nervous system (CNS) include:
(a) spinal cord and brain
(b) peripheral nerves and ganglia
(c) autonomic components only
(d) efferent components only
(e) none of the above

_____ 2. The cervical cord:
(a) contains a smaller amount of gray matter than other areas of the spinal cord
(b) is not enlarged relative to some other regions of the spinal cord
(c) supplies the shoulder girdle and upper limb
(d) only a and b are true
(e) a, b, and c are true

_____ 3. The spinal cord segments are:
(a) 35 in number
(b) always designated, or identified as "s.c." followed by a number
(c) associated on a one-to-one basis with a pair of dorsal root ganglia
(d) a nervous system method for transmitting motor nerve information
(e) none of the above

_____ 4. A spinal nerve:
(a) begins distal to the dorsal root ganglion
(b) contains only the dorsal rami
(c) contains sensory and motor roots
(d) are efferent nerves
(e) a and c are correct

_____ 5. The spinal meninges function to:
(a) protect the spinal cord
(b) stabilize the spinal cord
(c) absorb shocks to the spinal cord
(d) assist in delivering nutrients and oxygen to the spinal cord
(e) do all of the above

_____ 6. The meninges include:
(a) different layers for the brain and the spinal cord
(b) pia mater, dura mater, and venous sinuses from the outside inward
(c) pia mater, arachnoid, and dura mater from the inside outward
(d) two layers of transverse collagen fibers
(e) none of the above

_____ 7. The inner surface of the dura mater is separated from the next deeper layer by the:
(a) pia mater
(b) arachnoid
(c) subdural space
(d) subarachnoid space
(e) none of the above

_____ 8. The dura mater is the:
(a) innermost of the meninges
(b) toughest and thickest of the meninges
(c) outermost covering over the brain, but not the spinal cord
(d) meninx that forms the coccygeal ligament along with components of the filum terminale
(e) structure that does all of the above

_____ 9. The arachnoid:
  (a) forms the middle layer of the three meningeal layers of the brain and the spinal cord
  (b) has an outer layer of the epithelium, and an inner layer of collagen and elastic fibers
  (c) contains the CSF
  (d) all of the above are true
  (e) a and b from above are true

_____ 10. The pia mater:
  (a) covers the brain but not the spinal cord
  (b) is the delicate innermost meningeal layer
  (c) contains the space in which the CSF circulates
  (d) forms the filum terminale
  (e) b and d from above are true

_____ 11. Which of the following are true of the ventral rami of the spinal nerves?
  (a) they carry only motor information
  (b) they are smaller than the dorsal rami
  (c) they form plexi
  (d) they supply the intrinsic muscles of the back region
  (e) all of the above are true

_____ 12. The major nerve plexi of the body include:
  1. brachial   (a) 1,2,3,4,5,6,7,8
  2. cervical   (b) 2,5,7,8
  3. lumbosacral   (c) 1,2,3
  4. thoracic   (d) 3,5,6
  5. coccygeal   (e) 2,4,6
  6. dorsal
  7. ventral
  8. peripheral

_____ 13. Which of the following is/are true of the brachial plexus?
  (a) it innervates the shoulder girdle and the upper extremity
  (b) it has contributions from the ventral rami of spinal nerves C_5–T_1
  (c) the nerves forming this structure arise from nerve structures called cords
  (d) the main nerves are the musculocutaneous, radial, median, and ulnar
  (e) all of the above are true

_____ 14. The axillary nerve:
  (a) arises from the upper portion of the brachial plexus
  (b) arises from the medial cord of the brachial plexus
  (c) innervates the skin over the lateral surface of the hand
  (d) innervates the pronators of the forearm
  (e) is or does all of the above

_____ 15. The anterior muscles of the thigh are innervated by the:
  (a) femoral nerve
  (b) iliohypogastric nerve
  (c) genitofemoral nerve
  (d) obturator nerve
  (e) saphenous nerve

Fill in the Blank

16. The increased amount of gray matter in the lumbar region of the spinal cord is known as the _____ .

17. The tapered, conical region of the spinal cord inferior to the lumbar enlargement is the _____ .

18. The pia-arachnoid is connected to the dura mater by the _____ .

19. The _____ of the brachial plexus divide to form the nerves that innervate the arm and shoulder region.

20. The lateral, medial, and posterior cords of the brachial plexus are named with respect to the location of the _____ .

15 The Nervous System: The Brain and Cranial Nerves

Laboratory Review Questions

Multiple Choice

_____ 1. Which of the following are *not true* of the brain?
   (a) it is equally as complex as the spinal cord
   (b) it is more versatile than is the spinal cord
   (c) the neuronal pools are united by extremely complex interconnections
   (d) it has the greatest number of neuronal pools in the nervous system
   (e) no exceptions; all of the above are true

_____ 2. The cerebral hemispheres are separated by a space termed the:
   (a) coronal fissure
   (b) longitudinal fissure
   (c) sylvian fissure
   (d) lateral ventricles
   (e) none of the above

_____ 3. The structures that protect the brain from suffering an impact against the interior of the cranium are the:
   (a) denticulate ligaments
   (b) choroid plexi
   (c) cranial meninges
   (d) cranial ventricles
   (e) none of the above

_____ 4. Which of the following is true of the dura mater?
   (a) the outer layer is fused to the periosteum of the cranial bones
   (b) it consists of three layers
   (c) it contains the subarachnoid spaces
   (d) it follows precisely the contours of the brain tissue
   (e) none of the above

_____ 5. Which of the following are *not* true of the arachnoid?
   (a) it is the middle of the cranial meninges
   (b) it is the most (and only) delicate cranial meninx
   (c) it does not follow the underlying convolutions or sulci
   (d) it contains a meshwork of collagen and elastic fibers
   (e) it is linked to the pia mater

_____ 6. Cranial blood vessels flow under the:
   (a) dura mater
   (b) pia mater
   (c) arachnoid villi
   (d) arachnoid
   (e) none of the above

_____ 7. Which of the following are *not* true of the pia mater?
   (a) it adheres to the contours of the brain
   (b) processes of the astrocytes anchor it
   (c) it is avascular
   (d) it acts as a floor to support the cranial blood vessels
   (e) it is the innermost layer of the cranial meninges

_____ 8. Functions of the cerebrospinal fluid include all of the following *except*:
   (a) cushioning delicate neural structures
   (b) carrying electrical impulses to stimulate the neural cortex
   (c) supporting the brain
   (d) transporting nutrients to the brain
   (e) transporting waste products away from neural tissue

_____ 9. The choroid plexus:
   (a) is located on the entire surface of the brain-ventricle interface
   (b) is the site of production of the CSF
   (c) is the site of drainage of used CSF from neural tissues
   (d) is the site where the spinal nerves first enter the medulla
   (e) none of the above

_____ 10. The cortical surface of the cerebral hemispheres forms a series of elevated ridges called:
   (a) gyri
   (b) nuclei
   (c) sulci
   (d) lobes
   (e) none of the above

_____ 11. The groove between the frontal and parietal lobes of the brain is the:
   (a) longitudinal fissure
   (b) central sulcus
   (c) lateral sulcus
   (d) parieto-occipital sulcus
   (e) none of the above

_____ 12. The central sulcus separates which regions of the cerebrum?
   (a) the sensory and motor areas
   (b) the pyramidal cells and the frontal lobes
   (c) the parietal and occipital lobes
   (d) the temporal and insula lobes
   (e) none of the above

_____ 13. The fornix:
   (a) is a white matter tract that connects the parahippocampal gyrus with the thalamus
   (b) is a region present in only one brain hemisphere
   (c) has many fibers that end in the caudate nucleus
   (d) has many fibers that end in brain structures that control reflex movements associated with eating such as chewing, licking, and swallowing
   (e) is a part of the reticular formation

_____ 14. The structures of the diencephalon include the:
   (a) hypothalamus, but not the epithalamus
   (b) hypothalamus and thalamus
   (c) thalamus and mammillary bodies
   (d) fornix and hippocampus
   (e) a and c from above

_____ 15. The thalamus:
   (a) makes up the walls of the diencephalon
   (b) processes motor information to the spinal cord in a nucleus in the left or right side of the brain
   (c) contains centers that are involved with emotions and visceral processes
   (d) performs many voluntary functions
   (e) none of the above

Matching

Match each choice from the column on the left with the best response from the column on the right.

_____ 16. The smallest of the cranial nerves

_____ 17. Cranial nerves that move the eyeball

_____ 18. Provides sensory information from taste on the anterior 2/3 of the tongue and motor information from the pons

_____ 19. Special sensory nerve for balance, equilibrium and hearing

_____ 20. Special sensory nerves include

(a) facial nerve
(b) I, II, VII, VIII, IX
(c) trochlear nerve
(d) vestibulocochlear
(e) III, IV, VI

Fill in the Blank

21. The largest region of the brain, and the region where all conscious thought processes and intellectual functions originate is the _____ .

22. A brain region bounded anterosuperiorly by the area superior to the optic chiasma, and extending both to the posterior margins of the mammillary bodies posteriorly and to the pituitary inferiorly is the _____ .

16 The Nervous System:
Pathways and Higher-Order Functions

Laboratory Review Questions

Multiple Choice

_____ 1. Which of the following are *not true* of nerve pathways?
   (a) they consist of a chain of tracts and associated nuclei
   (b) no processing occurs along motor pathways; decisions are made in the brain and cannot be modified
   (c) there are several successive levels of processing of sensory and motor pathways
   (d) they relay sensory and motor information between the PNS and the brain
   (e) no exceptions; all of the above are true

_____ 2. Which of the following are *not true* of a sensory pathway?
   (a) only 1% of sensory input is processed in spinal cord tracts or the brain stem along the way to the higher centers
   (b) sensory somatic information reaches the primary sensory cortex in the cerebral hemispheres
   (c) the initial neuron activated in a sensory pathway is a first order neuron
   (d) sensory information conveyed by pathways crosses the corpus callosum to the other side of the brain, so the response is symmetrical
   (e) no exceptions; all of the above are true

_____ 3. Which is true of the motor pathways of the autonomic nervous system?
   (a) they control visceral effectors, such as smooth and cardiac muscle and glands
   (b) they show a neuronal organization identical to the somatic nervous system
   (c) they contain both upper and lower motor neurons
   (d) none of their neurons are located on the periphery
   (e) all of the above are true

_____ 4. The pyramidal system consists of descending motor tracts, which include all of the following *except*:
   (a) spinocerebellar tracts
   (b) corticobulbar tracts
   (c) lateral corticospinal tracts
   (d) anterior corticospinal tracts
   (e) no exceptions; all of the above are included

_____ 5. Processing centers of the extrapyramidal system (EPS) *exclude* which of the following?
   (a) the vestibular nuclei and the superior colliculus
   (b) the inferior colliculus and the mamillary body
   (c) the red nucleus and the reticular formation
   (d) the cerebral nuclei
   (e) no exclusions; all of the above are EPS processing centers

_____ 6. Outputs of the EPS processing centers target all of the following spinal cord tracts *except*:
   (a) vestibulospinal tracts
   (b) tectospinal tracts
   (c) rubrospinal tracts
   (d) reticulospinal tracts
   (e) no exceptions; all of the above are correct

_____ 7. Ascending sensory information is sent to the brain in a series of steps. At which of the following steps does information processing occur?
   (a) spinal cord to nucleus in the medulla
   (b) a nucleus in the medulla to a nucleus in the thalamus
   (c) a nucleus in the thalamus to the primary cortex
   (d) all of the above
   (e) no processing occurs at any of the steps listed above

_____ 8. Which of the following is true of the motor neurons controlled by simple reflexes?
   (a) they are the simplest reflexes, but require the most complex processing
   (b) the higher brain levels perform more elaborate processing
   (c) simple reflexes are controlled by nuclei in the spinal cord and brain stem
   (d) they have the greatest variability of response
   (e) none of the above

_____ 9. The most complicated involuntary motor patterns are controlled by the:
- (a) spinal cord and brain stem
- (b) cerebral nuclei and cerebellum
- (c) midbrain and hypothalamus
- (d) a and b from above are correct
- (e) b and c from above are correct

_____ 10. A specialized brain cortical region that has extensive connections to other cortical regions and other brain regions, such as the limbic system, is the:
- (a) hypothalamus (not cortex)
- (b) Broca's area
- (c) Wernicke's area
- (d) prefrontal cortex
- (e) none of the above

_____ 11. Damage to which region of the cerebral cortex would impair the ability of a person to regulate the pattern for breathing and vocalization required for speech?
- (a) none; no cerebral area is involved in these patterns
- (b) the general interpretive area
- (c) the primary motor cortex
- (d) Broca's area
- (e) none of the above is correct

_____ 12. Which of the following statements are true concerning the hemispheres of the cerebral cortex?
- (a) higher order centers in the two hemispheres have completely different, non-related functions
- (b) in the majority of people in the United States, the right hemisphere contains the general interpretive and speech centers
- (c) reading, writing, and speaking are processed in the left hemisphere
- (d) the left hemisphere is responsible for spatial relationships and analyses
- (e) none of the above are true

_____ 13. Identification of familiar objects by touch, smell, taste occurs in the:
- (a) left cerebral hemisphere
- (b) right cerebral hemisphere
- (c) both cerebral hemispheres
- (d) neither cerebral hemisphere
- (e) in the left hemisphere in most people living in the United States, but not all

_____ 14. Memory is:
- (a) accessing stored bits of information gathered through experience
- (b) located in the nucleus basalis
- (c) not possible to convert from short term to long term; the initial learning processes differ
- (d) all memories can be voluntarily retrieved, if the proper stimulus is recognized by the individual
- (e) I'm not certain I remember, but I think it is none of the above

_____ 15. Age-related anatomical changes in brain tissue:
- (a) are not reflected in similar functional changes
- (b) do not result in decreased neural processing changes
- (c) increase the speed of memory consolidation
- (d) slow reaction times
- (e) include none of the above

Fill in the Blank

16. The spinocerebellar pathway carries _____ information to the cerebellum for processing.

17. The corticospinal tracts that synapse on motor neurons in the anterior gray horns of the spinal cord are visible as they descend along the ventral surface of the medulla as a pair of thick bands, the _____ .

18. The most important and complex components of the EPS include the processing centers that provide the background patterns of movement involved in the performance of voluntary motor activities. These components are the

_____ .

19. The complex processing and integration of neural information from peripheral structures, visual information from the eyes and equilibrium-related sensations from the inner ear are performed by the _____ .

20. Brain shrinkage in older individuals has been attributed to _____ neurons.

17   The Nervous System: Autonomic Division

Laboratory Review Questions

Multiple Choice

_____ 1. Functions of the autonomic nervous system (ANS) *include*:
    (a) regulation of body temperature
    (b) coordination of cardiovascular and respiratory functions
    (c) coordination of digestive and excretory functions
    (d) regulation of reproductive functions
    (e) no exceptions; all of the above are included

_____ 2. In the ANS:
    (a) there is always a synapse between the CNS motor neuron and the peripheral effector organ
    (b) preganglionic fibers innervate the peripheral effector organs
    (c) preganglionic fibers are unmyelinated
    (d) motor neuron pathways synapse in the same pattern types as do those in the somatic nervous system
    (e) none of the above are true

_____ 3. Visceral motor neurons in the CNS:
    (a) are second-order (ganglionic) neurons
    (b) are postganglionic neurons
    (c) are not myelinated
    (d) send axons to synapse on peripherally located ganglionic neurons
    (e) are not any of the above

_____ 4. Postganglionic fibers:
    (a) are myelinated
    (b) carry impulses to the ganglion
    (c) innervate peripheral organs such as cardiac muscle, smooth muscle, glands, and adipose tissue
    (d) are the axons of first-order neurons
    (e) are or do all of the above

_____ 5. The major divisions of the autonomic nervous system (ANS) include:
    (a) the conscious and unconscious nervous system
    (b) pre and postganglionic fibers
    (c) the sympathetic and parasympathetic
    (d) the voluntary and involuntary
    (e) none of the above

_____ 6. The sympathetic division of the autonomic nervous system has an origin that is termed:
    (a) craniosacral
    (b) dorsoventral
    (c) thoracolumbar
    (d) pre and postganglionic
    (e) none of the above

_____ 7. The parasympathetic division of the autonomic nervous system has an origin that is termed:
    (a) craniosacral
    (b) peripheral only
    (c) dorsolumbar
    (d) cervicothoracic
    (e) none of the above

_____ 8. Which of the following ganglion types belong to the sympathetic division of the ANS?
    (a) preganglionic neurons
    (b) ganglionic neurons
    (c) paravertebral ganglia
    (d) prevertebral ganglia
    (e) all of the above belong to the sympathetic division of the ANS

_____ 9. Which of the following are true of the structures of the sympathetic chain?
  (a) each chain contains 3 cervical, 11 thoracic, 4 lumbar, and 4 sacral sympathetic ganglia
  (b) first-order sympathetic neurons are located in the cervical and thoracic regions only
  (c) these spinal nerves have white rami, but lack gray rami
  (d) about 8% of the axons of each spinal nerve are parasympathetic postganglionic fibers
  (e) none of the above are true

_____ 10. The parasympathetic division of the ANS includes:
  (a) preganglionic neurons in the brain stem
  (b) preganglionic neurons in the lumbar segments of the spinal cord
  (c) ganglionic neurons in the parasympathetic chain
  (d) neurons with a greater range of effects upon the target organs than have the neurons of the sympathetic division
  (e) none of the above

_____ 11. Cranial nerves that carry parasympathetic components are:
  (a) I, II, III, IV, V, VI
  (b) II, IV, VI, VII, X
  (c) VII, VIII, IX, XI, XII
  (d) III, VII, IX, X
  (e) none of the above combinations is correct

_____ 12. General functions of the parasympathetic division include:
  (a) pupillary constriction
  (b) secretion by digestive glands
  (c) secretion of hormones that promote nutrient absorption by peripheral cells
  (d) constriction of the respiratory passageways
  (e) all of the above

_____ 13. The division of the autonomic nervous system that increases activity levels in its target organs is the:
  (a) postganglionic or lumbosacral division
  (b) cervicothoracic division
  (c) sympathetic division
  (d) preganglionic division
  (e) none of the above

_____ 14. Brain regions that can affect the regulatory activities of the ANS include the:
  (a) hypothalamus
  (b) limbic system
  (c) thalamus
  (d) cerebral cortex
  (e) all of the above

_____ 15. If the greater splanchnic nerve is cut prior to its exit from the celiac ganglion, which of the following structures will be most affected?
  (a) stomach
  (b) spleen
  (c) kidney
  (d) adrenal gland
  (e) small intestine

Fill in the Blank

16. The cell bodies of ganglionic neurons are located in _____ .

17. Descending branches of the vagus nerve and thoracic splanchnic nerves leaving the sympathetic chain are contained within the _____ .

18. The parasympathetic outflow of the pelvic splanchnic nerves and sympathetic postganglionic fibers, from the inferior mesenteric ganglion, as well as sacral splanchnic nerves from the sacral sympathetic chain, are contained in the
_____ .

19. Smooth muscle contractions that propel materials and mix them with secretions in the digestive tract are controlled by actions of the _____ .

20. The increase in heart rate and the force of contractions coordinated in the medullary cardiac center is the
_____ reflex.

18 The Nervous System: General and Special Senses

Laboratory Review Questions

Multiple Choice

_____ 1. A sensory receptor:
   (a) cannot change sensitivity
   (b) may receive many different types of stimuli
   (c) is a specialized cell that monitors conditions in the body or the external environment
   (d) receives sensations from the CNS
   (e) none of the above are true

_____ 2. The special senses include all of the following *except*:
   (a) taste
   (b) olfaction
   (c) proprioception
   (d) vision
   (e) no exceptions; all of the above are included

_____ 3. All of the following are true of general sensory receptors *except*:
   (a) they are scattered throughout the body
   (b) they have a relatively simple structure
   (c) they provide information about the external environment
   (d) they provide information about the internal environment
   (e) no exceptions; all of the above are true

_____ 4. The olfactory organs consist of all of the following *except*:
   (a) an olfactory epithelium
   (b) specialized olfactory receptors within the epithelium
   (c) olfactory bulbs
   (d) basal cells
   (e) olfactory glands, which are located in the underlying lamina propria

_____ 5. Which of the following cranial nerves transmit sensory impulses from the taste buds?
   (a) I, II, IV
   (b) III, VI, VIII
   (c) VII, IX, XI
   (d) VII, IX, X
   (e) V, VII, IX, XI, XII

_____ 6. The three main anatomical regions into which the ear is divided are:
   (a) external ear, tympanum, stapes and malleus
   (b) stapes, malleus, incus and Organ of Corti
   (c) external ear, middle ear, and inner ear
   (d) ceruminous gland, cochlea, and utricle
   (e) none of the above are correct

_____ 7. The auditory ossicles consists of:
   (a) cochlea, saccule, and utricle
   (b) Organ of Corti
   (c) stapes, malleus, and incus
   (d) tympanum, stapedius, and tensor tympani
   (e) none of the above

_____ 8. The ear and the nasopharynx are connected by the pharyngotympanic tube, which connects the throat and the:
   (a) external ear
   (b) cochlea
   (c) middle ear
   (d) inner ear
   (e) none of the above

_____ 9. Structures of the external ear, in correct order from the outside to the inside, include:

1. tympanum
2. ceruminous glands
3. external auditory meatus
4. auricle
5. auditory tube
6. tensor tympani muscle

(a) 1,3,4,5,6
(b) 4,5,1,3,6
(c) 6,5,4,3,1
(d) 4,3,1
(e) 1,3,5

_____ 10. The semicircular canals and ducts include which of the following?
(a) dorsal and ventral
(b) lateral, middle, and medial
(c) anterior, posterior, and lateral
(d) spiral, upright, and reverse
(e) none of the above

_____ 11. The bony labyrinth, a shell of dense bone that surrounds and protects the membranous labyrinth, can be subdivided into:
(a) no subdivisions; it is all a single structure
(b) the perilymph, endolymph, saccule, and utricle
(c) cochlear duct and pharyngotympanic tube
(d) vestibule, semicircular canals, and the cochlea
(e) none of the above

_____ 12. Deep to the subcutaneous layer, the eyelids are supported by broad sheets of connective tissues, collectively called the:
(a) palpebrae
(b) tarsal plate
(c) chalazion
(d) medial canthus
(e) none of the above

_____ 13. The epithelium covering the inner surface of the eyelids and the outer surface of the eye is called the:
(a) conjunctiva
(b) lacrimal caruncle
(c) cornea
(d) lacrimal glands
(e) none of the above

_____ 14. The primary function of the lens of the eye is to:
(a) provide the coloring of the eye
(b) maintain the shape of the eye
(c) focus the visual image onto the optic disc
(d) focus the visual image on the retinal photoreceptors
(e) a and b from above are correct

_____ 15. Visual information from the retinas first arrives for processing at:
(a) the hippocampus
(b) the temporal lobe
(c) the occipital lobes
(d) the lateral geniculate nuclei
(e) none of the above

Fill in the Blank

16. General sensory receptors that provide information about the conditions inside the body are _____ .

17. The olfactory epithelium covers the inferior surface of the _____ , a portion of the _____ bone.

18. The taste receptors are clustered in individual _____ , each containing about 40 slender receptors.

19. The function of structures within the middle ear is to _____ and _____ sound waves to the inner ear.

20. The two optic nerves, one from each eye, reach the diencephalon at the _____ .

19 The Endocrine System

Laboratory Review Questions

Multiple Choice

_____ 1. Of the following statements, which best describes how the cells of the endocrine and nervous systems work together?
(a) provide widespread physiological effects throughout the body
(b) provide long lasting effects on a systemic basis
(c) monitor and adjust ongoing physiological activities
(d) affect target organs, which are restricted to nerve, gland, muscle, and fat cells
(e) provide gradual onset of the systemic effects

_____ 2. In general, the effects of the nervous system are:
(a) generalized with response to the environment
(b) opposite to those of the endocrine system
(c) short term and very specific
(d) particularly effective in regulating ongoing processes such as growth and development
(e) a and b from above are correct

_____ 3. The endocrine system:
(a) releases chemicals into the bloodstream for distribution throughout the body
(b) releases hormones that alter the metabolic activities of many different tissues and organs simultaneously
(c) produces effects that can last for hours, days, and even longer
(d) functions to control ongoing metabolic processes
(e) all of the above

_____ 4. The adrenal medulla is difficult to establish as either a nervous or an endocrine system structure exclusively because of all of the following reasons *except*:
(a) the adrenal medulla is a modified sympathetic ganglion
(b) adrenal neurons secrete epinephrine
(c) the systemic effects of the adrenal medulla are immediate
(d) adrenal neurons secrete norepinephrine
(e) no exceptions; all of the above are true

_____ 5. The anterior pituitary can be divided into two regions:
(a) neurohypophysis and infundibulum
(b) pars distalis and pars intermedia
(c) supraoptic and paraventricular nuclei
(d) adenohypophysis and neurohypophysis
(e) none of the above

_____ 6. Blood vessels that supply or drain the thyroid gland include all of the following *except* the:
(a) superior thyroid artery
(b) inferior thyroid artery
(c) superior and middle thyroid veins
(d) inferior thyroid veins
(e) no exceptions; all of the above supply or drain the thyroid gland

_____ 7. In addition to producing hormones, the pancreas also produces:
(a) digestive enzymes
(b) nothing else; this is the only function of the pancreas
(c) components of the white blood cells
(d) autonomic nervous stimulation of the digestive process
(e) none of the above

_____ 8. The alpha cells of the pancreas produce the hormone:
(a) insulin
(b) glucagon
(c) renin
(d) cortisol
(e) digestive enzymes

_____ 9. The _____ lies within the abdominopelvic cavity near the border between the stomach and the small intestine.
   (a) thymus gland
   (b) adrenal gland
   (c) pancreas
   (d) thyroid gland
   (e) ovary

_____ 10. Neurons of the supraoptic nuclei of the hypothalamus manufacture:
   (a) FSH
   (b) TSH
   (c) ADH
   (d) oxytocin
   (e) growth hormone

_____ 11. How does aging affect the function of the endocrine system?
   (a) it is relatively less affected than many other systems
   (b) the endocrine organs must produce increased amounts of hormones, because all types of receptors in the body become much less sensitive with age
   (c) the reproductive and endocrine systems are most affected by aging
   (d) the thyroid gland is the hormone-producing gland most affected by the aging process
   (e) none of the above are correct

_____ 12. The gland that produces growth hormone is the:
   (a) thymus
   (b) pituitary
   (c) thyroid
   (d) adrenal cortex
   (e) pancreas

_____ 13. How do nervous system functions affect those of the endocrine system?
   1. nervous stimulation causes the release of hormones
   2. the secretion of hormones may stimulate nervous responses
   3. high levels of hormones produced may inhibit nervous activity
   4. the release of hormones may inhibit nervous stimulation
   5. hormone activity may inhibit nervous activity and vice versa

   (a) 1,2,3,4,5
   (b) 1,3,5
   (c) 1,2,3,4
   (d) 3,4,5
   (e) 2,4

_____ 14. Which of the following organs secrete hormones?
   1. pancreas
   2. liver
   3. cerebrum
   4. adrenal cortex
   5. adrenal medulla
   6. thymus
   7. ovary
   8. testis
   9. thyroid
   10. pineal

   (a) 1,2,3,4,5,6,7,8,9,10
   (b) 1,4,5,6,7,8,9,10
   (c) 2,4,6,8,10
   (d) 1,3,5,7,9
   (e) 6,7,8,9,10

_____ 15. The zona glomerulosa of the adrenal gland produces:
   (a) androgens
   (b) glucocorticoids
   (c) mineralcorticoids
   (d) epinephrine
   (e) norepinephrine

Fill in the Blank

16. Peripheral cells sensitive to the presence of hormones are called _____ .

17. A slender stalk of neural tissue called the _____ connects the pituitary gland to the hypothalamus.

18. The two lobes of the thyroid gland are connected by the _____ .

19. The thyroid gland is composed of many _____ that produce and store thyroid hormone.

20. The _____ gland sits along the superior border of the kidney.

20 The Cardiovascular System: Blood

Laboratory Review Questions

Multiple Choice

_____ 1. To what type of tissue does blood belong?
  (a) muscle tissue
  (b) nervous tissue
  (c) connective tissue
  (d) epithelial tissue
  (e) blood does not fit into any of the typical categories

_____ 2. Functions of the blood include all of the following *except*:
  (a) filling vascular spaces to provide hydrostatic skeletal effects to increase the rigidity of the limbs
  (b) transportation of dissolved gases
  (c) transportation of metabolic wastes
  (d) delivery of hormones and enzymes
  (e) restricting fluid losses

_____ 3. The two main components of the blood include:
  (a) ions and proteins
  (b) lymph and lipids
  (c) plasma and formed elements
  (d) oxygen and carbon dioxide
  (e) none of the above

_____ 4. The most numerous formed elements in the blood are the:
  (a) platelets
  (b) white blood cells
  (c) proteins
  (d) red blood cells
  (e) none of the above

_____ 5. Formed elements of the blood include all of the following *except*:
  (a) plasma proteins
  (b) red blood cells
  (c) white blood cells
  (d) platelets
  (e) no exceptions; all of the above are correct

_____ 6. A formed element in the blood that is flattened, circular, lacks a nucleus, mitochondria, or ribosomes, and is red in color because of the presence of hemoglobin is a (an):
  (a) eosinophil
  (b) basophil
  (c) lymphocyte
  (d) erythrocyte
  (e) platelet

_____ 7. What metallic element or ion is essential to the normal function of a red blood cell in loading and transporting oxygen?
  (a) calcium
  (b) copper
  (c) potassium
  (d) manganese
  (e) iron

_____ 8. White blood cells:
  (a) carry oxygen and carbon dioxide as well as defend the body against infection
  (b) remove toxins, wastes, and abnormal or damaged cells from the body
  (c) are active mostly in the blood stream
  (d) are more abundant than the erythrocytes
  (e) produce blood clots

_____ 9. The most common white blood cell is a(an):
   (a) neutrophil
   (b) eosinophil
   (c) basophil
   (d) monocyte
   (e) lymphocyte

_____ 10. The white blood cell with the largest nucleus with respect to the size of its cytoplasm is a(an):
   (a) eosinophil
   (b) neutrophil
   (c) erythrocyte
   (d) basophil
   (e) lymphocyte

Matching
Match each of the statements in the left column with the appropriate term from the right column.

_____ 11. This is the blood matrix, which has a density only slightly greater than water.

_____ 12. Blood cells and cell fragments suspended in the matrix are called this.

_____ 13. Small packets of cytoplasm that contain enzymes and factors important in the blood clotting process in the matrix are called this.

_____ 14. This is the most common element in the matrix.

_____ 15. These are the components of the immune system in the blood matrix.

(a) white blood cells
(b) red blood cells
(c) formed elements
(d) platelets
(e) plasma

Fill in the Blank
16. The most important gas distributed to all cells of the body by the blood is _____ .
17. Blood cells and cell fragments that are suspended in the plasma are called _____ .
18. The protein in red blood cells that binds the oxygen is _____ .
19. The most noticeable visual difference between a red blood cell and a white blood cell seen under the microscope is that the red blood cell lacks a _____ .
20. In the adult, red blood cells are produced in areas of the _____ by stem cells.

21 The Cardiovascular System: The Heart

Laboratory Review Questions

Multiple Choice

_____ 1. The cavity that surrounds the heart and contains a small amount of serous lubricating fluid is the:
  (a) thoracic cavity
  (b) pleural cavity
  (c) diaphragmatic hiatus
  (d) pericardial cavity
  (e) none of the above

_____ 2. Which of the following do *not* apply to the pericardium?
  (a) it can be divided into fibrous and serous layers
  (b) the visceral pericardium is the outer layer
  (c) the parietal pericardium lines the inner surface of the sac that surrounds the heart
  (d   the pericardial sac is reinforced by a dense network of collagen fibers
  (e) no exceptions; all of the above apply to the pericardium

_____ 3. Functions of the fibrous skeleton of the heart include all of the following *except*:
  (a) stabilization of the positions of the muscle fibers and valves in the heart
  (b) provide physical support for the cardiac muscle fibers and myocardial blood vessels and nerves
  (c) controlling the distribution of the forces of muscle contraction
  (d) adding strength and preventing overexpansion of the heart
  (e) no exceptions; all of the above are true

_____ 4. Prominent muscular ridges, which run along the inner surface of the auricle and across the adjacent atrial wall, are the:
  (a) chordae tendineae
  (b) foramen ovale
  (c) papillary muscles
  (d) trabeculae carneae
  (e) none of the above

_____ 5. The two main branches of the right coronary artery are the:
  (a) circumflex branch and the left marginal branch
  (b) anterior interventricular branch and the left anterior descending artery
  (c) right marginal branch and the posterior interventricular branch
  (d) great and middle cardiac vessels or branches
  (e) anastomoses

_____ 6. Which of the following are true of the coronary arteries?
  (a) they branch off the pulmonary trunk
  (b) together, they supply part of the heart muscle with oxygen
  (c) they originate at the base of the aorta and are the first branch off this vessel
  (d) they open and close in pulsation with the contraction and relaxation of the heart ventricles
  (e) none of the above are true

_____ 7. The pacemaker of the heart that normally sets the beat is the:
  (a) atrioventricular node
  (b) sinoatrial node
  (c) purkinje fibers
  (d) bundle branches
  (e) pectinate muscles

_____ 8. Which heart structures are involved in assisting the function of the bicuspid and tricuspid valves as the ventricles contract?
  (a) the auricles, which supply an additional bolus of blood to each ventricle
  (b) the intercalated disks
  (c) the pectinate muscles
  (d) the chordae tendineae
  (e) none of the above

_____ 9. If the atrioventricular valves fail to close, what will happen to heart function?
   (a) nothing unusual will occur
   (b) blood will regurgitate from the ventricles to the atria
   (c) blood will be ejected too early from the ventricles into the pulmonary trunk and the ascending aorta
   (d) the foramen ovale will remain open for a much longer than is usual after birth
   (e) none of the above will occur

_____ 10. If the sinoatrial node is damaged, what will happen to the heartbeat?
   (a) it will be generated by the bundle branches, at a much lower rate
   (b) the heart will stop
   (c) the heart beat will increase in rate, but not in forcefulness
   (d) the atrioventricular node will take over setting the pace, at a speed somewhat slower than normal
   (e) it will become more forceful

Matching

Match each of the statements in the left column with the appropriate structure from the right column.

_____ 11. The right atrium receives blood from the

_____ 12. The coronary veins of the heart return blood to the

_____ 13. The posterior wall of the left atrium receives oxygenated blood from the lungs through the

_____ 14. The superior end of the right ventricle tapers to a cone-shaped pouch, called the

_____ 15. Blood leaves the left ventricle by passing through the aortic semilunar valve into the

(a) conus arteriosus

(b) ascending aorta

(c) coronary sinus

(d) superior and inferior vena cavae and coronary sinus

(e) left and right pulmonary veins

Fill in the Blank

16. The primary function of the pericardial fluid is to provide _____ between the pericardial membranes.

17. The extremely expandable portion of an atrium is called an _____ because it reminded early anatomists of the external ear.

18. The fossa ovalis in the adult heart marks the site of the _____ , an opening between the two atria of the heart prior to birth.

19. When it is closed, the _____ prevents blood from flowing from the left ventricle to the left atrium:

20. Blood passes into the lower extremity through the _____ artery.

22 The Cardiovascular System: Vessels and Circulation

Laboratory Review Questions

Multiple Choice

_____ 1. The elastic arteries that originate along the aortic arch include the:
(a) pulmonary arteries
(b) right subclavian artery and right common carotid artery
(c) brachiocephalic (innominate) artery, left common carotid artery, and left subclavian artery
(d) arteries listed in parts a and b
(e) none of the above

_____ 2. After passing from the thoracic cavity over the border of the first rib, the subclavian artery changes its name. The name changes continue through the arm and into the hand. In correct order these names are:

1. superficial palmar arch	(a) 1,2,3,4,5,6
2. deep palmar arch	(b) 6,5,4,3,2,1
3. brachial artery	(c) 1 or 4, 2, 3 or 5, 6
4. radial artery	(d) 5,3,4 or 6, 1 or 2
5. axillary artery	(e) none of the above orders is correct
6. ulnar artery	

_____ 3. The branches of the internal carotid artery include:

1. external carotid artery	(a) 1,2,3,4,5,6,7
2. common carotid artery	(b) 2,4,6
3. ophthalmic artery	(c) 1,3,4,5,7
4. anterior cerebral artery	(d) 1,3,5,6
5. maxillary artery	(e) 3,4,7
6. basilar artery	
7. middle cerebral artery	

_____ 4. The abdominal aorta bifurcates inferiorly to form:
(a) three inferior branches
(b) the internal iliac arteries
(c) the external iliac arteries
(d) the common iliac arteries
(e) none of the above

_____ 5. Which of the following do not empty directly into the inferior vena cava?
(a) hepatic veins
(b) lumbar veins
(c) superior mesenteric vein
(d) suprarenal veins
(e) no exceptions; all of them empty into the inferior vena cava

Matching

_____ 6. Artery that arises from the abdominal aorta and delivers blood to the pancreas, small intestine and most of the large intestine is the

_____ 7. Artery that arises about two inches superior to the terminal aorta and delivers blood to the terminal portions of the large intestine and rectum is the

_____ 8. Small arteries that arise on the posterior surface of the aorta and supply the spinal cord and the abdominal wall are the

_____ 9. Long, thin arteries, present only in males, that originate between the superior and inferior mesenteric arteries are the

_____ 10. Arteries that originate near the superior mesenteric arteries and supply the suprarenal glands are the

(a) testicular arteries
(b) superior mesenteric
(c) lumbar arteries
(d) suprarenal arteries
(e) inferior mesenteric

_____ 11. Veins that empty into a network of dural sinuses are the

_____ 12. Veins that drain the cerebral hemispheres are the

_____ 13. The sigmoid sinus, which penetrates the jugular foramen, leaves the skull as the

_____ 14. The cervical spinal cord and the posterior part of the skull is drained by the

_____ 15. The temporal and maxillary veins drain into the

(a) internal cerebral veins

(b) internal jugular vein

(c) superficial cerebral veins

(d) external jugular veins

(e) vertebral veins

Fill in the Blank

16. The blood vessel type that is present in the body in the greatest numbers is the _____ .

17. Blood leaves the heart in major vessels called the _____ and the _____ .

18. The arteries in the _____ circuit differ from those of the systemic circuit in that they carry deoxygenated blood.

19. The vertebral arteries enter the cranium at the foramen magnum where they fuse along the ventral surface of the medulla oblongata to form the _____ .

20. The blood supply to the superior surface of the muscular diaphragm that separates the thoracic and abdominopelvic cavities is provided by the _____ .

23 The Lymphatic System

Laboratory Review Questions

Multiple Choice

_____ 1. Lymphatic vessels originate in the:
   (a) spleen
   (b) bone marrow
   (c) venous system
   (d) peripheral tissues of the body
   (e) a and b from above

_____ 2. Which of the following is not a component of the lymphatic system?

   1. pancreas   (a) all of the above
   2. spleen   (b) none of the above
   3. lymphatic vessels   (c) 1,3,5
   4. thymus   (d) 2,3,5
   5. thoracic duct   (e) 1,4

_____ 3. Flow of lymph through lymphatic collecting vessels is slow, and is maintained by all of the following *except*:
   (a) contraction of skeletal muscles
   (b) arterial pulsations
   (c) contraction of the smooth muscle walls of the lymphatic vessels
   (d) lymphatic vessel valves
   (e) no exceptions; all of the above are correct

_____ 4. Lymph is:
   (a) excess tissue fluid
   (b) present in body tissues only when there is a pathological condition that allows it to escape
   (c) picked up by lymph vessels and returned to the great veins at the root of the neck
   (d) high in protein and other nutrient content
   (e) a and c from above are correct

_____ 5. Identify the correct order for the movement of lymph:

   1. lymph trunks   (a) 2,4,3,1,5
   2. lymph capillaries   (b) 1,2,3,4,5
   3. lymph nodes   (c) 4,2,5,3,1
   4. lymph collecting vessels   (d) 2,1,3,5
   5. lymph sacs   (e) none of the above orders is correct

_____ 6. Lymph nodes:
   (a) occur in all regions of the body
   (b) filter both blood and lymph
   (c) do not occur in the brain
   (d) receive lymph through the efferent vessels
   (e) manufacture lymph

_____ 7. The small organs that are a part of the lymphatic system and that are intimately associated with the lymph vessels are:
   (a) lymph nodes
   (b) cisterna chyli
   (c) spleen
   (d) suprarenal glands
   (e) none of the above

_____ 8. The largest of the organs of the lymphatic system is/are the:
   (a) lymph nodes in the groin
   (b) cisterna chyli
   (c) spleen
   (d) carotid bodies
   (e) kidneys

_____ 9. Tonsils located on the posterior aspect of the tongue are called the:
- (a) palatine tonsils
- (b) lingual tonsils
- (c) pharyngeal tonsils
- (d) esophageal tonsils
- (e) epiglottal tonsils

_____ 10. The bulges that give the lymphatic vessels the "string of beads" appearance are caused by:
- (a) periodic inflammation of the tissues that remove bacteria and viruses from other body tissues
- (b) closely spaced valves within the vessels
- (c) thickenings in the endothelial lining of the lymph vessels as evenly spaced intervals
- (d) regions of higher pressure within the lymph vessels, which cause the walls to bulge
- (e) none of the above

_____ 11. The thoracic duct:
- (a) carries lymph to the right brachiocephalic vein
- (b) carries lymph originating in tissues superior to the diaphragm
- (c) carries lymph from the right side of the lower body
- (d) joins the junction of the left subclavian vein and left internal jugular veins
- (e) a and b are correct

_____ 12. The thoracic duct:
- (a) drains superiorly through the medial region of the thoracic cavity
- (b) the thoracic duct travels parallel to the azygos vein, but not the hemiazygous vein
- (c) drains tissues inferior to the diaphragm
- (d) joins directly to the right lymphatic duct near the point where they enter the venous system
- (e) a and c from above are correct

_____ 13. The cisterna chyli is:
- (a) the point of origin of the right lymphatic duct
- (b) located superior to the diaphragm
- (c) receives some drainage directly from the azygous vein
- (d) the location where lymph is received from the lower abdomen, pelvis, and lower extremities
- (e) none of the above is correct

_____ 14. The spleen:
- (a) is the lymphatic organ that grows to its greatest size in the individual at puberty
- (b) is composed of a cortex and a medulla
- (c) lies tucked inside of the lesser curvature of the stomach
- (d) is attached to the stomach by the broad mesentery band, the gastrosplenic ligament
- (e) is or does all of the above

_____ 15. The shape of the spleen is:
- (a) rigid and unaffected by the surrounding tissues
- (b) is smooth and convex on the diaphragmatic surface
- (c) kidney-bean shaped, as is that of the kidney
- (d) is extremely changeable depending upon the amount of blood flowing within it
- (e) governed by all of the considerations listed above

Fill in the Blank

16. The major role of the lymphatic system is to guard the body against _____ and _____ .

17. The components of the lymphatic system that transport the lymph and begin in the peripheral tissues and end in the venous system are the _____ .

18. The fluid that is transported by the lymphatic vessels, similar in composition to blood plasma, but lower in protein concentration is the _____ .

19. Lymphatic capillaries are _____ in areas of loose connective tissue, that allow interstitial fluid to enter between the endothelial cells of their walls.

20. Lymphatic capillaries are present in almost every tissue of the body, with the exception of the _____ .

24 The Respiratory System

Laboratory Review Questions

Multiple Choice

_____ 1. Actions of the passageways of the respiratory system include all of the following *except*:
(a) filtering air
(b) warming air
(c) humidifying air
(d) accelerating air as it moves toward and away from the lungs to ensure a sufficient rate of air passage to permit gas exchange to occur
(e) no exceptions; all of the above are true

_____ 2. What is the effect of turbulence of the air passing between the vestibule and the nasal chamber?
(a) it promotes air filtration
(b) it decreases the volume of air entering, to prevent possible damage to delicate respiratory surfaces by too rapid a rate of airflow
(c) it allows extra time for warming and humidifying incoming air
(d) the effect is negligible
(e) a and c are correct

_____ 3. What is the function of the bony hard palate that forms a portion of the roof of the oral cavity?
(a) it acts as a vertical strut to separate the left and right nasal cavities
(b) it forms the floor of the nasal cavity and separates the oral and nasal cavities
(c) it stiffens and reinforces the soft palate
(d) it contains more ciliated epithelial cells than soft tissues because of the stiffness of the support it provides
(e) none of the above are correct

_____ 4. The openings to the pharyngotympanic tube are located in the:
(a) nasopharynx
(b) oropharynx
(c) laryngopharynx
(d) larynx
(c) nasal cavity

_____ 5. The palatine tonsils lie in the walls of the:
(a) nasopharynx
(b) oropharynx
(c) laryngopharynx
(d) larynx
(e) nasal cavity

_____ 6. The cartilage blocks in the walls of the secondary and tertiary bronchi:
(a) have a completely different function than do those of the tracheal rings, but not those of the bronchi
(b) support the bronchi and assist in keeping the lumens open
(c) are unusual among cartilaginous tissues in that they are highly vascularized
(d) assist directly in gas exchange by acting as baffles to direct the air flow
(e) are or do all of the above

_____ 7. The parietal pleura:
(a) covers the outer surfaces of the lungs
(b) extends into the fissures between the lung lobes
(c) covers the inner surfaces of the thoracic walls and extends over the diaphragm and the mediastinum
(d) are richly supplied with respiratory capillaries
(e) a and b from above are true

_____ 8. Pleural fluid is secreted by:
(a) the parietal pleura only
(b) both visceral and parietal pleura
(c) the mediastinum
(d) the visceral pleura only
(e) the respiratory membrane

_____ 9. The most important skeletal muscles involved in making respiratory movements include the:
   (a) serratus anterior and levator scapulae
   (b) diaphragm and external and internal intercostal muscles
   (c) rectus abdominis, external and internal oblique, and the transversus abdominis
   (d) intrinsic back muscles and the scalenus anterior and medius
   (e) all of the above

_____ 10. Contraction of the diaphragm:
   (a) decreases the volume of the thoracic cavity
   (b) increases the pressure in the thoracic cavity and pushes air out of the lungs
   (c) flattens the floor of the thoracic cavity thereby increasing its volume
   (d) depresses the ribs
   (e) does none of the above

_____ 11. Each of the following muscles can function in respiration *except* one. Identify the exception as the:
   (a) diaphragm
   (b) external intercostals
   (c) latissimus dorsi
   (d) sternocleidomastoid
   (e) scalenes

_____ 12. Which is greater?
   (a) the number of lobes in the right lung
   (b) the number of lobes in the left lung

_____ 13. What is the significance of the "C-shaped" cartilages that reinforce the tracheal rings?
   (a) these cartilages hold the trachea rigidly open at the same diameter at all times
   (b) the incomplete portion of the "C-shaped" cartilages are located at the posterior of the tracheal cartilages, to permit the esophagus to bulge anteriorly into the tracheal lumen in transient fashion to permit a large bolus to pass
   (c) they form a solid cartilaginous tube
   (d) the open region permits passage of the nerves that stimulate the muscles of the laryngeal region
   (e) none of the above is true

_____ 14. What is likely to happen if the epiglottis fails to fold over the glottis during the act of swallowing food or liquid?
   (a) air may enter the trachea
   (b) food or liquid may enter the trachea
   (c) the person may be induced to sneeze
   (d) nothing is likely to happen
   (e) none of the above is true

_____ 15. The following is a list of some of the structures of the respiratory tree:

   1. secondary (lobar) bronchi   The order in which air passes through
   2. bronchioles   the structures is:
   3. alveolar ducts   (a) 4, 1, 2, 7, 5, 3, 6
   4. primary bronchi   (b) 4, 1, 2, 5, 7, 3, 6
   5. respiratory bronchioles   (c) 1, 4, 2, 5, 7, 3, 6
   6. alveoli   (d) 1, 4, 2, 7, 5, 3, 6
   7. terminal bronchioles   (e) 2, 4, 1, 7, 5, 3, 6

Fill in the Blank

16. The lungs are divided into _____ that are separated by deep fissures.

17. The opening to the larynx is termed the _____ .

18. The cartilage that makes up most of the anterior and lateral surfaces of the larynx is the _____ .

19. The _____ are formed by the branching of the trachea within the mediastinum.

20. The actual site of gas exchange in the lungs occurs in the _____ .

25 The Digestive System

Laboratory Review Questions

Multiple Choice

_____ 1. Most of the digestive tract is lined by:
   (a) pseudostratified ciliated columnar epithelium
   (b) cuboidal epithelium
   (c) stratified squamous epithelium
   (d) ciliated columnar epithelium
   (e) simple columnar epithelium

_____ 2. Which of the following are true of the mesenteries?
   1. they are sheets of serous membrane
   2. they suspend portions of the digestive tract within the peritoneal cavity
   3. they are double sheets of peritoneal membrane
   4. they provide an access route to digestive structures for nerves and blood vessels, which travel
      along the outer surfaces of the membranes
   5. sensory receptors located in mesenteries provide proprioceptive information to the spinal cord regarding the
      digestive organs
   (a) all of the above are true
   (b) none of the above are true
   (c) 1, 3 and 5 are true
   (d) 1, 2 and 3 are true
   (e) 2, 4 and 5 are true

_____ 3. Which of the following is or are true of the greater omentum?
   (a) it is formed from the ventral mesentery of the peritoneal cavity
   (b) it forms a pouch that extends inferiorly, between the body wall and the anterior surface of the small intestine
   (c) it is formed of a double layer of double layered serous membrane
   (d) it is never the storage site for large amounts of fat
   (e) b and c from above are true

_____ 4. Gross movements of the tongue are performed by:
   (a) intrinsic tongue musculature
   (b) muscles innervated primarily by the hypoglossal nerve, Cranial Nerve XII
   (c) the hyoid apparatus, which moves and orients the tongue
   (d) the muscles of mastication
   (e) none of the above

_____ 5. What gland empties into the oral cavity at the level of the second upper molar?
   (a) submaxillary
   (b) submandibular
   (c) parotid
   (d) sublingual
   (e) vestibular

_____ 6. The esophagus:
   (a) is a hollow muscular tube that connects the pharynx to the stomach
   (b) always remains open
   (c) is reinforced by cartilaginous structures
   (d) is under both voluntary and autonomic nervous control
   (e) is or does all of the above

_____ 7. The bulge of the greater curvature of the stomach superior to the esophageal junction is known as:
   (a) pylorus
   (b) cardia
   (c) pyloric antrum
   (d) fundus
   (e) vestibule

_____ 8. Which of the following is *not* an accessory digestive organ?
  (a) salivary glands
  (b) spleen
  (c) liver
  (d) pancreas
  (e) gall bladder

_____ 9. Absorptive effectiveness of the small intestine is enhanced by:
  (a) plicae
  (b) villi
  (c) microvilli
  (d) intestinal movements
  (e) all of the above

_____ 10. Structures that unite to form the common bile duct include:

  1. hepatopancreatic sphincter   (a) 1, 2
  2. porta hepatica   (b) 1, 2, 3
  3. cystic duct   (c) 3, 4
  4. common hepatic duct   (d) 1, 3, 5
  5. duodenal papilla   (e) 2, 4

_____ 11. Which of the following is *not* a lobe of the liver?
  (a) caudate
  (b) left
  (c) right
  (d) quadrate
  (e) all of the above are lobes of the liver

_____ 12. The fusion of the hepatic duct and the cystic duct forms:
  (a) the hepatic portal vein
  (b) the porta hepatica
  (c) the common bile duct
  (d) the common pancreatic duct
  (e) the bile canaliculus

_____ 13. At the hepatic flexure, the colon becomes the:
  (a) ascending colon
  (b) transverse colon
  (c) descending colon
  (d) sigmoid colon
  (e) rectum

_____ 14. External pouches of the colon are known as:
  (a) the cecum
  (b) plicae
  (c) haustra
  (d) the taenia coli
  (e) rugae

_____ 15. Damage to which nerve would make it impossible for a person to protrude the tongue?
  (a) VII
  (b) IX
  (c) X
  (d) XII
  (e) a and b from above

Fill in the Blank

16. Lubricating fluid containing enzymes that break down carbohydrates is secreted by the _____ .

17. _____ stabilize the positions of the attached organs within the peritoneal cavity and prevent the intestines from becoming entangled during digestive movements of changes in body position.

18. A portion of the ventral mesentery that is retained in the adult, and occurs between the liver and the anterior abdominal wall, is the _____ .

19. The release of chyme from the stomach into the duodenum occurs by relaxation of the _____ .

20. The head of the pancreas lies within the loop formed by the _____ , a portion of the small intestine.

26 The Urinary System

Laboratory Review Questions

Multiple Choice

_____ 1. The urinary system interacts most closely with components of the:
(a) circulatory system
(b) respiratory system
(c) nervous system
(d) skeletal system
(e) muscular system

_____ 2. The urinary system includes all *but* which of the following?
(a) kidneys
(b) ureters
(c) urethra
(d) urinary bladder
(e) adrenal glands

_____ 3. Each of the following systems of the body is involved in the process of excretion to some degree except for one. Identify the exception.
(a) urinary system
(b) integumentary system
(c) digestive system
(d) endocrine system
(e) respiratory system

_____ 4. The ureters and urinary bladder are lined chiefly by:
(a) stratified squamous epithelium
(b) pseudostratified columnar epithelium
(c) simple cuboidal epithelium
(d) simple columnar epithelium
(e) transitional epithelium

_____ 5. The kidneys:
(a) are located in a position that is retroperitoneal
(b) are surrounded by a renal capsule
(c) are surrounded by a thick layer of adipose tissue
(d) are held in place by the renal fascia
(e) all of the above are true

_____ 6. The position of the kidneys in the abdominal cavity is maintained, in part, by the:
(a) overlying peritoneum
(b) diaphragm
(c) floating ribs
(d) osmotic pressure of the fluid in the ureters
(e) none of the above

_____ 7. Kidneys are often difficult to see without dissection because they are surrounded by a layer of fat. What is the significance of this fat?
(a) expansion for storage of additional urine once the bladder is full; it acts as a sponge
(b) cushioning or padding for protection from sudden jolts or other injuries
(c) no special significance; this fat stores energy as do other fat deposits in the abdominal cavity
(d) this fat cools the kidneys during active filtration
(e) a and d from above are correct

_____ 8. The urinary bladder, ureters, and kidneys are located:
(a) retroperitoneally
(b) anterior to the colon and pancreas
(c) medial to the aorta
(d) at the levels between T_{10} and L_1
(e) nowhere near any of the above structures

_____ 9. Each kidney receives blood directly from:
   (a) the descending aorta
   (b) branches of the gonadal arteries that branch off the celiac artery at the same place
   (c) the common iliac arteries
   (d) the internal iliac arteries
   (e) none of the above

_____ 10. The following is a list of the blood vessels that transport blood to the kidney:

1. afferent arteriole	The proper order in which blood passes
2. arcuate artery	through the blood vessels is:
3. interlobar artery	(a) 4,6,2,3,1,5,7,8
4. renal artery	(b) 4,3,2,6,1,5,7.8
5. glomerulus	(c) 4,3,2,6,7,5,1,8
6. interlobular artery	(d) 4,6,2,3,7.5.1.8
7. efferent arteriole	(e) 4,3,6,2,1,5,7,8
8. peritubular capillary	

_____ 11. The ureters:
   (a) are retroperitoneal
   (b) float freely within the abdominal cavity
   (c) take exactly the same path to the bladder in men and women
   (d) have specialized subdivisions called the urethrae
   (e) are or do none of the above

_____ 12. The expanded beginning of the ureter is the:
   (a) major calyx
   (b) minor calyx
   (c) hilus
   (d) renal pelvis
   (e) renal pyramids

_____ 13. A ligament that extends from the anterior and superior border of the bladder to the umbilicus is the:
   (a) round ligament
   (b) square ligament
   (c) urachus or median umbilical ligament
   (d) median umbilical ligaments
   (e) none of the above

_____ 14. The structure that stores urine in the urinary tract is the:
   (a) ureter
   (b) urethra
   (c) bladder
   (d) kidney
   (e) all of the above

_____ 15. The portions of the urethra in the male, from the bladder to the exterior, in correct order are:

1. urachus	(a) 1,3,5
2. penile urethra	(b) 2,4
3. dysuria	(c) 4,2,1
4. membranous urethra	(d) 5,4,2
5. prostatic urethra	(e) 1,2,3,4,5

Fill in the Blank

16. The renal medulla consists of six to eighteen distinct conical or triangular structures, called _____ .

17. The renal arteries give rise to the _____ arteries, which give rise to the interlobar arteries.

18. The triangular area bounded by the urethral openings and the entrance to the urethra constitutes the

   _____ .

19. The region surrounding the urethral opening, known as the neck of the urinary bladder, contains a muscular

   _____ .

20. Upon exiting the bladder, the urethra passes through the round, walnut-sized organ in males, called the

   _____ .

27 The Reproductive System

Laboratory Review Questions

Multiple Choice

_____ 1. The human reproductive system:
  (a) is as essential to the survival of an individual as all other anatomic systems
  (b) ensures the continued existence of the human species
  (c) remains functional throughout the life of an individual
  (d) stores but does not produce gametes
  (e) is or does none of the above

_____ 2. The reproductive system includes the following structures:
  (a) gonads
  (b) ducts that receive and transport the gametes
  (c) accessory glands and organs that secrete fluids for the reproductive process
  (d) external genitalia
  (e) all of the above

_____ 3. The following is a list of structures of the male reproductive tract:
  1. ductus deferens
  2. urethra
  3. ejaculatory duct
  4. epididymis
  The order in which sperm passes through these structures from the testes to the penis is:
  (a) 1,3,4,2
  (b) 4,3,1,2
  (c) 4,1,2,3
  (d) 4,1,3,2
  (e) 1,4,3,2

_____ 4. The function of the dartos and cremaster muscles is:
  (a) to attach the penis to the body wall
  (b) to produce erections
  (c) to regulate the temperature of the testes
  (d) to help the testes descend prior to birth
  (e) to move sperm along the ductus deferens

_____ 5. For erection and ejaculation to occur:
  (a) there must be sufficient blood hydrostatic pressure
  (b) the parasympathetic and sympathetic branches of the nervous system must be properly functioning
  (c) the urinary sphincters must be closed
  (d) the level of nutrients supplied by the digestive system must be sufficient for sperm formation and glandular secretions
  (e) all of the above

_____ 6. Contraction of the bulbospongiosus and ischiocavernosus muscles:
  (a) causes wrinkling of the scrotal sac
  (b) assists in erection
  (c) propels sperm through the penile urethra
  (d) moves sperm through the ductus deferens
  (e) moves the testes closer to the body cavity

_____ 7. The ovaries are supported and stabilized in position by the:
  (a) roof of the pelvic cavity
  (b) ovarian blood vessels and the ovarian hilum
  (c) mesovarium and other ligaments
  (d) uterus
  (e) none of the above

_____ 8. The vagina:
    (a) serves as a passageway for the elimination of menstrual fluids
    (b) receives the penis during coitus
    (c) holds spermatozoa prior to their passage to the uterus
    (d) forms the lower portion of the birth canal
    (e) all of the above

_____ 9. The vulva includes:
    (a) mons pubis
    (b) labia majora
    (c) labia minora
    (d) clitoris
    (e) all of the above

_____ 10. Jane is an avid jogger and trains extensively. She has trimmed down so that she is now under her weight by 25%. Consequently, she has very little body fat. One effect of low body fat in some women is:
    (a) have heavy menstrual flows
    (b) double ovulate
    (c) be amenorrheic
    (d) have painful menstrual cramps
    (e) show elevated levels of FSH

Fill in the Blank

11. Most of the body of the penis consists of three masses of _____ .

12. The internal orifice connects the uterine cavity to the _____ .

13. Sperm cells are produced by the process of _____ .

14. The _____ is the part of the spermatozoa that contains the chromosomes.

15. The _____ contains the ductus deferens, spermatic arteries/veins, nerves, and lymphatics that service the testes.

16. The _____ is an extensive mesentery that encloses the ovaries, uterine tubes, and uterus.

17. The _____ ligaments arise on the lateral margins of the uterus and extend anteriorly, passing through the inguinal canals to the base of the external genitalia.

18. The finger-like projections of the infundibulum are known as _____ .

19. The _____ is the inferior constricted portion of the uterus that projects into the vagina.

20. The _____ is the layer of the uterus that provides mechanical protection and nutritional support for the developing embryo.

Index

Transitional epithelium, 14
Transverse:
    cervical nerve, 204
    colon, 380
    jugular vein, 310
    mesocolon, 366, 381
    plane, 5
    sections, 5
    sinus, 217
Transversus abdominis muscle, 122, 123, 134
Transversus thoracis muscle, 122, 123
Trapezium bone, 74
Trapezius group, 130
Trapezius muscle, 144, 145, 180, 184
Trapezoid bone, 74
Triangular bone, 74
Triceps brachii muscle, 137, 149
    lateral head, 137, 150
    long head, 147, 150, 151
    medial head, 150
Triceps muscle, 187, 307
Tricuspid valve, 300
Trigeminal nerve, 233, 234
Trigone, 400
Triiodothyronine, 279
Triquetrum bone, 74
Trochanters, 36
    greater, 190
    lesser, 190
Trochlea, 36
Trochlear nerve, 233, 234
True ribs, 65
True vocal cords, 348, 360
Tuber cinereum, 227
Tubercle, 36
Tuberosity, 36
Tunica albuginea, 417, 419-20
Tunica externa, 314-15
Tunica interna, 314-15
Tunica media, 314-15
Tunica vaginalis, 409, 427
Tympanic duct, 263, 264
Tympanic membrane, 260
Typical tooth, 371

U

Ulna, 187
    bones of, 73-74
    markings, 73
Ulnar, 319
    artery, 307
    collateral ligament, 91, 93
    nerve, 187, 206
    vein, 328
Umbilical:
    arteries, 309, 332
    vein, 332
Umbilicus, 185
Unipolar neurons, 195
Upper arm, surface anatomy, 187
Upper extremities, 186
    muscles of, 144
    skeletal system, 68
    surface anatomy, 186-89
    venous return from, 328
Upper respiratory tract, 344
Ureteral openings, 400
Ureters, 391, 392, 399-400, 401, 404
    microscopic identification of, 401-402
Urethra, 391, 400, 403, 404
Urethral meatus, external, 400
Urethral orifice, 428
Urethral sphincter:
    external, 400

internal, 400
Urethral sphincter muscle:
    female, 126
    male, 126
Urinary bladder, 250, 391, 392, 400, 402, 404
    microscopic identification of, 402
        urethra, microscopic identification of, 403
Urinary system, 391-405
    identification of components of, 393-94
    kidney:
        circulation pattern in, 396-97
        nephron of, 398-400
        specimen dissection, 396
        structure, 394-95
    ureters, 399-400
        microscopic identification of, 401-2
    urethra, 400
        microscopic identification of, 403
    urinary bladder, 400
        microscopic identification of, 402
    urine, organs responsible for conduction/storage of, 399-401
Urine, 391, 399-401
Urogenital regions, perineum, 425
Urogenital sinus, 429
Urogenital triangle, 127
Uterine:
    body, 429
    cavity, 422
    cervix, 423
    horn, 429
    tubes, 406, 419, 420, 429
        microscopic identification of, 421
    wall, 424
Uterosacral ligaments, 422
Uterus, 406, 422
    microscopic identification of, 424
Utricle, 263
Uvea, 270
Uvula, 347, 368

V

Vagal trunks, 372
Vagina, 406, 422-23, 429
    microscopic identification of, 424-25
Vaginal:
    canal, 422
    orifice, 422, 425
    rugae, 422
Vagus nerve, 234, 235, 258, 372
Valves, heart, 302-3
Vascular tunic structures, 270-71
Vas deferens, *See* Ductus deferens
Vasoconstriction, 315
Vasodilation, 315
Vastus intermedius muscle, 141, 165, 166
Vastus lateralis muscle, 141, 165, 166, 191
Vastus medialis muscle, 141, 165, 166, 191
Veins, 297, 317, 325, 329, 331
    microscopic, 314-16
    See also specific vessels by name
Vena cava:
    inferior, 297, 310
    superior, 297, 310
Venous arch, dorsal, 193
Ventral, 4
    body cavity, 6
    nerve roots, 200
    ramus, 200

roots, 202
    thoracic vein, 310
Ventricles, 220-21, 294, 297
    brain, 221
    fourth, 221
    lateral, 221
    left, 297
    right, 297
    third, 221
Venules, 314
Vermiform appendix, 380
Vermis, 229
Vertebrae, 57
    cervical, 59-61
    lumbar, 63
    sacral, 64
    thoracic, 62
Vertebral, 184, 319
    arch, 57
    artery, 307
    canal, 201
    column, 39, 57-58
    veins, 310, 327
Vertebrochondral ribs (false), 65
Vertebrosternal ribs (true), 65
Vesicouterine pouch, 418
Vessels, 332
Vestibular:
    complex, 260
    duct, 263, 264
    nerve, 223, 260
Vestibule, 346, 368, 387, 422, 425
Vestibulocochlear nerve, 233, 234, 260, 264
Vestibulospinal tract, 241
Villus, 378
Visceral, 295
    pericardium, 295
    peritoneum, 389
    pleura, 351, 360
    surface, 341, 382
Vision, 252
Vitreous body, 270-71
Vitreous chamber, 271
Vocal folds, 348
Volkmann's canals, 32
Vomer bone, 53
Vulva, 425, 429

W

White blood cells (WBCs), 290
White matter, 202, 225
White pulp, 341
Whole blood, 290
Wormian bones, 34
Wrist:
    articulations of, 93-94
    bones of, 74-75
    surface anatomy, 187

X

Xiphihumeralis, 135
Xiphoid process, 183

Z

Zona fasciculata, 283, 285
Zona glomerulosa, 283, 285
Zona reticularis, 283, 285
Zygomatic arch, 180
Zygomatic bone, 54, 180
Zygomaticus major muscle, 109
Zygomaticus minor muscle, 109

Photo Credits

Unless noted otherwise below, all photographs are courtesy of Ralph Hutchings

Chapter 3
3-1	Wards Natural Science Establishment Inc.
3-2	Wards Natural Science Establishment Inc.
3-3	Michael J. Timmons
3-4	Frederic H. Martini
3-5a,b	Frederic H. Martini
3-6	Frederic H. Martini
3-7	CNRI/Science Photo Library
3-8	Frederic H. Martini
3-9	Frederic H. Martini
3-10	Wards Natural Science Establishment Inc.
3-11	Science Source/Photo Researchers
3-12	Frederic H. Martini
3-13	Wards Natural Science Establishment Inc.
3-14	John D. Cunningham/Visuals Unlimited
3-15	Bruce Iverson/Visuals Unlimited
3-16	Frederic H. Martini
3-17	Robert Brons/Biological Photo Service
3-18	Science Source/Photo Researchers
3-19	Ed Reschke/Peter Arnold, Inc.
3-20	Frederic H. Martini
3-21	G. W. Willis/Biological Photo Service
3-22	Frederic H. Martini
3-23	G. W. Willis/Biological Photo Service
3-24	Frederic H. Martini

Chapter 4
4-1, AIR 4-1	Wehr/Custom Medical Stock Photo
4-2a	Manfred Kage/Peter Arnold, Inc.
4-2b,c	Frederic H. Martini
4-3	Wards Natural Science Establishment Inc.
4-4	John D. Cunningham/Visuals Unlimited
AIR 4-2	Elias-Paul's & Peters, Amenta, HISTOLOGY: AND HUMAN MICROANATOMY, 5/E, Puccin Publishers, 1987.

Chapter 7
7-18b	Photo Researchers

Chapter 8
8-1a	Wards Natural Science Establishment Inc.
8-5	Wards Natural Science Establishment Inc.
8-6b	Patrick M. Timmons
8-9	Wards Natural Science Establishment Inc.
8-12	Wards Natural Science Establishment Inc.

8-15, AIR 8-1	Patrick M. Timmons
8-17	Wards Natural Science Establishment Inc.

Chapter 9
9-1	Wards Natural Science Establishment Inc.
9-2	Fred Hossler/Visuals Unlimited

Chapter 10
10-10	Mentor Networks, Inc.

Chapter 11
11-8	Custom Medical Stock Photo
11-9	Mentor Networks, Inc.

Chapter 12
12-1	Mentor Networks, Inc.
12-2	Mentor Networks, Inc.
12-3	Mentor Networks, Inc.
12-4	Mentor Networks, Inc.
12-5	Custom Medical Stock Photo
12-6	Mentor Networks, Inc.
12-7	Mentor Networks, Inc.
12-8	Mentor Networks, Inc.
12-9	Mentor Networks, Inc.
12-10	Custom Medical Stock Photo
12-11	Mentor Networks, Inc.
12-12	Mentor Networks, Inc.
12-13	Mentor Networks, Inc.
12-14	Mentor Networks, Inc.

Chapter 13
13-1, AIR 13-1	Phototake

Chapter 14
14-3	Michael J. Timmons

Chapter 15
15-3	Wards Natural Science Establishment Inc.
15-9	Michael J. Timmons
15-13	Wards Natural Science Establishment Inc.

Chapter 18
18-2c	Frederic H. Martini
18-6a,b	Frederic H. Martini
18-8b	Lennart Nilsson, BEHOLD MAN, Little, Brown, & Co., 1973.
18-11	Wards Natural Science Establishment Inc.
18-18	Michael J. Timmons
18-19	Frederic H. Martini

Chapter 19
19-2	Manfred Kage/Peter Arnold, Inc.
19-4a,b	Frederic H. Martini
19-6	Martin Rotker/Phototake
19-9	Frederic H. Martini
19-11	Wards Natural Science Establishment Inc.

Chapter 20
20-1, AIR 20-1	Ed Reschke/Peter Arnold, Inc.

Chapter 21

All cadaver photos in Chapter 21 courtesy of Ralph Hutchings

21-1 left	Phototake
21-1 right	Peter Arnold, Inc.

Chapter 22
22-1a,b	Michael J. Timmons
22-1c	Wards Natural Science Establishment Inc.

Chapter 23
23-3	Frederic H. Martini
23-5	Wards Natural Science Establishment Inc.
23-7a,b	Frederic H. Martini
23-8	Frederic H. Martini

Chapter 24
24-2	Frederic H. Martini
24-6	John D. Cunningham/Visuals Unlimited
24-12	Wards Natural Science Establishment Inc.

Chapter 25
25-2	Phototake
25-6	Frederic H. Martini
25-9	Wards Natural Science Establishment Inc.
25-12	Wards Natural Science Establishment Inc.
25-14a	John D. Cunningham/Visuals Unlimited
25-14b left	Michael J. Timmons
25-14b right	G. W. Willis/Biological Photo Service
25-15	Wards Natural Science Establishment Inc.
25-19	Wards Natural Science Establishment Inc.
25-21	Wards Natural Science Establishment Inc.

Chapter 26
26-11	Wards Natural Science Establishment Inc.
26-12	G. W. Willis/Biological Photo Service
26-13	Frederic H. Martini

Chapter 27
27-5	Dr. Robert Kelley, University of New Mexico
27-6	Wards Natural Science Establishment Inc.
27-7	David M. Phillips/Visuals Unlimited
27-10	Frederic H. Martini
27-11	Frederic H. Martini
27-12	Frederic H. Martini
27-15	Wards Natural Science Establishment Inc.
27-19	Frederic H. Martini
27-20	Frederic H. Martini
27-23	Wards Natural Science Establishment Inc.
27-24	Michael J. Timmons

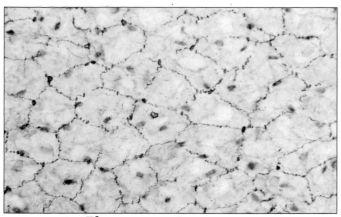

Plate 1: Simple squamous
epithelium (mesothelium) × 156

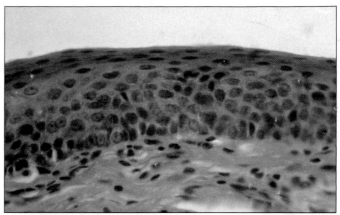

Plate 2: Stratified squamous epithelium,
tongue × 171

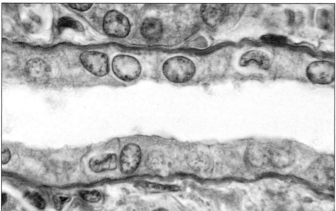

Plate 3: Simple cuboidal epithelium,
kidney tubule × 1724

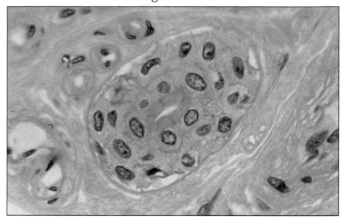

Plate 4: Stratified cuboidal epithelium,
sweat gland duct × 647

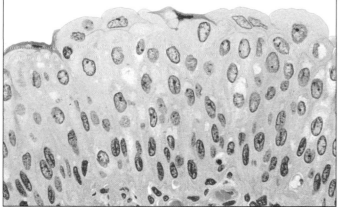

Plate 5: Transitional epithelium,
urinary bladder × 465

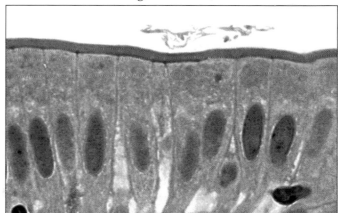

Plate 6: Simple columnar epithelium,
small intestine × 632

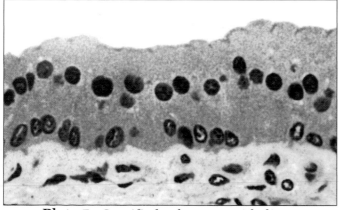

Plate 7: Stratified columnar epithelium,
salivary gland × 292

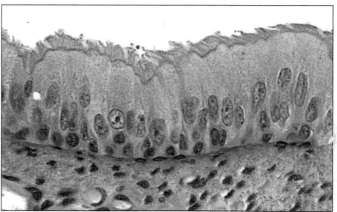

Plate 8: Pseudostratified (ciliated) columnar
epithelium, trachea × 308

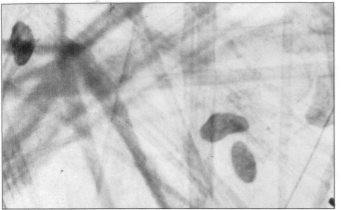

Plate 9: Loose (areolar) connective tissue, peritoneum × 630

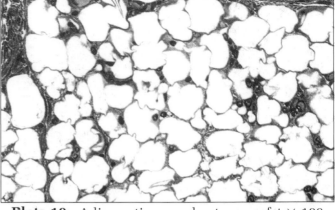

Plate 10: Adipose tissue, subcutaneous fat × 109

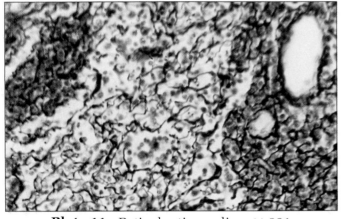

Plate 11: Reticular tissue, liver × 291

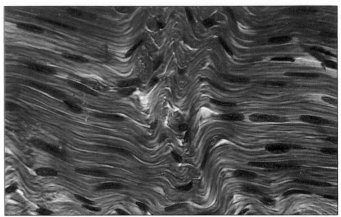

Plate 12: Dense regular connective tissue, tendon × 356

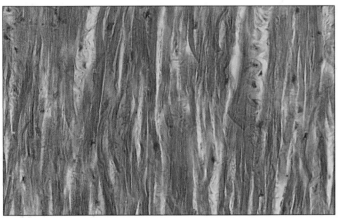

Plate 13: Elastic tissue, intervertebral ligaments × 930

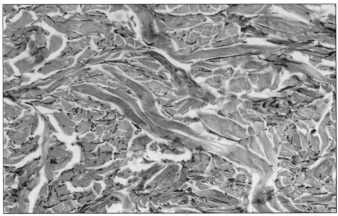

Plate 14: Dense irregular connective tissue, dermis of skin × 115

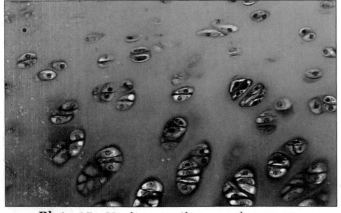

Plate 15: Hyaline cartilage, trachea × 481

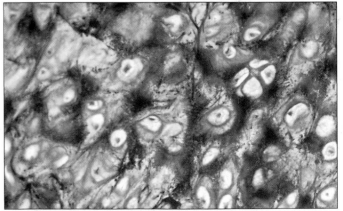

Plate 16: Elastic cartilage, external ear (pinna) × 338

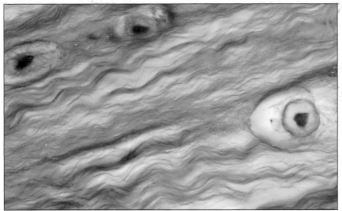

Plate 17: Fibrocartilage, intervertebral disc x 792

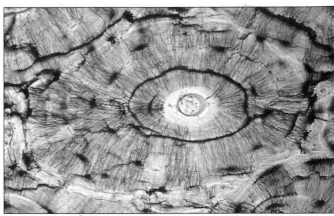

Plate 18: Compact bone, femur

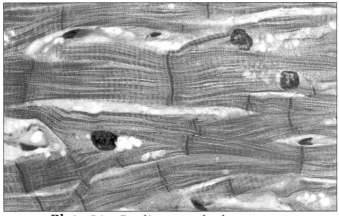

Plate 19: Skeletal muscle tissue x 102

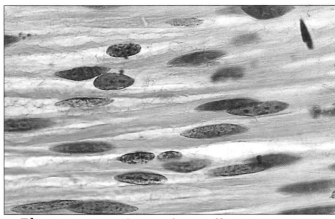

Plate 20: Smooth muscle, small intestine x 300

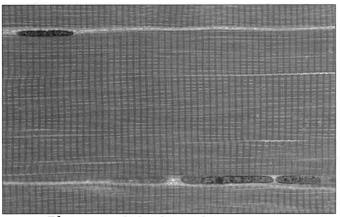

Plate 21: Cardiac muscle, heart x 469

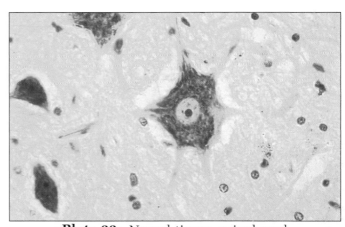

Plate 22: Neural tissue, spinal cord

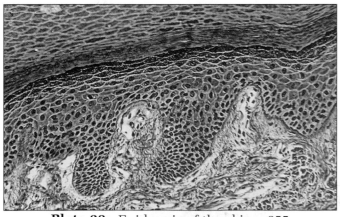

Plate 23: Epidermis of the skin x 255

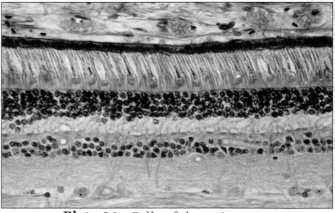

Plate 24: Cells of the retina x 52

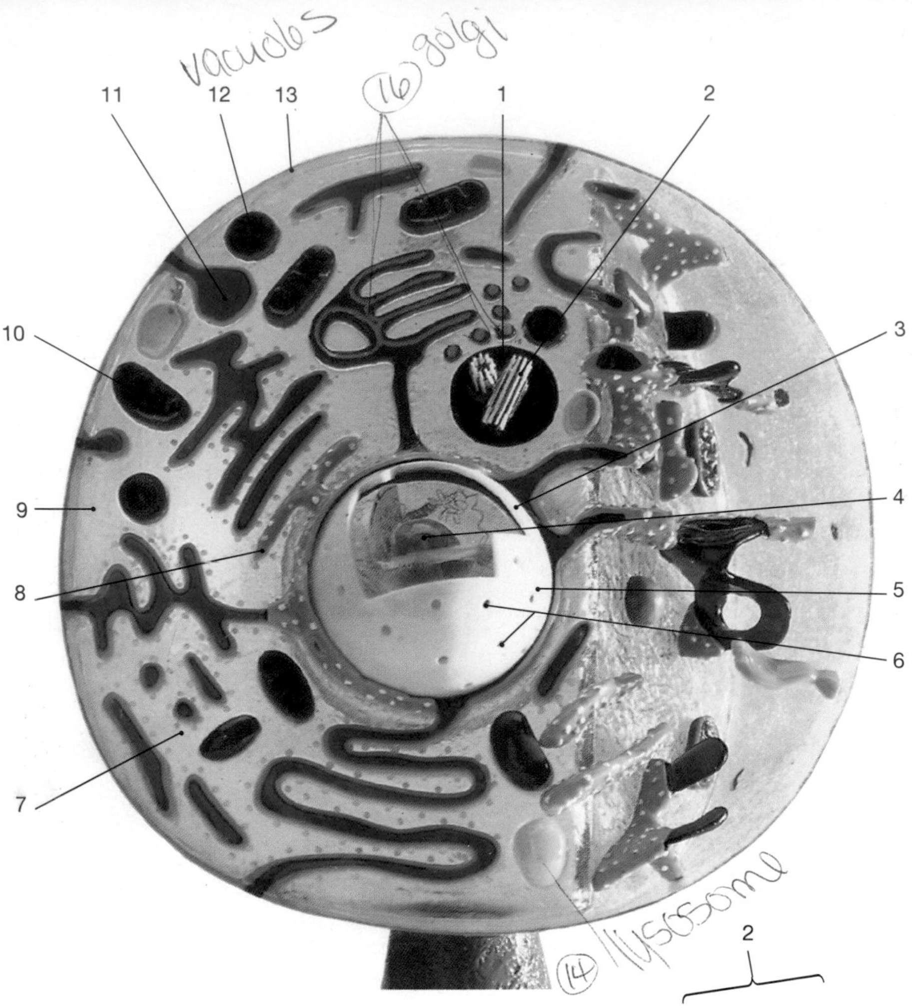

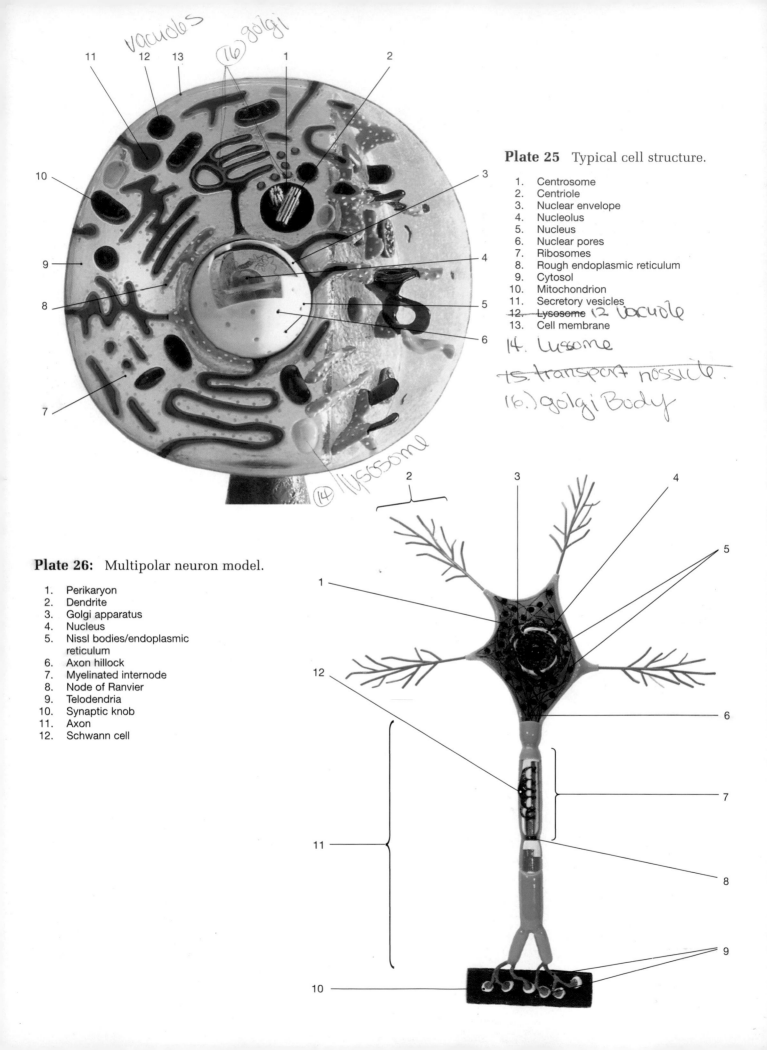

Plate 25 Typical cell structure.

1. Centrosome
2. Centriole
3. Nuclear envelope
4. Nucleolus
5. Nucleus
6. Nuclear pores
7. Ribosomes
8. Rough endoplasmic reticulum
9. Cytosol
10. Mitochondrion
11. Secretory vesicles
~~12. Lysosome~~ 12 vacuole
13. Cell membrane

14. Lysome

~~15. transport vessicle.~~

16.) golgi Body

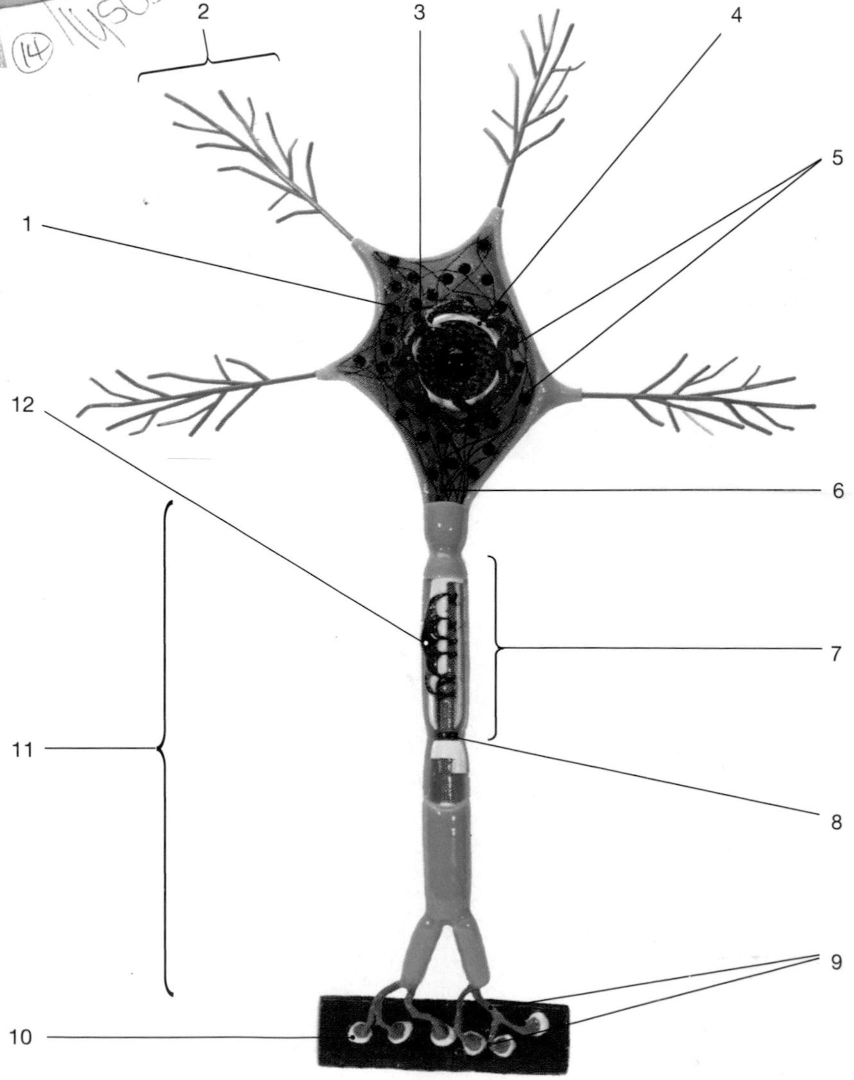

Plate 26: Multipolar neuron model.

1. Perikaryon
2. Dendrite
3. Golgi apparatus
4. Nucleus
5. Nissl bodies/endoplasmic reticulum
6. Axon hillock
7. Myelinated internode
8. Node of Ranvier
9. Telodendria
10. Synaptic knob
11. Axon
12. Schwann cell

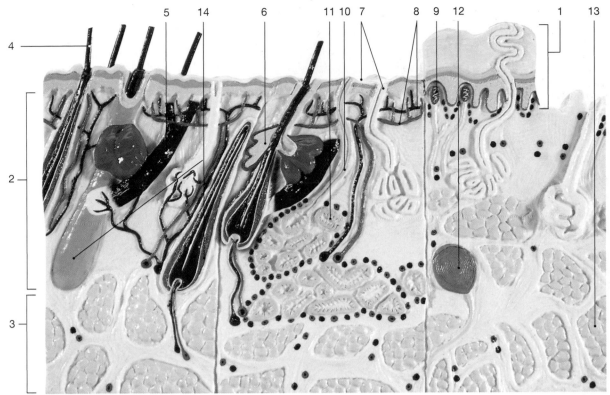

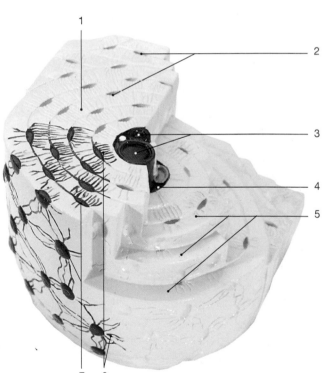

Plate 27:
Skin, general view

1. Epidermis
2. Dermis
3. Hypodermis (subcutaneous layer)
4. Hair shaft
5. Arrector pili
6. Sebaceous gland
7. Pore of sweat gland
8. Artery/Vein (cutaneous plexus)
9. Meissner's corpuscle
10. Sweat gland duct
11. Sweat gland
12. Pacinian corpuscle
13. Adipose cells
14. Hair follicle

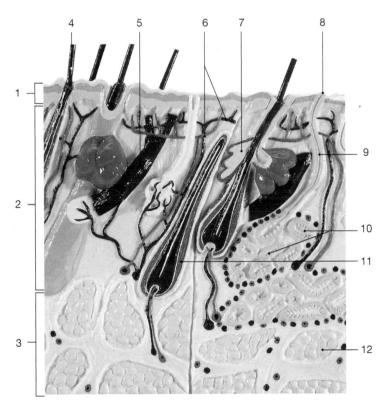

Plate 28: Detail of hair follicles and glands of the skin.

1. Epidermis
2. Dermis
3. Hypodermis (subcutaneous layer)
4. Hair shaft
5. Arrector pili
6. Artery/Vein (cutaneous plexus)
7. Sebaceous gland
8. Pore of sweat gland
9. Sweat gland duct
10. Sweat gland
11. Hair follicle
12. Adipose cells

Plate 29: Model of an osteon.

1. Matrix
2. Lacunae
3. Blood vessels
4. Central canal (Haversian)
5. Concentric lamellae
6. Canaliculi
7. Osteocyte

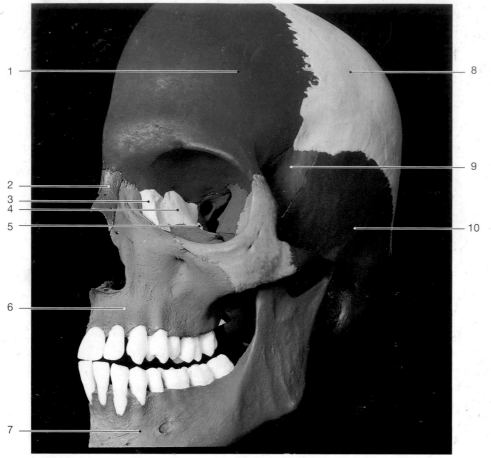

Plate 30: Skull, anterior and lateral views.

1. Frontal bone
2. Nasal bone
3. Lacrimal bone
4. Ethmoid bone
5. Palatine bone
6. Maxilla
7. Mandible
8. Parietal bone
9. Sphenoid bone
10. Temporal bone

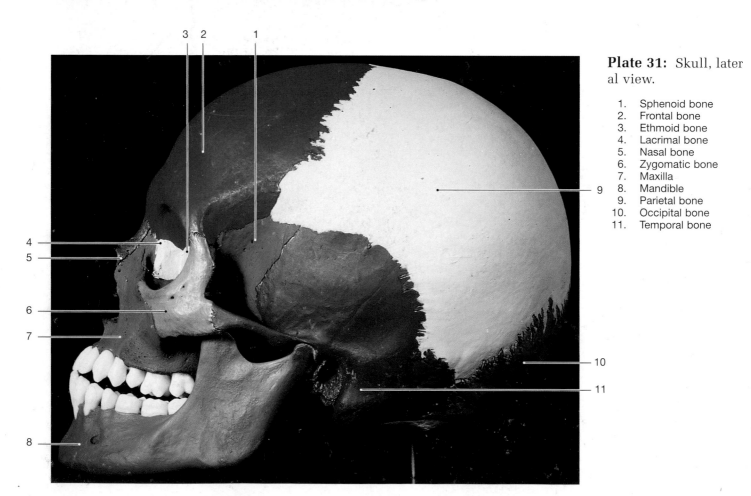

Plate 31: Skull, lateral view.

1. Sphenoid bone
2. Frontal bone
3. Ethmoid bone
4. Lacrimal bone
5. Nasal bone
6. Zygomatic bone
7. Maxilla
8. Mandible
9. Parietal bone
10. Occipital bone
11. Temporal bone

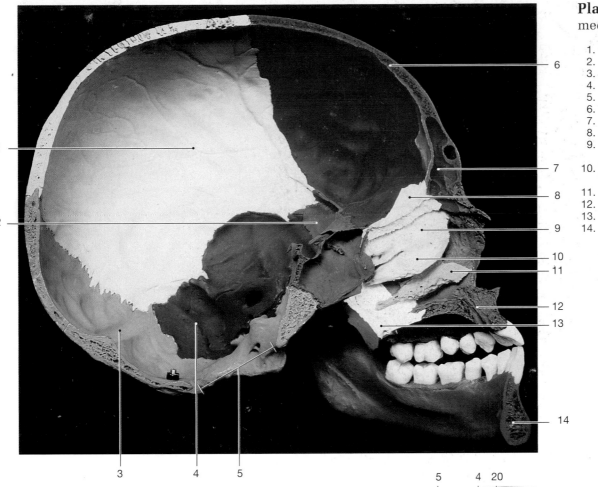

Plate 33: Skull, horizontal section.

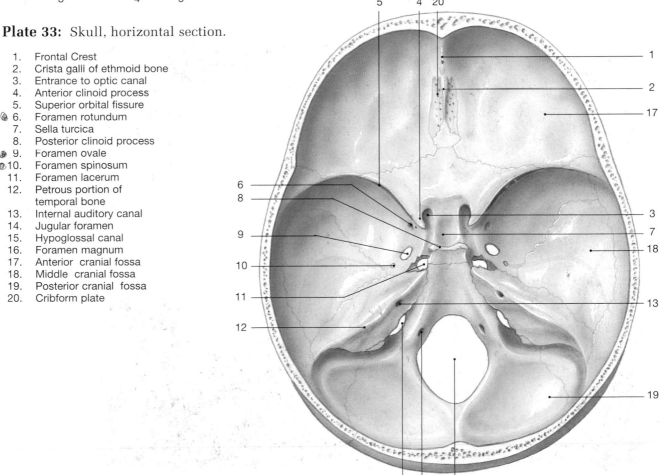

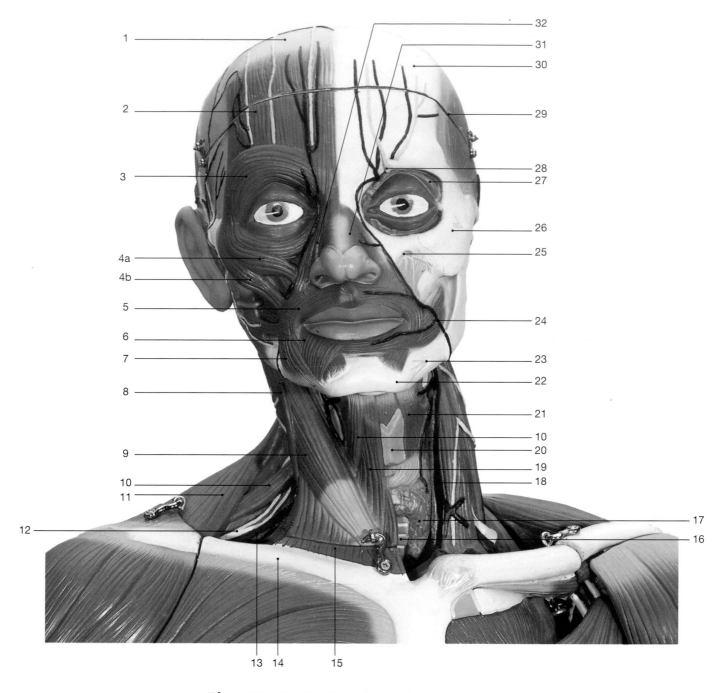

Plate 34: Head and neck muscles, anterior view

1. Galea aponeurotica
2. Frontalis
3. Orbicularis oculi
4a. Zygomaticus minor
4b. Zygomaticus major
5. Orbicularis oris
6. Depressor labii inferioris
7. Depressor anguli oris
8. External jugular vein
9. Sternocleidomastoid
10. Omohyoid
11. Trapezius

12. Brachial plexus
13. Subclavian artery
14. Clavicle
15. Platysma (cut)
16. Trachea
17. Thyroid gland
18. Common carotid artery
19. Sternohyoid
20. Thyroid cartilage
21. Thyrohyoid
22. Mandible
23. Mental nerve

24. Facial artery
25. Infraorbital nerve
26. Zygomatic bone
27. Lacrimal gland
28. Supraorbital nerve
29. Temporalis
30. Frontal bone
31. Nasalis
32. Levator labi superioris

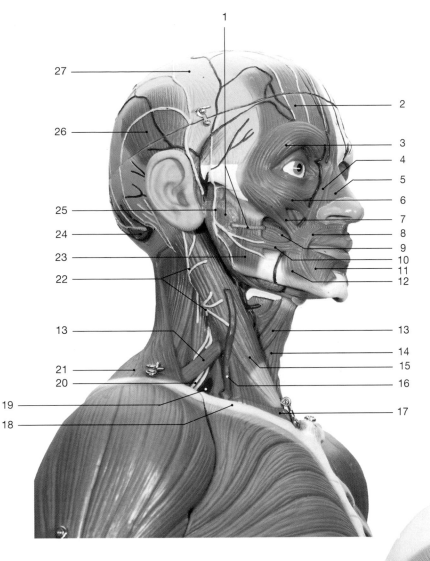

Plate 35: Head, neck and shoulder muscles, lateral view as seen from the right

1. Parotid gland and duct
2. Frontalis
3. Orbicularis oculi
4. Levator labi superioris
5. Nasalis
6. Zygomaticus minor
7. Zygomaticus major
8. Orbicularis oris
9. Buccinator
10. Risorius
11. Depressor labii inferiouis
12. Depressor anguli oris
13. Omohyoid
14. Sternohyoid
15. Sternocleidomastoid
16. External jugular vein
17. Platysma (cut)
18. Clavicle
19. Subclavian artery
20. Brachial plexus
21. Trapezius
22. Ansa cervicalis
23. Masseter
24. Occipitalis
25. Facial nerve
26. Temporalis
27. Galea aponeuratica

Plate 36: Head and neck muscles, lateral view as seen from the right

1. Frontal bone
2. Nasalis
3. Zygomatic bone
4. Facial artery
5. Thyrohyoid ligament
6. Superior thyroid artery
7. Thyrohyoid
8. Thyroid cartilage
9. Common carotid artery
10. Thyroid gland
11. Brachiocephalic veins
12. Temporalis
13. Superficial temporal artery
14. Internal carotid artery
15. External carotid artery
16. Carotid sinus
17. Splenius
18. Longissimus capitis
19. Sternohyoid
20. Levator scapulae
21. Brachial plexus
22. Subclavian artery
23. Clavicle
24. 1st Rib

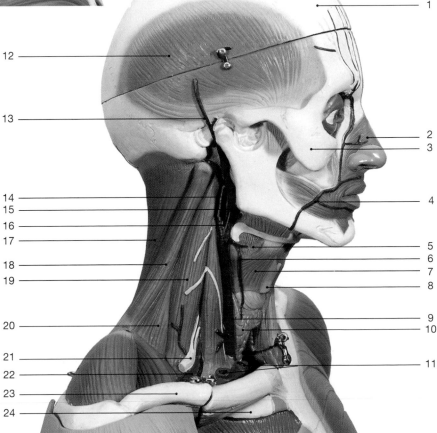

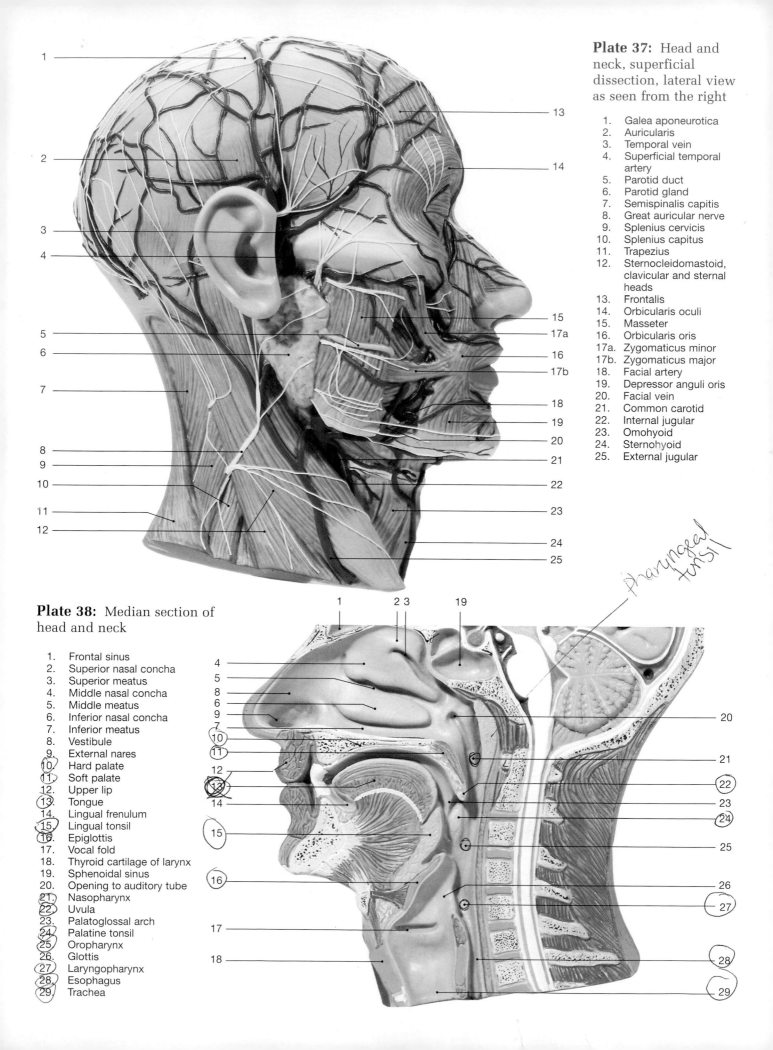

Plate 37: Head and neck, superficial dissection, lateral view as seen from the right

1. Galea aponeurotica
2. Auricularis
3. Temporal vein
4. Superficial temporal artery
5. Parotid duct
6. Parotid gland
7. Semispinalis capitis
8. Great auricular nerve
9. Splenius cervicis
10. Splenius capitus
11. Trapezius
12. Sternocleidomastoid, clavicular and sternal heads
13. Frontalis
14. Orbicularis oculi
15. Masseter
16. Orbicularis oris
17a. Zygomaticus minor
17b. Zygomaticus major
18. Facial artery
19. Depressor anguli oris
20. Facial vein
21. Common carotid
22. Internal jugular
23. Omohyoid
24. Sternohyoid
25. External jugular

pharyngeal tonsil

Plate 38: Median section of head and neck

1. Frontal sinus
2. Superior nasal concha
3. Superior meatus
4. Middle nasal concha
5. Middle meatus
6. Inferior nasal concha
7. Inferior meatus
8. Vestibule
9. External nares
10. Hard palate
11. Soft palate
12. Upper lip
13. Tongue
14. Lingual frenulum
15. Lingual tonsil
16. Epiglottis
17. Vocal fold
18. Thyroid cartilage of larynx
19. Sphenoidal sinus
20. Opening to auditory tube
21. Nasopharynx
22. Uvula
23. Palatoglossal arch
24. Palatine tonsil
25. Oropharynx
26. Glottis
27. Laryngopharynx
28. Esophagus
29. Trachea

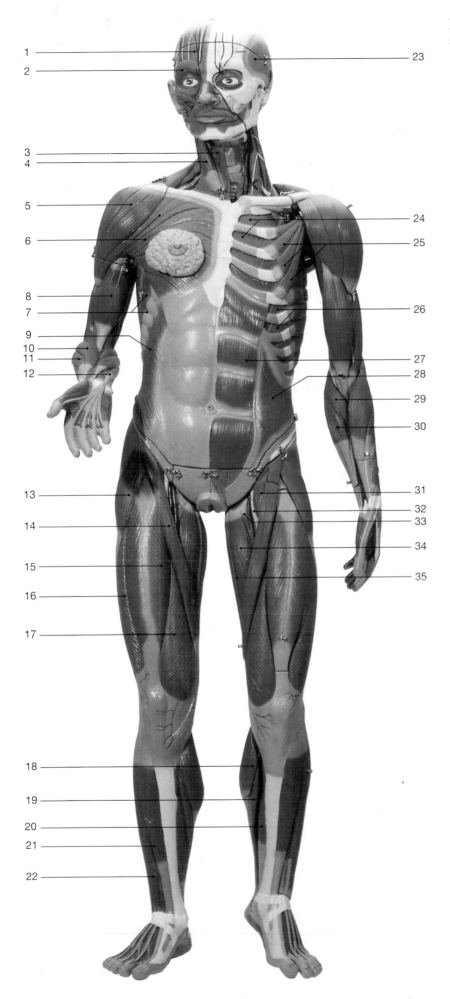

Plate 39: Superficial muscles, full anterior view of torso.

1. Frontalis
2. Orbicularis oculi
3. Sternohyoid
4. Sternocleidomastoid
5. Deltoid
6. Pectoralis major
7. Serratus anterior
8. Biceps brachii
9. External oblique
10. Brachioradialis
11. Extensor carpi radialis
12. Palmaris longus
13. Tensor fasciae latae
14. Sartorius
15. Rectus femoris
16. Vastus lateralis
17. Vastus medialis
18. Gastrocnemius
19. Soleus
20. Flexor digitorum longus
21. Tibialis anterior
22. Extensor digitorum longus
23. Temporalis
24. Internal intercostal
25. Pectoralis minor
26. External intercostal
27. Rectus abdominis
28. Internal oblique
29. Pronator teres
30. Flexor carpi radialis
31. Iliopsoas
32. Femoral artery, vein and nerve
33. Pectineus
34. Adductor longus
35. Gracilis

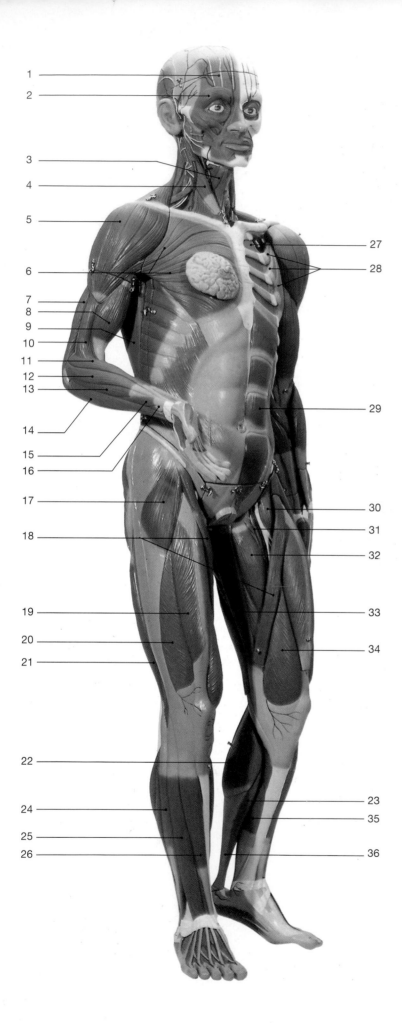

Plate 40: Superficial muscles, full anterior and lateral views of torso.

1. Frontalis
2. Orbicularis oculi
3. Omohyoid
4. Sternocleidomastoid
5. Deltoid
6. Pectoralis major
7. Triceps brachii
8. Biceps brachii
9. Latissimus dorsi
10. Brachialis
11. Brachioradialis
12. Extensor carpi radialis longus
13. Extensor carpi radialis brevis
14. Extensor digitorum
15. Abductor pollicis longus
16. Extensor pollicis brevis
17. Tensor fasciae latae
18. Sartorius
19. Rectus femoris
20. Vastus lateralis
21. Biceps femoris
22. Gastrocnemius
23. Soleus
24. Peroneus longus
25. Extensor digitorum longus
26. Tibialis anterior
27. Internal intercostal
28. Pectoralis minor
29. Rectus abdominis
30. Iliopsoas
31. Pectineus
32. Adductor longus
33. Gracilis
34. Vastus medialis
35. Flexor digitorum longus
36. Calcaneal tendon

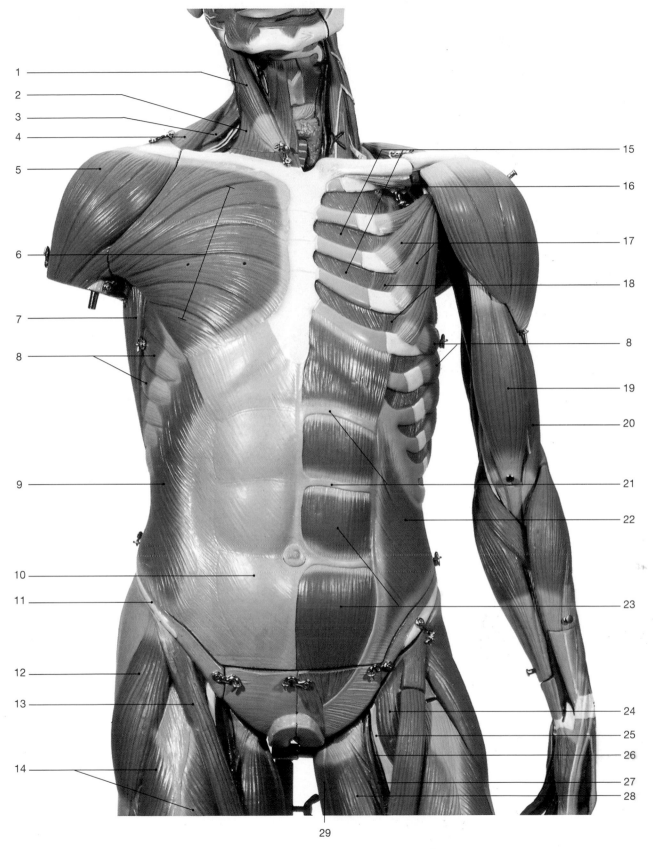

Plate 41: Superficial and deep muscles of the neck, shoulder, abdomen, and upper thigh, anterior view.

1. Sternocleidomastoid
2. Sternohyoideus
3. Omohyoid
4. Trapezius
5. Deltoid
6. Pectoralis major
7. Latissimus dorsi
8. Serratus anterior
9. External oblique
10. Rectus sheath
11. Iliac crest
12. Tensor fasciae latae
13. Sartorius
14. Rectus femoris
15. Internal intercostals
16. Subclavius
17. Pectoralis minor
18. External intercostals
19. Biceps brachii
20. Triceps, lateral head
21. Tendinous inscriptions
22. Internal oblique
23. Rectus abdominis
24. Iliopsoas
25. Femoral nerve
26. Femoral artery
27. Femoral vein
28. Adductor longus
29. Gracilis

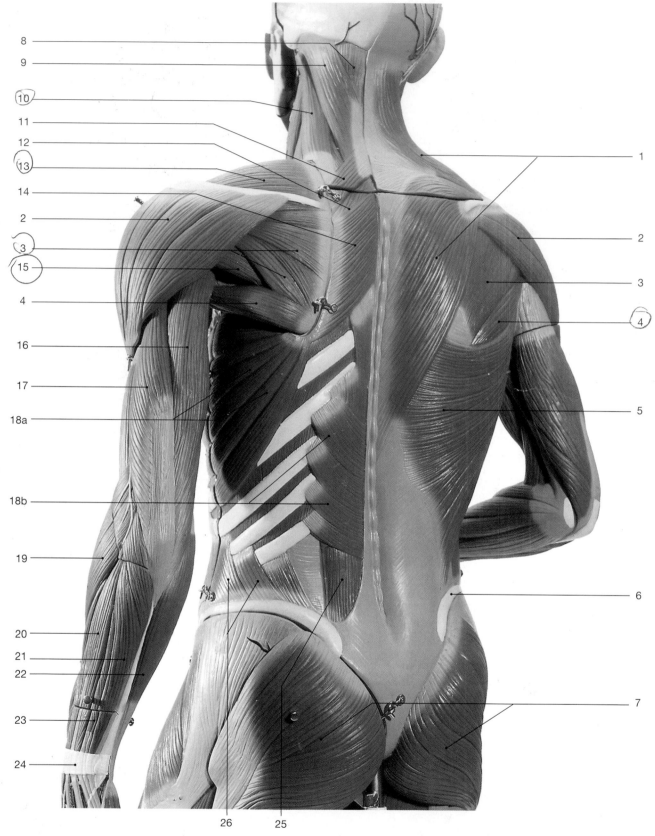

Plate 42: Superficial and deep muscles of the neck, shoulder, arm, back and gluteal regions, posterior view.

1. Trapezius
2. Deltoid
3. Infraspinatus
4. Teres major
5. Latissimus dorsi
6. Iliac crest
7. Gluteus maximus
8. Semispinalis capitis
9. Splenius

10. Levator scapulae
11. Rhomboideus minor
12. Middle scalene
13. Supraspinatus
14. Rhomboideus major
15. Teres minor
16. Triceps brachii, long head
17. Triceps brachii, lateral head

18a. Serratus anterior
18b. Serratus posterior
19. Extensor carpi radialis longus
20. Extensor digitorum
21. Extensor carpi ulnaris
22. Flexor carpi ulnaris
23. Extensor carpi radialis brevis

24. Extensor retinaculum
25. Erector spinae group
26. Oblique muscles

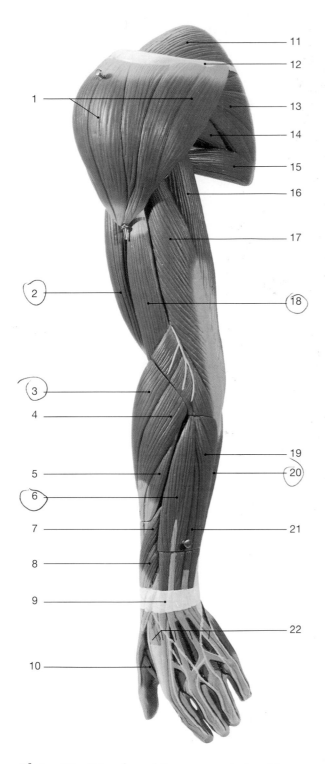

Plate 43: Muscles of the arm and shoulder, lateral view.

1. Deltoid
2. Biceps brachii
3. Brachioradialis
4. Extensor carpi radialis longus
5. Extensor carpi radialis brevis
6. Extensor digitorum
7. Abductor pollicis longus
8. Extensor pollicis brevis
9. Extensor retinaculum
10. 1st Dorsal interosseous
11. Supraspinatus
12. Scapular spine
13. Infraspinatus
14. Teres minor
15. Teres major
16. Triceps brachii, long head
17. Triceps brachii, lateral head
18. Brachialis
19. Extensor carpi ulnaris
20. Flexor carpi ulnaris
21. Extensor digiti minimi
22. Tendons of extensor muscles

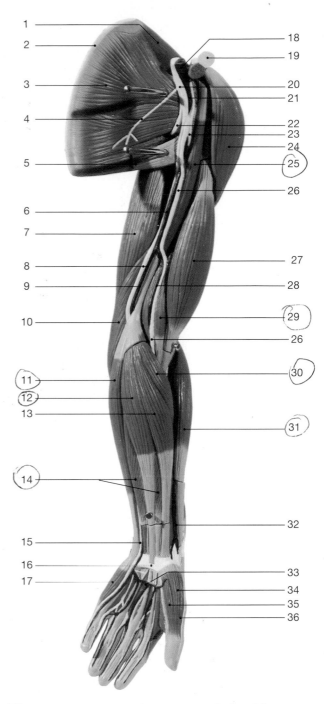

Plate 44: Muscles of the arm and shoulder, posterior view.

1. Supraspinatus
2. Medial border of scapula
3. Subscapularis
4. Thorocodorsal nerve
5. Teres major
6. Brachial artery
7. Triceps brachii, long head
8. Ulnar nerve
9. Superior ulnar collateral artery
10. Triceps brachii, medial head
11. Flexor carpi ulnaris
12. Palmaris longus
13. Flexor carpi radialis
14. Flexor digitorum superficialis
15. Ulnar artery and nerve
16. Flexor retinaculum
17. Abductor digiti minimi
18. Subclavian artery
19. Clavicle
20. Medial cord of brachial plexus
21. Axillary artery
22. Axillary nerve
23. Musculocutaneous nerve
24. Deltoid
25. Coracobrachialis
26. Median nerve
27. Biceps brachii
28. Radial collateral artery
29. Brachialis
30. Pronator teres
31. Brachioradialis
32. Radial artery
33. Superficial palmar arterial arch
34. Abductor pollicis brevis
35. Flexor pollicis brevis
36. Opponens pollicis

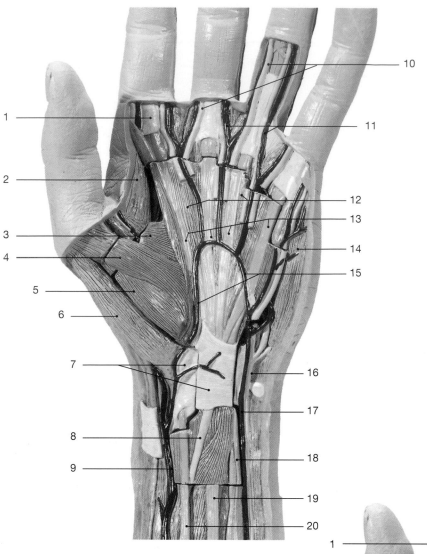

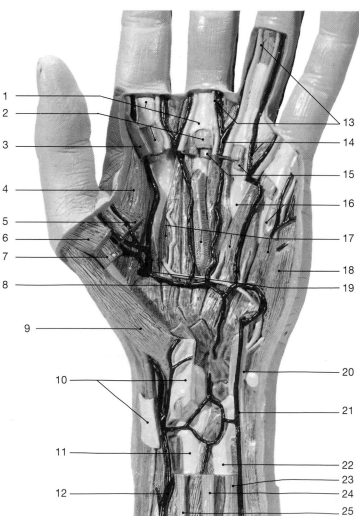

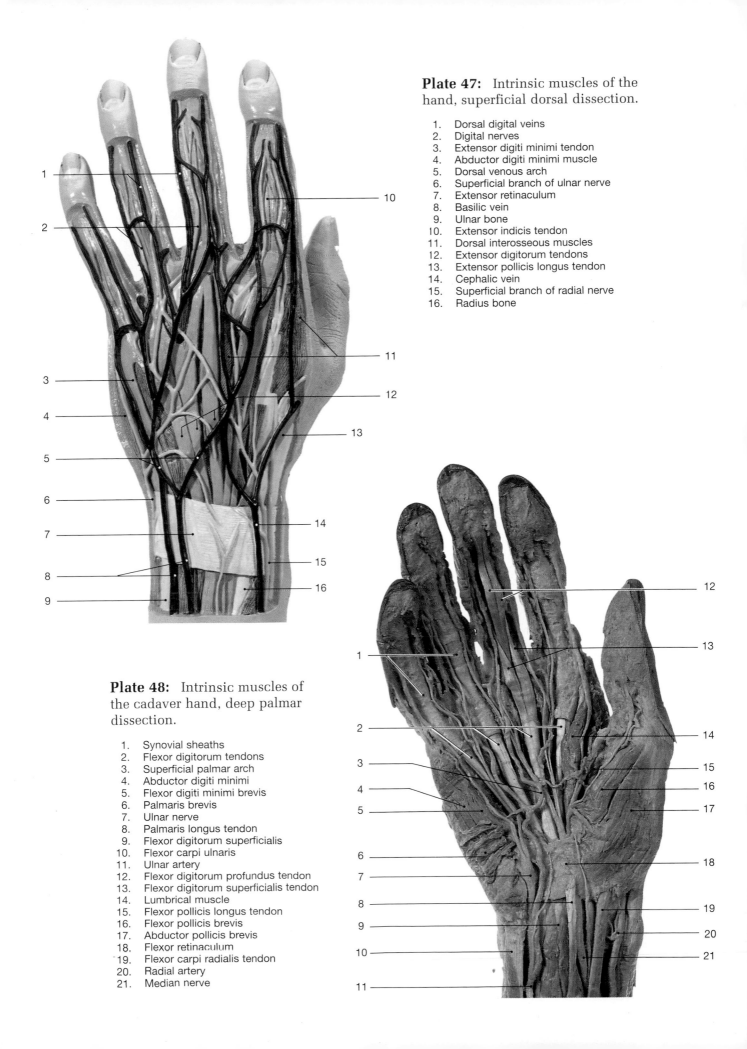

Plate 47: Intrinsic muscles of the hand, superficial dorsal dissection.

1. Dorsal digital veins
2. Digital nerves
3. Extensor digiti minimi tendon
4. Abductor digiti minimi muscle
5. Dorsal venous arch
6. Superficial branch of ulnar nerve
7. Extensor retinaculum
8. Basilic vein
9. Ulnar bone
10. Extensor indicis tendon
11. Dorsal interosseous muscles
12. Extensor digitorum tendons
13. Extensor pollicis longus tendon
14. Cephalic vein
15. Superficial branch of radial nerve
16. Radius bone

Plate 48: Intrinsic muscles of the cadaver hand, deep palmar dissection.

1. Synovial sheaths
2. Flexor digitorum tendons
3. Superficial palmar arch
4. Abductor digiti minimi
5. Flexor digiti minimi brevis
6. Palmaris brevis
7. Ulnar nerve
8. Palmaris longus tendon
9. Flexor digitorum superficialis
10. Flexor carpi ulnaris
11. Ulnar artery
12. Flexor digitorum profundus tendon
13. Flexor digitorum superficialis tendon
14. Lumbrical muscle
15. Flexor pollicis longus tendon
16. Flexor pollicis brevis
17. Abductor pollicis brevis
18. Flexor retinaculum
19. Flexor carpi radialis tendon
20. Radial artery
21. Median nerve

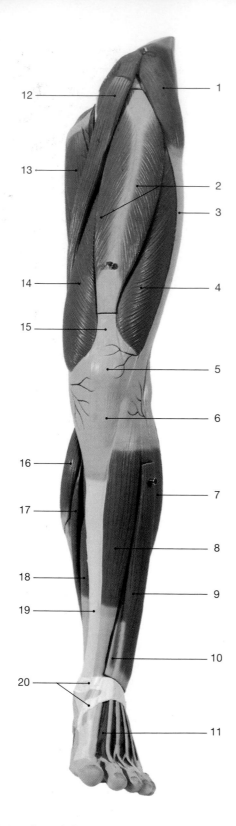

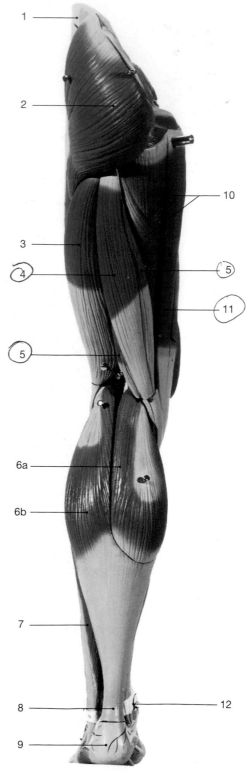

Plate 49: Muscles of the leg, anterior view.

1. Tensor fasciae latae
2. Rectus femoris
3. Iliotibial tract
4. Vastus lateralis
5. Patella
6. Patellar ligament
7. Peroneus longus
8. Tibialis anterior
9. Extensor digitorum
10. Extensor hallucis longus
11. Extensor hallucis brevis

12. Sartorius
13. Adductor longus
14. Vastus medialis
15. Quadriceps tendon
16. Gastrocnemius
17. Soleus
18. Flexor digitorum longus
19. Tibia
20. Inferior extensor retinaculum

Plate 50: Muscles of the leg, posterior view.

1. Gluteal aponeurosis over gluteus medius
2. Gluteus maximus
3. Biceps femoris, long head
4. Semitendinosus
5. Semimembranosus
6a. Gastrocnemius, medial head
6b. Gastrocnemius, lateral head
7. Peroneus longus
8. Calcaneal tendon
9. Calcaneus bone
10. Adductor magnus
11. Gracilis
12. Inferior extensor retinaculum

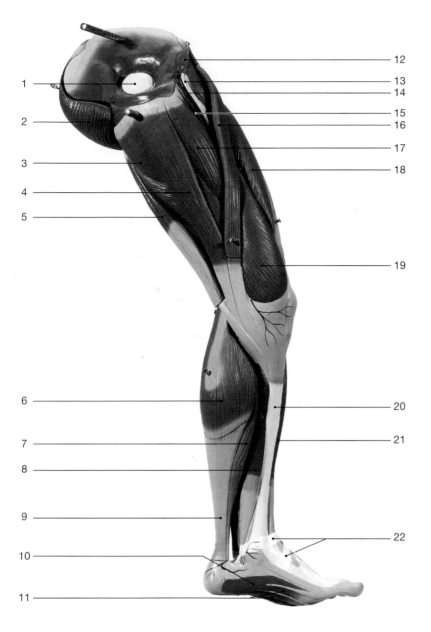

Plate 51: Muscles of the leg, medial view.

1. Head of femur
2. Gluteus maximus
3. Adductor magnus
4. Gracilis
5. Semimbranosus
6. Gastrocnemius, medial head
7. Soleus
8. Flexor digitorum longus
9. Calcaneal tendon
10. Abductor hallucis
11. Flexor hallucis brevis
12. Iliopsoas
13. Femoral nerve
14. Femoral artery
15. Femoral vein
16. Sartorius
17. Adductor longus
18. Rectus femoris
19. Vastus medialis
20. Tibia
21. Tibialis anterior
22. Inferior extensor retinaculum

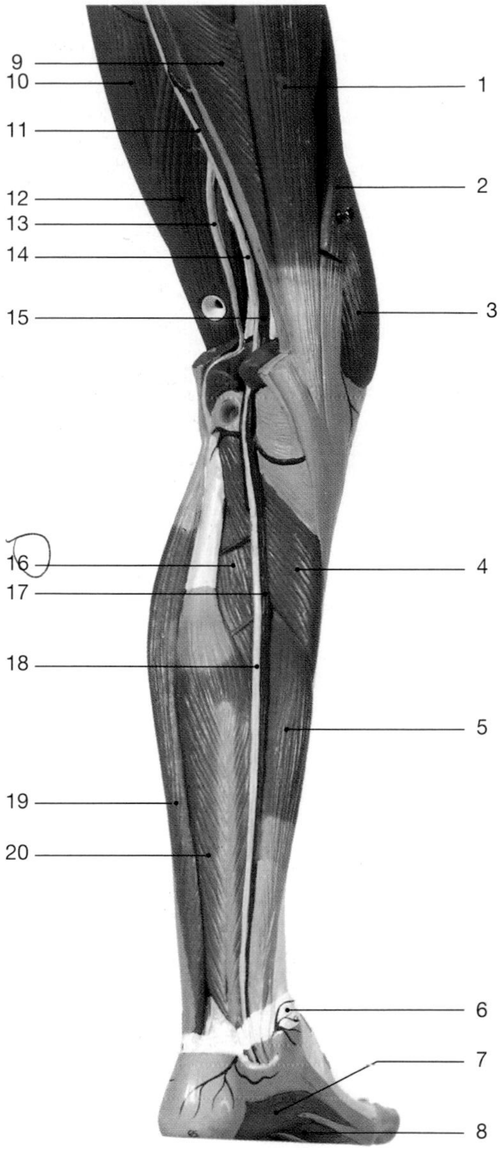

Plate 52: Muscles of the leg, as viewed from medial and posterior.

1. Gracilis
2. Sartorius
3. Vastus medialis
4. Popliteus
5. Flexor digitorum longus
6. Inferior extensor retinaculum
7. Abductor hallucis
8. Flexor hallucis brevis
9. Adductor magnus
10. Vastus lateralis
11. Sciatic nerve
12. Biceps femoris
13. Common peroneal nerve
14. Tibial nerve
15. Popliteal artery
16. Tibialis posterior
17. Posterior tibial artery
18. Tibial nerve
19. Flexor digitorum longus
20. Flexor hallucis longus

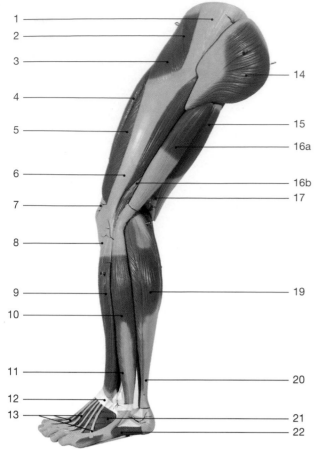

Plate 53: Muscles of the leg, lateral view.

1. Gluteal aponeurosis over gluteus medius
2. Gluteus medius
3. Tensor fasciae latae
4. Rectus femoris
5. Vastus lateralis
6. Iliotibial tract
7. Quadriceps tendon
8. Patellar ligament
9. Extensor digitorum longus
10. Peroneus longus
11. Peroneus brevis
12. Inferior extensor retinaculum
13. Tendons of extensor digitorum longus
14. Gluteus maximus
15. Semitendinosus

16a. Biceps femoris, long head
16b. Biceps femoris, short head
17. Semimembranosus
18. Gastrocnemius, lateral head
19. Soleus
20. Calcaneal tendon
21. Extensor digitorum brevis
22. Abductor digiti minim

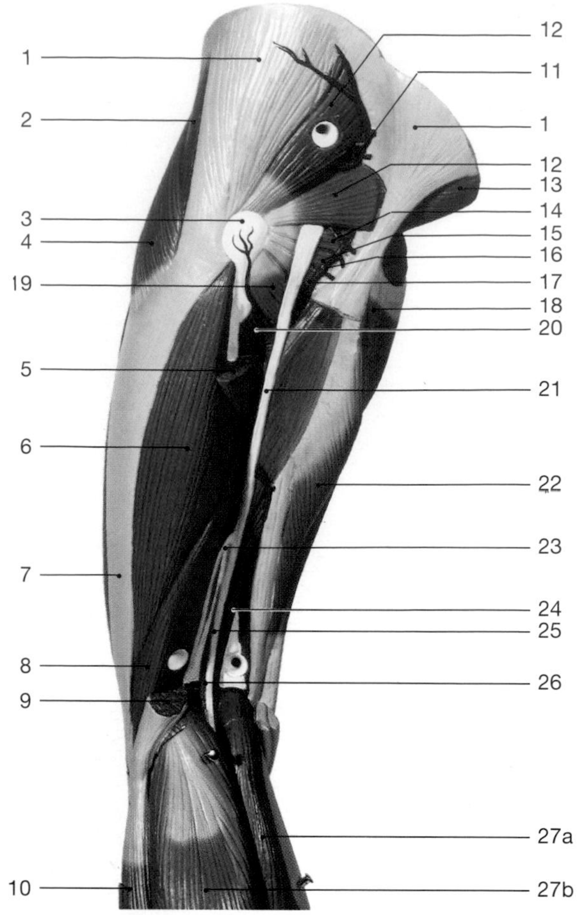

Plate 54: Muscles, vessels, and nerves of thigh, as viewed from lateral and posterior.

1. Gluteal aponeurosis over gluteus medius
2. Gluteus medius
3. Greater trochanter
4. Tensor fasciae latae
5. Gluteus maximus (cut)
6. Vastus lateralis
7. Iliotibial tract
8. Biceps femoris, short head
9. Biceps femoris, long head (cut)
10. Peroneus longus
11. Superior gluteal artery
12. Piriformis
13. Gluteus maximus (cut)
14. Superior gemellus
15. Internal obturator
16. Inferior gluteal artery
17. Inferior gemellus
18. Adductor magnus
19. Quadratus femoris
20. Medial femoral circumflex artery, deep branch
21. Sciatic nerve
22. Semimembranosus
23. Common peroneal nerve
24. Popliteal artery
25. Tibial nerve
26. Popliteal vein
27a. Gastrocnemius, medial head
27b. Gastrocnemius, lateral head

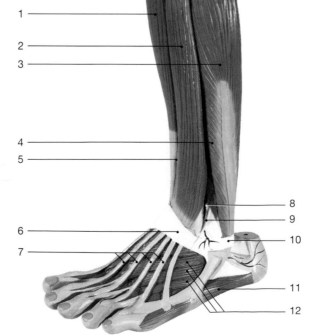

Plate 55: Muscles of the leg and foot, lateral view.

1. Tibialis anterior
2. Extensor digitorum longus
3. Peroneus longus
4. Peroneus brevis
5. Extensor hallucis longus
6. Inferior extensor retinaculum
7. Extensor digitorum longus tendons
8. Peroneal artery
9. Superficial peroneal nerve
10. Superior extensor retinaculum
11. Abductor digiti minimi
12. Extensor digitorum brevis

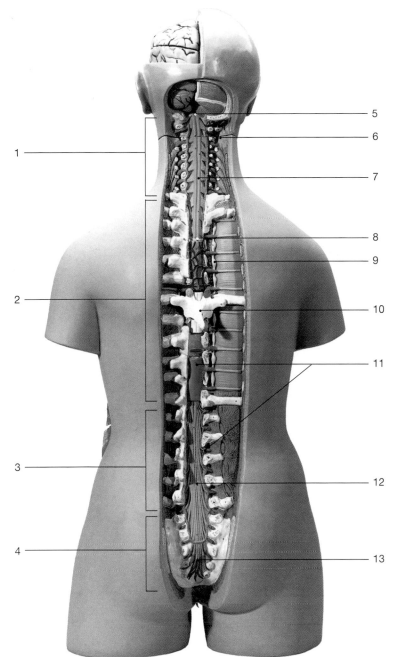

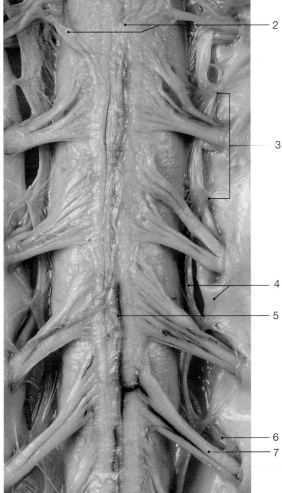

Plate 56: Torso dissected to show spinal cord in situ, posterior view.

1. Cervical spinal nerves
2. Thoracic spinal nerves
3. Lumbar spinal nerves
4. Sacral spinal nerves
5. Spinal cord emerging from foramen magnum
6. Vertebral artery
7. Cervical enlargement
8. Arachnoid
9. Sympathetic trunk
10. 12th Thoracic vertebra
11. Dura mater
12. Cauda equina
13. Filum terminale

Plate 57: The spinal cord and spinal meninges, ventral view.

1. Anterior median fissure
2. Pia mater
3. Denticulate ligaments
4. Arachnoid and dura mater (reflected)
5. Spinal blood vessel
6. Dorsal root of sixth cervical nerve
7. Ventral root of sixth cervical nerve

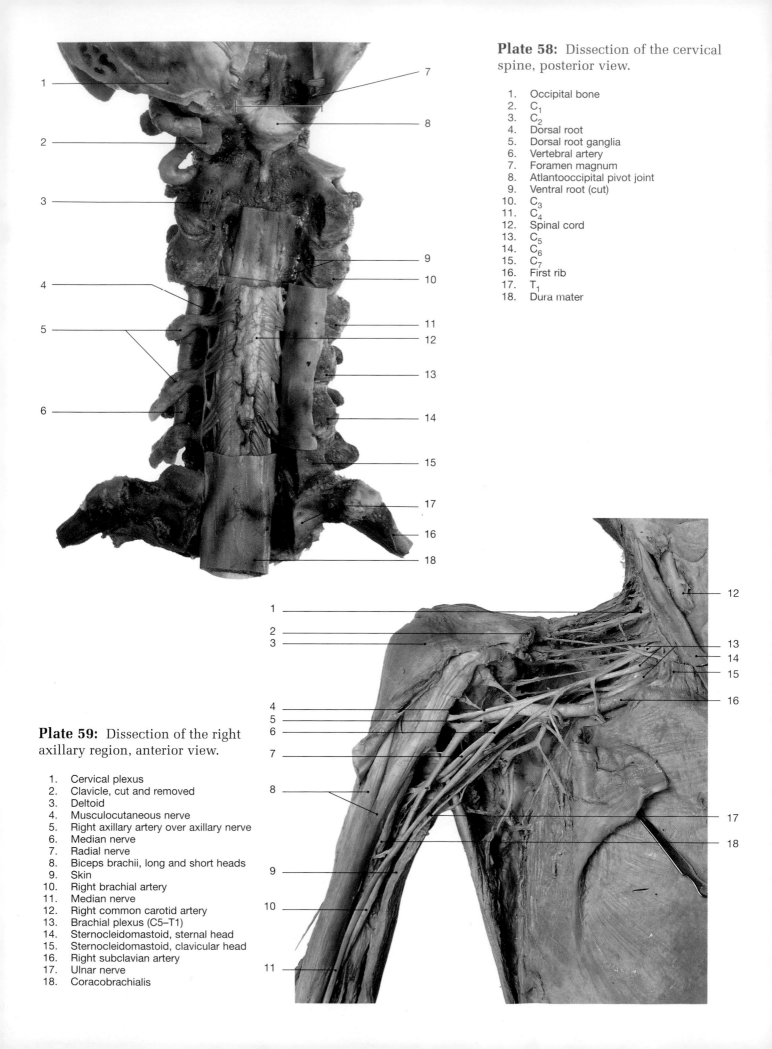

Plate 58: Dissection of the cervical spine, posterior view.

1. Occipital bone
2. C_1
3. C_2
4. Dorsal root
5. Dorsal root ganglia
6. Vertebral artery
7. Foramen magnum
8. Atlantooccipital pivot joint
9. Ventral root (cut)
10. C_3
11. C_4
12. Spinal cord
13. C_5
14. C_6
15. C_7
16. First rib
17. T_1
18. Dura mater

Plate 59: Dissection of the right axillary region, anterior view.

1. Cervical plexus
2. Clavicle, cut and removed
3. Deltoid
4. Musculocutaneous nerve
5. Right axillary artery over axillary nerve
6. Median nerve
7. Radial nerve
8. Biceps brachii, long and short heads
9. Skin
10. Right brachial artery
11. Median nerve
12. Right common carotid artery
13. Brachial plexus (C5–T1)
14. Sternocleidomastoid, sternal head
15. Sternocleidomastoid, clavicular head
16. Right subclavian artery
17. Ulnar nerve
18. Coracobrachialis

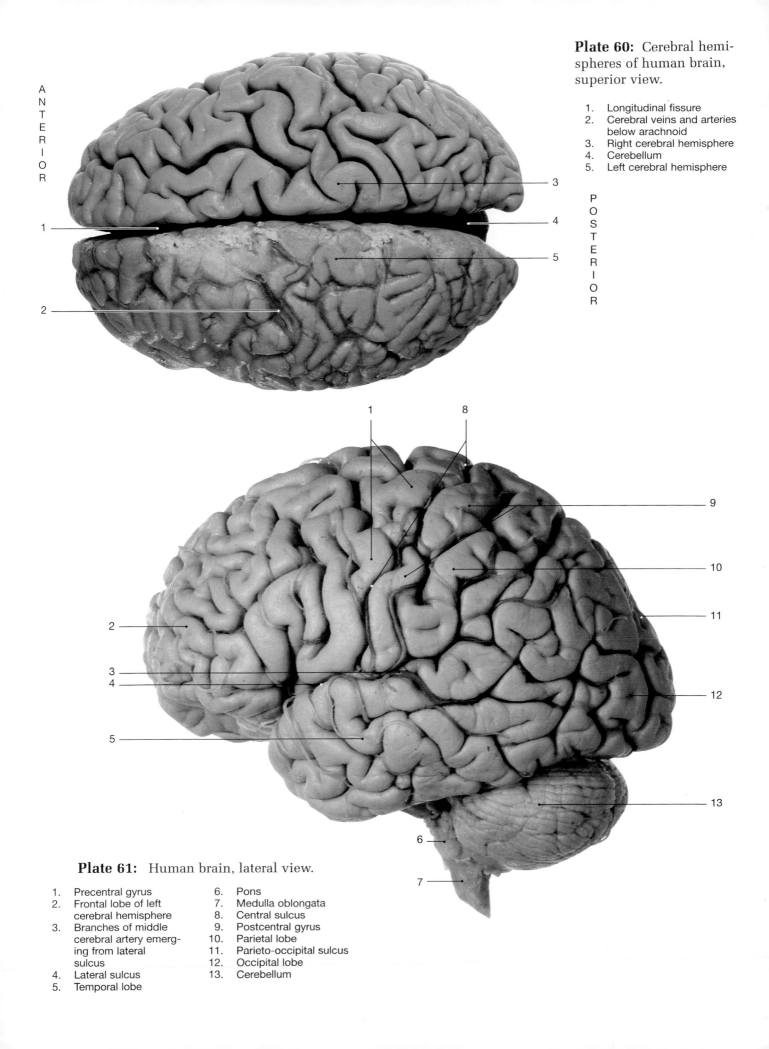

ANTERIOR

POSTERIOR

Plate 60: Cerebral hemispheres of human brain, superior view.

1. Longitudinal fissure
2. Cerebral veins and arteries below arachnoid
3. Right cerebral hemisphere
4. Cerebellum
5. Left cerebral hemisphere

Plate 61: Human brain, lateral view.

1. Precentral gyrus
2. Frontal lobe of left cerebral hemisphere
3. Branches of middle cerebral artery emerging from lateral sulcus
4. Lateral sulcus
5. Temporal lobe
6. Pons
7. Medulla oblongata
8. Central sulcus
9. Postcentral gyrus
10. Parietal lobe
11. Parieto-occipital sulcus
12. Occipital lobe
13. Cerebellum

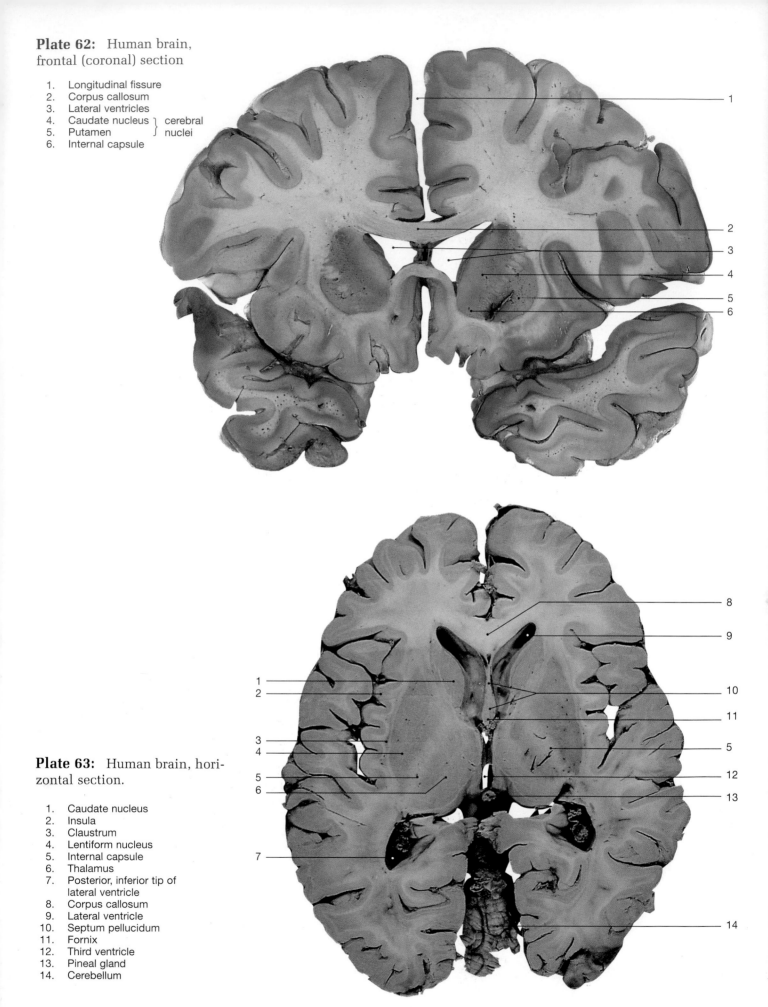

Plate 62: Human brain, frontal (coronal) section

1. Longitudinal fissure
2. Corpus callosum
3. Lateral ventricles
4. Caudate nucleus ⎫ cerebral
5. Putamen ⎭ nuclei
6. Internal capsule

Plate 63: Human brain, horizontal section.

1. Caudate nucleus
2. Insula
3. Claustrum
4. Lentiform nucleus
5. Internal capsule
6. Thalamus
7. Posterior, inferior tip of lateral ventricle
8. Corpus callosum
9. Lateral ventricle
10. Septum pellucidum
11. Fornix
12. Third ventricle
13. Pineal gland
14. Cerebellum

Plate 64:
Human brain, midsagittal section.

1. Central sulcus
2. Precentral gyrus
3. Corpus callosum
4. Septum pellucidum (separating lateral ventricles)
5. Interventricular foramen
6. Frontal lobe
7. Anterior commissure
8. Mamillary body
9. Optic chiasma
10. Temporal lobe
11. Pons
12. Medulla oblongata
13. Postcentral gyrus
14. Cingulate gyrus
15. Fornix
16. Thalamus
17. Pineal gland
18. Hypothalamus
19. Superior colliculus
20. Inferior colliculus
21. Mesencephalic aqueduct
22. Fourth ventricle
23. Cerebellum

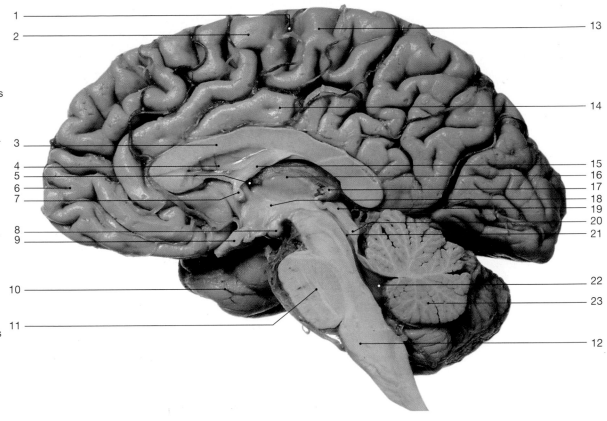

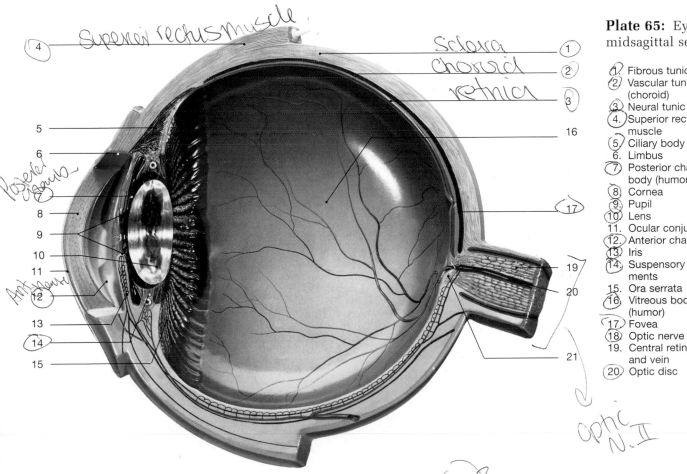

Plate 65: Eye, midsagittal section.

1. Fibrous tunic (sclera)
2. Vascular tunic (choroid)
3. Neural tunic (retina)
4. Superior rectus muscle
5. Ciliary body
6. Limbus
7. Posterior chamber body (humor)
8. Cornea
9. Pupil
10. Lens
11. Ocular conjunctiva
12. Anterior chamber
13. Iris
14. Suspensory ligaments
15. Ora serrata
16. Vitreous body (humor)
17. Fovea
18. Optic nerve
19. Central retinal artery and vein
20. Optic disc

(Handwritten annotations on eye diagram: "Superior rectus muscle", "Sclera", "Choroid", "retina", "Posterior chamber", "Anterior", "Optic N. II", "18")

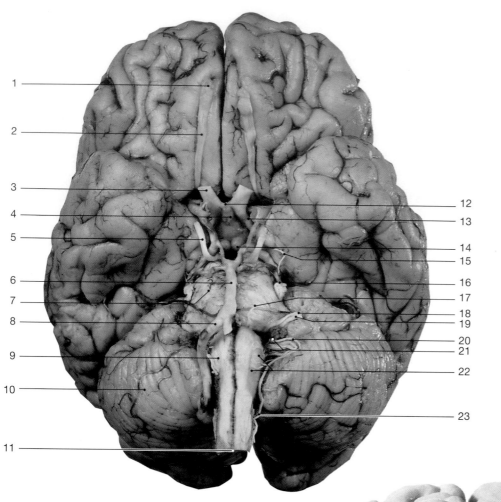

Plate 66: Human brain, inferior view showing cranial nerves and blood vessels.

1. Olfactory bulb
2. Olfactory tract
3. Optic nerve (II)
4. Infundibulum
5. Oculomotor nerve (III)
6. Basilar artery
7. Pons
8. Vertebral artery
9. Medulla oblongata
10. Cerebellum
11. Spinal cord
12. Optic chiasma
13. Optic tract
14. Mamillary body
15. Trochlear nerve (IV)
16. Trigeminal nerve (V)
17. Abducens nerve (VI)
18. Facial nerve (VII)
19. Vestibulocochlear nerve (VIII)
20. Glossopharyngeal nerve (IX)
21. Vagus nerve (X)
22. Hypoglossal nerve (XII)
23. Accessory nerve (XI)

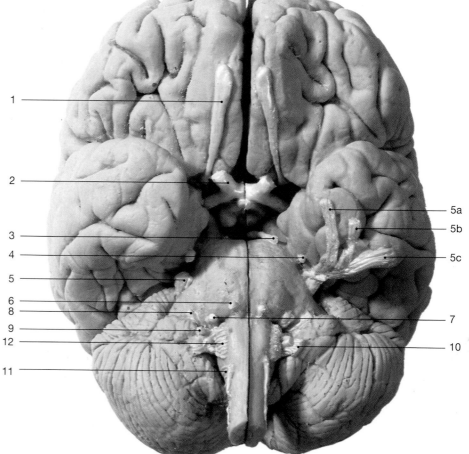

Plate 67: Brain model, inferior view.

1. Olfactory tract
2. Optic nerve (II)
3. Oculomotor nerve (III)
4. Trochlear nerve (IV)
5. Trigeminal nerve (V)
5a. Opthalmic branch ⎫
5b. Maxillary branch ⎬ Trigeminal nerve
5c. Mandibular branch ⎭
6. Abducens nerve (VI)
7. Facial nerve (VII)
8. Vestibulocochlear nerve (VIII)
9. Glossopharyngeal nerve (IX)
10. Vagus nerve (X)
11. Spinal accessory nerve (XI)
12. Hypoglossal nerve (XII)

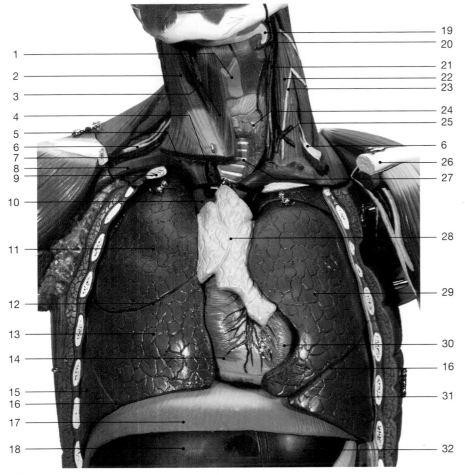

Plate 68: Thoracic cavity of torso exposed to show heart, lungs and great vessels.

1. Larynx (thyroid prominence)
2. Sternocleidomastoid
3. Omohyoid, sternothyroid and sternohyoid
4. Common carotid artery, right
5. Subclavian vein
6. Brachial plexus
7. Axillary artery, right
8. Axillary vein, right
9. Rib, cut
10. Superior vena cava
11. Right lung, superior lobe
12. Horizontal fissure
13. Right lung, middle lobe
14. Right ventricle
15. Right lung, inferior lobe
16. Oblique fissure
17. Diaphragm
18. Liver, right lobe
19. Hyoid bone
20. Superior thyroid artery
21. Common carotid artery, left
22. Anterior scalene
23. Phrenic nerve, left
24. Thyroid gland
25. Thyrocervical trunk
26. Clavicle, cut
27. Trachea
28. Thymus
29. Left lung, superior lobe
30. Left ventricle
31. Left lung, inferior lobe
32. Liver, left lobe

Plate 69: Torso thoracic cavity showing heart in situ with lungs dissected to show bronchi.

1. Larynx (thyroid prominence)
2. Trachea
3. Common carotid artery, right
4. Subclavian vein, right
5. Axillary vein
6. Brachiocephalic (innominate) vein
7. Right lung, superior lobe
8. Superior vena cava
9. Superior lobar bronchus, right
10. Middle lobar bronchus, right
11. Right lung, middle lobe
12. Inferior lobar bronchus, right
13. Right lung, inferior lobe
14. Pulmonary arteries and veins, right
15. Inferior vena cava
16. Ventricles, right and left
17. Diaphragm covered with diaphragmatic pleura
18. External carotid artery, left
19. Internal carotid artery, left
20. Carotid sinus
21. Superior thyroid artery
22. Common carotid artery, left
23. Thyroid gland
24. Thyrocervical trunk
25. Internal jugular vein, left
26. Brachial plexus
27. External jugular vein, left
28. Axillary artery, left
29. Subclavian vein, left
30. Brachiocephalic (innominate) vein, left
31. Aortic arch
32. Left lung, superior lobe
33. Pulmonary artery, left
34. Superior lobar bronchus, left
35. Pulmonary trunk
36. Inferior lobar bronchus, left
37. Left lung, inferior lobe
38. Auricles, right and left
39. Apex

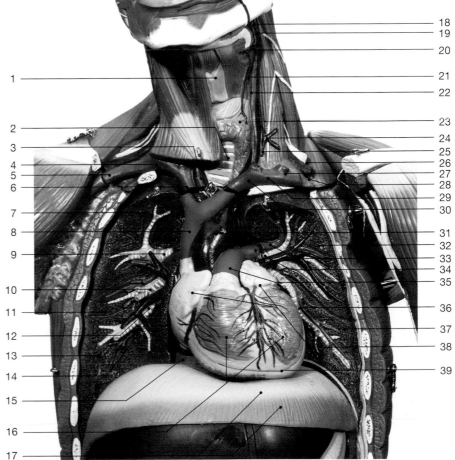

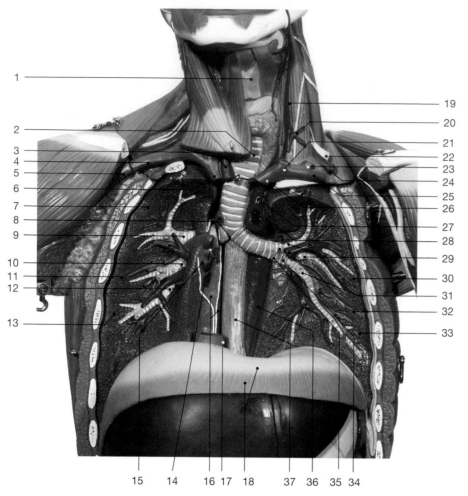

Plate 70: Thoracic cavity of torso with heart removed to show trachea and lungs dissected to display the major bronchi.

1. Larynx (thyroid prominence)
2. Trachea
3. Common carotid artery, right
4. Subclavian vein, right
5. Axillary vein, right
6. Brachiocephalic (innominate) vein
7. Right lung, superior lobe
8. Carina
9. Superior lobar bronchus, right
10. Middle lobar bronchus, right
11. Right lung, middle lobe
12. Inferior lobar bronchus, right
13. Right lung, inferior lobe
14. Right pulmonary artery
15. Right pulmonary vein
16. Visceral (mediastinal) pleura, cut edge
17. Inferior vena cava
18. Diaphragm covered with diaphragmatic pleura
19. Common carotid artery, left
20. Thyrocervical trunk
21. Internal jugular vein, left
22. Brachial plexus
23. External jugular vein, left
24. Axillary artery, left
25. Subclavian vein, left
26. Brachiocephalic (innominate) vein, left
27. Subclavian artery, left
28. Aortic arch
29. Superior lobar bronchus, left
30. Left lung, superior lobe
31. Inferior lobar bronchus, left
32. Oblique fissure
33. Left lung, inferior lobe
34. Left pulmonary vein
35. Left pulmonary artery
36. Thoracic aorta
37. Esophagus

Plate 71: Torso with heart and right lung removed to show thoracic cavity structures.

1. Larynx (thyroid prominence)
2. Common carotid artery, right
3. Axillary artery, right
4. Axillary vein
5. Brachiocephalic (innominate) vein, right
6. Trachea
7. Internal thoracic vein
8. Carina
9. Azygos vein
10. Intercostal vein, artery, and nerve
11. Parietal pleura (right lung removed)
12. Esophagus
13. Inferior vena cava
14. Diaphragm covered with diaphragmatic pleura
15. Kidney, right
16. Common carotid artery, left
17. Thyrocervical trunk
18. Brachial plexus
19. Internal jugular vein, left
20. Axillary artery, left
21. External jugular vein, left
22. Subclavian vein, left
23. Brachiocephalic (innominate) vein, left
24. Subclavian artery, left
25. Aortic arch
26. Left lung, superior lobe
27. Superior lobar bronchus, left
28. Inferior lobar bronchus, left
29. Left primary bronchus
30. Left lung, inferior lobe
31. Thoracic aorta
32. Visceral pleura covering base of left lung
33. Abdominal aorta
34. Kidney, right

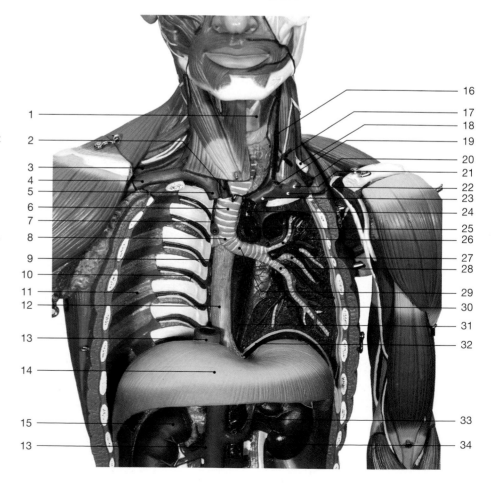

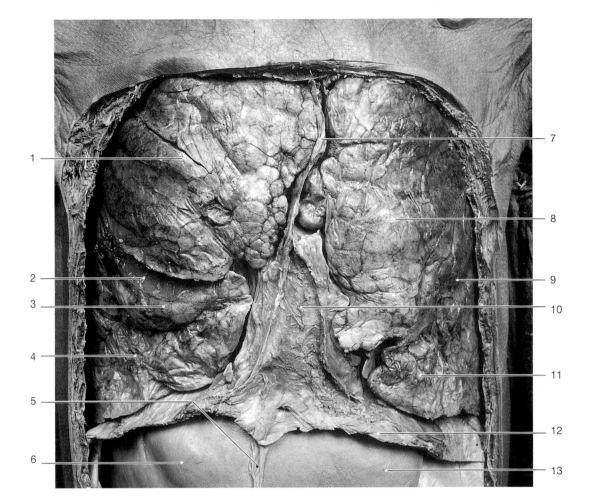

Plate 72:
Thoracic cavity
showing heart and
lungs in situ.

1. Superior lobe, right lung
2. Horizontal fissure
3. Middle lobe, right lung
4. Inferior lobe, right lung
5. Falciform ligament
6. Liver, right lobe
7. Boundary between right and left pleural cavities
8. Superior lobe, left lung
9. Oblique fissure
10. Fibrous layer of pericardium
11. Inferior lobe, left lung
12. Cut edge of diaphragm
13. Liver, left lobe

Plate 73: Heart and
lungs isolated and dis-
sected, anterior view.

1. Trachea
2. Right common carotid
3. Right subclavian artery
4. Brachiocephalic trunk
5. Superior lobe, right lung
6. Superior vena cava
7. Right primary bronchus
8. Pulmonary arteries, right
9. Pulmonary veins, right
10. Middle lobe, right lung
11. Inferior lobe, right lung
12. Right atrium
13. Marginal branch of right coronary artery
14. Left brachiocephalic vein
15. Aortic arch (cut)
16. Left primary bronchus
17. Superior lobe, left lung
18. Pulmonary trunk
19. Left ventricle
20. Inferior lobe, left lung
21. Apex of heart
22. Anterior interventricular artery
23. Right ventricle

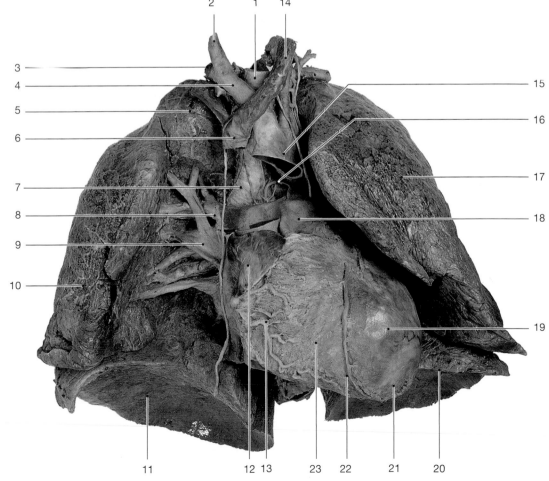

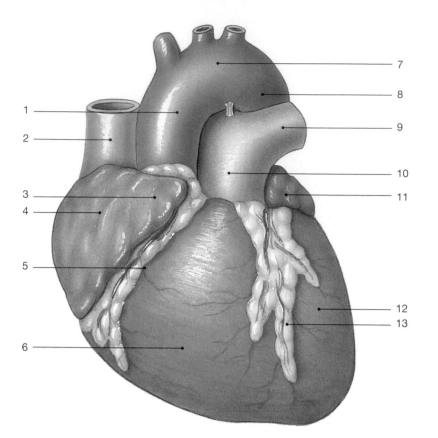

Plate 74: Superficial anatomy of the heart, anterior view.

1. Ascending aorta
2. Superior vena cava
3. Auricle of right atrium
4. Right atrium
5. Coronary sulcus
6. Right ventricle
7. Aortic arch
8. Descending aorta
9. Left pulmonary artery
10. Pulmonary trunk
11. Auricle of left atrium
12. Left ventricle
13. Anterior interventricular sulcus

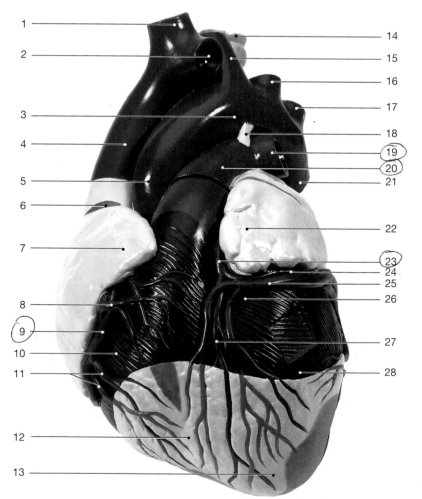

Plate 75: Heart, anterior view.

1. Right innominate (brachiocephalic) vein
2. Left innominate (brachiocephalic) vein
3. Aortic arch
4. Superior vena cava
5. Ascending aorta
6. Sinoatrial node
7. Auricle of right atrium
8. Anterior cardiac vein
9. Right coronary artery
10. Right ventricle
11. Marginal branch of right coronary artery and small cardiac vein
12. Fat in epicardium
13. Apex
14. Trachea
15. Innominate (brachiocephalic) artery
16. Left common carotid artery
17. Left subclavian artery
18. Ligamentum arteriosum
19. Left pulmonary artery
20. Pulmonary trunk
21. Descending thoracic aorta
22. Auricle of left atrium
23. Left coronary artery
24. Circumflex branch of left coronary artery
25. Great cardiac vein
26. Anterior interventricular sulcus
27. Anterior interventricular (descending) branch of left coronary artery
28. Myocardium of left ventricle

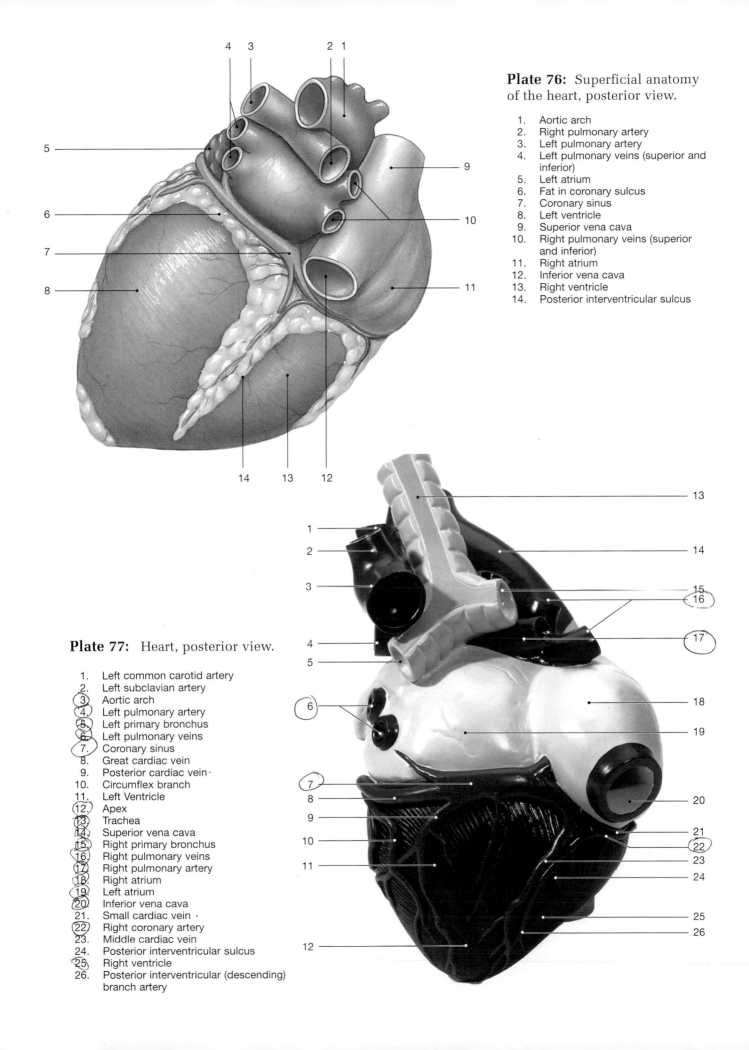

Plate 76: Superficial anatomy of the heart, posterior view.

1. Aortic arch
2. Right pulmonary artery
3. Left pulmonary artery
4. Left pulmonary veins (superior and inferior)
5. Left atrium
6. Fat in coronary sulcus
7. Coronary sinus
8. Left ventricle
9. Superior vena cava
10. Right pulmonary veins (superior and inferior)
11. Right atrium
12. Inferior vena cava
13. Right ventricle
14. Posterior interventricular sulcus

Plate 77: Heart, posterior view.

1. Left common carotid artery
2. Left subclavian artery
3. Aortic arch
4. Left pulmonary artery
5. Left primary bronchus
6. Left pulmonary veins
7. Coronary sinus
8. Great cardiac vein
9. Posterior cardiac vein·
10. Circumflex branch
11. Left Ventricle
12. Apex
13. Trachea
14. Superior vena cava
15. Right primary bronchus
16. Right pulmonary veins
17. Right pulmonary artery
18. Right atrium
19. Left atrium
20. Inferior vena cava
21. Small cardiac vein·
22. Right coronary artery
23. Middle cardiac vein
24. Posterior interventricular sulcus
25. Right ventricle
26. Posterior interventricular (descending) branch artery

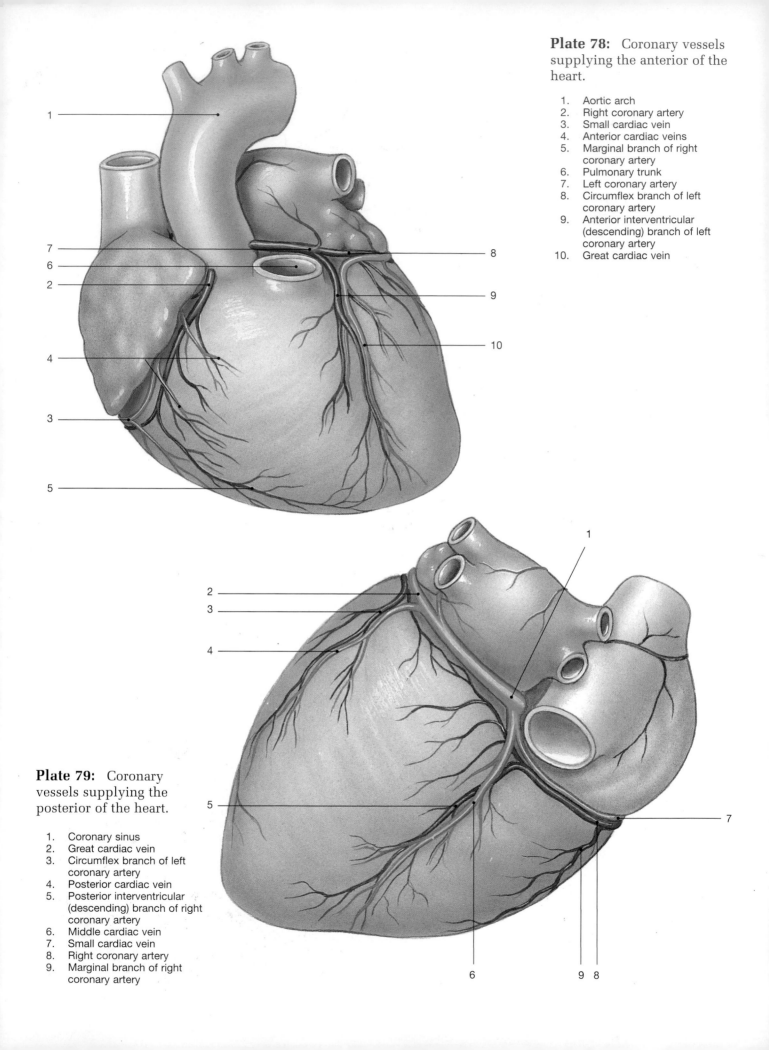

Plate 78: Coronary vessels supplying the anterior of the heart.

1. Aortic arch
2. Right coronary artery
3. Small cardiac vein
4. Anterior cardiac veins
5. Marginal branch of right coronary artery
6. Pulmonary trunk
7. Left coronary artery
8. Circumflex branch of left coronary artery
9. Anterior interventricular (descending) branch of left coronary artery
10. Great cardiac vein

Plate 79: Coronary vessels supplying the posterior of the heart.

1. Coronary sinus
2. Great cardiac vein
3. Circumflex branch of left coronary artery
4. Posterior cardiac vein
5. Posterior interventricular (descending) branch of right coronary artery
6. Middle cardiac vein
7. Small cardiac vein
8. Right coronary artery
9. Marginal branch of right coronary artery

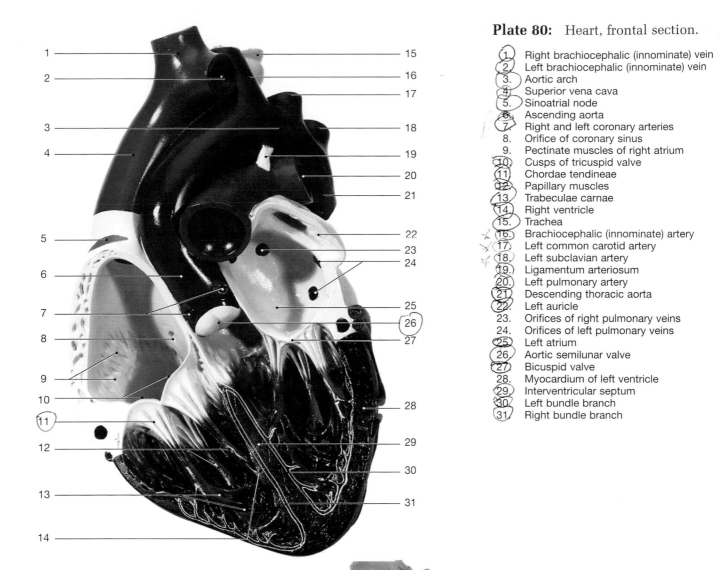

Plate 80: Heart, frontal section.

1. Right brachiocephalic (innominate) vein
2. Left brachiocephalic (innominate) vein
3. Aortic arch
4. Superior vena cava
5. Sinoatrial node
6. Ascending aorta
7. Right and left coronary arteries
8. Orifice of coronary sinus
9. Pectinate muscles of right atrium
10. Cusps of tricuspid valve
11. Chordae tendineae
12. Papillary muscles
13. Trabeculae carnae
14. Right ventricle
15. Trachea
16. Brachiocephalic (innominate) artery
17. Left common carotid artery
18. Left subclavian artery
19. Ligamentum arteriosum
20. Left pulmonary artery
21. Descending thoracic aorta
22. Left auricle
23. Orifices of right pulmonary veins
24. Orifices of left pulmonary veins
25. Left atrium
26. Aortic semilunar valve
27. Bicuspid valve
28. Myocardium of left ventricle
29. Interventricular septum
30. Left bundle branch
31. Right bundle branch

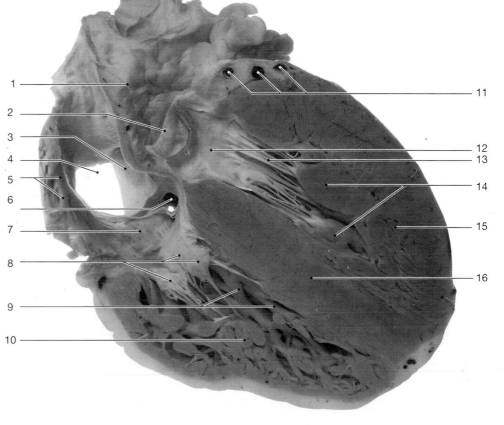

Plate 81: Heart dissected to show internal features and valves, sectional view.

1. Ascending aorta
2. Cusp of aortic valve
3. Fossa ovalis
4. Inferior vena cava
5. Pectinate muscles
6. Coronary sinus
7. Right atrium
8. Cusps of right AV (tricuspid) valve
9. Trabeculae carneae
10. Right ventricle
11. Left coronary artery branches and great cardiac vein
12. Cusp of left AV (bicuspid) valve
13. Chordae tendineae
14. Papillary muscles
15. Left ventricle
16. Interventricular septum

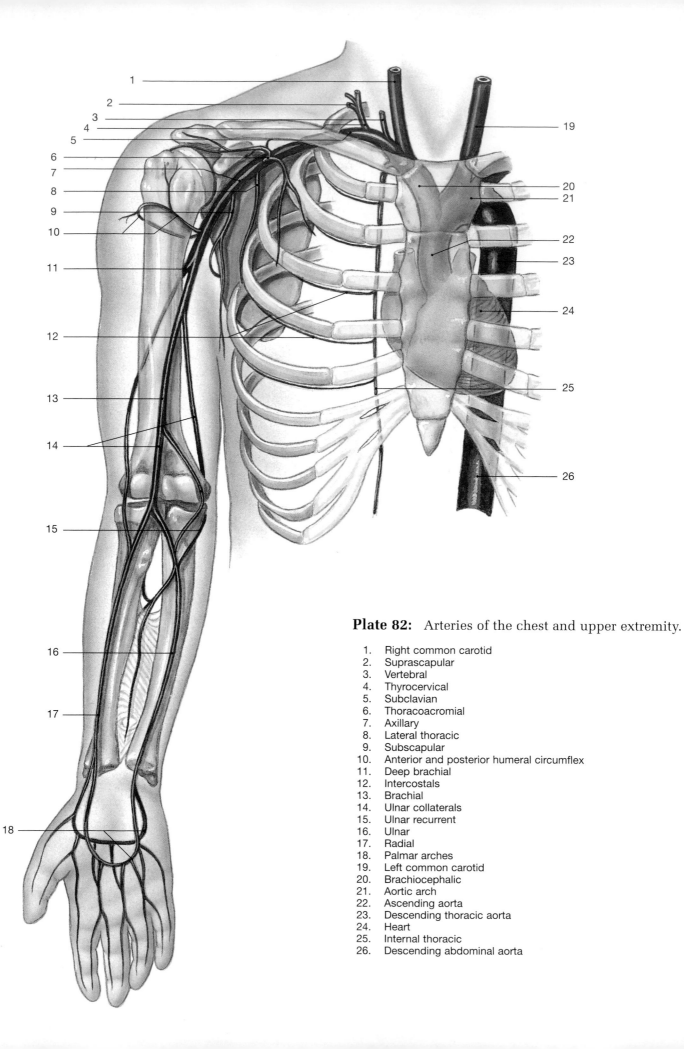

Plate 82: Arteries of the chest and upper extremity.

1. Right common carotid
2. Suprascapular
3. Vertebral
4. Thyrocervical
5. Subclavian
6. Thoracoacromial
7. Axillary
8. Lateral thoracic
9. Subscapular
10. Anterior and posterior humeral circumflex
11. Deep brachial
12. Intercostals
13. Brachial
14. Ulnar collaterals
15. Ulnar recurrent
16. Ulnar
17. Radial
18. Palmar arches
19. Left common carotid
20. Brachiocephalic
21. Aortic arch
22. Ascending aorta
23. Descending thoracic aorta
24. Heart
25. Internal thoracic
26. Descending abdominal aorta

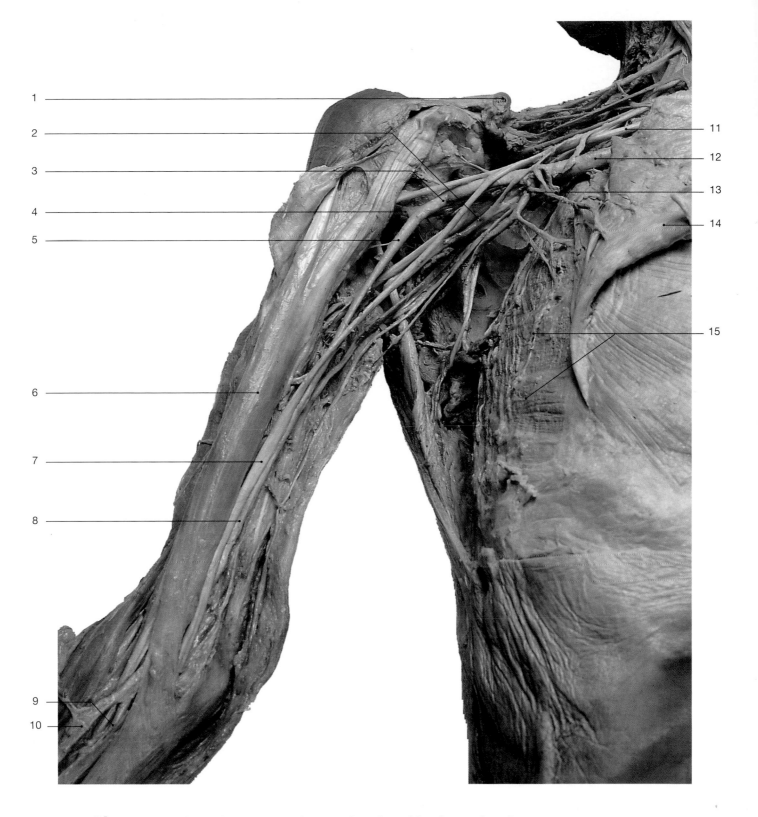

Plate 83: Right axillary region dissected to show blood vessel and nerves, anterior view.

1. Clavicle (cut)
2. Axillary artery
3. Posterior cord of brachial plexus
4. Deep brachial artery
5. Brachial artery
6. Biceps brachii
7. Median nerve
8. Brachial artery
9. Ulnar artery
10. Radial artery
11. Medial cord of brachial plexus
12. Subclavian artery
13. Subscapular artery
14. Pectoralis major (cut and reflected)
15. Serratus anterior

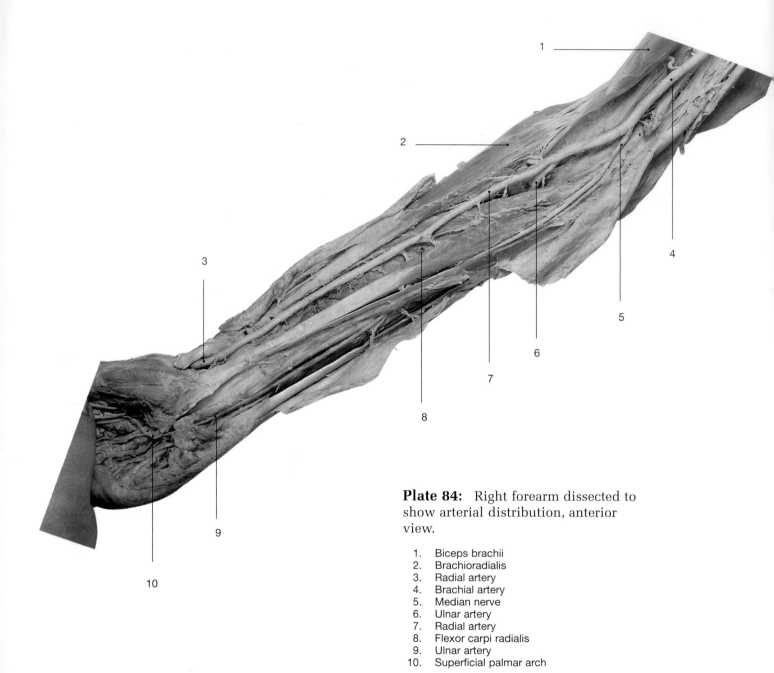

Plate 84: Right forearm dissected to show arterial distribution, anterior view.

1. Biceps brachii
2. Brachioradialis
3. Radial artery
4. Brachial artery
5. Median nerve
6. Ulnar artery
7. Radial artery
8. Flexor carpi radialis
9. Ulnar artery
10. Superficial palmar arch

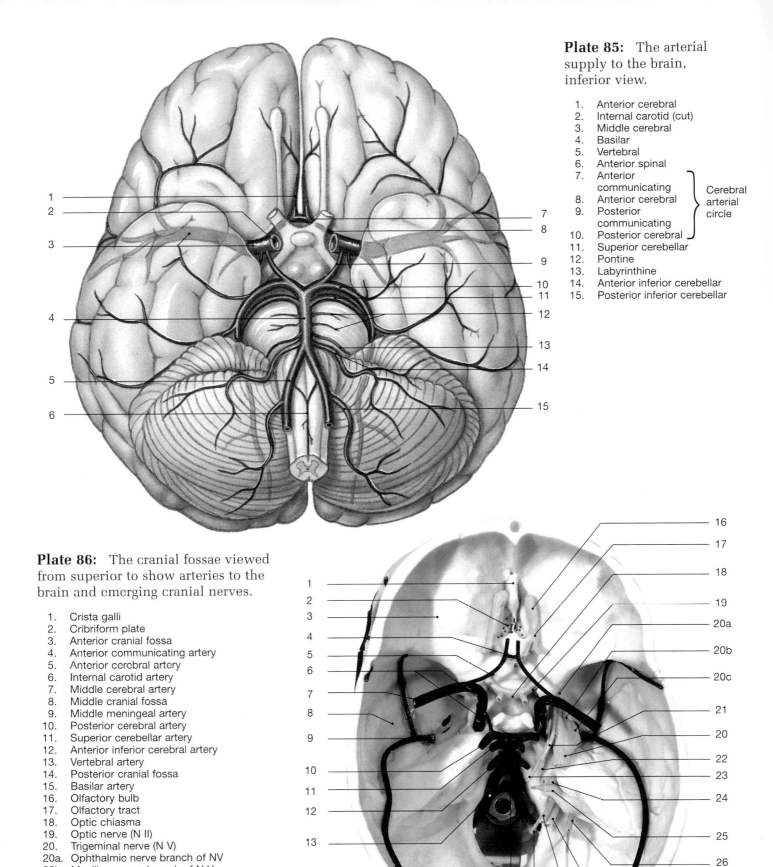

Plate 85: The arterial supply to the brain, inferior view.

1. Anterior cerebral
2. Internal carotid (cut)
3. Middle cerebral
4. Basilar
5. Vertebral
6. Anterior spinal
7. Anterior communicating ⎫
8. Anterior cerebral ⎬ Cerebral arterial circle
9. Posterior communicating ⎪
10. Posterior cerebral ⎭
11. Superior cerebellar
12. Pontine
13. Labyrinthine
14. Anterior inferior cerebellar
15. Posterior inferior cerebellar

Plate 86: The cranial fossae viewed from superior to show arteries to the brain and emerging cranial nerves.

1. Crista galli
2. Cribriform plate
3. Anterior cranial fossa
4. Anterior communicating artery
5. Anterior cerebral artery
6. Internal carotid artery
7. Middle cerebral artery
8. Middle cranial fossa
9. Middle meningeal artery
10. Posterior cerebral artery
11. Superior cerebellar artery
12. Anterior inferior cerebral artery
13. Vertebral artery
14. Posterior cranial fossa
15. Basilar artery
16. Olfactory bulb
17. Olfactory tract
18. Optic chiasma
19. Optic nerve (N II)
20. Trigeminal nerve (N V)
20a. Ophthalmic nerve branch of NV
20b. Maxillary nerve branch of N V
20c. Mandibular nerve branch of N V
21. Oculomotor nerve (N III)
22. Trochlear nerve (N IV)
23. Abducens nerve (N VI)
24. Facial nerve (N VII)
25. Vestibulocochlear nerve (N VIII)
26. Hypoglossal (N XII)
27. Vagus (N X)
28. Glossopharyngeal (N IX)
29. Spinal accessory (Accessory) nerve (N XI)

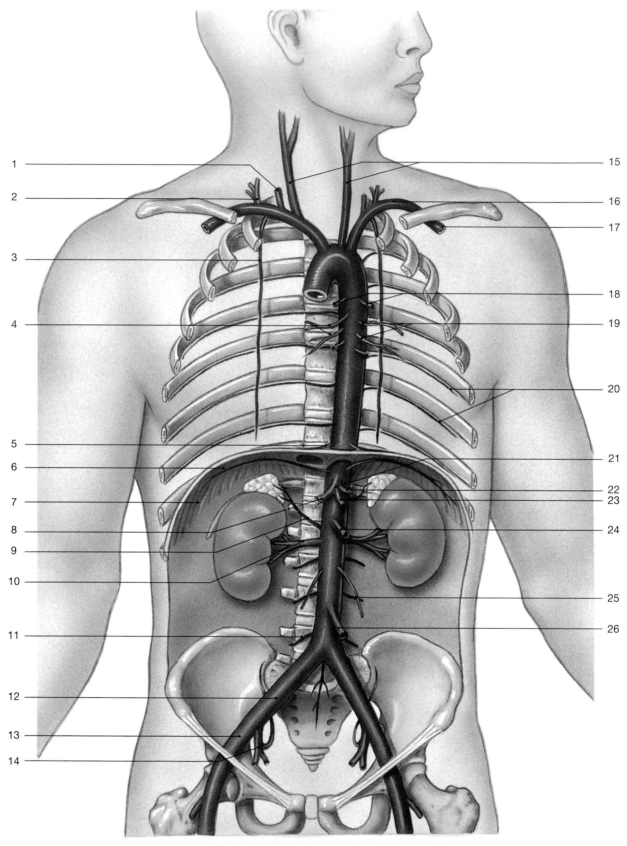

Plate 87: Major arteries of the trunk.

1.	Vertebral	10.	Renal	19.	Mediastinals
2.	Thyrocervical	11.	Lumbar	20.	Intercostals
3.	Internal thoracic	12.	Common iliac	21.	Celiac trunk
4.	Pericardials	13.	External iliac	22.	Left gastric
5.	Superior phrenic	14.	Internal iliac	23.	Splenic
6.	Inferior phrenic	15.	Common carotid	24.	Superior mesenteric
7.	Diaphragm	16.	Subclavian	25.	Gonadal
8.	Common hepatic	17.	Axillary	26.	Inferior mesenteric
9.	Suprarenal	18.	Bronchials		

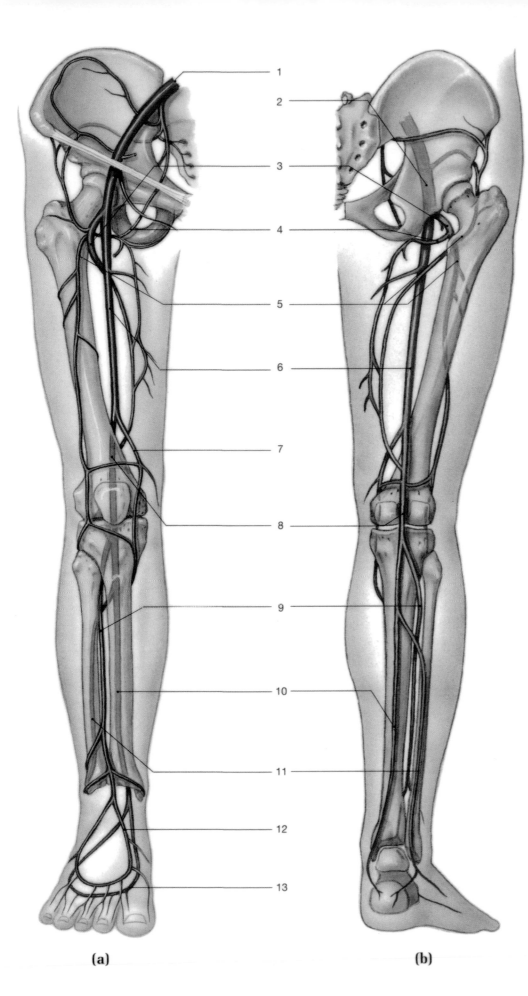

Plate 88: Major arteries of the lower extremity, (a) anterior and (b) posterior views.

1. Common iliac
2. External iliac
3. Deep femoral
4. Medial femoral circumflex
5. Lateral femoral circumflex
6. Femoral
7. Descending genicular
8. Popliteal
9. Anterior tibial
10. Posterior tibial
11. Peroneal
12. Dorsalis pedis
13. Plantar arch

(a)

(b)

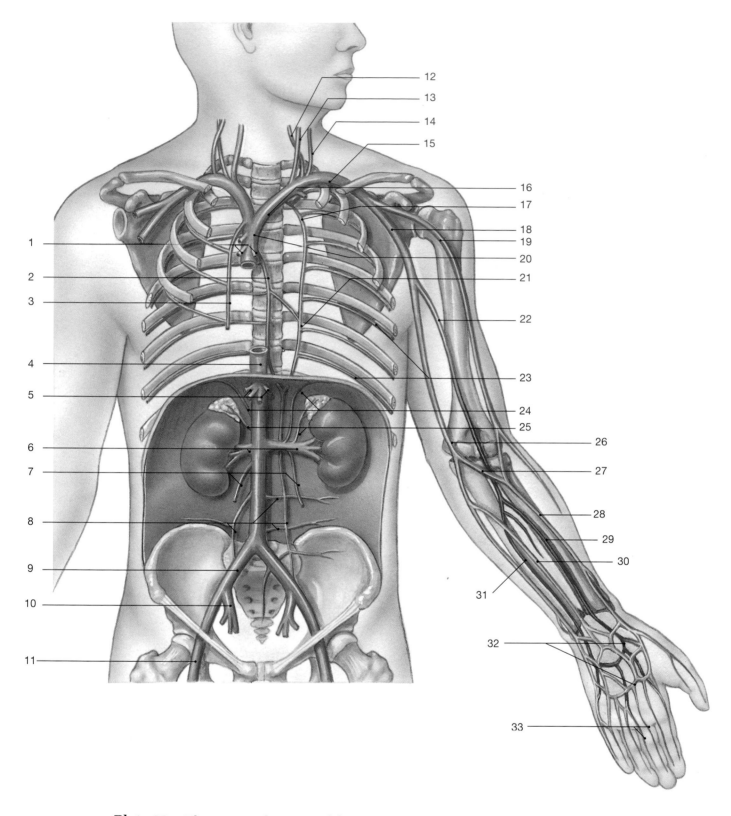

Plate 89: The venous drainage of the upper extremities, chest, and abdomen.

1. Mediastinals
2. Azygos
3. Internal thoracic
4. Inferior vena cava
5. Hepatics
6. Renals
7. Gonadals
8. Lumbars
9. Common iliac
10. Internal iliac
11. External iliac

12. Vertebral
13. Internal jugular
14. External jugular
15. Subclavian
16. Brachiocephalic
17. Highest intercostal
18. Axillary
19. Cephalic
20. Superior vena cava
21. Hemiazygos
22. Brachial

23. Intercostals
24. Phrenics
25. Suprarenals
26. Basiliac
27. Median cubital
28. Cephalic
29. Radial
30. Median antebrachial
31. Ulnar
32. Palmar venous arches
33. Digital veins

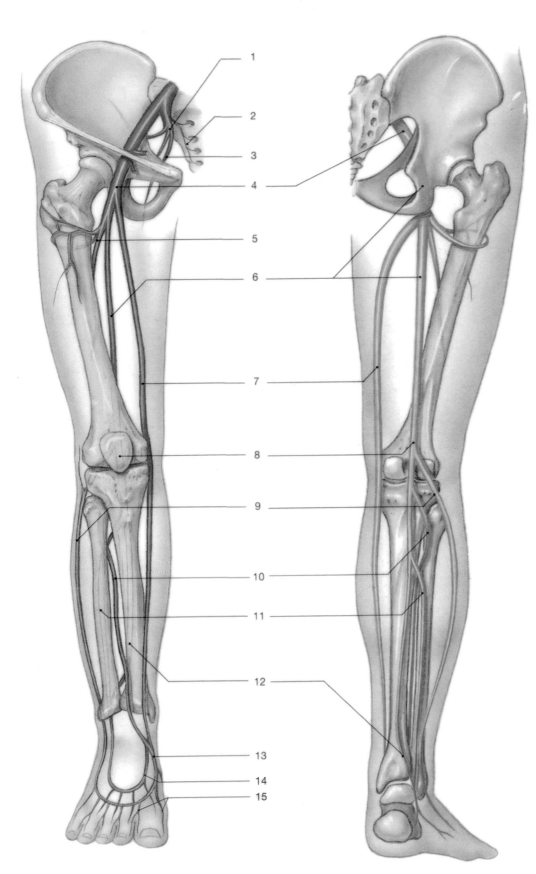

Plate 90: The venous drainage of the lower extremity, (a) anterior and (b) posterior views.

1. Internal iliac
2. Sacral
3. Obturator
4. External iliac
5. Deep femoral
6. Femoral
7. Great saphenous
8. Popliteal
9. Small saphenous
10. Anterior tibial
11. Peroneal
12. Posterior tibial
13. Dorsal venous arch
14. Plantar venous arch
15. Digitals

(a)   (b)

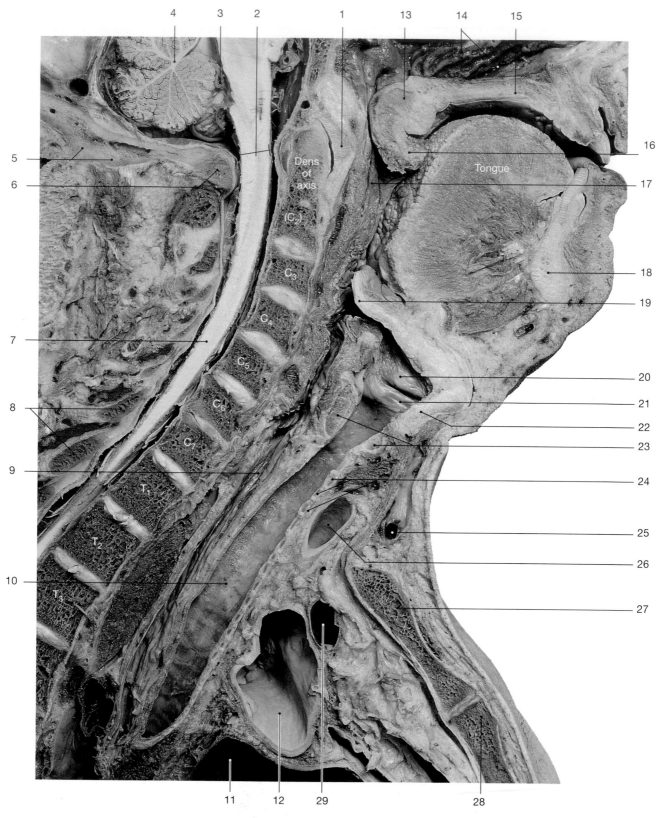

Plate 91: Sectional anatomy of the neck.

1. Atlas (C₁) (anterior arch)
2. Foramen magnum
3. Choroid plexus
4. Arbor vitae of cerebellum
5. External occipital crest
6. Atlas (C₁) (posterior arch)
7. Spinal cord
8. Spinous processes
9. Esophagus
10. Trachea

11. Pleural cavity
12. Aortic arch
13. Soft palate
14. Nasal conchae
15. Hard palate
16. Uvula
17. Oropharynx
18. Mandible
19. Epiglottis
20. Ventricular fold

21. Vocal fold
22. Thyroid cartilage
23. Cricoid cartilage
24. Tracheal cartilage
25. External jugular vein
26. Right common carotid artery
27. Manubrium of sternum
28. Body of sternum
29. Left brachiocephalic vein

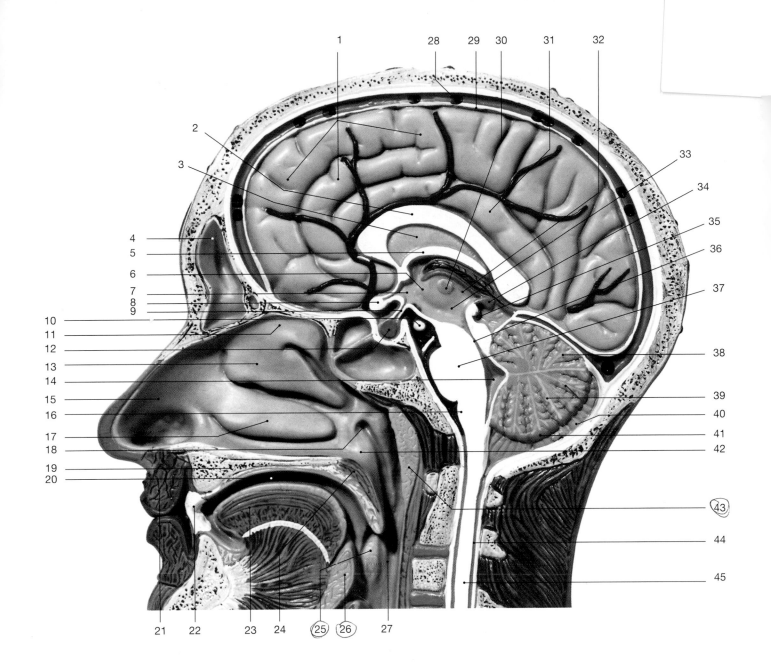

Plate 92: Midsagittal section of the head.

1.	Cerebrum	16.	Medulla oblongata	31.	Limbic lobe (cingulate gyrus)
2.	Corpus callosum	17.	Inferior nasal concha	32.	Epithalamus
3.	Septum pellucidum	18.	Opening to auditory tube	33.	Third ventricle
4.	Frontal sinus	19.	Hard palate	34.	Pineal gland
5.	Fornix	20.	Oral cavity	35.	Tectum
6.	Thalamus	21.	Upper lip	36.	Mesencephalic aqueduct
7.	Hypothalamus	22.	Incisive tooth	37.	Pons
8.	Optic chiasma	23.	Tongue	38.	Cerebellum
9.	Infundibulum	24.	Soft palate	39.	Arbor vitae
10.	Mamillary body	25.	Palatine tonsil	40.	Folia
11.	Superior nasal concha	26.	Lingual tonsil	41.	Cerebellar cortex
12.	Hypophysis (pituitary gland)	27.	Oropharynx	42.	Nasopharynx
13.	Middle nasal concha	28.	Arachnoid villus	43.	Pharyngeal tonsil
14.	Fourth ventricle	29.	Subarachnoid space	44.	Central canal of spinal cord
15.	Nasal vestibule	30.	Intermediate mass	45.	Spinal cord

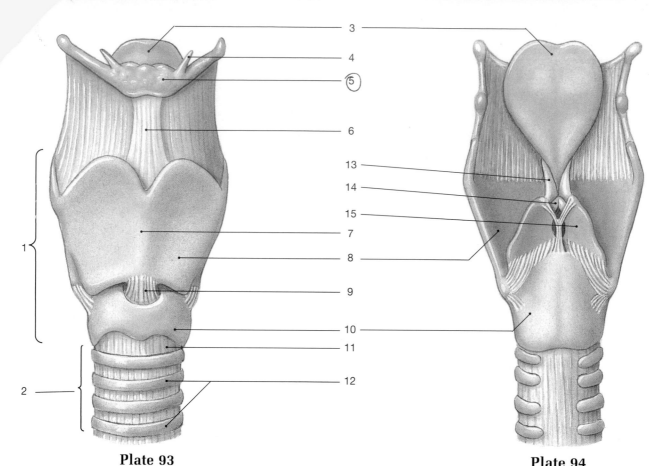

Plate 93

Plate 94

Plate 93: Larynx, anterior view.
Plate 94: Larynx, posterior view.

1. Larynx
2. Trachea
3. Epiglottis
4. Lesser cornu
5. Hyoid bone
6. Extrinsic (thyrohyoid) ligament
7. Laryngeal prominence
8. Thyroid cartilage
9. Intrinsic (cricothyroid) ligament
10. Cricoid cartilage
11. Extrinsic (cricotracheal) ligament
12. Tracheal cartilages
13. Ventricular fold
14. Vocal fold
15. Arytenoid cartilage

Plate 95: Larynx, sagittal section of the intact larynx.

1. Epiglottis
2. Thyroid cartilage
3. Ventricular fold — *false*
4. Vocal fold — *true*
5. Cricothyroid ligament
6. Tracheal cartilages
7. Hyoid bone
8. Thyrohyoid membrane
9. Superior horn of thyroid cartilage
10. Corniculate cartilage
11. Arytenoid cartilage
12. Cricoid cartilage
13. Cricotracheal ligament

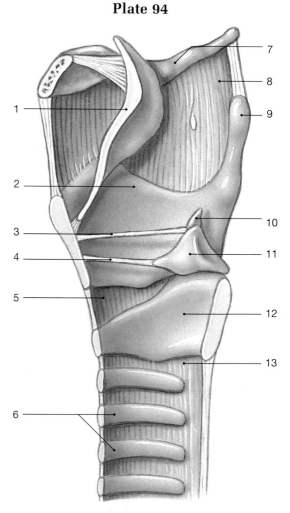

Plate 95

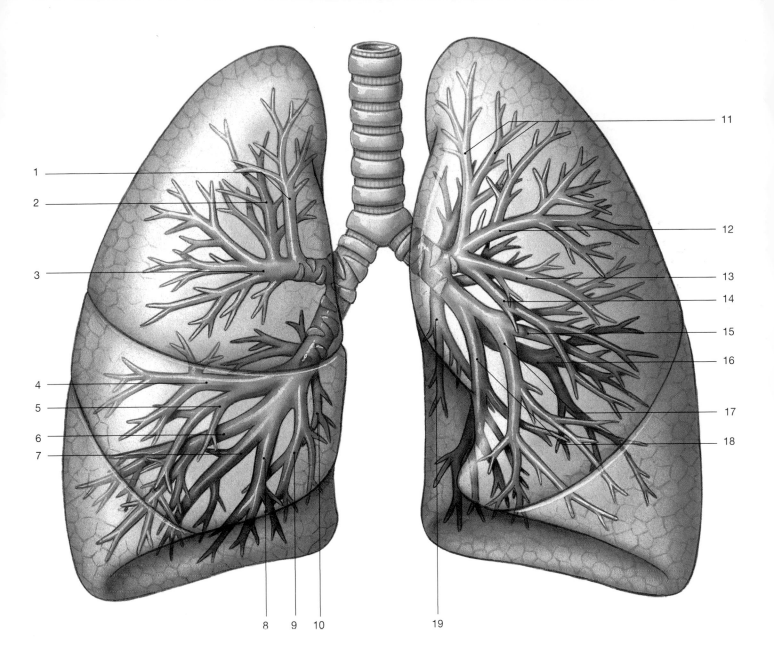

Plate 96: Gross anatomy of the lungs, showing the bronchial tree, its divisions, and the following bronchopulmonary segments.

Segments of Right Superior Lobe
1. Apical
2. Posterior
3. Anterior

Segments of Right Middle Lobe
4. Lateral
5. Medial

Segments of Right Inferior Lobe
6. Anterior basal
7. Lateral basal
8. Posterior basal
9. Medial basal
10. Superior

Segments of Left Superior Lobe
11. Apical and posterior (fused)
12. Anterior
13. Superior lingular
14. Inferior lingular

Segments of Left Inferior Lobe
15. Anterior basal
16. Lateral basal
17. Medial basal
18. Posterior basal
19. Superior

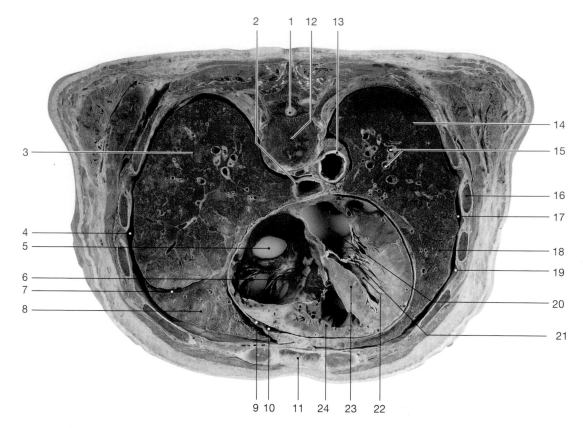

Plate 97: Lungs and heart in thoracic cavity as seen in transverse section.

1. Spinal cord	9. Parietal (fibrous) pericardium	17. Left pleural cavity
2. Esophagus	10. Pericardial cavity	18. Left lung, superior lobe
3. Right lung, middle lobe	11. Body of sternum	19. Parietal pleura
4. Right pleural cavity	12. Body of vertebra	20. Left atrium
5. Inferior vena cava	13. Descending aorta	21. Bicuspid valve
6. Right atrium	14. Left lung, inferior lobe	22. Papillary muscle of left ventricle
7. Oblique fissure	15. Bronchi	23. Interventricular septum
8. Right lung, inferior lobe	16. Rib (cut)	24. Right ventricle

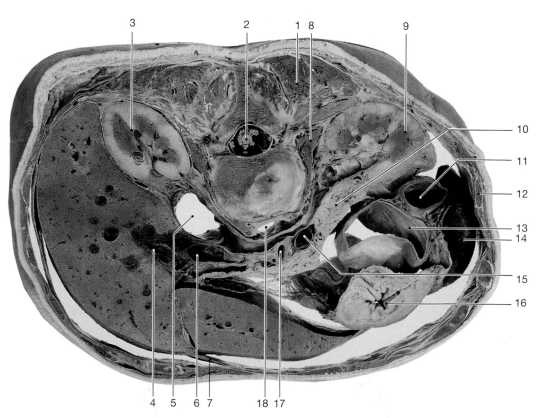

Plate 98: Abdominal organs in abdominopelvic cavity viewed in transverse section through upper abdomen.

1. Erector spinae
2. Tip of spinal cord
3. Right kidney
4. Hepatic duct
5. Inferior vena cava
6. Hepatic portal vein
7. Falciform ligament
8. Quadratus lumborum
9. Left kidney
10. Pancreas
11. Descending colon
12. 12th rib
13. Transverse colon
14. Spleen
15. Splenic vein
16. Stomach
17. Superior mesenteric artery
18. Abdominal aorta

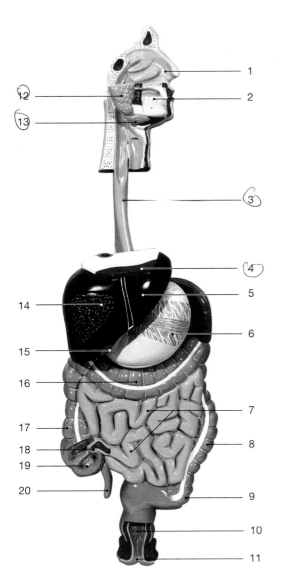

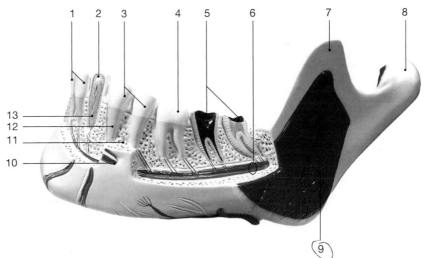

Plate 100: Mandible exposed to show teeth.

1. Incisors
2. Cuspid (canine)
3. Bicuspids (premolars)
4. Crown, first molar tooth
5. 2nd and 3rd molars with dental caries
6. Alveolar artery, vein, and nerve
7. Coronoid processs
8. Condylar process
9. Masseter muscle
10. Mental foramen with mental vessels emerging
11. Alveolar socket
12. Root of tooth
13. Dentin of tooth

Plate 99: The digestive system.

1. Nasal cavity
2. Oral cavity
3. Esophagus
4. Diaphragm
5. Left liver lobe
6. Stomach
7. Small intestine (jejunum and ileum)
8. Descending colon
9. Sigmoid colon
10. Rectum
11. Anus
12. Parotid salivary gland
13. Submandibular salivary gland
14. Right liver lobe
15. Gallbladder
16. Transverse colon
17. Ascending colon
18. Ileocecal valve
19. Cecum
20. Vermiform appendix

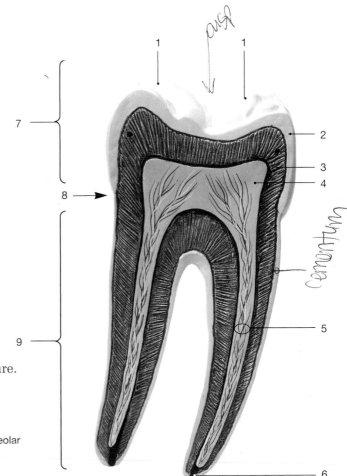

Plate 101: Adult tooth
sectioned to show structure.

1. Occlusal surface
2. Enamel
3. Dentin
4. Pulp cavity
5. Root canal containing alveolar blood vessels
6. Apical foramen
7. Crown
8. Neck
9. Root

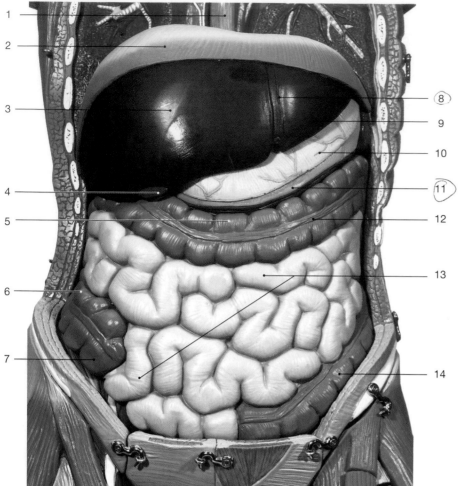

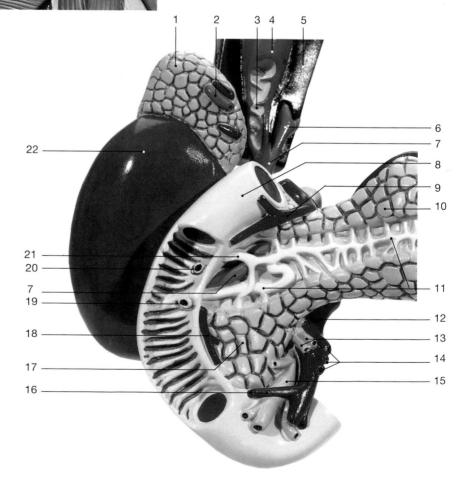

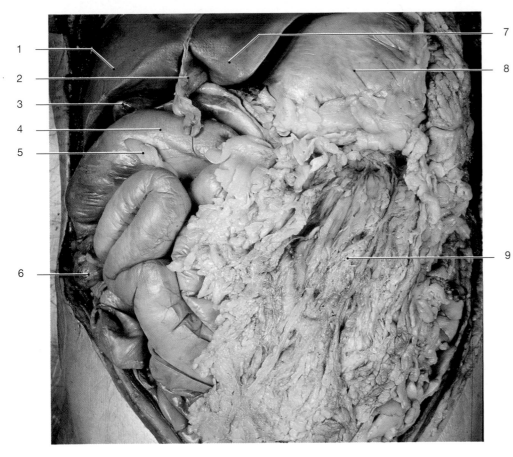

Plate 104: Abdominopelvic viscera and greater omentum.

1. Liver (right lobe)
2. Falciform ligament
3. Gallbladder
4. Transverse colon
5. Epiploic appendage
6. Ascending colon
7. Liver (left lobe)
8. Stomach
9. Greater omentum

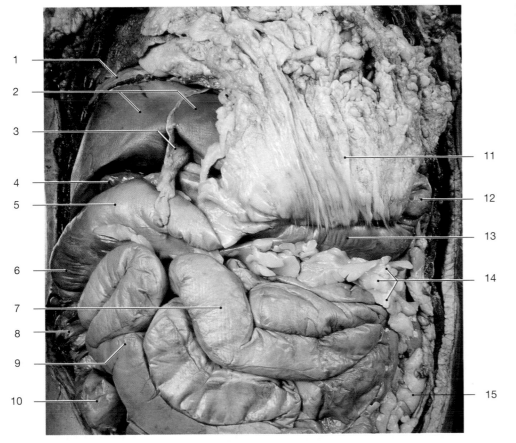

Plate 105: Greater omentum reflected to expose abdominopelvic viscera.

1. Diaphragm
2. Left and right lobes of liver
3. Falciform ligament
4. Gallbladder
5. Transverse colon
6. Right colic (hepatic) flexure
7. Jejunum
8. Ascending colon
9. Ileum
10. Cecum
11. Greater omentum (reflected)
12. Left colic (splenic) flexure
13. Transverse colon
14. Epiploic appendages
15. Descending colon

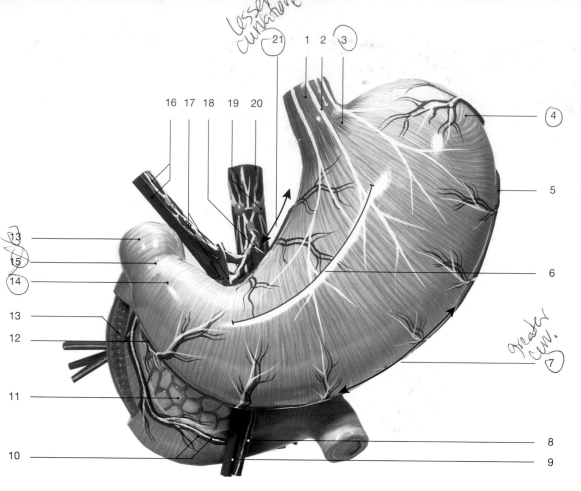

lesser curvature

Plate 106: Gross anatomy of the stomach, showing blood vessels and relation to duodenum and pancreas.

1. Esophagus
2. Left vagus nerve
3. Cardia
4. Fundus
5. Left gastroepiploic artery
6. Body
7. Greater curvature
8. Superior mesenteric artery
9. Superior mesenteric vein
10. Anterior inferior pancreaticoduodenal artery and vein
11. Pancreas, head portion
12. Right gastroepiploic artery
13. Duodenum
14. Pylorus
15. Pyloric valve
16. Common hepatic artery and vein
17. Right gastric artery and vein
18. Celiac trunk
19. Celiac ganglion
20. Abdominal aorta
21. Lesser curvature

Greater curv.

Plate 107: Internal anatomy of the stomach.

1. Esophagus
2. Cardiac orifice
3. Cardia
4. Fundus
5. Mucosa layer
6. Oblique muscle layer
7. Circular muscle layer
8. Longitudinal muscle layer
9. Body
10. Pancreas, tail portion
11. Rugae
12. Duodenum
13. Pancreas, head portion
14. Pylorus
15. Pyloric outlet
16. Pyloric sphincter

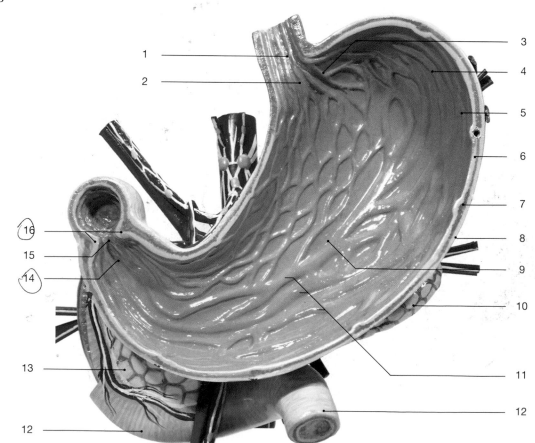

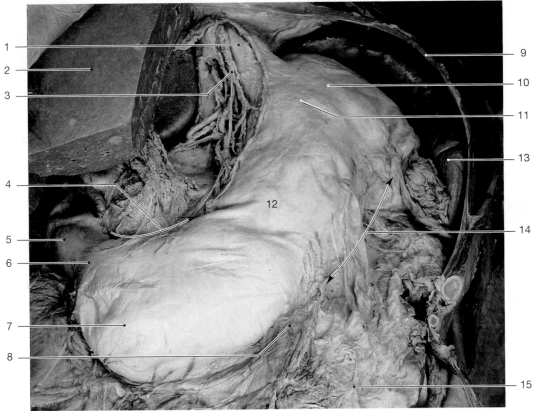

Plate 108: Left lobe of liver and lesser omentum removed to show relation of stomach to liver and spleen.

1. Esophagus
2. Right lobe of liver
3. Vagus nerve (N X)
4. Lesser curvature
5. Duodenum
6. Pyloric sphincter
7. Pylorus
8. Left gastroepiploic vessels
9. Diaphragm
10. Fundus
11. Cardia
12. Body
13. Spleen
14. Greater curvature with greater omentum attached
15. Greater omentum

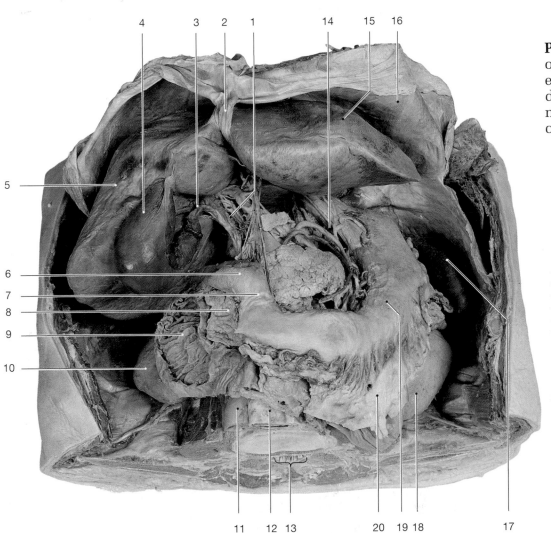

Plate 109: Greater omentum removed to expose liver, gallbladder, stomach, duodenum, pancreas, and colon.

1. Common bile duct
2. Falciform ligament
3. Cystic duct
4. Gallbladder
5. Right lobe of liver
6. Duodenum
7. Pylorus
8. Pancreas, head portion
9. Transverse colon, cut to expose mucosa
10. Right kidney
11. Inferior vena cava
12. Aorta
13. Cauda equina
14. Left gastric artery
15. Left lobe of liver
16. Diaphragm
17. Spleen
18. Left kidney
19. Stomach (body)
20. Greater omentum

Plate 110: Transverse colon reflected to show branches of the superior mesenteric artery.

1. Right lobe of liver
2. Mesentery proper
3. Ascending colon
4. Cecum
5. Ileum
6. Vermiform appendix
7. Transverse colon
8. Middle colic artery
9. Superior mesenteric artery
10. Ileocolic artery
11. Jejunal arteries
12. Jejunum
13. Descending colon

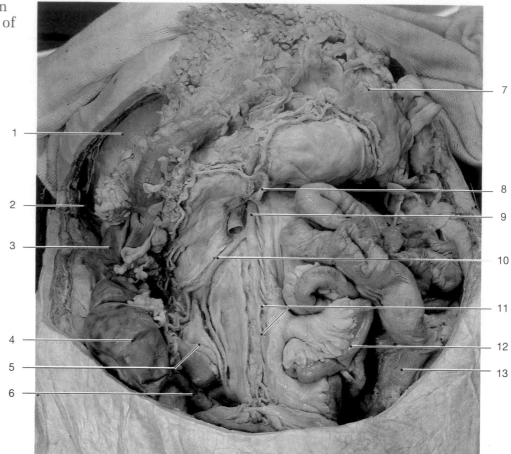

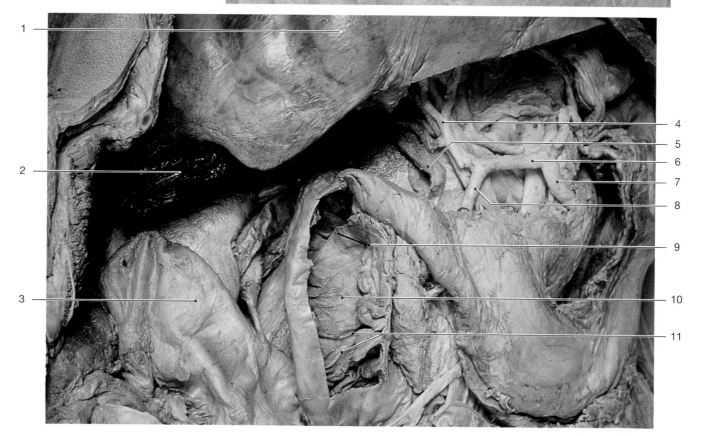

Plate 111: Structures and vessels of the duodenal region.

1. Right lobe of liver
2. Gallbladder
3. Transverse colon
4. Hepatic artery
5. Common bile duct
6. Common hepatic artery
7. Splenic artery
8. Gastroduodenal artery
9. Probe in entrance to duodenal papilla
10. Duodenum (cut open to expose detail of mucosa)
11. Plicae

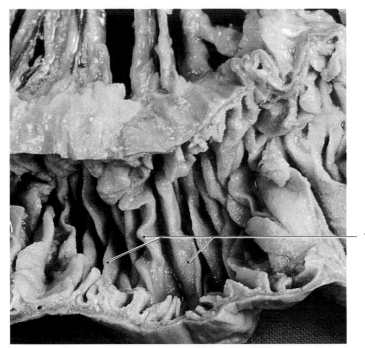

Plate 112: Jejunum dissected to show internal structure.

1. Plicae

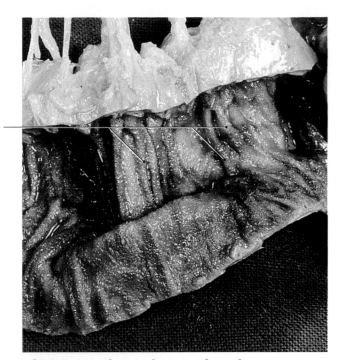

Plate 113: Ileum dissected to show internal structure.

1. Plicae

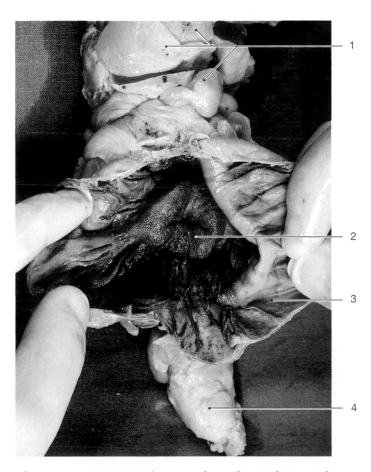

Plate 114: Cecum dissected to show ileocecal valve.

1. Epiploic appendages
2. Ileocecal valve
3. Cecum (cut open)
4. Veriform appendix

Plate 115: Liver
and ligaments,
superior view.

1. Inferior vena cava
2. Coronary ligament
3. Right lobe
4. Falciform ligament
5. Pericardium
6. Diaphragm
7. Left lobe

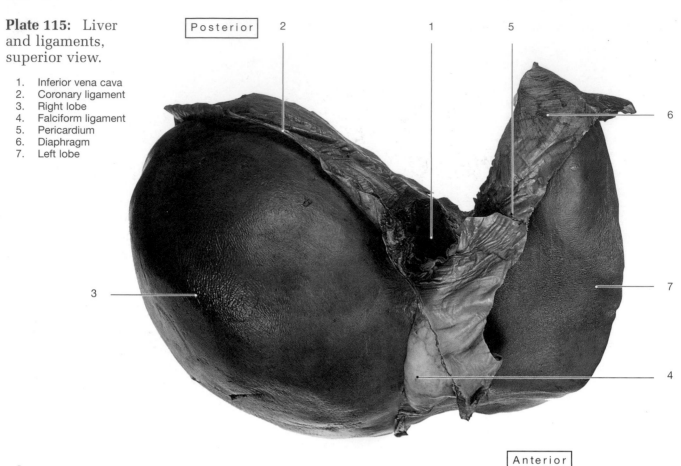

Posterior

Anterior

Plate 116: Liver, superior
view.

1. Right lobe of liver
2. Coronary ligament
3. Inferior vena cava
4. Caudate lobe
5. Left lobe of liver
6. Falciform ligament

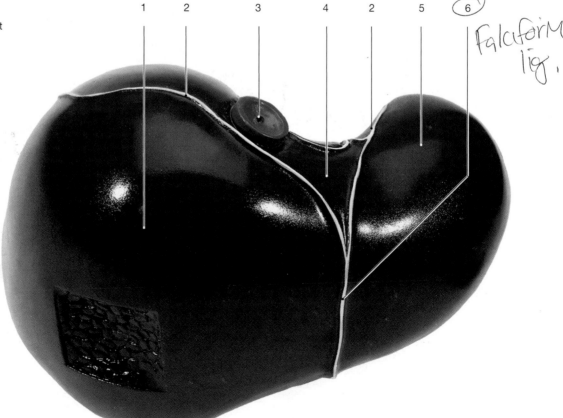

Falciform
lig.

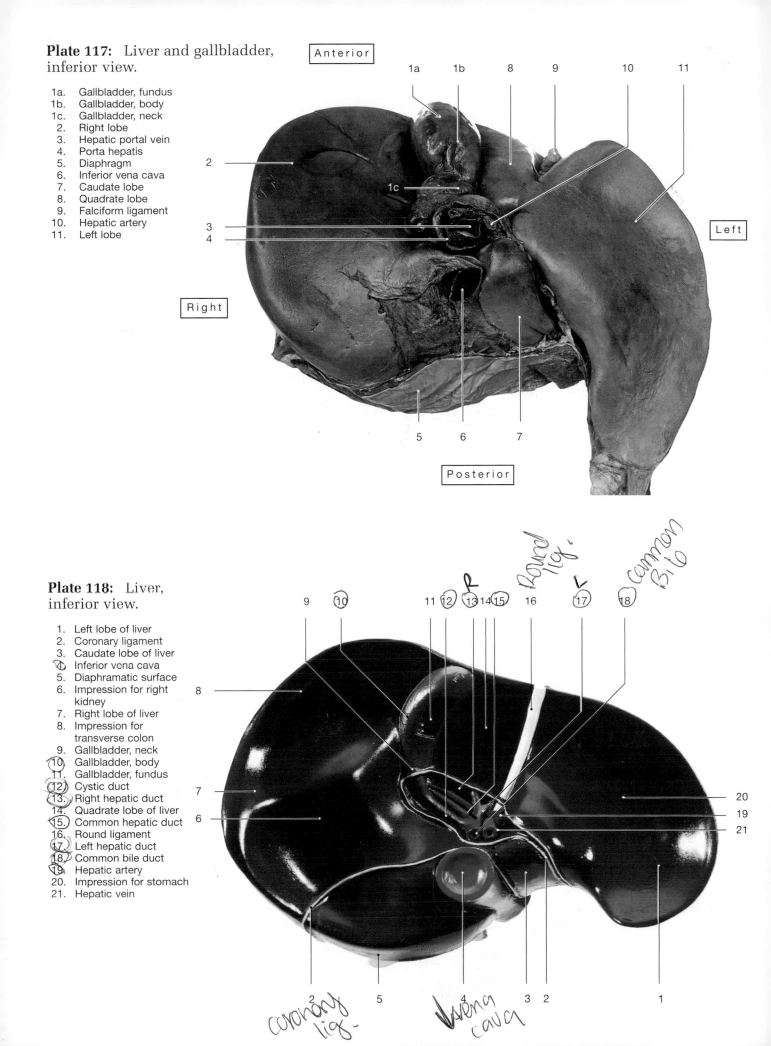

Plate 117: Liver and gallbladder, inferior view.

1a. Gallbladder, fundus
1b. Gallbladder, body
1c. Gallbladder, neck
2. Right lobe
3. Hepatic portal vein
4. Porta hepatis
5. Diaphragm
6. Inferior vena cava
7. Caudate lobe
8. Quadrate lobe
9. Falciform ligament
10. Hepatic artery
11. Left lobe

Anterior

1a 1b 8 9 10 11

2

1c

3
4

Right

Left

5 6 7

Posterior

Plate 118: Liver, inferior view.

1. Left lobe of liver
2. Coronary ligament
3. Caudate lobe of liver
4. Inferior vena cava
5. Diaphramatic surface
6. Impression for right kidney
7. Right lobe of liver
8. Impression for transverse colon
9. Gallbladder, neck
10. Gallbladder, body
11. Gallbladder, fundus
12. Cystic duct
13. Right hepatic duct
14. Quadrate lobe of liver
15. Common hepatic duct
16. Round ligament
17. Left hepatic duct
18. Common bile duct
19. Hepatic artery
20. Impression for stomach
21. Hepatic vein

9 10   11 12 13 14 15 16   17   18

Round lig. R L common Bile

8

7

6

20
19
21

coronary lig.   5   4 Iveng cava   3 2   1

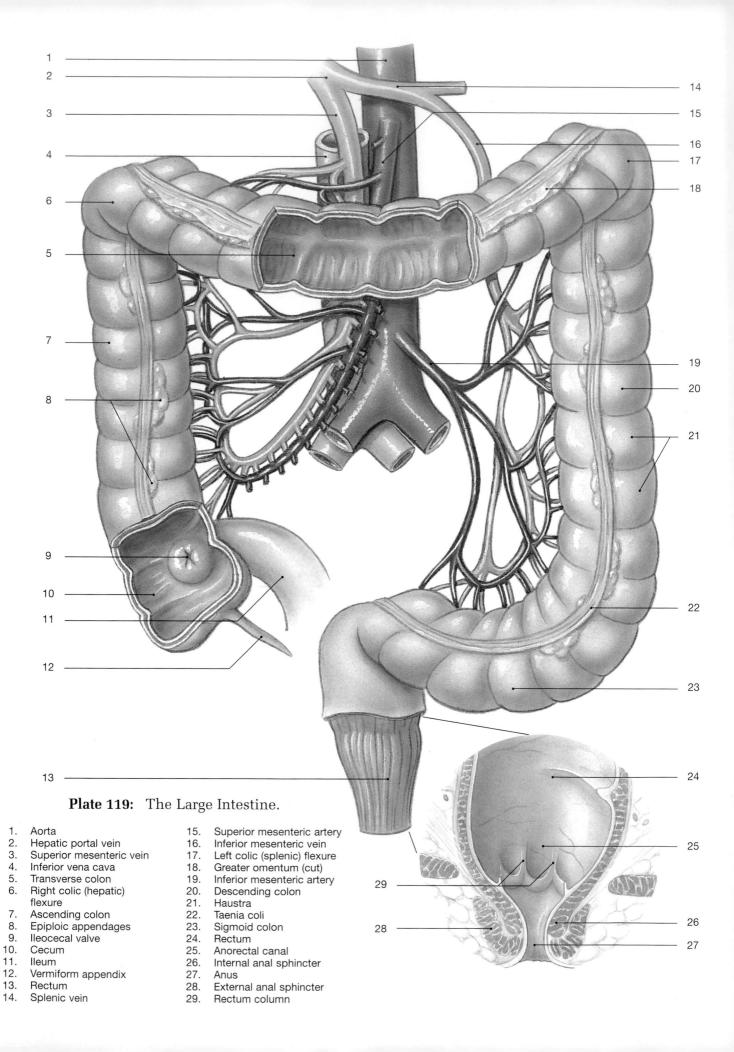

Plate 119: The Large Intestine.

1. Aorta
2. Hepatic portal vein
3. Superior mesenteric vein
4. Inferior vena cava
5. Transverse colon
6. Right colic (hepatic) flexure
7. Ascending colon
8. Epiploic appendages
9. Ileocecal valve
10. Cecum
11. Ileum
12. Vermiform appendix
13. Rectum
14. Splenic vein

15. Superior mesenteric artery
16. Inferior mesenteric vein
17. Left colic (splenic) flexure
18. Greater omentum (cut)
19. Inferior mesenteric artery
20. Descending colon
21. Haustra
22. Taenia coli
23. Sigmoid colon
24. Rectum
25. Anorectal canal
26. Internal anal sphincter
27. Anus
28. External anal sphincter
29. Rectum column

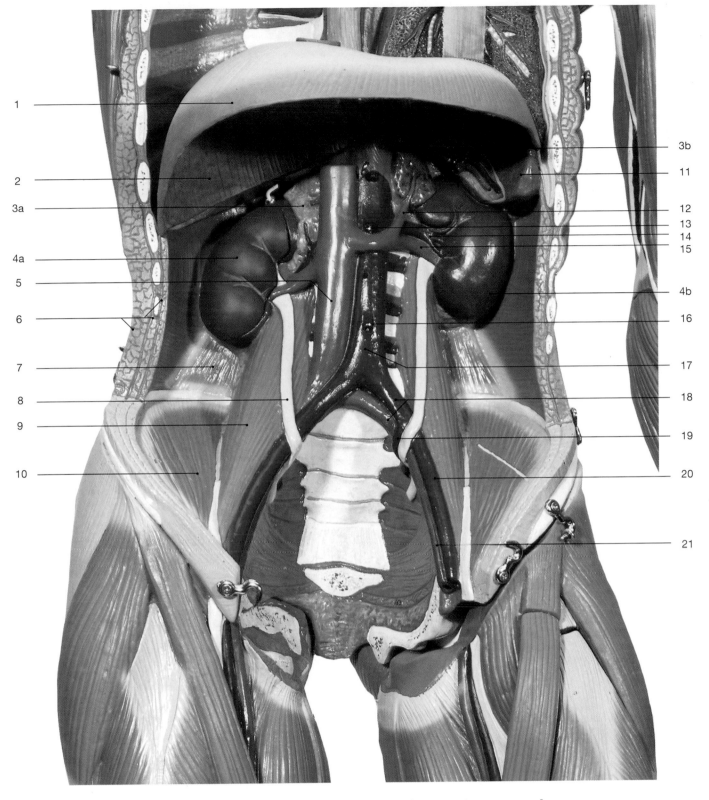

Plate 120: Viscera of the abdominopelvic cavity removed to expose the structures of the urinary system and associated blood vessels.

1. Diaphragm
2. Liver, right lobe, cut
3a Adrenal (suprarenal) gland, right
3b Adrenal (suprarenal) gland, left
4a Kidney, right
4b Kidney, left
5 Inferior vena cava
6. Oblique muscles, cut

7. Quadratus lumborum muscle
8. Ureter, right
9. Psoas major muscle
10. Iliacus muscle
11. Spleen
12. Suprarenal vein, left (middle)
13. Superior mesenteric artery
14. Renal artery, left

15. Renal vein, left
16. Inferior mesenteric artery
17. Abdominal aorta
18. Common iliac artery and vein, left
19. Internal iliac artery, left
20. External iliac artery, left
21. External iliac vein, left

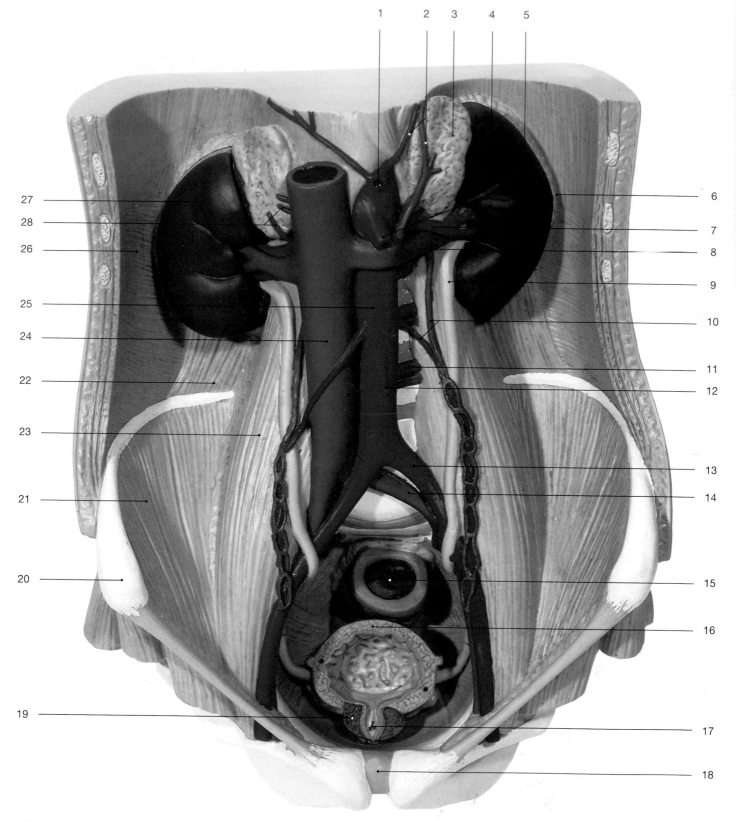

Plate 121: Viscera of male abdominopelvic cavity removed to show urinary and reproductive systems structures.

1. Celiac trunk
2. Left phrenic artery and vein
3. Left adrenal (suprarenal) gland
4. Left kidney
5. Superior mesenteric artery
6. Perirenal fat
7. Left renal artery
8. Left renal vein
9. Left ureter
10. Left testicular artery and vein
11. Lumbar artery
12. Inferior mesenteric artery
13. Common iliac artery, left
14. Common iliac vein, left
15. Rectum
16. Urinary bladder
17. Prostatic urethra
18. Pubic symphysis
19. Prostate
20. Anterior superior iliac spine
21. Iliacus muscle
22. Quadratus lumborum muscle
23. Psoas major muscle
24. Inferior vena cava
25. Abdominal aorta
26. Transverse abdominis muscle
27. Right kidney
28. Right suprarenal artery and vein

Plate 122: Gross structure of the kidney, frontal section.

1. Renal pyramids
2. Medulla
3. Interlobar arteries
4. Segmental artery
5. Renal artery
6. Renal vein
7. Segmental veins
8. Hilus
9. Ureter
10. Renal capsule
11. Cortex
12. Renal column
13. Renal pelvis
14. Interlobar veins
15. Minor calyx
16. Major calyx
17. Renal papilla
18. Arcuate artery and vein

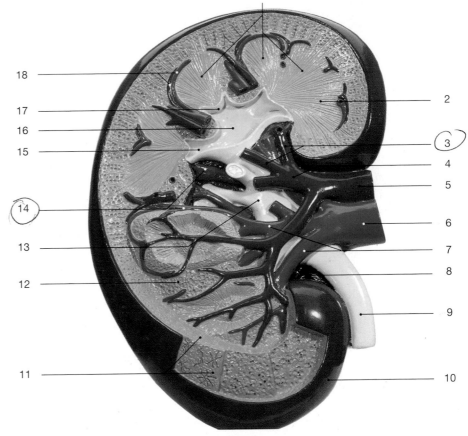

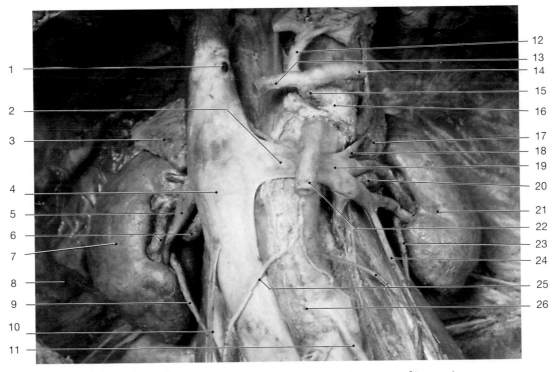

Plate 123: The urinary system as seen in gross dissection.

1. A hepatic vein (stump)
2. Left renal vein
3. Right suprarenal gland
4. Inferior vena cava
5. Right renal vein
6. Right renal artery
7. Right kidney
8. Peritoneum
9. Right ureter
10. Right gonadal vein
11. Inferior mesenteric artery
12. Left gastric artery
13. Common hepatic artery
14. Splenic artery
15. Coeliac trunk
16. Coeliac ganglion
17. Left suprarenal gland
18. Left suprarenal vein
19. Left renal vein
20. Left renal artery
21. Left kidney
22. Superior mesenteric artery
23. Left ureter
24. Left gonadal vein
25. Gonadal arteries
26. Abdominal aorta

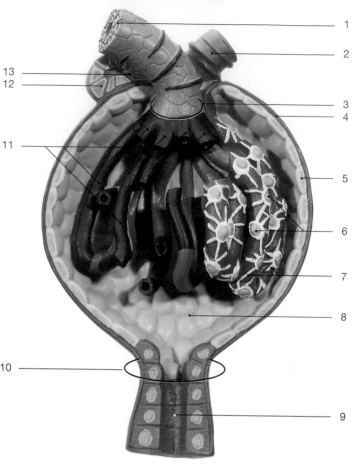

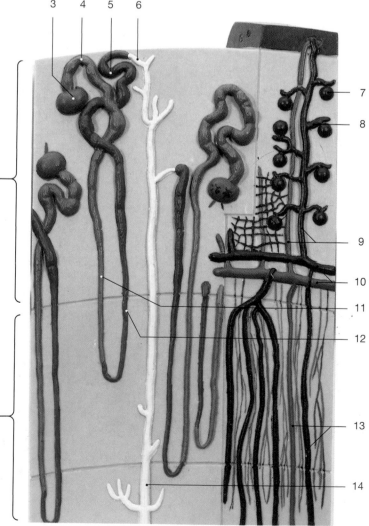

Plate 124: The renal corpuscle.

1. Afferent arteriole
2. Efferent arteriole
3. Vascular pole
4. Renal corpuscle
5. Capsular epithelium
6. Podocytes
7. Pedicels
8. Capsular space
9. Proximal convoluted tubule
10. Tubular pole
11. Glomerulus
12. Distal convoluted tubule
13. Juxtaglomerular apparatus

Plate 125: A cortical nephron.

1. Cortex
2. Medulla
3. Renal corpuscle
4. Proximal convoluted tubule
5. Distal convoluted tubule
6. Collecting tubule
7. Glomerulus
8. Afferent arteriole
9. Interlobular artery and vein
10. Arcuate artery and vein
11. Descending limb of loop of Henle
12. Ascending limb of loop of Henle
13. Interlobar artery and vein
14. Collecting duct

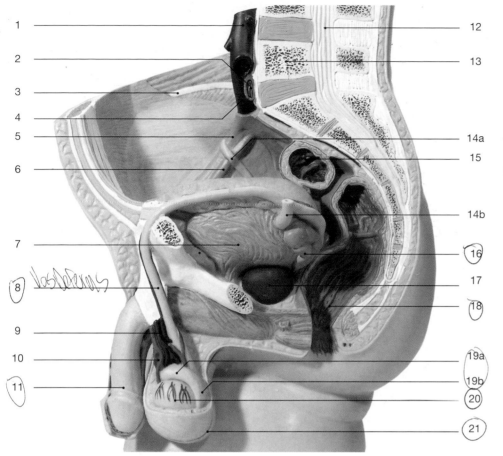

Plate 126: The male pelvis, sagittal section.

1. Abdominal aorta
2. Common iliac vein
3. Iliac crest
4. Common iliac artery
5. External iliac artery
6. External iliac vein
7. Urinary bladder
8. Ductus deferens
9. Testicular artery
10. Pampiniform plexus
11. Penis, shaft
12. Spinal cord, cauda equina
13. Lumbar vertebra (body)
14a. Ureter, right
14b. Ureter, left
15. Sigmoid colon, cut
16. Seminal vesicle
17. Prostate gland
18. Rectum
19a. Epididymis, head
19b. Epididymis, tail
20. Testis
21. Scrotum

(handwritten note near 8: "Vas deferens")

Plate 127: The male pelvis as seen in gross dissection, sagittal section.

1. Sigmoid colon (cut)
2. Rectus abdominis muscle
3. Urinary bladder
4. Pubic symphysis
5. Penis
6. Penile urethra and corpus spongiosum
7. Corpus cavernosum
8. Epididymis head
9. Testis
10. Rectum
11. Internal urethral orifice
12. Prostatic urethra
13. Prostate gland
14. Ejaculatory duct
15. Membranous urethra
16. Penile urethra
17. Bulbospongiosus muscle
18. Ductus deferens
19. Scrotum

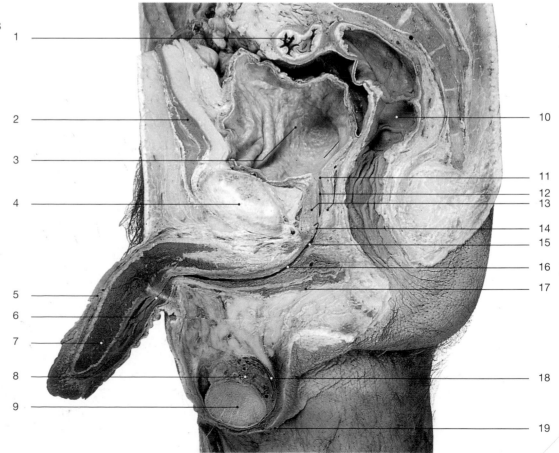

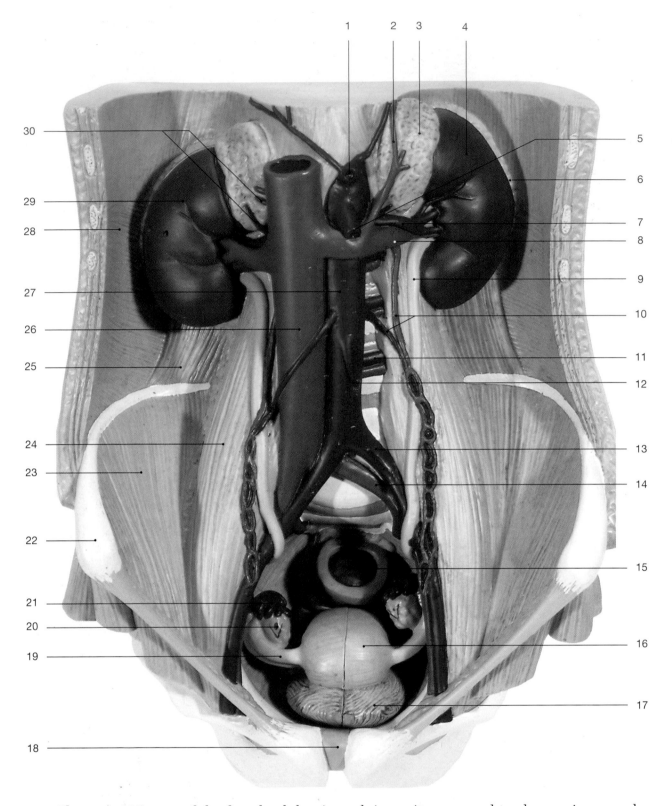

Plate 128: Viscera of the female abdominopelvic cavity removed to show urinary and reproductive system structures.

1. Celiac trunk
2. Left phrenic artery and vein
3. Left adrenal (suprarenal) gland
4. Left kidney
5. Superior mesenteric artery
6. Perirenal fat
7. Left renal artery
8. Left renal vein
9. Left ureter
10. Ovarian artery and vein

11. Lumbar artery
12. Inferior mesenteric artery
13. Common iliac artery
14. Common iliac vein
15. Rectum, cut
16. Uterus
17. Urinary bladder
18. Pubic symphysis
19. Uterine tube
20. Ovary

21. Fimbriae
22. Anterior iliac superior spine
23. Iliacus muscle
24. Psoas major muscle
25. Quadratus lumborum muscle
26. Inferior vena cava
27. Abdominal aorta
28. Transverse abdominis muscle
29. Right kidney
30. Right suprarenal artery and vein

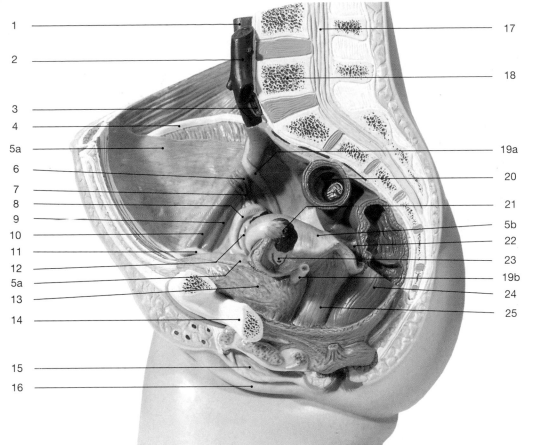

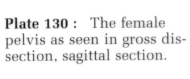

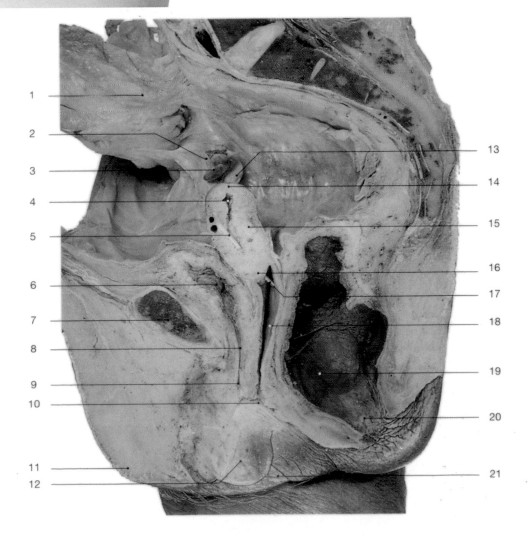

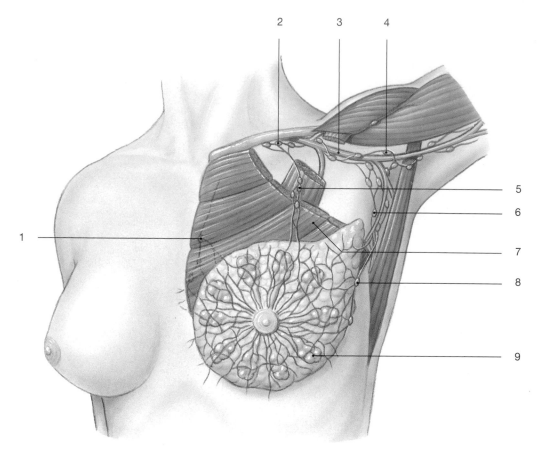

Plate 131: Lymphatic drainage of the upper extremity in the female.

1. Parasternal lymph node
2. Subclavian lymph node
3. Axillary vein
4. Axillary lymph nodes
5. Central lymph node
6. Subscapular lymph node
7. Pectoralis major muscle (cut)
8. Pectoral lymph node
9. Mammary gland

Plate 132: Mammary gland as seen in dissection.

1. Parasternal lymph nodes
2. Lymphatic duct
3. Pectoralis major muscle
4. Pectoral fat pad
5. Areola
6. Nipple
7. Lactiferous sinus
8. Lactiferous ducts
9. External oblique muscle
10. Connective tissue
11. Mammary glands
12. Suspensory ligaments
13. Rib
14. Intercostal muscle
15. Pectoralis minor muscle

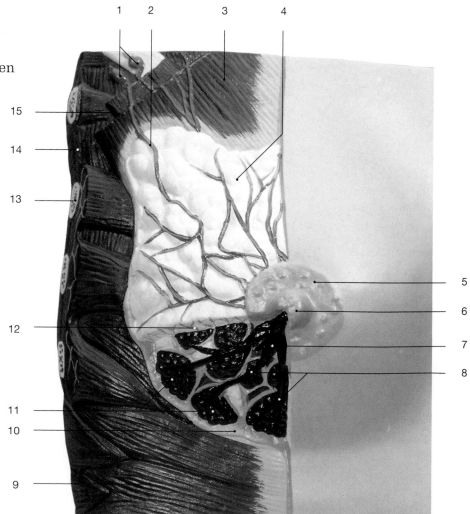